AC
6-65

WITHDRAWN

TEXTBOOK OF POLYMER SCIENCE

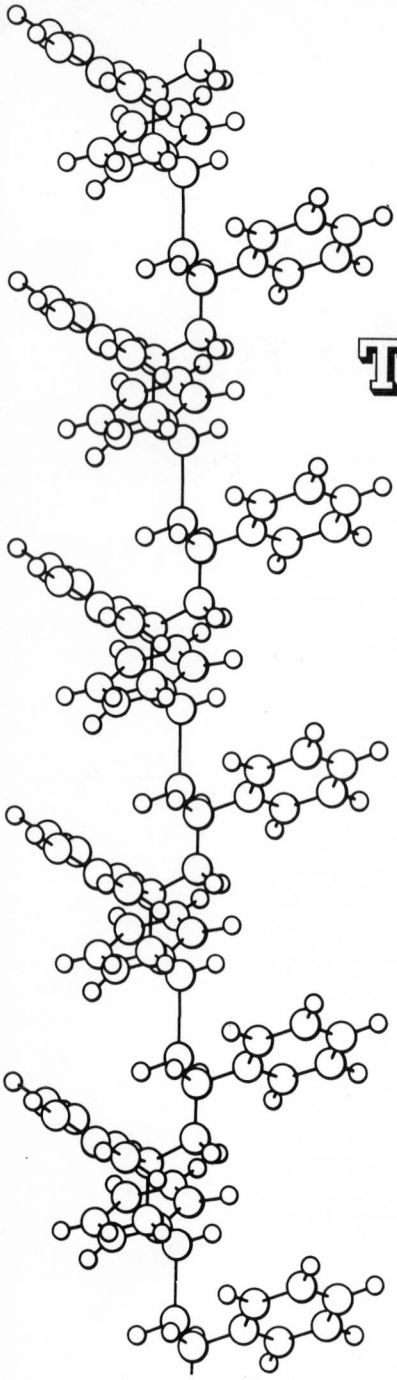

TEXTBOOK OF

Fred W. Billmeyer, Jr.

POLYMER SCIENCE

Professor of Chemistry, University of Delaware

Research Associate, Plastics Department

E. I. du Pont de Nemours & Co., Inc.

and Sometime Visiting Professor

of Chemical Engineering

Massachusetts Institute of Technology

Interscience Publishers, a Division of John Wiley and Sons

New York and London

SECOND PRINTING, MARCH, 1964

Copyright © 1962 by John Wiley & Sons, Inc.

All Rights Reserved
This book or any part thereof
must not be reproduced in any form
without the written permission of the publisher.

Library of Congress Catalog Card Number: 62-18350
Printed in the United States of America

29,581

QD
281
.P6
B43
1962

Preface

Like its predecessor, *Textbook of Polymer Chemistry* (1957), this book is written as a textbook in the chemistry and physics of high polymers. It includes no original work, but is an attempt to collect in one location and to classify some of the more interesting and important information about polymeric substances. It is based entirely on previously published information.

In view of the rapidly increasing size of the polymer literature,* it is not surprising that an early and extensive revision of the previous volume became necessary. Its accomplishment at this time was made possible by the generosity of the Du Pont Company and of the Massachusetts Institute of Technology, which extended to the author the privilege of writing the revised edition while preparing and teaching courses in the chemistry and physics of polymers in the Department of Chemical Engineering at the Institute, during the academic year 1960–61.

It is hoped that this textbook will be useful for first courses in polymer chemistry at the senior or graduate level. It requires of the student some familiarity with organic and physical chemistry, elementary physics, and mathematics through elementary calculus. No prior knowledge of high polymers is assumed. It is intended that the book be expanded in lectures and supplemented by reading assignments taken from the works listed under General References at the end of each section. These details are left to the discretion of the instructor. Like its predecessor, *Textbook of Polymer Chemistry*, the book may, hopefully, also prove useful to the chemists and engineers working in the polymer field.

FRED W. BILLMEYER, JR.

Wilmington, Delaware
August, 1962

* E. H. Immergut, paper presented at the International Symposium on Macromolecular Chemistry, Montreal, Canada, July, 1961; *J. Polymer Sci.* 1962 (in press).

12.75
6-3-65

Acknowledgments

The assistance of my many colleagues, at the Massachusetts Institute of Technology, at the University of Delaware, and in the Du Pont Company, who have reviewed and commented upon the manuscript, has been most valuable. The manuscript was typed in major part by Mrs. Nathan R. Hopton, whose assistance I acknowledge with thanks.

The preparation of the manuscript was supported by a grant made to the Massachusetts Institute of Technology by the Ford Foundation for the purpose of aiding in the improvement of engineering education. This support is gratefully acknowledged.

Finally, I am deeply grateful to my wife Annette, whose presence has made this undertaking, like all those we share, a joy and pleasure.

FRED W. BILLMEYER, JR.

Wilmington, Delaware
August, 1962

Contents

Part I Introduction ***1***

Chapter 1 The Science of Large Molecules 3

A Basic Concepts of Polymer Science 3
B History of Macromolecular Science 11
C Molecular Forces and Chemical Bonding in Polymers 15

Part II Structure and Properties of Polymers ***23***

Chapter 2 Polymer Solutions 25

A Criteria for Polymer Solubility 25
B Conformations of Dissolved Polymer Chains 27
C Thermodynamics of Polymer Solutions 30
D Phase Separation in Polymer Solutions 40
E Fractionation of Polymers 47
F Swelling of Crosslinked Polymers 52

Chapter 3 Measurement of Molecular Weight and Size 53

A End Group Analysis 53
B Colligative Property Measurement 55
C Light Scattering 64
D Ultracentrifugation 73
E Solution Viscosity and Molecular Size 79
F Polyelectrolytes 86

Chapter 4 Analysis and Testing of Polymers 88

A Analytical Chemistry of Polymers 88
B Infrared Spectroscopy 90
C X-Ray Diffraction Analysis 97
D Nuclear Magnetic and Electron Paramagnetic Resonance Spectroscopy 100
E Calorimetric and Thermal Analysis 105
F Light and Electron Microscopy 108
G Physical Testing 109

Chapter 5 Morphology and Order in Crystalline Polymers 117

A Configurations of Polymer Chains 117
B Crystal Structures of Polymers 124
C Morphology of Polymer Single Crystals 140
D Structure of Polymers Crystallized from the Melt 145
E Crystallization Kinetics 154
F Thermodynamics of Crystallization and Degree of Crystallinity 157
G Orientation and Drawing 164

Chapter 6 Rheology and the Mechanical Properties of Polymers 172

A Viscous Flow 173
B Kinetic Theory of Rubber Elasticity 180
C Viscoelasticity 188
D The Glassy State and the Glass Transition 198
E The Mechanical Properties of Crystalline Polymers 205

Chapter 7 Polymer Structure and Physical Properties 208

A Basic Determinants of Polymer Properties 208
B Properties Involving Large Deformations 209
C Transitions 212
D Mechanical Properties of Solid Polymers 221
E Other Properties 226
F Property Requirements and Polymer Utilization 230

Part III Polymerization **235**

Chapter 8 Step-Reaction (Condensation) Polymerization 237

A Classification of Polymers and Polymerization Mechanisms 237
B Chemistry of Stepwise Polymerization 239
C Kinetics and Statistics of Linear Stepwise Polymerization 245
D Polyfunctional Step-Reaction Polymerization 253
E Ring-Scission Polymerization 259

Chapter 9 Radical Chain (Addition) Polymerization 262

A Chemistry of Vinyl Polymerization 262
B Laboratory Methods in Vinyl Polymerization 270
C Steady-State Kinetics of Vinyl Radical Polymerization 272
D Absolute Reaction Rates 280
E Molecular Weight and Its Distribution 283
F Thermochemistry of Chain Polymerization 286

Chapter 10 Nonradical Chain (Addition) Polymerization 291

A Chemistry of Nonradical Chain Polymerization 291
B Cationic Polymerization 293
C Anionic Polymerization 297
D Coordination Polymerization 302
E Polymerization with Supported Metal Oxide Catalysts 307

Chapter 11 Copolymerization 310

A Kinetics of Copolymerization 310
B Composition of Copolymers 319
C Chemistry of Copolymerization 326
D Block and Graft Copolymers 333

Chapter 12 Polymerization Conditions and Polymer Reactions 338

A Polymerization in Homogeneous Systems 338
B Polymerization in Heterogeneous Systems 340
C Diene and Divinyl Polymerization 348
D Chemical Reactions of Polymers 353
E Degradation of Polymers 354
F Radiation Chemistry of Polymers 358

Part IV Properties of Commercial Polymers **361**

Chapter 13 Olefin Polymers 363

A Low-Density (Branched) Polyethylene 363
B High-Density (Essentially Linear) Polyethylene 369
C Polypropylene 371
D Other Olefin Polymers 374

Chapter 14 Polymers Derived from Dienes 377

A Natural Rubber and Other Polyisoprenes 378
B Polymers Derived from Butadiene 383
C Polyisobutylene and Butyl Rubber 388
D Polychloroprene (Neoprene) 392

Chapter 15 Vinyl and Vinylidene Polymers 394

A Polystyrene and Related Polymers 394
B Acrylic Polymers 400
C Poly(vinyl Esters) and Derived Polymers 406
D Chlorine-Containing Polymers 412
E Miscellaneous Vinyl and Related Polymers 416

Chapter 16 Fluorocarbon Polymers 419

A Polytetrafluoroethylene and Other Fully Fluorinated Polymers 419
B Poly(vinyl Fluoride), Poly(vinylidene Fluoride), and Related Polymers 425
C Polymers Based on Chlorotrifluoroethylene 426
D Other Fluorocarbon Polymers 427

Chapter 17 Heterochain Thermoplastics 429

A Polyamides 429
B Polyesters 434
C Polyurethans 437
D Polypeptides 439
E Ether and Acetal Polymers 442
F Polysulfides 445
G Polycarbonates 446

Chapter 18 Cellulosic Polymers 448

A Native Cellulose 449
B Regenerated Cellulose 452
C Cellulose Esters 456
D Other Polymers Related to Cellulose 461

Chapter 19 Thermosetting Resins 463

A Phenolic Resins 463
B Amino Resins 470
C Unsaturated Polyester Resins 472
D Epoxy Resins 475
E Urethan Foams 477
F Silicone Polymers 479
G Miscellaneous Thermosetting Resins 483

Part V Polymer Processing **485**

Chapter 20 Plastics Technology 487

A Molding 487
B Other Processing Methods 492
C Fillers 498
D Plasticizers and Other Additives 499
E Tables of Plastics Properties 506

Chapter 21 Fiber Technology 507

A Textile and Fabric Properties 508
B Spinning 513
C Fiber After-treatments 520
D Table of Fiber Properties 525

Chapter 22 Elastomer Technology 526

A Compounding and Elastomer Properties 526
B Vulcanization 533
C Reinforcement 541
D Table of Elastomer Properties 547

Appendixes **549**

I. List of Symbols 549
II. Table of Physical Constants 558
III. Trademarks, Generic Names, and Manufacturers 559

Author Index **571**

Subject Index **585**

I

Introduction

1

The Science of Large Molecules

A. Basic Concepts of Polymer Science

Almost half a century ago, Wolfgang Ostwald (2)* coined the phrase "the land of neglected dimensions" to describe the range of sizes between molecular and macroscopic within which occur most colloidal particles. The term "neglected dimensions" might have been applied equally well to the world of polymer molecules, the high-molecular-weight compounds so important to man and his modern technology. It was not until the third decade of this century that the science of high polymers began to emerge, and the major growth of the technology of these materials came even later. Yet today polymer dimensions are neglected no more, for industries associated with polymeric materials employ more than a third of all American chemists and chemical engineers.

The science of macromolecules is divided between biological and nonbiological materials. Each is of great importance. Biological polymers form the very foundation of life and intelligence, and provide much of the food on which man exists. This book, however, is concerned with the chemistry, physics, and technology of nonbiological polymers. These are primarily the synthetic materials used for plastics, fibers, and elastomers, but a few naturally occurring polymers, such as rubber, wool, and cellulose, are included. Today, these substances are truly indispensable to mankind, being essential to his clothing, shelter, transportation, and communication, as well as to the conveniences of modern living.

A *polymer* is a large molecule built up by the repetition of small, simple chemical units. In some cases the repetition is linear, much as a chain is

* Parenthetical numbers refer to items in the bibliography following the immediate section.

built up from its links. In other cases the chains are *branched* or interconnected to form *three-dimensional networks*. The *repeat unit* of the polymer is usually equivalent or nearly equivalent to the *monomer*, or starting material from which the polymer is formed. Thus (Table 1-1) the repeat unit of poly(vinyl chloride) is $—CH_2CHCl—$; its monomer is vinyl chloride, $CH_2{=}CHCl$.

TABLE 1-1. Some linear high polymers, their monomers, and their repeat units

Polymer	Monomer	Repeat Unit
Polyethylene	$CH_2{=}CH_2$	$—CH_2CH_2—$
Poly(vinyl chloride)	$CH_2{=}CHCl$	$—CH_2CHCl—$
Polyisobutylene	$CH_2{=}C(CH_3)_2$	$—CH_2—C(CH_3)_2—$
Polystyrene*	$CH_2{=}CH$	$—CH_2—CH—$
Polycaprolactam (6 nylon)	$H—N(CH_2)_5C(=O)—OH$	$—N(CH_2)_5C(=O)—$
Polyisoprene (natural rubber)	$CH_2{=}CH—C(CH_3){=}CH_2$	$—CH_2CH{=}C(CH_3)—CH_2—$

* By convention, the symbol is used throughout to represent the benzene ring, double bonds being omitted.

The length of the polymer chain is specified by the number of repeat units in the chain. This is called the *degree of polymerization* (DP). The molecular weight of the polymer is the product of the molecular weight of the repeat unit and the degree of polymerization. Using poly(vinyl chloride) as an example, a polymer of degree of polymerization 1000 has a molecular weight of $63 \times 1000 = 63{,}000$. Most high polymers useful for plastics, rubbers, or fibers have molecular weights between 10,000 and 1,000,000.

Unlike many products whose structure and reactions were well known before their industrial application, some polymers were produced on an

industrial scale long before their chemistry or physics was studied. Empiricism in recipes, processes, and control tests was usual.

Gradually the study of polymer properties began. Almost all were first called anomalous because they were so different from the properties of low-molecular-weight compounds. It was soon realized, however, that polymer molecules are many times larger than those of ordinary substances. The presumably anomalous properties of polymers were shown to be normal for such materials, as the consequences of their size were included in the theoretical treatments of their properties.

Primary chemical bonds along polymer chains are entirely satisfied. The only forces between molecules are secondary bond forces of attraction, which are weak relative to primary bond forces. The high molecular weight of polymers allows these forces to build up enough to impart excellent strength, dimensional stability, and other mechanical properties to the substances.

Polymerization processes The processes of polymerization were divided by Carothers and Flory (Flory, and Mark and Whitby, gen. refs.) into two groups known as *condensation* and *addition* polymerization or, in more precise terminology (Chapter 8*A*), *step-reaction* and *chain-reaction* polymerization.

Condensation or *step-reaction polymerization* is entirely analogous to condensation in low-molecular-weight compounds. In polymer formation the condensation takes place between two polyfunctional molecules to produce one larger polyfunctional molecule, with the possible elimination of a small molecule such as water. For example, a first step in the formation of the polyester poly(ethylene adipate) is the reaction:

$$\mathrm{HOCH_2CH_2OH} + \mathrm{HO\underset{\underset{\mathrm{O}}{\|}}{C}(CH_2)_4\underset{\underset{\mathrm{O}}{\|}}{C}OH} \longrightarrow \mathrm{HOCH_2CH_2O\underset{\underset{\mathrm{O}}{\|}}{C}(CH_2)_4\underset{\underset{\mathrm{O}}{\|}}{C}OH} + \mathrm{H_2O}$$

The reaction continues until all of one of the reagents is used up:

$$x\mathrm{HOCH_2CH_2OH} + x\mathrm{HO\underset{\underset{\mathrm{O}}{\|}}{C}(CH_2)_4\underset{\underset{\mathrm{O}}{\|}}{C}OH} \longrightarrow \mathrm{H}\left(\mathrm{OCH_2CH_2O\underset{\underset{\mathrm{O}}{\|}}{C}(CH_2)_4\underset{\underset{\mathrm{O}}{\|}}{C}}\right)_x\mathrm{OH} + (2x-1)\mathrm{H_2O}$$

The equilibrium of this reaction can be shifted at will at high temperatures by controlling the amounts of the reactants and products.

Addition or *chain-reaction polymerization* involves chain reactions in which the chain carrier may be an ion or a reactive substance with one unpaired electron called a *free radical.* A free radical is usually formed by the decomposition of a relatively unstable material called an *initiator*:

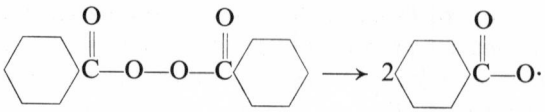

initiator free radical

The free radical is capable of reacting to open the double bond of a vinyl monomer and add to it, with an electron remaining unpaired:

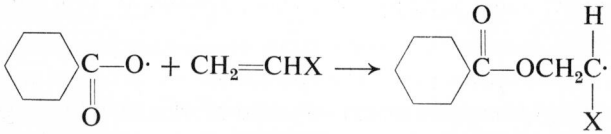

free radical vinyl monomer growing polymer chain

In a very short time (usually a few seconds or less) many more monomers add successively to the growing chain. Finally two free radicals react to annihilate each other's growth activity and form one or more polymer molecules:

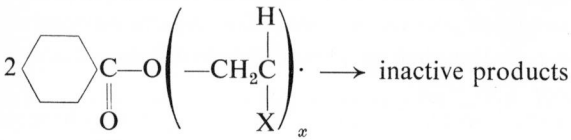

growing chains polymer

With some exceptions, polymers made in chain reactions often contain only carbon atoms in the main chain (*homochain polymers*), whereas polymers made in step reactions may have other atoms, originating in the monomer functional groups, as part of the chain (*heterochain polymers*).

Molecular weight and its distribution In both chain and stepwise polymerization, the length of a chain is determined by purely random events. In step reactions, the chain length is determined by the local availability of reactive groups at the ends of the growing chains. In radical polymerization, chain length is determined by the time during which the chain grows before it diffuses into the vicinity of a second free radical and the two react.

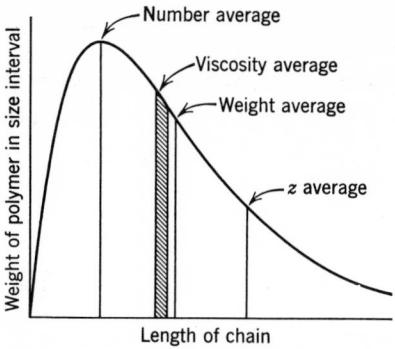

Fig. 1-1. Distribution of molecular weights in a typical polymer (1).

In either case, the polymeric product contains molecules having many different chain lengths. For some types of polymerization the resulting *distribution of molecular weights* can be calculated statistically. It can be illustrated by plotting the weight of polymer of a given size against the chain length or molecular weight (Fig. 1-1).

Since a distribution of molecular weights exists in any finite sample of polymer, the experimental measurement of molecular weight can give only an average value. Several different averages are important. For example, some methods of molecular weight measurement in effect count the number of molecules in a known mass of material. Through knowledge of Avogadro's number this information leads to the *number-average molecular weight* $\bar{M}_n$ of the sample. For typical polymers the number average lies near the peak of the weight-distribution curve or the most probable molecular weight.

In other experiments, such as sedimentation in an ultracentrifuge, the contribution of a molecule to the observed effect is a function of its mass. Heavy molecules are favored in the averaging process; a *weight-average molecular weight* $\bar{M}_w$ results. $\bar{M}_w$ is equal to or greater than $\bar{M}_n$. The ratio $\bar{M}_w/\bar{M}_n$ is sometimes used as a measure of the breadth of the molecular weight distribution. Values of $\bar{M}_w/\bar{M}_n$ for typical polymers range from 1.5–2.0 to 20–50.

The molecular weight averages shown in Fig. 1-1 are defined mathematically in Chapter 3.

Branched and network polymers In contrast to the linear-chain molecules discussed so far, some polymers have branched chains, often as a result of side reactions during polymerization (Fig. 1-2*a*). The term *branching* implies that the individual molecules are still discrete. Regularly repeating side groups which are a part of the monomer structure are not

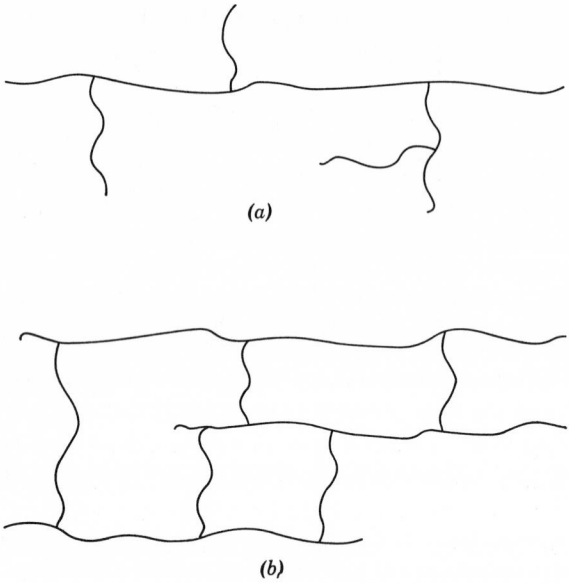

Fig. 1-2. Schematic representation of (*a*) branched and (*b*) network polymers.

considered branches. Thus, poly(butyl methacrylate) is classed as a linear polymer:

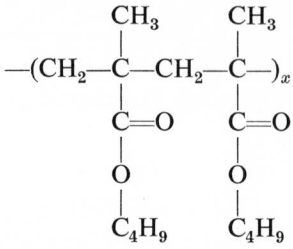

In still other cases *crosslinked* or *network structures* are formed (Fig. 1-2*b*), as in the use of monomers containing more than two reactive groups in stepwise polymerization. If, e.g., glycerol is substituted for ethylene glycol in the reaction with a dibasic acid, a three-dimensional network polymer results.

In commercial practice such reactions may take place during the fabrication of articles made with *thermosetting* resins. The crosslinked network extending throughout the final article is stable to heat and cannot be made to flow or melt. In contrast, most linear polymers can be made to soften and take on new shapes by the application of heat and pressure. They are said to be *thermoplastic*.

The texture of polymers The geometrical arrangement of the atoms in a polymer chain can be divided conveniently into two categories:

a. Arrangements fixed by the chemical bonding in the molecule, such as *cis* and *trans* isomers, or *d*- and *l*-forms. Throughout this book, such arrangements are described as *configurations.* The configuration of a polymer chain cannot be altered unless chemical bonds are broken and reformed.

b. Arrangements arising from rotation about single bonds. These arrangements, including the manifold forms that the polymer chain may have in solution, are described as *conformations.*

In dilute solution, where the polymer chain is surrounded by small molecules, or in the melt, where it is in an environment of similar chains, the polymer molecule is in continual motion because of its thermal energy, assuming in rapid succession many different *conformations.* As a polymer melt is cooled, this motion becomes restricted, and one of several things may happen (Fig. 1-3). The material may become hard, strong, and stiff if *crystallinity* develops. If the polymer cannot crystallize, it may be either hard, stiff, and brittle or limp and flexible at room temperature.

The major factor determining whether a polymer can crystallize is the occurrence of successive units in the chain in a *configuration* of geometrical regularity. If all the substituents on each chain carbon atom are small or all are alike, as in polyethylene, crystallinity is highly developed. If, however, some of the substituents are bulky groups, as in polystyrene, the material can crystallize only if they are arranged in one of several ordered, or tactic, configurations (Chapter 5*A*). The achievement of

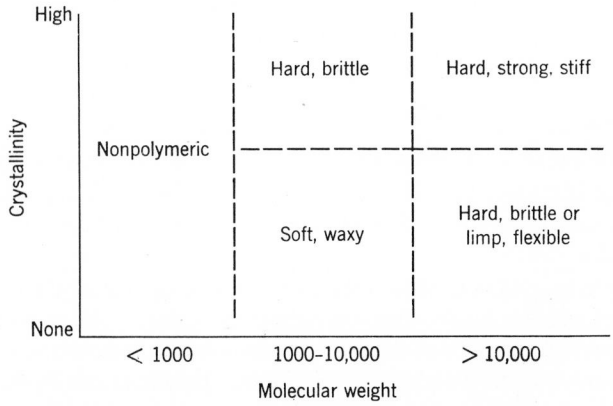

Fig. 1-3. Classification of the expected properties of materials on the basis of molecular weight and crystallinity.

stereospecific polymerizations (Chapter 10*D*) leading to such ordered configurations is a major triumph in polymer chemistry.

The properties of an amorphous (noncrystalline) polymer are determined by the degree of restriction of molecular motion of its chains. Below temperatures where viscous flow is possible, the motion of chain segments of a few (10–30) carbon atoms is still possible. As the polymer is cooled still more, it undergoes a transition reminiscent of a second-order transition,* at which the slope of the volume-temperature curve changes. This is not a thermodynamic second-order transition and is preferably called the *glass transition* (Chapter 6*D*).

If the polymer is above the glass transition at room temperature, it is typically limp and flexible. In this state it may show the phenomenon of delayed elastic response or *viscoelasticity* characteristic of polymers.

An amorphous polymer below its glass transition at room temperature is characterized by stiffness, hardness, and form stability, and often by transparency and high optical clarity. However, it is likely to be brittle. This brittleness arises from the fact that molecular motion is restricted to very short chain segments or side groups, and the material cannot deform enough to withstand the high stress.

In contrast to an amorphous polymer, a highly crystalline plastic like 6 nylon is much stronger and tougher. The classical explanation for the improved mechanical properties of a crystalline polymer involved a composite single-phase structure (Chapter 5*D*) in which polymer chains passed through several crystallites and surrounding amorphous regions, holding the polymer together and conferring strength and stiffness. It is now becoming clear that the classical picture of the crystalline-amorphous structure of polymers must be drastically modified. The implications of this knowledge constitute one of the most fascinating developments in polymer chemistry today. The current concept of a lamellar, folded-chain structure of solid polymers is developed in Chapter 5.

GENERAL REFERENCES

Flory, P. J., *Principles of Polymer Chemistry*, Cornell University Press, Ithaca, N.Y., 1953, Chaps. II-1, II-2.

* In thermodynamics, a *first-order transition* involves a discontinuity in a primary thermodynamic quantity such as volume or heat content. A plot of such a quantity versus temperature shows a discontinuous change at a *first-order transition temperature*. Similarly, a *second-order transition* involves a discontinuous change in a secondary thermodynamic quantity such as expansion coefficient or heat capacity—the temperature derivative of a primary quantity. At a *second-order transition temperature* there is a discontinuity in the slope of a plot of a primary quantity, or in the ordinate of a plot of a secondary quantity, against temperature.

Mark, H., and G. S. Whitby, eds., *Collected Papers of Wallace Hume Carothers on High Polymeric Substances*, Interscience Publishers, New York, 1940.
Melville, Sir Harry, *Big Molecules*, The Macmillan Co., New York, 1958.
Scientific American issue on Giant Molecules, **197,** #3 (September, 1957).
Stille, J. K., *Introduction to Polymer Chemistry*, John Wiley & Sons, New York, 1962.

BIBLIOGRAPHY

1. McGrew, F. C., *J. Chem. Educ.* **35,** 178 (1958).
2. Ostwald, W., *Die Welt der Vernachlässigten Dimensionen*, T. Steinkopff, Dresden, 1914.

B. History of Macromolecular Science

Early investigations

Natural polymers Natural polymers have been utilized throughout the ages. Since his beginning man has been dependent upon animal and vegetable matter for sustenance, shelter, warmth, and other requirements and desires. Natural resins and gums have been used for thousands of years. Asphalt was utilized in pre-Biblical times; amber was known to the ancient Greeks; and gum mastic was used by the Romans.

About a century ago the unique properties of natural polymers were recognized. The term *colloid* was proposed to distinguish them as a class from materials which could be obtained in crystalline form. The concept was later broadened to that of the "colloidal state of matter," which was considered to be like the gaseous, liquid, and solid states. Although useful for describing many colloidal substances, such as gold sols and soap solutions, the concept of a reversibly attainable colloidal state of matter has no validity.

The hypothesis that colloidal materials are very high in molecular weight is also quite old, but before the work of Raoult and van't Hoff in the 1880's no suitable methods were available for estimating molecular weights. When experimental methods did become available, molecular weights ranging from 10,000 to 40,000 were obtained for such substances as rubber, starch, and cellulose nitrate. The existence of large molecules implied by these measurements was not accepted by the chemists of the day for two reasons.

First, true macromolecules were not distinguished from other colloidal substances which could be obtained in noncolloidal form as well. When a material of well-known structure was seen in the colloidal state, its

apparent high molecular weight was considered erroneous. Thus it was assumed that Raoult's solution law did not apply to any material in the colloidal state. Second, coordination complexes and the association of molecules were often used to explain polymeric structures in terms of physical aggregates of small molecules.

For example, the empirical formula C_5H_8 was found for rubber as early as 1826, and isoprene was obtained on destructive distillation of the polymer in 1860. The presence of the repeating unit

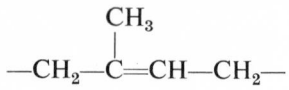

was demonstrated in the early 1900's. At that time rubber was thought to consist of short sequences of this unit arranged in either chains or cyclic structures. Obvious difficulties regarding end groups, which could not be found chemically, favored ring structures, leading to the concept of the rubber molecule being a ring like dimethylcycooctadiene. Large numbers of these were considered held together by "association" to give the colloidal material:

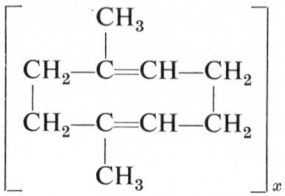

Synthetic polymers In the search by the early organic chemists for pure compounds in high yields, many polymeric substances were discovered and as quickly discarded as oils, tars, or undistillable residues. A few of these materials, however, attracted interest. Poly(ethylene glycol) was prepared about 1860; the individual polymers with degree of polymerization up to 6 were isolated and their structures correctly assigned. The concept of extending the structure to very high molecular weights by continued condensation was understood.

Other condensation polymers were prepared in succeeding decades. As the molecular aggregation theories gained in popularity, structures involving small rings held together by secondary bond forces were often assigned to these products.

Some vinyl polymers were also discovered. Styrene was polymerized as early as 1839, isoprene in 1879, and methacrylic acid in 1880. Again cyclic structures held together by "partial valences" were assigned.

The rise of polymer science

Acceptance of the existence of macromolecules Acceptance of the macromolecular hypothesis came about in the 1920's, largely because of the efforts of Staudinger (8), who received the Nobel prize in 1953 for his championship of this viewpoint. In 1920 he proposed long-chain formulas for polystyrene, rubber, and polyoxymethylene. His extensive investigations of the latter polymers left no doubt as to their long-chain nature. More careful molecular weight measurements substantiated Staudinger's conclusions, as did x-ray studies showing structures for cellulose and other polymers which were compatible with chain formulas. The outstanding series of investigations by Carothers (2) beginning in 1929 supplied quantitative evidence substantiating the macromolecular viewpoint.

The problem of end groups One deterrent to the acceptance of the macromolecular theory was the problem of the ends of the long-chain molecules. Since the degree of polymerization of a typical polymer is several hundred, chemical methods for detecting end groups were at first not successful. Staudinger (9) suggested that no end groups were needed to saturate terminal valences of the long chains; they were considered to be unreactive because of the size of the molecules. Large ring structures were also hypothesized (10), and this concept was popular for many years. Not until Flory (3) elucidated the mechanism for chain-reaction polymerization did it become clear that the ends of long-chain molecules consist of normal, satisfied valence structures. The presence and nature of end groups have since been investigated in detail by chemical methods (1, 4, 7).

Molecular weight and its distribution Staudinger (10) was among the first to recognize the large size of polymer molecules, and to utilize the dependence on molecular weight of a physical property, such as dilute solution viscosity (11), for determining polymer molecular weights. He also understood clearly that synthetic polymers are polydisperse (10). A few years later, Lansing and Kraemer (5) distinguished unmistakably among the various average molecular weights obtainable experimentally.

Configurations of polymer chain atoms Staudinger's name is also associated with the first studies (12) of the configuration of polymer chain atoms. He showed that the phenyl groups in polystyrene are attached to alternate chain carbon atoms. This regular head-to-tail configuration has since been established for most vinyl polymers. The mechanism for producing branches in normally linear vinyl polymers was introduced by Flory (3), but such branches were not adequately identified and characterized for another decade (see Chapter 3*E*). Natta (6) first recognized the presence of stereospecific regularity in vinyl polymers.

Rubber The modern plastics industry began with the utilization of natural rubber for erasers and in rubberized fabrics a few years before Goodyear's discovery of vulcanization in 1839. In the next decade the rubber industry arose both in England and in the United States. In 1851 hard rubber, or ebonite, was patented and commercialized.

Derivatives of cellulose Cellulose nitrate, or nitrocellulose, discovered in 1838, was successfully commercialized by Hyatt in 1870. His product, Celluloid, cellulose nitrate plasticized with camphor, could be formed into a wide variety of useful products by the application of heat and pressure. Nitrocellulose found application in the manufacture of explosives, photographic film, synthetic fibers (Chardonnet silk), airplane dopes, automobile lacquers, and automobile safety glass. Nitrocellulose in turn has been superseded in almost all these uses by more stable and more suitable polymers.

Cellulose acetate, discovered in 1865, was not utilized commercially for several decades because of the low solubility and lack of dyeability of the early cellulose triacetate products. In the early 1900's the manufacture of less highly substituted, more readily soluble compositions opened the way for the commercial development of acetate rayon fibers and cellulose acetate plastics.

Later still, processes were developed whereby cellulose itself could be dissolved and reprecipitated by chemical treatment. These led to the production of viscose rayon fiber and cellophane film.

Synthetic polymers The oldest of the purely synthetic plastics is the family of phenol-formaldehyde resins, of which Baekeland's Bakelite was the first commercial product. Small-scale production of phenolic resins and varnishes was begun in 1907.

The first commercial use of styrene was in synthetic rubbers made by copolymerization with dienes in the early 1900's. Polystyrene was produced commercially in Germany about 1930 and successfully in the United States in 1937. Large-scale production of vinyl chloride-acetate resins began in the early 1930's also.

Thus the last quarter-century has seen the development of all but a handful of the wide variety of synthetic polymers now in common use.

GENERAL REFERENCES

Flory, P. J., *op. cit.*, Chap. I.
Mark, H., and G. S. Whitby, eds., *loc. cit.*

McGrew, F. C., *J. Chem. Educ.* **35,** 178 (1958).
Purves, C. B., Chap. IIIA in E. Ott, H. M. Spurlin, and M. W. Grafflin, eds., *Cellulose and Cellulose Derivatives*, 2nd ed., Part I, Interscience Publishers, New York, 1954.

BIBLIOGRAPHY

1. Bevington, J. C., H. W. Melville, and R. P. Taylor, *J. Polymer Sci.* **12,** 449 (1954).
2. Carothers, W. H., *J. Am. Chem. Soc.* **51,** 2548 (1929); *Chem. Revs.* **8,** 353 (1931).
3. Flory, P. J., *J. Am. Chem. Soc.* **59,** 241 (1937).
4. Joyce, R. M., W. E. Hanford, and J. Harmon, *J. Am. Chem. Soc.* **70,** 2529 (1948).
5. Lansing, W. D., and E. O. Kraemer, *J. Am. Chem. Soc.* **57,** 1369 (1935).
6. Natta, G., *et al.*, *J. Am. Chem. Soc.* **77,** 1708 (1955); G. Natta, *J. Polymer Sci.* **16,** 143 (1955).
7. Price, C. C., R. W. Kell, and E. Krebs, *J. Am. Chem. Soc.* **64,** 1103 (1942).
8. Staudinger, H., *Ber.* **53,** 1073 (1920).
9. Staudinger, H., *Helv. Chim. Acta* **8,** 67 (1925).
10. Staudinger, H., *Ber.* **61,** 2427 (1928).
11. Staudinger, H., and W. Heuer, *Ber.* **63,** 222 (1930).
12. Staudinger, H., and A. Steinhofer, *Ann.* **517,** 35 (1935).

C. Molecular Forces and Chemical Bonding in Polymers

The nature of the bonds which hold atoms together in molecules is explained by quantum mechanics in terms of an atom consisting of a small nucleus, concentrating the mass and positive charge, surrounded by clouds or shells of electrons relatively far away. It is among the outermost, more loosely bound electrons, called *valence electrons*, that chemical reactions and primary bond formation take place.

Primary bonds

Ionic bond The most stable electronic configuration for all but very light atoms is a complete outer shell of eight electrons, called an *octet*. This structure may be obtained by the donation of an electron by one atom to another:

$$\mathrm{Na}\cdot + \cdot\underset{\cdot\cdot}{\overset{\cdot\cdot}{\mathrm{Cl}}}: \longrightarrow \mathrm{Na}^{+} + :\underset{\cdot\cdot}{\overset{\cdot\cdot}{\mathrm{Cl}}}:^{-}$$

This results in electrostatic charges on the atoms, creating the attractive forces of these *ionic bonds*. Ionic bonds lead to the buildup of large salt crystals. These bonds are not usually found in macromolecular substances.

Covalent bond These bonds are formed when one or more pairs of valence electrons are shared between two atoms, again resulting in stable electronic shells:

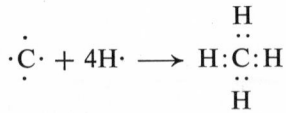

The *covalent bond* is the predominant bond in polymers.

Coordinate bond This bond is similar to the covalent bond in that electrons are shared to produce stable octets; but in the coordinate bond both of the shared electrons come from one atom, called the *donor*. Addition compounds of boron trichloride are common examples:

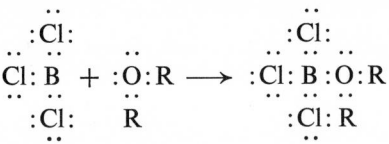

where R is an organic group. The donor atom (O) effectively loses an electron and carries a positive charge, whereas the *acceptor* (B) becomes negative. Thus, the *coordinate* or *semipolar bond* has properties between those of the ionic and the covalent bonds.

TABLE 1-2. Typical primary bond lengths and energies

Bond	Bond Length, A	Dissociation Energy, kcal/mole
C—C	1.54	83
C=C	1.34	146
C—H	1.10	99
C—N	1.47	73
C≡N	1.15	213
C—O	1.46	86
C=O	1.21	179
C—F*	1.32–1.39	103–123
C—Cl	1.77	81
N—H	1.01	93
O—H	0.96	111
O—O	1.32	35

* The bond length decreases and the dissociation energy increases as additional fluorine atoms are substituted on the same carbon atom (1).

Metallic bonds In the *metallic bond*, which does not occur in polymers, the number of valence electrons is far too small to provide complete shells for all the atoms. The resulting bonds involve the concept of positively charged atoms embedded in a permeating "gas" of electrons free to move about at will. The features of high electrical conductivity, nonsaturation of valence, and lack of direction in the bonding forces are accounted for. The metallic bond is perhaps the least understood of all the primary bond types.

Typical primary-bond distances and energies From studies of the positions of atoms in molecules and the energetics of molecular formation and dissociation, it is possible to assign typical energies and lengths to primary bonds. Table 1-2 lists some of these properties of interest in polymeric systems. The angles between successive single bonds involving the atomic arrangements usual in polymers range between 105° and 113°, not far from the tetrahedral angle of 109°28′.

Secondary bond forces

Experience shows that, even after all the primary valences within covalent molecules are saturated, there are still forces acting between the molecules. These are generally known as *secondary valence* or *intermolecular forces*. The first quantitative statements about them were made by van der Waals in his studies of the equation of state of real gases; hence they are frequently called *van der Waals forces*.

Dipole forces (3) When different atoms in a molecule carry equal and opposite electric charges, the molecule is said to be *polar* or to have a *dipole moment*. At large distances such a molecule acts like an electrically neutral system, but at molecular distances the charge separation becomes significant and leads to a net intermolecular force of attraction. The magnitude of the interaction energy depends on the mutual alignment of the dipoles. Molecular orientation of this sort is always opposed by thermal agitation; hence the dipole force is strongly dependent upon temperature.

Induction forces (2) A polar molecule also influences surrounding molecules which do not have permanent dipoles. The electric field associated with a dipole causes slight displacements of the electrons and nuclei of surrounding molecules, which lead to induced dipoles. The intermolecular force between the permanent and induced dipoles is called the *induction force*. The ease with which the electronic and nuclear

displacements are made is called the *polarizability* of the molecule. The energy of the induction force is always small and is independent of temperature.

Dispersion forces (4) The existence of intermolecular forces in nonpolar materials, plus the small temperature dependence of intermolecular forces even where the dipole effect is known to far outweigh the induction effect, suggests the presence of a third type of intermolecular force. All molecules have time-varying dipole moments which average out to zero and which arise from different instantaneous configurations of the electrons and nuclei. These fluctuations lead to perturbations of the electronic clouds of neighboring atoms and give rise to attractive forces called *dispersion forces.* They are present in all molecules and make up a major portion of the intermolecular forces unless very strong dipoles are present. In nonpolar materials only the dispersion forces exist. They are independent of temperature.

Interrelation of intermolecular forces The energy of the intermolecular attractive forces varies as the inverse sixth power of the intermolecular distance. As with primary bond forces, repulsion arises when the atoms approach more closely than the equilibrium bond distance of 3–5 A. The energy of typical secondary bond attractive forces is 2–10 kcal/mole, divided among the three secondary bond types according to the polarizability and dipole moment of the bonding molecules.

The hydrogen bond The bond in which a hydrogen atom is associated with two other atoms is particularly important in many polymers, including proteins, and is held by many to be essential to life processes. Since the classical concepts of chemical bonding allow hydrogen to form only one covalent bond, the hydrogen bond can be considered electrostatic or ionic in character. This model does not, however, account for all the properties of the hydrogen bond; it is appealing to consider the bond covalent in some cases. The hydrogen bond occurs between two functional groups in the same or different molecules. The hydrogen is usually attached to an acidic group (a proton donor), typically a hydroxyl, carboxyl, amine, or amide group. The other group must be basic, usually oxygen as in carbonyls, ethers, or hydroxyls; nitrogen as in amines and amides; and occasionally halogens. Some typical hydrogen bond lengths and energies are given in Table 1-3. The association of such polar liquid molecules as water, alcohols, and hydrofluoric acid, the formation of dimers of simple organic acids, and important structural effects in polar polymers such as nylon, cellulose, and proteins, are due to hydrogen bonding.

TABLE 1-3. Typical hydrogen bond lengths and energies

Bond	Bond Length, A*	Dissociation Energy, kcal/mole
C—H - - - N	...	3
O—H - - - N	2.8	...
O—H - - - O	2.6–2.8	3–6
O—H - - - Cl	3.1	...
N—H - - - N	3.1	3–5
N—H - - - O	2.9–3.0	4
N—H - - - Cl	3.2	...
N—H - - - F	2.8	...
F—H - - - F	2.4	7

* A to B distance in A—H - - - B.

Intermolecular forces and physical properties

Secondary bond forces are not of great importance in the formation of stable chemical compounds. They lead rather to the aggregation of separate molecules into solid and liquid phases. As a result many physical properties such as volatility, viscosity, surface tension and frictional properties, miscibility, and solubility are determined largely by intermolecular forces.

The *cohesive energy* is the total energy necessary to remove a molecule from a liquid or solid to a position far from its neighbors. This is approximately equal to the heat of vaporization or sublimation at constant volume and can be estimated from thermodynamic data. The cohesive energy per unit volume, sometimes called the *specific cohesive energy* or *cohesive energy density*, and its variation with molecular structure, illustrate the effects of intermolecular forces on the physical properties of matter.

Volatility and molecular weight The tendency of a molecule to volatilize from its liquid is a function of its total translational energy and therefore of the temperature. The boiling point depends on the relation of the translational energy to the cohesive energy, and thus is a function of molecular weight in a homologous series. At high molecular weights the total cohesive energy per molecule becomes greater than primary bond energy, and the molecules decompose before they volatilize. This point is reached at molecular weights far below those of typical polymers.

The melting point is also related to the cohesive energy, but here another

important factor comes into play. This is the influence of molecular order or entropy. In thermodynamic terms changes of state take place only when the free energy* change in the process

$$\Delta F = \Delta H - T\Delta S$$

is favorable, and the energy term ΔH may easily be outweighed by the entropy term $T\Delta S$ whenever a radical change in molecular configurations occurs in the process. Thus in general a high boiling point is associated with a high melting point, but the relation between melting point and molecular structure is fairly complicated. Symmetrical molecules, which have low entropies of fusion, melt at higher temperatures than do similar but less symmetrical molecules.

Effect of polarity A molecule containing strongly polar groups exerts correspondingly strong attractive forces on its neighbors. This is reflected in higher boiling and melting points and other manifestations of higher cohesive energy density.

Miscibility and solubility These properties are also determined by the intermolecular forces. The thermal effect on mixing or solution is the difference between the cohesive energy of the mixture and that of the individual pure components. Again entropy considerations are important, but in general a negative heat of mixing favors solubility and a positive heat of mixing favors immiscibility. The intermolecular forces therefore lead directly to the solubility law of "like dissolves like."

Intermolecular forces and polymer types Table 1-4 lists the cohesive energy densities of some typical polymers. From these data may be drawn some generalities regarding the dependence of polymer properties on secondary bond forces.

If the intermolecular forces are small and the cohesive energy is low, and the molecules have relatively flexible chains, they comply readily to applied stresses and have properties usually associated with elastomers. Somewhat higher cohesive energy densities, accompanied in some cases by bulky side groups giving stiffer chains, are characteristic of typical plastics. If the cohesive energy is higher still, the materials exhibit the high resistance to stress, high strength, and good mechanical properties typical of fibers, especially where molecular symmetry is favorable for the formation of ordered crystalline regions (Chapter 5).

Symmetry and crystallizability provide some exceptions, as in the case of

* This book follows the American convention of defining the (Gibbs) *free energy* as $F = H - TS$, and the *work content* or Helmholtz free energy as $A = E - TS$.

***TABLE 1-4. Cohesive energy densities of linear polymers* (5, 6)**

Polymer	Repeat Unit	Cohesive Energy Density, cal/cm^3
Polyethylene	$—CH_2CH_2—$	62
Polyisobutylene	$—CH_2C(CH_3)_2—$	65
Polyisoprene	$—CH_2C(CH_3)=CHCH_2—$	67
Polystyrene	$—CH_2CH(C_6H_5)—$	74
Poly(methyl methacrylate)	$—CH_2C(CH_3)(COOCH_3)—$	83
Poly(vinyl acetate)	$—CH_2CH(OCOCH_3)—$	88
Poly(vinyl chloride)	$—CH_2CHCl—$	91
Poly(ethylene terephthalate)	$—CH_2CH_2OCOC_6H_4COO—$	114
Poly(hexamethylene adipamide)	$—NH(CH_2)_6NHCO(CH_2)_4CO—$	185
Polyacrylonitrile	$—CH_2CHCN—$	237

polyethylene. Despite its weak cohesive forces, its simple, highly symmetrical structure enhances its tendency to crystallize. As a result, polyethylene has the characteristics of a plastic or fiber.

GENERAL REFERENCES

Cottrell, T. L., *The Strengths of Chemical Bonds*, 2nd ed., Butterworth's Scientific Publications, London, 1958.

Ketelaar, J. A. A., *Chemical Constitution*, Elsevier Publishing Co., New York, 1953.

Pauling, L., *The Nature of the Chemical Bond*, 3rd ed., Cornell University Press, Ithaca, N.Y., 1960.

Pimentel, G. C., and A. L. McClellan, *The Hydrogen Bond*, W. H. Freeman and Co., San Francisco, Calif., 1960.

BIBLIOGRAPHY

1. Bryant, W. M. D., *J. Polymer Sci.* **56,** 277 (1962).
2. Debye, P., *Physik. Z.* **21,** 178 (1920); **22,** 302 (1921).
3. Keesom, W. H., *Physik. Z.* **22,** 129, 643 (1921); **23,** 225 (1922).
4. London, F., *Trans. Faraday Soc.* **33,** 8 (1937); F. London and R. Eisenschitz, *Z. Physik* **60,** 491 (1930).
5. Small, P. A., *J. Appl. Chem.* **3,** 71 (1953).
6. Walker, E. E., *J. Appl. Chem.* **2,** 470 (1952).

II

Structure and Properties of Polymers

2

Polymer Solutions

A. Criteria for Polymer Solubility

The solution process Dissolving a polymer is a slow process occurring in two stages. First, solvent is slowly imbibed into the polymer to produce a swollen gel. In the second stage, the gel gradually disintegrates into a true solution. Only the second stage is materially speeded by agitation. If the polymer is crosslinked by primary valence bonds or strong hydrogen bonds, or is highly crystalline, only swelling may take place.

Empirical rules of solubility Solubility relations in polymer systems are more complex than those among low-molecular-weight compounds, because of the size of the polymer molecules, the viscosity of their solutions, and the possible presence of a crystalline phase. There has nevertheless been some success in applying to polymers empirical solubility rules derived from the behavior of small molecules. Some of these generalities are as follows.

a. Chemical and structural similarity favors solubility. Solution takes place when the attractive forces of the solvent molecules for the solute molecules outweigh those between pairs of solvent molecules and between pairs of solute molecules. The solvent-solute attractions are greatest when the two molecules have similar polarity.

b. Solubility decreases as the molecular weight of the solute increases. This is the basis for the fractionation of polymers according to molecular weight.

c. Solubility decreases as the melting point of the solute increases. Crystalline polymers are relatively insoluble and often dissolve only at temperatures slightly below their crystalline melting points.

Solubility parameters Solubility occurs when the free energy of mixing

$$\Delta F = \Delta H - T\,\Delta S$$

is negative. Since the entropy of mixing is positive, the sign of ΔF is determined by the sign and magnitude of the heat of mixing, ΔH. If there is a positive interaction between polymer and solvent, ΔH is negative and solution occurs. This is usually the case with polar molecules, especially if hydrogen bonds can form with the solvent. If dispersion forces only are involved, solution may or may not occur, depending on the magnitude of ΔH. For endothermic mixing, the heat of mixing per unit volume is given (Hildebrand and Scott, gen. ref.) by

$$\Delta H = v_1 v_2 (\delta_1 - \delta_2)^2$$

where v is volume fraction, and subscripts 1 and 2 refer to solvent and polymer, respectively. The quantity δ^2 is the cohesive energy density, or for

***TABLE 2-1. Typical values of the solubility parameter* δ (1, 2, 3)**

Solvent	δ_1, (cal/cm^3)$^{1/2}$	Polymer	δ_2, (cal/cm^3)$^{1/2}$
n-Hexane	7.24	Polyethylene	7.9
Carbon tetrachloride	8.58	Polystyrene	8.6
2-Butanone	9.04	Poly(methyl methacrylate)	9.1
Benzene	9.15	Poly(vinyl chloride)	9.5
Chloroform	9.24	Poly(ethylene terephthalate)	10.7
Acetone	9.71	66 nylon	13.6
Methanol	14.5	Polyacrylonitrile	15.4

small molecules the energy of vaporization per unit volume. It is generally true that, in the absence of interactions, solution will not occur if $\delta_1 - \delta_2$ is large, but the converse is not always true, probably because specific interaction effects dominate. Some values of the parameter δ are given in Table 2-1. For polymers, they are the square roots of the cohesive energy densities in Table 1-4.

In contrast to the above considerations of the thermodynamics of dissolution of polymers, the rate of this step depends primarily on how rapidly the polymer and the solvent diffuse into one another (4). Solvents promoting rapid solubility are usually small, compact molecules, but these kinetically good solvents need not be thermodynamically good as well. Mixtures of a kinetically good and a thermodynamically good liquid are often very powerful and rapid polymer solvents.

GENERAL REFERENCES

Allen, P. W., Chap. 1 in P. W. Allen, ed., *Techniques of Polymer Characterization*, Academic Press, New York, 1959.

Hildebrand, J. H., and R. L. Scott, *Solubility of Non-Electrolytes*, 3rd ed., Reinhold Publishing Corp., New York, 1950.
Walker, E. E., "Solubility," Chap. 13 in R. Hill, ed., *Fibres from Synthetic Polymers*, Elsevier Publishing Co., New York, 1953.

BIBLIOGRAPHY

1. Bristow, G. M., and W. F. Watson, *Trans. Faraday Soc.* **54,** 1731, 1742 (1958).
2. Small, P. A., *J. Appl. Chem.* **3,** 71 (1953).
3. Walker, E. E., *J. Appl. Chem.* **2,** 470 (1952).
4. Ueberreiter, K., and F. Asmussen, *J. Polymer Sci.* **57**, 187 (1962); F. Asmussen and K. Ueberreiter, *ibid.*, p. 199.

B. Conformations of Dissolved Polymer Chains

As specified in Chapter 1, those arrangements of the polymer chain differing by reason of rotations about single bonds are termed *conformations.* In solution, a polymer molecule is a randomly coiling mass most of whose conformations occupy many times the volume of its segments alone. The average density of segments within a dissolved polymer molecule is of the order of 0.01 g/cm^3. The size of the molecular coil is very much influenced by the polymer-solvent interaction forces. In a thermodynamically "good" solvent, where polymer-solvent contacts are favored, the coils are relatively extended. In a "poor" solvent the converse is true. It is the purpose of this section to describe the conformational properties of both real and ideal polymer chains.

The random chain In contrast to the fully extended or *contour length* of the chain, the size of a randomly coiling polymer molecule is commonly designated either by the root-mean-square distance between its ends, $(\overline{r^2})^{1/2}$, or by its *radius of gyration*, the root-mean-square distance of the elements of the chain from its center of gravity, $(\overline{s^2})^{1/2}$. For linear polymers the mean-square end-to-end distance and the square of the radius of gyration are simply related: $\overline{r^2} = 6\overline{s^2}$.

A simple model of a polymer chain consists of a series of x links of length l joined in a linear sequence with no restrictions on the angles between successive bonds. The probability W that such an array has a given end-to-end distance r can be calculated by the classical random flight method (1, 5). The most important result of the calculation is that the end-to-end distance is proportional to the square root of the number of links:

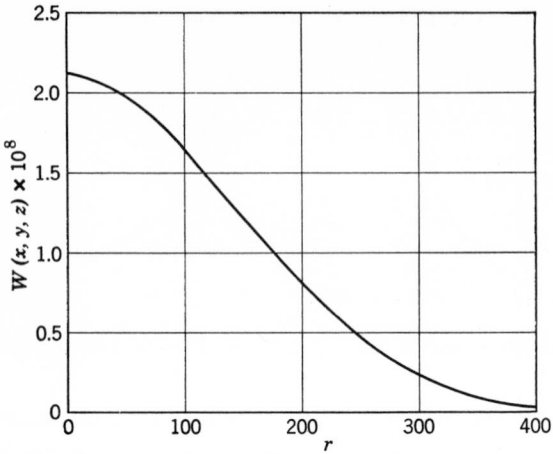

Fig. 2-1. Gaussian distribution of end-to-end distances in a random coil of 10^4 links each 2.5 A long. $W(x, y, z)$ is expressed in A^{-3} and r in A.

$(\overline{r_0^2})^{1/2} = lx^{1/2}$. (The subscript zero indicates the *unperturbed* or random flight end-to-end distance.) Thus $(\overline{r_0^2})^{1/2}$ is proportional to $M^{1/2}$, or $\overline{r_0^2}/M$ is a characteristic property of the polymer chain structure, independent of molecular weight or the surroundings (i.e., solvent).

The distribution of end-to-end distances over the space coordinates $W(x, y, z)$ is given by the Gaussian distribution function shown graphically in Fig. 2-1. This is the density distribution of end points and shows that, if one end of the chain is taken at the origin, the probability is highest of finding the other end in a unit volume near the origin. This probability decreases continuously with increasing distance from the origin. On the other hand, the probability of finding a chain end within the volume of a spherical shell between distances r and $r + dr$ from the origin has a maximum, as shown in Fig. 2-2, corresponding to the fact that $(\overline{r_0^2})^{1/2}$ has a nonzero value. It should be noted that the Gaussian distribution does not fall to zero at large extensions and so must fail to describe the actual conformations of the chain where r nears the contour length. Over most of the range of interest for the dilute solution properties of polymers this is of little consequence; better approximations are available where needed, as in the treatment of rubber elasticity (Chapter 6*B*).

The equivalent random chain (4) The problem of restrictions on bond angles and restricted rotation about bonds in real polymer chains is simply handled in principle. It is sufficient to replace the actual bond length l and number of segments x by a smaller number of hypothetical longer elements so chosen that the contour length of the chain remains unchanged.

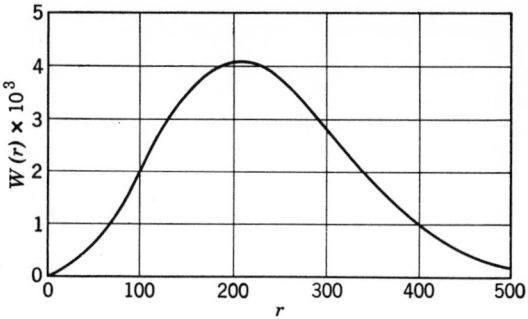

Fig. 2-2. Radial distribution of end-to-end distances for the coil of Fig. 2-1. $W(r)$ is expressed in A^{-1}.

If the number of segments is still large enough for the statistical treatment to apply, none of the conclusions above is altered. This is true for all but a few cases of interest: some unusually stiff chains such as cellulose and its derivatives are not long enough at ordinary molecular weights to exhibit Gaussian behavior (2, 3).

Detailed calculation from theory of end-to-end distances for actual polymers has so far not met with outstanding success, because of difficulties in evaluating hindrances to free rotation about the bonds. Actual values of polymer chain dimensions must still be determined experimentally.

Nonlinear chains Since branched chains possess a multiplicity of ends, it is more suitable to speak in terms of the radius of gyration $(\overline{s^2})^{1/2}$. A branched molecule occupies a smaller volume than a linear one with the same number of segments, i.e., the same molecular weight. It is convenient to express this diminution of size as a factor $g = \overline{s^2}$ (branched)$/\overline{s^2}$ (linear), which can be calculated statistically for various degrees and types of branching (6). For random branching, five trifunctional branch points per molecule reduce g to about 0.70. The change in size is the basis of a method of measuring branching (Chapter 3*E*).

The excluded volume The random flight calculation allows many conformations excluded to actual chains because different links approach each other closer than the average thickness of the real chain. These longer-range intramolecular interactions occur within a volume, occupied by one molecule, from which all others are effectively excluded. The calculation of the deviation from random flight statistics within this *excluded volume* remains a major problem for which a definitive solution has not been found. It is clear, however, that more extended conformations are favored; real polymer chains are in general expanded by a factor α such that $(\overline{r^2})^{1/2} = \alpha(\overline{r_0^2})^{1/2}$. The calculation of α is treated in Section *C*.

GENERAL REFERENCES

Bueche, F., *Physical Properties of Polymers*, Interscience Publishers, a Division of John Wiley & Sons, New York, 1962, Chap. 1.

Casassa, E. F., *Ann. Rev. Phys. Chem.* **11,** 477 (1960).

Flory, P. J., *Principles of Polymer Chemistry*, Cornell University Press, Ithaca, N.Y., 1953, Chap. X.

Tanford, C., *Physical Chemistry of Macromolecules*, John Wiley & Sons, New York, 1961, Secs. 3, 4.

Tompa, H., *Polymer Solutions*, Academic Press, New York, 1956, Chap. 8.

BIBLIOGRAPHY

1. Chandrasekhar, S., *Revs. Modern Phys.* **15,** 1 (1943).
2. Doty, P., N. S. Schneider, and A. Holtzer, *J. Am. Chem. Soc.* **75,** 754 (1953).
3. Kratky, O., and G. Porod, *Rec. trav. chim.* **68,** 1106 (1949).
4. Kuhn, W., *Kolloid Z.* **76,** 258 (1936); **87,** 3 (1939).
5. Rayleigh, Lord, *Phil. Mag.* **37,** 321 (1919).
6. Zimm, B. H., and W. H. Stockmayer, *J. Chem. Phys.* **17,** 1301 (1949).

C. Thermodynamics of Polymer Solutions*

The behavior of polymers toward solvents is characteristic and different from that of low-molecular-weight substances. The size and conformations of dissolved polymer molecules require special theoretical treatment to explain their solution properties. Conversely, it is possible to obtain information about the size and shape of polymer molecules from studies of their solution properties.

The first group of properties of interest includes those depending upon equilibrium between two phases, one or both of which is a solution of the polymer. In this section are discussed situations in which one phase is pure solvent; in Section *D*, solubility phenomena in which both phases contain polymer.

* The following books on thermodynamics are recommended as general references for review of this subject:

Glasstone, S., *Thermodynamics for Chemists*, D. Van Nostrand Co., Princeton, N. J., 1947.

Lewis, G. N., and M. Randall, *Thermodynamics*, revised by K. S. Pitzer and L. Brewer, 2nd ed., McGraw-Hill Book Co., New York, 1961.

MacDougall, F. H., *Physical Chemistry*, 3rd ed., The Macmillan Co., New York, 1952.

Tompa, H., *op. cit.*, Chaps. 2, 3.

Wall, F. T., *Chemical Thermodynamics*, W. H. Freeman and Co., San Francisco, Calif., 1958.

Thermodynamics of simple liquid mixtures

From the condition for equilibrium between two phases may be derived relations such as that for the free energy of dilution of a solution:

$$\Delta F_A = kT \ln (p_A/p_A^0) \tag{2-1}$$

where ΔF is the free energy of dilution resulting from the transfer of one molecule of liquid A from the pure liquid state with vapor pressure p_A^0 to a large amount of solution with vapor pressure p_A. This expression is written in terms of one of the *colligative properties* of solutions: vapor pressure lowering, freezing point depression, boiling point elevation, and osmotic pressure. The values of these properties cannot be related to the composition of the system by pure thermodynamic reasoning. It is necessary to know the type of variation of one property with concentration; the variation of the others may then be deduced.

Ideal solutions In the simplest type of mixing, the molecules of components A and B have roughly the same size and shape and similar force fields. They may then form an *ideal solution*, defined as one in which Raoult's law is obeyed. This law states that the partial vapor pressure of each component in the mixture is proportional to its mole fraction. Therefore

$$p_A = p_A^0 [N_A/(N_A + N_B)] = p_A^0 n_A \tag{2-2}$$

where n denotes mole fraction, and Eq. 2-1, for example, becomes

$$\Delta F_A = kT \ln n_A \tag{2-3}$$

The total free energy of mixing is

$$\begin{aligned} \Delta F &= N_A \Delta F_A + N_B \Delta F_B \\ &= kT(N_A \ln n_A + N_B \ln n_B) \end{aligned} \tag{2-4}$$

The conditions for ideal mixing imply that the heat of mixing $\Delta H = 0$, i.e., the components mix without change in energy. Since $\Delta F = \Delta H - T\Delta S$, the entropy of mixing is given by

$$\Delta S = -k(N_A \ln n_A + N_B \ln n_B) \tag{2-5}$$

which is positive for all compositions, so that, by the second law, spontaneous mixing occurs in all proportions.

Other types of mixing In practice few liquid mixtures obey Raoult's law. Three types of deviations are distinguished:

a. "Athermal" solutions, in which $\Delta H = 0$ but ΔS is no longer given by Eq. 2-5.

b. "Regular" solutions, in which ΔS has the ideal value but ΔH is finite.

c. "Irregular" solutions, in which both ΔH and ΔS deviate from the ideal values.

It is usually found in systems of similar-sized molecules that ΔS is nearly ideal when $\Delta H = 0$; therefore athermal solutions are nearly ideal. Many mixtures are found for which ΔH is finite, however. Such cases arise when the intermolecular force fields around the two types of molecule are different. Expressions for the heat of mixing may be derived in terms of the cohesive energy density of these force fields (Section *A*).

Entropy and heat of mixing of polymer solutions

Deviations from ideal behavior Polymer solutions invariably exhibit large deviations from Raoult's law except at extreme dilutions, where ideal behavior is approached as an asymptotic limit. At concentrations above a few per cent, deviations from ideality are so great that the ideal law is of little value for predicting or correlating the thermodynamic properties of polymer solutions. Even if the mole fraction is replaced with the volume fraction, in view of the different sizes of the polymer and solvent molecules, there is not a good correlation with the experimental results (Fig. 2-3).

Entropy of mixing Deviations from ideality in polymer solutions arise largely from small entropies of mixing. These are not abnormal but are the natural result of the large difference in molecular size between the two components. They can be interpreted in terms of a simple molecular model. The molecules in the pure liquids and the mixture are assumed to be representable without serious error by a lattice. A two-dimensional representation of such an arrangement for nonpolymer liquids is shown in Fig. 2-4. Whereas the molecules of a pure component can be arranged in only one way on such a lattice, assuming that they cannot be distinguished from one another, the molecules of a mixture of two components can be arranged on a lattice in a large but calculable number of ways, W. By the Boltzmann relation the entropy of mixing $\Delta S = k \ln W$. Equation 2-5 results for the case of molecules which can replace one another indiscriminately on the lattice.

It is assumed that the polymer molecules consist of a large number x of chain segments of equal length, flexibly joined together. Each link occupies one lattice site, giving the arrangement of Fig. 2-5. The solution is assumed to be concentrated enough that the occupied lattice sites are distributed at

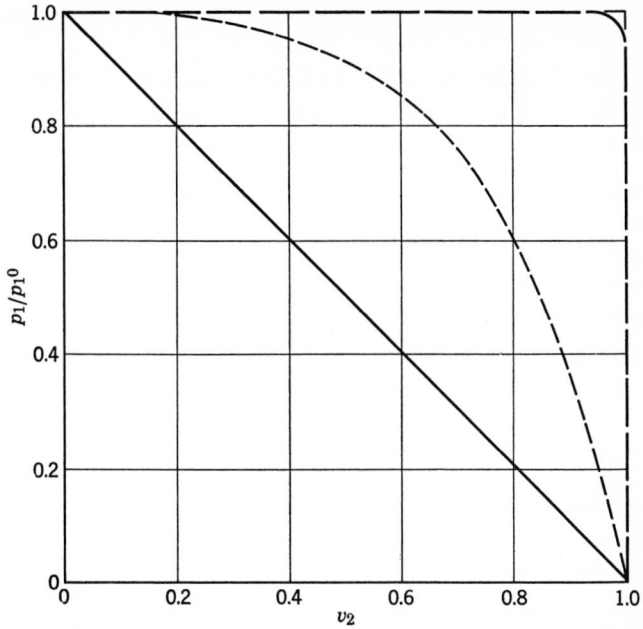

Fig. 2-3. Deviations of polymer solutions from ideality. The activity p_1/p_1^0 of the solvent is plotted against the volume fraction v_2 of the polymer. Dotted curve is experimental data for rubber in benzene (8). Dashed curve is Raoult's law; solid line is the approximation that the activity of the solvent equals its volume fraction.

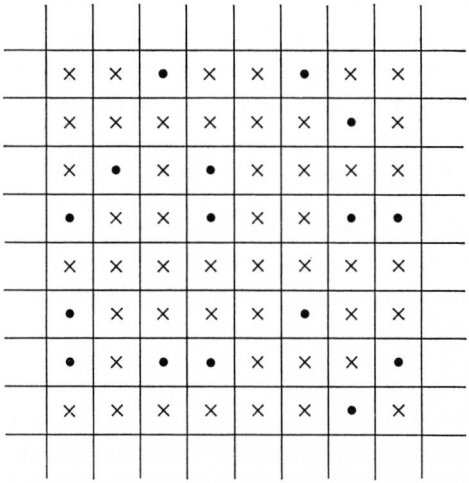

Fig. 2-4. Two-dimensional representation of molecules of nonpolymer liquids distributed over a lattice.

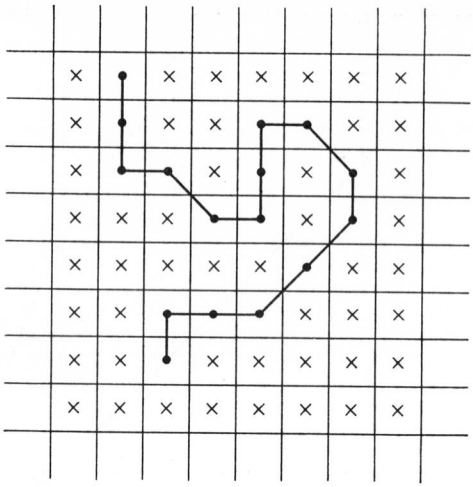

Fig. 2-5. Two-dimensional representation of a polymer molecule located in the liquid lattice.

random rather than lying in well-separated regions of x occupied sites each. It can now be seen qualitatively why the entropy of mixing of polymer solutions is small compared to that with normal solutes. There are fewer ways in which the same number of lattice sites can be occupied by polymer segments: fixing one segment at a lattice point severely limits the number of sites available for the adjacent segment. The approximate calculation of W for such a model is due separately to Flory (4) and Huggins (9); their results (which differ only in minor detail) are known as the *Flory-Huggins theory* of polymer solutions. The entropy of mixing is analogous to that given in Eq. 2-5 for simple liquids: for polymer solutions,

$$\Delta S = -k(N_1 \ln v_1 + N_2 \ln v_2) \tag{2-6}$$

where subscript 1 denotes the solvent and 2 the polymer, and v_1 and v_2 are *volume fractions* defined as

$$\begin{aligned} v_1 &= N_1/(N_1 + xN_2) \\ v_2 &= xN_2/(N_1 + xN_2) \end{aligned} \tag{2-7}$$

Heat and free energy of mixing The heat of mixing of polymer solutions is analogous to that of ordinary solutions:

$$\Delta H = \chi_1 kTN_1 v_2 \tag{2-8}$$

where χ_1 characterizes the interaction energy per solvent molecule divided by kT. Combining Eqs. 2-6 and 2-8 gives the Flory-Huggins expression for the free energy of mixing of a polymer solution with normal heat of mixing:

$$\Delta F = kT(N_1 \ln v_1 + N_2 \ln v_2 + \chi_1 N_1 v_2) \tag{2-9}$$

From this expression may be derived many useful relations involving experimentally obtainable quantities. For example, the partial molar free energy of mixing is

$$\Delta\bar{F}_1 = kT\left[\ln(1 - v_2) + \left(1 - \frac{1}{x}\right)v_2 + \chi_1 v_2^2\right] \tag{2-10}$$

and from it is obtained the osmotic pressure

$$\pi = -(kT/V_1)\left[\ln(1 - v_2) + \left(1 - \frac{1}{x}\right)v_2 + \chi_1 v_2^2\right] \tag{2-11}$$

where V_1 is the molecular volume of the solvent. If the logarithmic term in Eq. 2-11 is expanded and only low powers of v_2 are retained, the working equation for molecular weight determination by osmotic pressure measurement (Eq. 2-21) is obtained.

Experimental results with polymer solutions

The first system for which accurate experimental results were compared with the Flory-Huggins theories was that of rubber in benzene (6, 7). Free

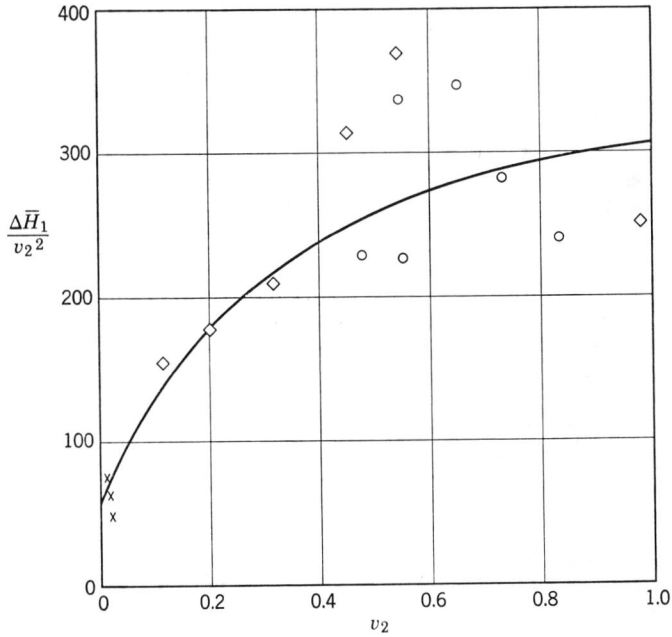

Fig. 2-6. Dependence of heat of dilution on volume fraction of polymer for rubber in benzene (6, 7.)

energies for the system were calculated from vapor pressure measurements by Eq. 2-1, and heats of solution were measured in a calorimeter. The predicted concentration dependence of the partial molar heat of dilution

$$\Delta \bar{H}_1 = kT\chi_1 v_2^2 \tag{2-12}$$

was not observed (Fig. 2-6). Despite this fact, the entropy of mixing, calculated from the heat and free energy, was in fair agreement with theory except for the dilute solution region (Fig. 2-7).

The properties of other systems (2, 11, 12) do not in general conform as well to the predictions of the theory. For example, the anticipated independence of χ_1 of v_2 is usually not observed (Fig. 2-8).

In addition to these shortcomings, which later modifications to the theory (10) have not completely overcome, application to dilute polymer solutions

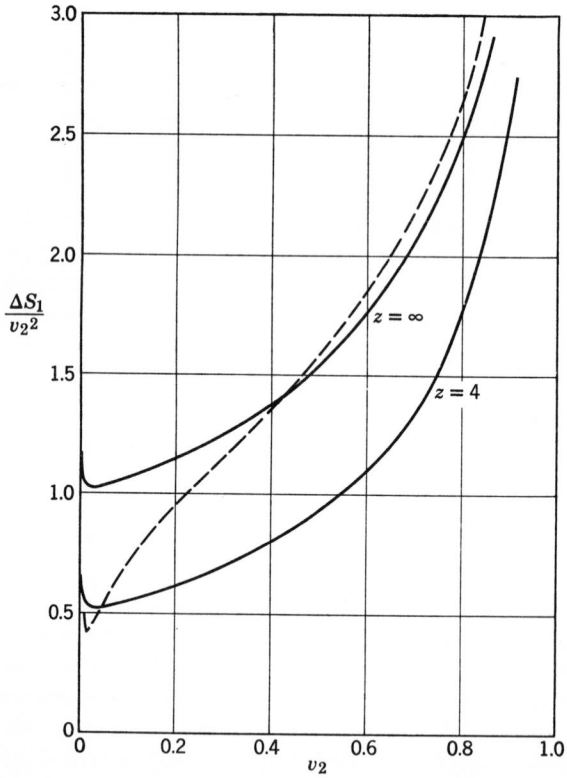

Fig. 2-7. Dependence of entropy of mixing on volume fraction of polymer for rubber in benzene (7). Dashed curve, experimental; solid curves, theoretical for various values of lattice parameter z corresponding to different values of χ_1.

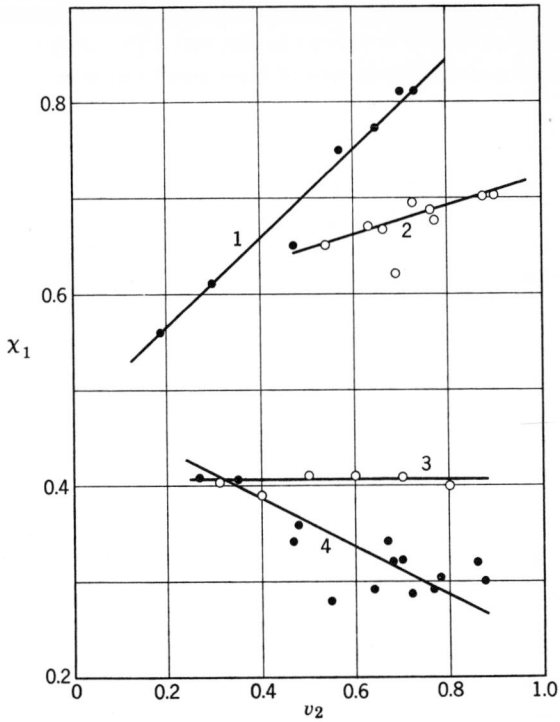

Fig. 2-8. Experimentally observed variation of χ_1 with concentration. Curve 1, poly-(dimethyl siloxane) in benzene (12); curve 2, polystyrene in benzene (2); curve 3, rubber in benzene (6, 7); curve 4, polystyrene in toluene (2).

requires a different approach, as described below. Nevertheless, the Flory-Huggins theory represents a major improvement over classical ideal solution theory for application to polymer solutions.

Dilute solutions

Flory-Krigbaum theory The lattice model used in the Flory-Huggins treatment neglects the fact that a very dilute polymer solution must be discontinuous in structure, consisting of domains or clusters of polymer chain segments separated on the average by regions of polymer-free solvent. Flory and Krigbaum (5) assume such a model, in which each cloud of segments is approximately spherical, with a density which is a maximum at the center and decreases in an approximately Gaussian function with distance from the center. Within the volume occupied by the segments of one molecule, all other molecules tend to be excluded. Within such an

excluded volume (Section *B*) occur long-range intramolecular interactions whose thermodynamic functions can be derived. The partial molar heat content, entropy, and free energy of these interactions are:

$$\begin{aligned} \Delta\bar{H}_1 &= kT\kappa_1 v_2^2 \\ \Delta\bar{S}_1 &= k\psi_1 v_2^2 \\ \Delta\bar{F}_1 &= kT(\kappa_1 - \psi_1)v_2^2 \end{aligned} \tag{2-13}$$

where, by comparison with the expansion of Eq. 2-10,

$$\kappa_1 - \psi_1 = \chi_1 - \tfrac{1}{2} \tag{2-14}$$

It is convenient to define as a parameter the *Flory temperature* Θ such that $\Theta = \kappa_1 T/\psi_1$. It follows that

$$\psi_1 - \kappa_1 = \psi_1(1 - \Theta/T) = \tfrac{1}{2} - \chi_1 \tag{2-15}$$

and that at the temperature $T = \Theta$ the partial molar free energy due to polymer-solvent interactions is zero and deviations from ideal solution behavior vanish. The excluded volume becomes smaller as the solvent becomes poorer and vanishes at $T = \Theta$, where the molecules interpenetrate one another freely with no net interactions. At temperatures below Θ they attract one another and the excluded volume is negative. If the temperature is much below Θ, precipitation occurs (Section *D*).

The extended conformation of a polymer chain may also be considered as determined by an equilibrium between an expansion force due to the excluded volume and a contraction force arising when the chain segments are expanded into less probable conformations. The free energy of the latter situation is analogous to that arising in the consideration of the rubberlike elasticity of a polymer network (Chapter 6*B*). The chain is considered extended linearly by a factor α over its dimension as calculated from random flight statistics. The actual root-mean-square end-to-end distance is given by $(\overline{r^2})^{1/2} = \alpha(\overline{r_0^2})^{1/2}$. The elastic part of the free energy is given by

$$\Delta F_{el} = kT[\tfrac{3}{2}(\alpha^2 - 1) - \ln \alpha^3] \tag{2-16}$$

At equilibrium the derivative with respect to α of the sum of the free energies equivalent to Eqs. 2-13 and 2-16 is zero. This leads to the evaluation of α in terms of thermodynamic quantities:

$$\alpha^5 - \alpha^3 = 2C_m\psi_1\left(1 - \frac{\Theta}{T}\right)M^{1/2} \tag{2-17}$$

where C_m lumps together numerical and molecular constants.

This equation, which involves the assumption of validity of theories involving intermolecular interactions (Eq. 2-16) as well as intramolecular interactions (Eq. 2-13), must be accepted with some reservations. It predicts that α increases without limit with increasing molecular weight; since $\overline{r_0^2}$ is proportional to M, $\overline{r^2}/M$ should increase with M. It also leads to the fact that, at $T = \Theta$, $\alpha = 1$ and molecular dimensions are unperturbed by intramolecular interactions. Since α depends on the entropy parameter ψ, it is larger in better solvents.

Extensions of the Flory-Krigbaum treatment The most far-reaching extensions of Flory and Krigbaum's dilute solution theory involve treating the statistical mechanics of polymer solutions in terms of molecular distribution functions, as first carried out by Zimm (14). This treatment evaluates the excluded volume u as the binary cluster integral, involving a correlation function describing the dependence of interactions between pairs of segments on distance from the center of the cluster of segments. The expansion factor α is given (15) as an infinite series in powers of a parameter

$$z = (3\pi\overline{r_0^2}/2)^{3/2}ux^2 \tag{2-18}$$

For small z (i.e., near $T = \Theta$) (3),

$$\alpha^2 = 1 + 1.33z + 1.74z^2 + \cdots \tag{2-19}$$

where higher terms, though clearly important, are exceedingly difficult to evaluate. Both this treatment and that of Flory lead to equations of the type

$$\alpha^5 - \alpha^3 = Cz \tag{2-20}$$

but the numerical value of C is in doubt by a factor of about 2.

The *second virial coefficient* A_2 is related to the coefficient of v_2^2 in the expansion of Eq. 2-11,

$$\pi = \frac{kT}{V_1}\left(\frac{v_2}{x} + (\tfrac{1}{2} - \chi_1)v_2^2 + \cdots\right) \tag{2-21}$$

and is defined (1, 13) as

$$A_2 = \frac{\bar{v}_2^2}{N_0V_1}(\tfrac{1}{2} - \chi_1)\,\mathrm{f}(z) \tag{2-22}$$

consistent with Eq. 3-5. Here $\bar{v}_2$ is the specific volume of the polymer. The contribution of the Flory treatment, $\frac{1}{2} - \chi_1$, is independent of molecular weight; f(z) decreases slowly with increasing M. However, a serious discrepancy exists between the predicted and the observed dependence of A_2 on M (and T). This represents a major shortcoming of existing theories of polymer solutions [see (15); and Casassa, Stockmayer, gen. refs.].

GENERAL REFERENCES

Casassa, E. F., *Ann. Rev. Phys. Chem.* **11,** 477 (1960).
Flory, P. J., *op. cit.*, Chap. XII.
Grimley, T. B., "The Theory of High Polymer Solutions," pp. 1–36 in J. C. Robb and F. W. Peaker, eds., *Progress in High Polymers*, Vol. 1, Academic Press, New York, 1961.
Stockmayer, W. H., *Makromol. Chem.* **35,** 54 (1960).
Tompa, H., *op. cit.*, Chap. 4.

BIBLIOGRAPHY

1. Albrecht, A. C., *J. Chem. Phys.* **27,** 1002 (1957).
2. Bawn, C. E. H., R. F. J. Freeman, and A. R. Kamaliddin, *Trans. Faraday Soc.* **46,** 677 (1950).
3. Fixman, M., *J. Chem. Phys.* **23,** 1656 (1955).
4. Flory, P. J., *J. Chem. Phys.* **10,** 51 (1942)
5. Flory, P. J., and W. R. Krigbaum, *J. Chem. Phys.* **18,** 1086 (1950).
6. Gee, G., *J. Chem. Soc.* **1947,** 280.
7. Gee, G., and W. J. C. Orr, *Trans. Faraday Soc.* **42,** 507 (1946).
8. Gee, G., and L. R. G. Treloar, *Trans. Faraday Soc.* **38,** 147 (1942).
9. Huggins, M. L., *J. Phys. Chem.* **46,** 151 (1942); *J. Am. Chem. Soc.* **64,** 1712 (1942); *Ann. N. Y. Acad. Sci.* **43,** 1 (1942).
10. Huggins, M. L., *Physical Chemistry of High Polymers*, John Wiley & Sons, New York, 1958.
11. Krigbaum, W. R., and D. O. Geymer, *J. Am. Chem. Soc.* **81,** 1859 (1959).
12. Newing, M. J., *Trans. Faraday Soc.* **46,** 613 (1950).
13. Yamakawa, H., *J. Phys. Soc. Japan* **13,** 87 (1958); M. Kurata and H. Yamakawa, *J. Chem. Phys.* **29,** 311 (1958).
14. Zimm, B. H., *J. Chem. Phys.* **14,** 164 (1946).
15. Zimm, B. H., W. H. Stockmayer, and M. Fixman, *J. Chem. Phys.* **21,** 1716 (1953); see also M. Kurata, W. H. Stockmayer, and A. Roig, *ibid.*, **33,** 151 (1960); M. Kurata and W. H. Stockmayer, paper presented at the International Symposium on Macromolecular Chemistry, Montreal, Can., July, 1961.

D. Phase Separation in Polymer Solutions

This section is concerned with the equilibrium between two liquid phases, both of which contain amorphous polymer and one or more solvents. The treatment of cases involving a crystalline polymer phase is more complex and is given in Chapter 5*E*.

When the temperature of a polymer solution is lowered, the solvent becomes thermodynamically poorer. Finally a temperature is reached below which polymer and solvent are no longer miscible in all proportions.

At a lower temperature the mixture separates into two phases. Phase separation may also be brought about by the addition of a nonsolvent liquid to the solution. In these cases, an upper consolute temperature is reached. Rarely, phase separation occurs on raising the temperature until a lower consolute temperature is reached (6, 9).

Binary polymer-solvent systems The condition for equilibrium between two phases in a binary system is that the partial molar free energy of each component be equal in each phase. This condition corresponds to the requirement that the first and second derivatives of $\Delta\bar{F}_1$ (Eq. 2-10) with respect to v_2 be zero. Application of this condition leads to the critical concentration at which phase separation first appears:

$$v_{2c} = \frac{1}{1 + x^{1/2}} \simeq \frac{1}{x^{1/2}} \tag{2-23}$$

This is a rather small volume fraction; for a typical polymer ($x \simeq 10^4$), $v_{2c} \simeq 0.01$. The critical value of χ_1 is given by

$$\chi_{1c} = \frac{(1 + x^{1/2})^2}{2x} \simeq \tfrac{1}{2} + \frac{1}{x^{1/2}} \tag{2-24}$$

The critical value of χ_1 exceeds $\frac{1}{2}$ by a small increment depending on molecular weight and at infinite molecular weight equals $\frac{1}{2}$. The temperature at which phase separation begins is given by

$$\frac{1}{T_c} = \frac{1}{\Theta}\left[1 + \frac{1}{\psi_1}\left(\frac{1}{x^{1/2}} + \frac{1}{2x}\right)\right] \simeq \frac{1}{\Theta}\left(1 + \frac{C}{M^{1/2}}\right) \tag{2-25}$$

where C is a constant for the polymer-solvent system. Thus $1/T_c$ (°K) varies linearly with the reciprocal square root of molecular weight. The Flory temperature Θ is the critical miscibility temperature in the limit of infinite molecular weight.

The qualitative features of the theory are in agreement with experiment. The dependence of precipitation temperature on polymer concentration is shown in Fig. 2-9. The critical values T_c and v_{2c} correspond to the maxima in the curves. In the neighborhood of the critical point the polymer concentration in the dilute phase is extremely small. The phenomenon of the coexistence of two liquid phases, one of which is a dilute solution and the other nearly pure solvent, is called *coacervation* (2) (Section *E*).

The dependence of T_c upon molecular weight is shown in Fig. 2-10. The curves are accurately linear, and the Flory temperatures Θ obtained from the intercepts agree within experimental error ($<1°$) with those derived from osmotic measurements (5), taking Θ to be the temperature where A_2 is zero. Precipitation measurements on a series of fractions offer perhaps the best method of determining Θ.

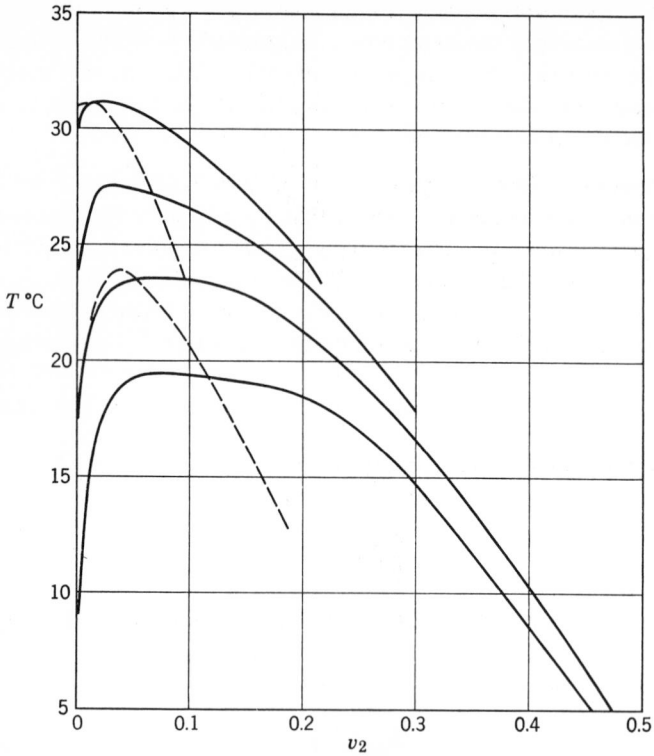

Fig. 2-9. Phase diagram showing precipitation temperature as a function of polymer concentration for fractions of polystyrene in diisobutyl ketone (7). Higher-molecular-weight fractions precipitate at higher temperatures. Solid lines, experimental; dotted lines, theoretical.

Ternary systems Although a thorough discussion of ternary systems (Flory, Tompa, gen. refs.) is outside the scope of this book, a few cases have particular interest. The most commonly encountered system is that of polymer, solvent, and nonsolvent. The phase relations are conveniently displayed in the usual ternary diagram (Fig. 2-11). The position of the *binoidal curve*, along which two phases are in equilibrium, depends upon molecular weight as indicated. The limiting critical point at infinite molecular weight is the analog of Θ in a two-component system. The ternary system is characterized by three interaction parameters χ_{ij}, one for each pair of components. With appropriate values of χ_{ij}, solubility may be achieved for some polymers in mixtures of liquids, neither of which is a solvent for that polymer. A well-known example is the solubility of cellulose acetate in mixtures of ethanol and chloroform but not in either pure liquid (1). In contrast, mixtures of two different polymer types in a single

solvent almost invariably separate into two phases, each of which contains practically all of one of the polymers. An example is rubber and polystyrene in benzene (3). The source of this general incompatibility lies in the low entropy of mixing (Eq. 2-6) arising from the small numbers of polymer molecules involved.

Multicomponent systems The theory of phase separation in systems comprising a heterogeneous polymer in a single solvent is developed with the simplifying assumption that the interaction constants χ_{ij} for all j between 0 and x are identical and equal to χ_{1j}; only the size parameter x varies from one molecular species to another. In theory all details of the system could be calculated from a knowledge of the size distribution of the polymer and the equilibrium conditions. In practice this would be an enormous task; but the principal point of interest, namely the efficiency with which molecular species are separated between the two phases, is simply derived. It is given by the equation

$$\ln (v_x'/v_x) = \sigma x \tag{2-26}$$

where v_x' and v_x are, respectively, the concentrations of species x in the precipitated or more concentrated phase (primed) and the dilute phase

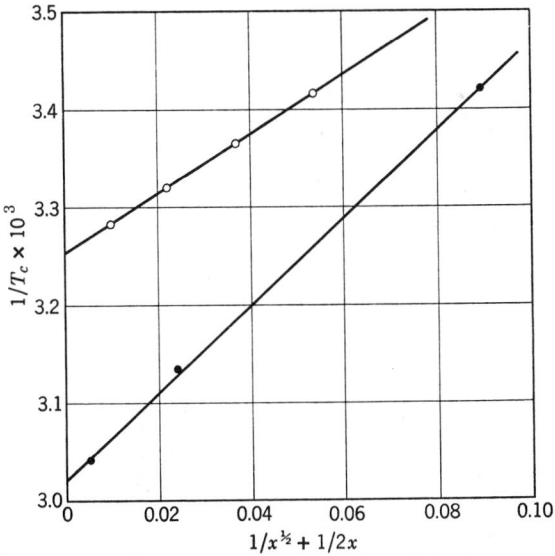

Fig. 2-10. Dependence of critical temperature for phase separation on molecular weight for (○) polystyrene in cyclohexane, and (●) polyisobutylene in diisobutyl ketone (7).

(unprimed), and

$$\sigma = v_2\left(1 - \frac{1}{\bar{x}_n}\right) - v_2'\left(1 - \frac{1}{\bar{x}_n'}\right) + \chi_1[(1 - v_2)^2 - (1 - v_2')^2] \quad (2\text{-}27)$$

Here v_2' and v_2 are the total polymer concentrations in the two phases, and $\bar{x}_n$ is the number average of x. The parameter σ thus depends in a complicated way on the relative amounts of all the polymer species in each phase. The following conclusions may be drawn from the theory:

a. A part of any given polymer species is always present in each phase: in fact, every species is actually more soluble in the precipitated phase, i.e., $v_x' > v_x$ for all values of x.

b. Since σ depends on the concentrations of all species and on the details of their weight distribution, the result of a fractionation cannot be predicted in advance unless the details of the distribution are known. (Although the Flory-Huggins theory (4) predicts that only the number average $\bar{x}_n$ appears

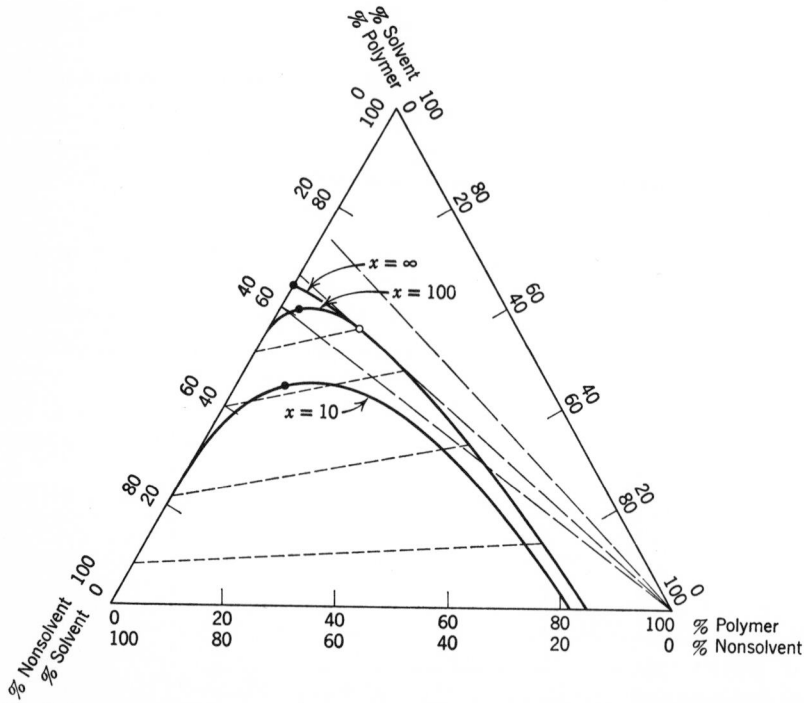

Fig. 2-11. Ternary phase diagram (11) for $\chi_{12} = \chi_{13} = 1.5$ and $\chi_{23} = 0$ and the indicated values of x. Dotted lines are tie lines for $x = 100$, connecting points representing the compositions of pairs of phases in equilibrium. Dashed lines represent constant solvent-nonsolvent ratios. ●, critical points; ○, precipitation threshold for $x = 100$.

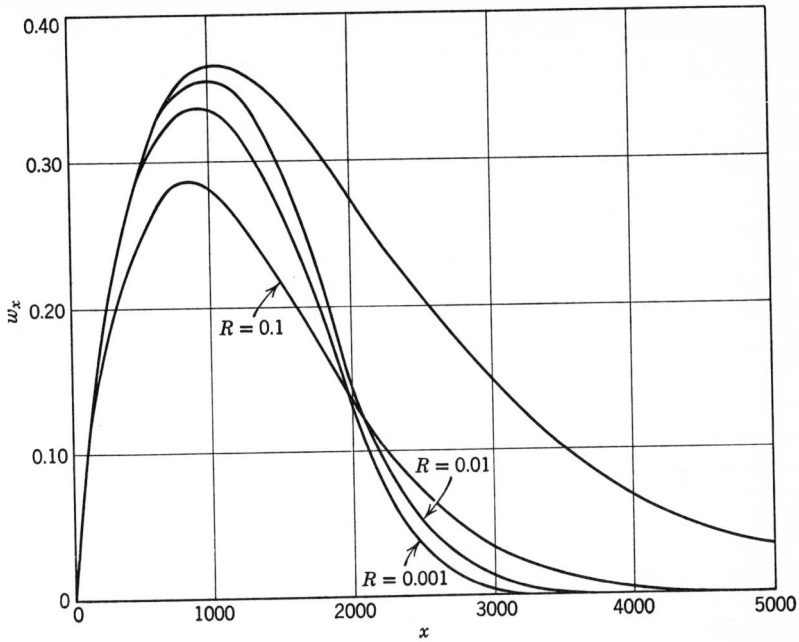

Fig. 2-12. Calculated separation of a polymer having the molecular weight distribution shown in the upper curve with $R = 0.1$, 0.01, and 0.001, f_x chosen to be $\frac{1}{2}$ at $x = 2000$ in each case (Flory, gen. ref.).

in σ, a more complete analysis (10) shows that σ depends upon higher moments of the distribution as well.)

c. Fractionation is significantly more efficient in very dilute solutions: if $R = v'/v$, the ratio of the volumes of the precipitated and dilute phases, the fraction f_x of the constituent x in the dilute phase is

$$f_x = 1/(1 + Re^{\sigma x}) \tag{2-28}$$

and that in the precipitated phase is

$$f_x' = Re^{\sigma x}/(1 + Re^{\sigma x}) \tag{2-29}$$

If $R \ll 1$, most of the smaller species remain in the dilute phase because of its greater volume. The larger partition factor v_x'/v_x for the higher-molecular-weight species causes them to appear selectively in the precipitated phase despite its smaller volume.

The resulting inefficient nature of the fractionation process is illustrated in Figs. 2-12 and 2-13. Figure 2-12 shows, in the three lower curves, the

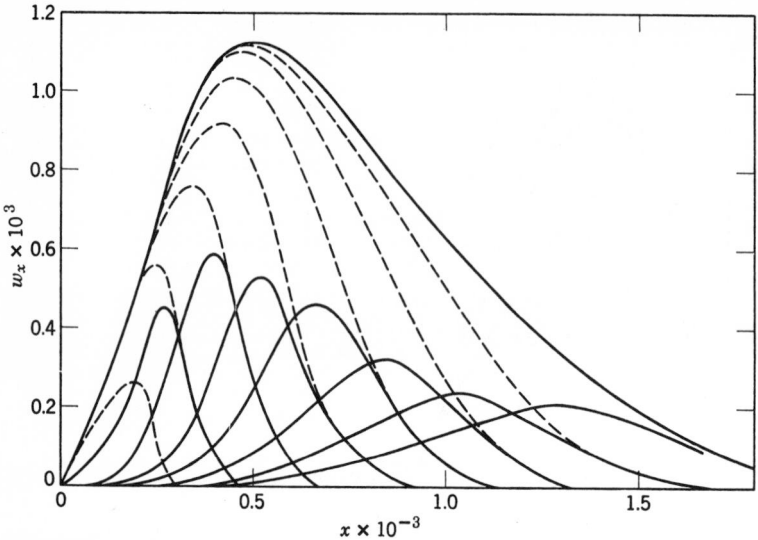

Fig. 2-13. Calculated distributions of a series of eight fractions separated from the distribution of the upper curve [Flory, gen. ref., data of (8)]. $R = 0.001$ in each case. Dashed curves represent the polymer remaining in the dilute phase after each successive precipitation.

distribution of polymer species left in the dilute phase after precipitation at various values of R. The initial distribution is shown in the upper curve. The increase in efficiency as R becomes smaller is apparent. Figure 2-13 shows the overlapping distribution curves resulting even when R is unusually small. It is clear that separation into fractions which are close to monodisperse is a most difficult and time-consuming process. It should be noticed that the distribution curves of the fractions in Fig. 2-13 become narrower at lower molecular weights.

The foregoing discussion refers to fractionation by cooling from a single solvent. Similar considerations apply to fractionation at constant temperature in a solvent-nonsolvent system. A difference arises in that the solvent composition is different in the two phases. This difference should lead to more efficient fractionation in solvent-precipitant mixtures. Experimental evidence for such an effect has not been reported.

GENERAL REFERENCES

Flory, P. J., *op. cit.*, Chap. XIII.
Tompa, H., *op. cit.*, Chap. 7.
Voorn, M. J., *Fortschr. Hochpolym.-Forsch.* (*Advances in Polymer Science*) **1**, 192 (1959).

BIBLIOGRAPHY

1. Bamford, C. H., and H. Tompa, *Trans. Faraday Soc.* **46,** 310 (1950).
2. Bungenberg deJong, H. G., *Kolloid Z.* **79,** 223 (1937).
3. Dobry, A., and F. Boyer-Kawenoki, *J. Polymer Sci.* **2,** 90 (1947).
4. Flory, P. J., *J. Chem. Phys.* **12,** 425 (1944).
5. Krigbaum, W. R., unpublished work, cited in Flory, gen. ref., p. 547.
6. Malcolm, G. N., and J. S. Rowlinson, *Trans. Faraday Soc.* **53,** 921 (1957).
7. Schultz, A. R., and P. J. Flory, *J. Am. Chem. Soc.* **74,** 4760 (1952).
8. Schulz, G. V., *Z. physik. Chem.* **B46,** 137 (1940); **B47,** 155 (1940).
9. Silberberg, A., J. Eliassaf, and A. Katchalsky, *J. Polymer Sci.* **23,** 259 (1957).
10. Stockmayer, W. H., *J. Chem. Phys.* **17,** 588 (1949).
11. Tompa, H., *Trans. Faraday Soc.* **45,** 1142 (1949).

E. Fractionation of Polymers

Techniques for fractionating polymers are conveniently divided into preparative and analytical methods (Hall, gen. ref.). In preparative fractionation, discrete fractions are obtained for further study. The fractions are usually small (milligram to gram amounts), although separation to 200-gram fractions has been performed (20). In analytical fractionation, discrete fractions are usually not isolated.

If fractionation on the basis of molecular weight is desired, it is essential that the polymer-rich phase be amorphous rather than crystalline, and that other parameters influencing the separation, such as chain branching, be controlled.

Fractionation alone yields no information about the distribution of molecular weights in a sample, but must be accompanied by simultaneous or subsequent analytical measurement of the molecular weights of the fractions.

Preparative fractionation by precipitation

Fractionation by addition of nonsolvent (1, 10) Fractionation by precipitation from a solvent-nonsolvent mixture is carried out by adding the nonsolvent to a dilute solution of the polymer until a slight turbidity develops at the temperature of fractionation. To ensure the establishment of equilibrium, the mixture may be warmed until it is homogeneous and allowed to cool slowly back to the required temperature, which should thereafter be carefully maintained. The precipitated phase is allowed to settle to a coherent layer, and the supernatant phase is removed. A further

increment of precipitant is added and the process repeated. The polymer is isolated from the precipitated phase, which may still be relatively dilute (perhaps 10% polymer).

Refractionation is often used to achieve better separation. A more efficient procedure is to restrict the initial experiments to very dilute solutions, or to refine each initial fraction by a single reprecipitation at higher dilution, returning the unprecipitated polymer from this separation to the main solution before the next fraction is removed.

The solvent and the precipitant should be chosen so that precipitation occurs over a relatively wide range of solvent composition, yet is complete before too high a ratio of precipitant to solvent is reached. Other important considerations are the stability and volatility of the liquids, and their ability to form a highly swollen, mobile gel phase.

Other precipitation methods Fractionation by *solvent volatilization* (2, 15) involves precipitation through removal of the solvent component by evaporation. Large volumes of solution are avoided, but the solvent must have higher volatility than the precipitant. Fractional precipitation by *cooling* (27) offers the simplicity of a single solvent and operation at fixed volume, but has the disadvantage that many polymers cannot be completely precipitated by cooling alone.

Preparative fractionation by solution

Extraction techniques In these methods, polymer is placed in contact with a liquid of limited solvent power. The lower-molecular-weight species predominate in the supernatant layer at equilibrium and are removed. The treatment is repeated, using liquids of progressively higher solvent power. In direct *sequential extraction* (17) the extraction liquids are applied directly to the solid polymer, whereas in *coacervation* (13) the polymer phase is prepared by precipitation from solution and is thus highly swollen. Alternatively, in *film extraction* (12) the polymer is precipitated as a very thin film on a supporting foil to ensure rapid equilibration with the extractants.

Elution techniques Considerable success has been achieved in *gradient elution* techniques (9, 11, 18, 19, 25), in which the polymer is precipitated onto the finely divided support material of a column. Polymer fractions are then eluted from the column with progressively better solvent liquids. The technique is amenable to continuous operation. The method of polymer deposition is important, suggesting (18) that the efficiency of the subsequent extraction is improved if the polymer is selectively precipitated, with the lower-molecular-weight species precipitating last. A modification

of the technique is *temperature gradient elution* (3, 25) in which a temperature gradient is applied to the column, so that each species undergoes a series of solution and precipitation steps as it is carried down the column by the gradient of solvent composition.

Other preparative techniques

Preparative fractionation methods which have not yet been widely used include *distribution* (26), in which polymer species are distributed between two immiscible solvents; *diffusion* (24), where the higher rate of diffusion of low-molecular-weight species from polymer into solvent is utilized; *dialysis* (14), depending on controlled passage of polymer species through porous membranes; *chromatographic absorption* (4), where separation is achieved because of the molecular weight dependence of the absorption of polymer molecules on a substrate; and *zone melting* (23), where the polymer migrates through a column of frozen solvent as portions of the column are successively melted and refrozen.

Determination of molecular weight distribution

Information about the distribution of molecular weights in a polymer is often the goal of a fractionation experiment. Such information requires

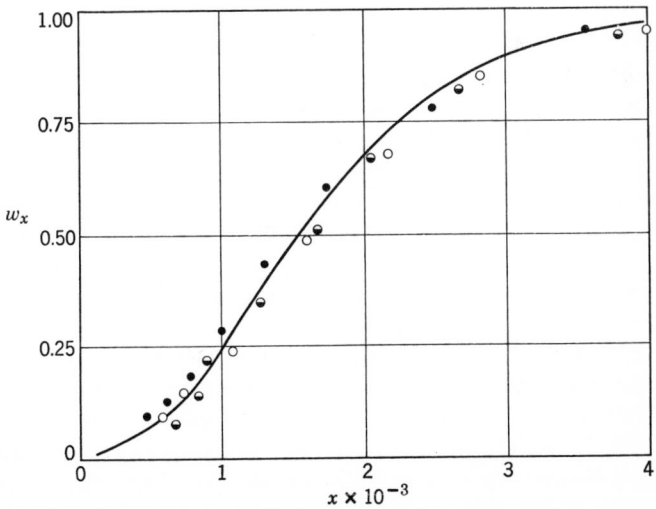

Fig. 2-14. Integral molecular weight distribution curve for polystyrene (3). The curve is drawn according to theory, while the experimental points are for three different fractionations of one polymer, using the temperature gradient elution technique.

measurement of the weight and molecular weight of each fraction. For convenience, intrinsic viscosity is often used for the molecular weight measurement, but this may lead to erroneous results for branched polymers (5).

The data are treated (16) by plotting against molecular weight the cumulative or integral distribution curve of the combined weight of all fractions having molecular weight up to and including M. A typical curve of this type is shown in Fig. 2-14. The differential distribution curve is obtained by differentiation. Experimental errors, indicated by the scatter of points in the integral curve, are magnified in this process. Only the gross features of the differential curve, such as the position of the peak and the breadth of the curve, are experimentally significant. Unless the fractionation is performed with extreme care, spurious features, such as double peaks in the differential distribution curve, may appear.

Analytical fractionation techniques

Since experimental and theoretical restrictions limit the amount of detailed information obtainable from preparative fractionation and subsequent construction of the distribution curve, methods for obtaining such information with less effort have been proposed. Perhaps the most sophisticated is *ultracentrifugation* (Chapter 3*D*), which can also serve as a preparative technique.

In *summative fractionation* (6, 8), a series of polymer solutions is used. In each, part of the polymer is precipitated, the point of fractionation being varied throughout the series. Weight fractions w_x and molecular weights M_x of the precipitates are obtained; the product $w_x M_x$ may be related to the *integral* of the cumulative distribution curve. Since the two differentiations required to obtain the differential distribution would introduce intolerably large errors into the resulting differential curve, the data are analyzed to provide only information equivalent to the position of the maximum and breadth of the molecular weight distribution curve. The *gel volume* method (7) simplifies the experiments still further at the cost of complicating their interpretation. In a summative fractionation, the volume of swollen precipitate is used as a measure of its weight, and the amount of nonsolvent added as a measure of the molecular weight of the precipitate. If these two functions are properly calibrated, the operation of the method becomes very simple; the extra work involved in the calibration limits its usefulness to cases where a large number of similar samples is to be studied. The data obtained are similar to those of the summative method.

Still another method (21, 22) based on similar calibrations is *turbidimetric*

titration. A precipitant is added slowly to a stirred polymer solution, and the turbidity due to precipitated polymer is measured by the decrease in the intensity of a transmitted beam of light.

Finally, it should be mentioned that careful measurements of molecular weight by two absolute methods giving different averages, such as osmotic pressure and light scattering, give about as many data as can be reliably obtained from a conventional fractionation.

GENERAL REFERENCES

Cragg, L. H., and H. Hammerschlag, *Chem. Revs.* **39,** 79 (1946).

Guzmán, G. M., "Fractionation of High Polymers," pp. 113–183 in J. C. Robb and F. W. Peaker, eds., *op. cit.*

Hall, R. W., "The Fractionation of High Polymers," Chap. 2 in P. W. Allen, ed., *op. cit.*

BIBLIOGRAPHY

1. Alfrey, T., A. Bartovics, and H. Mark, *J. Am. Chem. Soc.* **65,** 2319 (1943).
2. Badgely, W., V. J. Frilette, and H. Mark, *Ind. Eng. Chem.* **37,** 227 (1945).
3. Baker, C. A., and R. J. P. Williams, *J. Chem. Soc.* **1956,** 2352.
4. Bannister, D. W., C. S. G. Phillips, and R. J. P. Williams, *Anal. Chem.* **26,** 1451 (1954).
5. Billmeyer, F. W., Jr., *J. Am. Chem. Soc.* **75,** 6118 (1953).
6. Billmeyer, F. W., Jr., and W. H. Stockmayer, *J. Polymer Sci.* **5,** 121 (1950).
7. Boyer, R. F., *J. Polymer Sci.* **8,** 73 (1952); **9,** 197 (1952).
8. Broda, A., T. Niwinska, and S. Polowinski, *J. Polymer Sci.* **32,** 343 (1958).
9. Desreux, V., *Rec. trav. chim.* **68,** 789 (1949); V. Desreux and M. C. Spiegels, *Bull. soc. chim. Belges* **59,** 476 (1950).
10. Flory, P. J., *J. Am. Chem. Soc.* **65,** 372 (1943).
11. Francis, P. S., R. C. Cooke, and J. H. Elliott, *J. Polymer Sci.* **31,** 453 (1958).
12. Fuchs, O., *Makromol. Chem.* **5,** 245 (1951); **7,** 259 (1952).
13. Green, J. H. S., and M. F. Vaughan, *Chem. & Ind.* **1958,** 829.
14. Green, J. H. S., H. T. Hookway, and M. F. Vaughan, *Chem. & Ind.* **1958,** 862.
15. Goldberg, A. I., W. P. Hohenstein, and H. Mark, *J. Polymer Sci.* **2,** 503 (1947).
16. Hosemann, R., and W. Schramek, *J. Polymer Sci.* **59,** 13, 29, 51 (1962).
17. Howlett, F., and A. R. Urquhart, *J. Textile Inst.* **37,** T89 (1946).
18. Kenyon, A. S., and I. O. Salyer, *J. Polymer Sci.* **43,** 427 (1960).
19. Krigbaum, W. R., and J. E. Kurz, *J. Polymer Sci.* **41,** 275 (1959).
20. Merz, E. H., and R. W. Raetz, *J. Polymer Sci.* **5,** 587 (1950).
21. Morey, D. R., and J. W. Tamblyn, *J. Appl. Phys.* **16,** 419 (1945).
22. Oth, A., *Bull. soc. chim. Belges* **58,** 285 (1959).
23. Peaker, F. W., and J. C. Robb, *Nature* **182,** 1591 (1958).
24. Rosenberg, J. L., and C. O. Beckmann, *J. Colloid Sci.* **3,** 483 (1948).
25. Schneider, N. S., J. D. Loconti, and L. G. Holmes, *J. Appl. Polymer Sci.* **5,** 354 (1961); N. S. Schneider, *Anal. Chem.* **33,** 1829 (1961).
26. Schulz, G. V., and E. Nordt, *J. prakt. Chem.* **155,** 115 (1940).
27. Wesslau, H., *Makromol. Chem.* **20,** 111 (1956).

F. Swelling of Crosslinked Polymers

A crosslinked polymer cannot dissolve completely but may swell to many times its original volume by absorbing a liquid with which it is in contact. Swelling occurs for the same reason that a linear polymer dissolves: the addition of solvent affords an increase in entropy. The swollen gel is in fact an elastic rather than a viscous solution. The tendency toward swelling is opposed by an elastic retractive force arising as the chains between network junctions are forced to assume elongated conformations. The equilibrium between the two forces is closely analogous to an osmotic equilibrium, with the network structure acting as solute, osmotic membrane, and pressure-generating device. A gel previously swollen with pure solvent "deswells" to a smaller equilibrium volume when placed in contact with a polymer solution. The change in volume measures the change in activity of the solution and thus the number-average molecular weight and second virial coefficient of the polymer in the solution. With proper calibration, it can be used to determine these quantities (1). The method has been relatively unpopular because of difficulties in making and handling the gels. Alternatively, the swelling of the gel in pure solvent may be used to measure the second virial coefficient of the crosslinked polymer for that solvent (2).

BIBLIOGRAPHY

1. Boyer, R. F., *J. Chem. Phys.* **13,** 363 (1945).
2. Boyer, R. F., and R. S. Spencer, *J. Polymer Sci.* **3,** 97 (1948).

3

Measurement of Molecular Weight and Size

The molecular weights of polymers can be determined by chemical or physical methods of functional group analysis, by measurement of the colligative properties, light scattering, or ultracentrifugation, or by measurement of dilute solution viscosity. All these methods except the last are, in principle, absolute: molecular weights can be calculated without reference to calibration by another method. Dilute solution viscosity, however, is not a direct measure of molecular weight. Its value lies in the simplicity of the technique and the fact that it can be related empirically to molecular weight for many systems.

With the exception of some types of end group analysis, all molecular weight methods require solubility of the polymer, and all involve extrapolation to infinite dilution or operation in a Θ-solvent in which ideal solution behavior is attained.

A. End Group Analysis

Molecular weight determination through group analysis requires that the polymer contain a known number of determinable groups per molecule. The long-chain nature of polymers limits such groups to end groups. Thus the method is usually referred to as *end group analysis.* Since methods of end group analysis count the number of molecules in a given weight of sample, they yield the number-average molecular weight (Section *B*) for polydisperse materials. The methods become insensitive at high molecular weight, as the fraction of end groups becomes too small to be measured

with precision. Although published procedures often indicate that loss of precision occurs at molecular weights above 25,000, there is no doubt that the adoption of more sensitive techniques, such as gas chromatography or mass spectrometry to estimate reaction products, can raise this limit by at least a factor of 10. Similar considerations apply to the use of infrared spectroscopy to measure end groups.

The discussion of end group methods is conveniently divided to cover condensation and addition polymers because of the difference in the types of end groups usually found.

Condensation polymers End group analysis in condensation polymers usually involves chemical methods of analysis for functional groups. Carboxyl groups in polyesters (10) and in polyamides (12) are usually titrated directly with base in an alcoholic or phenolic solvent, while amino groups in polyamides may be titrated with acid under similar conditions (12). Hydroxyl groups are usually determined by reacting them with a titratable reagent (4, 9), but infrared spectroscopy has been used (5, 13). The chemical methods are often limited by insolubility of the polymer in solvents suitable for the titrations.

Addition polymers No general procedures can be given for end group analysis in addition polymers because of the variety of type and origin of the end groups. When the polymerization kinetics is well known, analysis may be made for initiator fragments containing identifiable functional groups (1, 2), elements (11), or radioactive atoms (3); or for end groups arising from transfer reactions with solvent (6, 7); or for unsaturated end groups in linear polyethylene and poly-α-olefins, as in the infrared spectroscopic analysis for vinyl groups in polypropylene (8).

GENERAL REFERENCE

Price, G. F., "Techniques of End-Group Analysis," Chap. 7 in P. W. Allen, ed., *Techniques of Polymer Characterization*, Academic Press, New York, 1959.

BIBLIOGRAPHY

1. Bartlett, P. D., and S. G. Cohen, *J. Am. Chem. Soc.* **65,** 543 (1943).
2. Baxendale, J. H., M. G. Evans, and G. S. Park, *Trans. Faraday Soc.* **42,** 155 (1946).
3. Bevington, J. C., H. W. Melville, and R. P. Taylor, *J. Polymer Sci.* **12,** 449 (1954); **14,** 463 (1954).
4. Conix, A., *Makromol. Chem.* **26,** 226 (1958).
5. Daniels, W. W., and R. E. Kitson, *J. Polymer Sci.* **33,** 161 (1958).

6. Joyce, R. M., W. E. Hanford, and J. Harmon, *J. Am. Chem. Soc.* **70,** 2529 (1948).
7. Mayo, F. R., *J. Am. Chem. Soc.* **65,** 2324 (1943).
8. Natta, G., and coworkers, *Chim. e ind.* (*Milan*) **38,** 124 (1956).
9. Ogg, C. L., W. L. Porter, and C. O. Willits, *Ind. Eng. Chem., Anal. Ed.* **17,** 394 (1945).
10. Pohl, H. A., *Anal. Chem.* **26,** 1614 (1954).
11. Price, C. C., and B. E. Tate, *J. Am. Chem. Soc.* **65,** 517 (1943).
12. Waltz, J. E., and G. B. Taylor, *Anal. Chem.* **19,** 448 (1947).
13. Ward, I. M., *Trans. Faraday Soc.* **53,** 1406 (1957).

B. Colligative Property Measurement

The relations between the colligative properties and molecular weight for infinitely dilute solutions rest upon the fact that the activity of the solute in a solution becomes equal to its mole fraction as the solute concentration becomes sufficiently small. It can be shown that the activity of the solvent must equal its mole fraction under these conditions, and it follows that the depression of the activity of the solvent by a solute is equal to the mole fraction of the solute.

The colligative property methods are based on vapor pressure lowering, boiling point elevation (*ebulliometry*), freezing point depression (*cryoscopy*), and the osmotic pressure (*osmometry*). The working equations for these methods are derived from Eq. 2-10. When Boltzmann's constant is replaced by the gas constant throughout to convert to a molar basis, the equations for the methods of interest for polymer solutions are:

$$\lim_{c\to 0} \frac{\Delta T_b}{c} = \frac{RT^2}{\rho\,\Delta H_v}\frac{1}{M}$$

$$\lim_{c\to 0} \frac{\Delta T_f}{c} = -\frac{RT^2}{\rho\,\Delta H_f}\frac{1}{M} \tag{3-1}$$

$$\lim_{c\to 0} \frac{\pi}{c} = \frac{RT}{M}$$

where ΔT_b, ΔT_f, and π are the boiling point elevation, freezing point depression, and osmotic pressure, respectively, ρ is the density of the solvent, ΔH_v and ΔH_f are the latent heats of vaporization and fusion of the solvent per gram, and c is the solute concentration in grams per cubic centimeter.

The relative applicability of the colligative methods to polymer solutions is demonstrated in Table 3-1. It is clear that direct measurement of vapor pressure lowering in polymer solutions is unrewarding. While temperature

differences of the order of 1×10^{-3}°C can currently be measured with considerable precision, the larger effect observed in the osmotic experiment suggests that this technique is the most fruitful for polymer solutions.

TABLE 3-1. Colligative properties of a solution of a polymer of $M = 20{,}000$ at $c = 0.01$ g/cm³

Property	Value
Vapor pressure lowering	4×10^{-3} mm Hg
Boiling point elevation	1.3×10^{-3}°C
Freezing point depression	2.5×10^{-3}°C
Osmotic pressure	15 cm solvent

Although other considerations, such as the perfection of available osmotic membranes, are important, the osmotic method has been more widely used than other colligative techniques for polymer systems.

Number-average molecular weight

Since typical polymers consist of mixtures of many molecular species, molecular weight methods always yield average values. The measurement of the colligative properties in effect counts the number of moles of solute per unit weight of sample. This number is the sum over all molecular species of the number of moles N_i of each species present:

$$\sum_{i=1}^{\infty} N_i$$

The total weight w of the sample is similarly the sum of the weights of each molecular species,

$$w = \sum_{i=1}^{\infty} w_i = \sum_{i=1}^{\infty} N_i M_i \tag{3-2}$$

The average molecular weight given by these methods is known as the *number-average molecular weight* $\bar{M}_n$. By the definition of molecular weight as weight of sample per mole,

$$\bar{M}_n = w \Big/ \sum_{i=1}^{\infty} N_i = \sum_{i=1}^{\infty} M_i N_i \Big/ \sum_{i=1}^{\infty} N_i \tag{3-3}$$

Equations 3-1 may then be put in the form applicable to heterogeneous solutes:

$$\lim_{c\to 0} \frac{\Delta T_b}{c} = \frac{RT^2}{\rho\,\Delta H_v}\frac{1}{\bar{M}_n}$$

$$\lim_{c\to 0} \frac{\Delta T_f}{c} = -\frac{RT^2}{\rho\,\Delta H_f}\frac{1}{\bar{M}_n} \tag{3-4}$$

$$\lim_{c\to 0} \frac{\pi}{c} = \frac{RT}{\bar{M}_n}$$

End group methods as well as colligative methods give the number-average molecular weight. The number average is very sensitive to changes in the weight fractions of low-molecular-weight species, and relatively insensitive to similar changes for high-molecular-weight species.

Concentration dependence of the colligative properties

For convenience, the concentration dependence of the colligative properties is developed in terms of the osmotic pressure; the results are directly applicable to cryoscopy and ebulliometry by appropriate substitution in Eq. 3-4. By Eq. 2-11 or 2-21, a general expression for the osmotic pressure at a finite concentration is

$$\frac{\pi}{RTc} = (A_1 + A_2c + A_3c^2 + \cdots)$$

$$\frac{\pi}{RTc} = \frac{1}{\bar{M}_n}(1 + \Gamma c + g\Gamma^2c^2 + \cdots) \tag{3-5}$$

where $\Gamma = A_2/A_1$, and g is a slowly varying function of the polymer-solvent interaction with values near zero for poor solvents and near 0.25 for good solvents (12, 21). In most cases, the term in c^2 may be neglected; in the rare instances where dependence on c^2 is significant, it may be convenient to take $g = 0.25$ and write

$$\frac{\pi}{RTc} = \frac{1}{\bar{M}_n}\left(1 + \frac{\Gamma}{2}c\right)^2 \tag{3-6}$$

In terms of the polymer-solvent interaction constant χ_1 of the Flory-Huggins theory, the osmotic pressure is given by

$$\frac{\pi}{RTc} = \frac{1}{M_2} + \frac{\rho_1}{M_1\rho_2^2}\left(\frac{1}{2} - \chi_1\right)c + \cdots \tag{3-7}$$

where subscript 1 indicates the solvent, and 2 the polymer.

As Eqs. 3-5 and 3-7 suggest, it is usual to plot π/c vs. c. In general a straight line results whose intercept at $c = 0$ is $A_1 = 1/\bar{M}_n$ and whose slope is the second virial coefficient A_2 and allows evaluation of the polymer-solvent interaction constant χ_1. If the solvent is good enough or the concentration high enough that the c^2 term is significant, the points may deviate from a straight line. In such cases, it is useful to plot $(\pi/RTc)^{1/2}$ vs. c as suggested by Eq. 3-6. Typical data are discussed under *Osmometry*.

Vapor pressure lowering

The *isopiestic* method, in which solutions of a polymer and a standard substance are joined through a common vapor space, has been applied (15, 23) to polymers of molecular weights up to about 20,000. *Isothermal distillation* occurs until the vapor pressures over the two solutions are equal; in the limit at $c = 0$ the equilibrium concentrations in the solutions are proportional to the solute molecular weights. The time required to attain equilibrium may be many days or many weeks, however, and the requirements for temperature constancy are stringent. The method has therefore seen little application.

Ebulliometry

In the boiling point elevation experiment (8, 10, 16, 18) the boiling point of the polymer solution is compared directly with that of the (condensing) pure solvent in a vessel known as an ebulliometer. Sensing devices include differential thermometers (10) and multijunction thermocouples or thermistors arranged in a Wheatstone bridge circuit (8, 16, 18). It is customary to calibrate the ebulliometer with a substance of known molecular weight, e.g., octacosane ($M = 396$) or tristearin ($M = 892$).

Although with care accurate molecular weights can be obtained up to $\bar{M}_n = 30{,}000$ or more, the ebulliometric experiment occasionally suffers from a limitation in the tendency of polymer solutions to foam on boiling. Not only may this lead to unstable operation, but also the polymer may concentrate in the foam because of its higher surface, thus rendering uncertain the actual concentration of the solution.

Cryoscopy

The freezing point depression, or cryoscopic, method (1, 22) is similar to the ebulliometric technique in several respects. The preferred temperature-sensing element is the thermistor used in a bridge circuit, but the freezing

points of solvent and solution are compared sequentially. Calibration with a substance of known molecular weight is customary. Although the limitations of the method appear to be somewhat less severe than those of ebulliometry, care must be taken that supercooling is controlled. The use of a nucleating agent (11) to provide controlled crystallization of the solvent is helpful in this respect. Reliable results are obtained for molecular weights at least as high as 30,000.

Osmometry

The success of the osmotic experiment depends on the availability of a *semipermeable membrane* through which solvent molecules pass freely, but solute molecules cannot pass. Existing membranes only approximate ideal semipermeability, and the chief limitation of the osmotic method is the diffusion of low-molecular-weight species through the membrane. As currently practiced, the method is reliable and accurate only for experiments in which diffusion is absent, such as measurement of unfractionated polymers with $\bar{M}_n > 50{,}000$, or polymers from which low-molecular-weight species have been removed by fractionation or extraction, with $\bar{M}_n > 20{,}000$. The upper limit to molecular weights which can be measured by the osmotic method is about 1,000,000, set by the precision with which small osmotic heights can be read.

Osmometers An osmometer is a device that holds a solution and the solvent in compartments separated by a membrane. A connection is made to a capillary to allow the osmotic pressure to be balanced by a hydrostatic pressure, or to a source of external pressure for the same purpose.

Two types of osmometer have been used extensively. One, the block osmometer (4, 5), is a relatively large and bulky metal instrument. The membrane area is large and the solution volume small, advantages for rapid equilibration. More popular are relatively small and simple osmometers based on the Zimm-Myerson design (24), in which two membranes are held against a glass solution cell by means of perforated metal plates, as shown in Fig. 3-1. The assembled instrument is suspended in a large tube partly filled with solvent. In the Stabin modification (19), the membranes are supported on both sides by channeled metal plates to ensure their rigidity. Advantages of these osmometers are small size and cost, making multiple installations feasible, the possibility of complete immersion in constant-temperature baths, and ease of filling and adjusting osmotic height.

Membranes Osmotic membranes fall into two classes: (*a*) nonswellable inorganic substances such as porous glass and metal; and (*b*) swellable

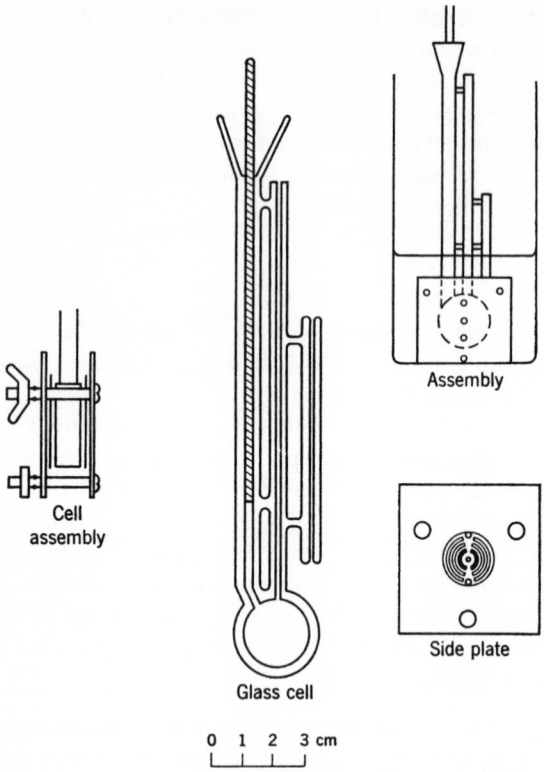

Fig. 3-1. Diagram of the Zimm-Myerson osmometer (24). Typical diameter for the measuring and reference (solvent cell) capillary is 0.5–1 mm. The closure of the filling tube is a 2-mm metal rod. A mercury seal is used at the top to ensure tightness.

organocolloidal materials such as cellulose and rubber. Although some theories of membrane action suggest that a "sieve" theory is not adequate and that the membrane material must interact with or be swollen by the solvent, the successful use of glass membranes has been reported (3).

Common organic membrane materials (9) include collodion (nitrocellulose of 11–13.5% nitrogen); regenerated cellulose, made by denitration of collodion (14); gel cellophane which has never been allowed to dry after manufacture (this material is available from commercial cellophane manufacturers in several thicknesses); bacterial cellulose, made by the action of certain strains of bacteria (13); rubber; poly(vinyl alcohol); polyurethans; poly(vinyl butyral); and polychlorotrifluoroethylene. Of these, gel cellophane is probably most widely used at present.

Permeability of membranes The serious defect of available membrane materials in allowing the diffusion of low-molecular-weight species as well

***TABLE 3-2. Cryoscopic and osmotic molecular weights for extracted and unextracted branched polyethylenes* (1, 11)**

Type of Sample	Molecular Weight by	
	Osmometry	Cryoscopy
Extracted to remove low-molecular-weight species	50,000	45,000
	37,000	31,000
	31,000	25,000
	28,000	21,000
Not extracted	31,400	19,100
	30,300	10,700
	26,600	13,300

as solvent was treated theoretically by Staverman (20). Theory and experiment show that the apparent osmotic pressure is always less than the true osmotic pressure and falls with time. Extrapolation back to zero time still gives too low an osmotic pressure and hence too high a molecular weight; however, the correct osmotic pressure may be calculated from the data from independent dialysis and diffusion experiments (6). In principle, measurement of the osmotic pressure after a very long time should give the true molecular weight of the nondiffusing species, but it seems likely that equilibrium is difficult to reach.

The osmotic method therefore measures only those polymer species larger than a certain minimum size. At the risk of oversimplification, the summations in Eq. 3-3 may be considered to begin at some molecular species j, where j is greater than 1. The result is that the osmotic molecular weight

$$\bar{M}_{os} = \left(\sum_{i=j}^{\infty} M_i N_i\right) \Big/ \left(\sum_{i=j}^{\infty} N_i\right) \tag{3-8}$$

may be different from and larger than the number-average molecular weight obtained from cryoscopy or ebulliometry. The magnitude of the error and its dependence on the type of polymer specimen are illustrated in Table 3-2 (1, 11).

Typical data Figures 3-2 and 3-3 show typical osmotic data. In both cases concentrations were low enough so that dependence of π/RTc on c^2 was negligible. The parallelism of the lines in Fig. 3-2, representing samples of different molecular weight in the same solvent, illustrates the usual result that the polymer-solvent interactions are a slowly varying function of molecular weight. The constant intercept of the lines in Fig. 3-3 obtained

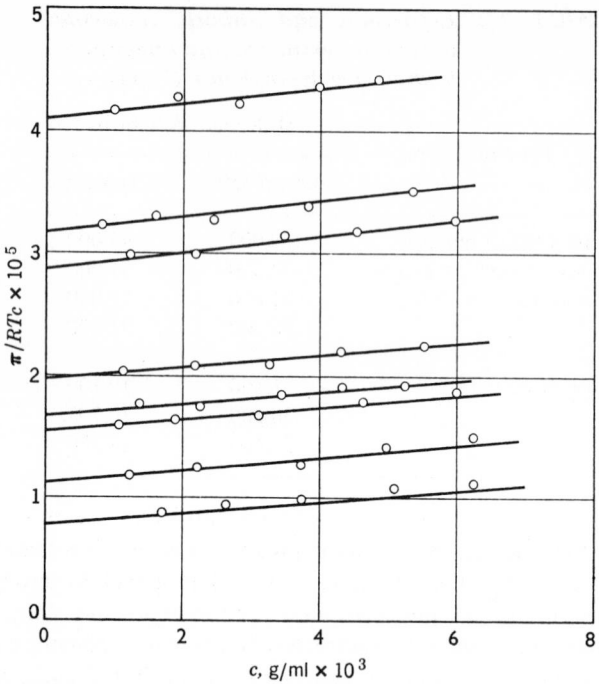

Fig. 3-2. Plots of $\pi/RTc = 1/\bar{M}_n$ vs. c for cellulose acetate fractions in acetone solution (2).

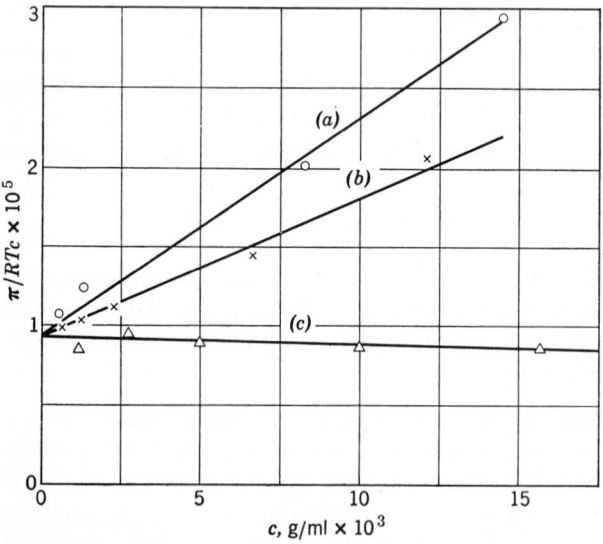

Fig. 3-3. Plot of $\pi/RTc = 1/\bar{M}_n$ vs. c for nitrocellulose in (*a*) acetone, (*b*) methanol, and (*c*) nitrobenzene (7).

for one sample in several different solvents illustrates the independence of $(\pi/RTc)_{c=0} = 1/\bar{M}_n$ of solvent type and the range of A_2 or χ_1 (17) which can be obtained with a given polymer and a range of solvents.

The advantage of selecting a poor solvent for osmotic measurements in order to minimize the error in extrapolation is evident. At the Flory temperature Θ the slope of π/RTc vs. c is zero up to concentrations of several per cent.

GENERAL REFERENCES

Bonnar, R. U., M. Dimbat, and F. H. Stross, *Number-Average Molecular Weights*, Interscience Publishers, New York, 1958.

Flory, P. J., *Principles of Polymer Chemistry*, Cornell University Press, Ithaca, N.Y., 1953, Chap. VII-1.

Hookway, H. T., "Number-Average Molecular Weights by Osmometry," Chap. 3 in P. W. Allen, ed., *op. cit.*

Lehrle, R. S., "Ebulliometry Applied to Polymer Solutions," pp. 37–111 in J. C. Robb and F. W. Peaker, eds., *Progress in High Polymers*, Vol. 1, Academic Press, New York, 1961.

Rushman, D. F., "Other Methods for the Determination of Number-Average Molecular Weights," Chap. 4 in P. W. Allen, ed., *op. cit.*

Tompa, H., *Polymer Solutions*, Academic Press, New York, 1956, Chap. 6.

Wagner, R. H., and L. D. Moore, Jr., "Determination of Osmotic Pressure," Chap. XV in A. Weissberger, ed., *Technique of Organic Chemistry*, Vol. I, *Physical Methods of Organic Chemistry*, 3rd ed., Part I, Interscience Publishers, New York, 1959.

BIBLIOGRAPHY

1. Ashby, C. E., J. S. Reitenour, and C. F. Hammer, *J. Am. Chem. Soc.* **79,** 5086 (1957).
2. Badgley, W. J., and H. Mark, "Osmometry and Viscometry of Polymer Solutions," pp. 75–113 in R. E. Burk and O. Grummitt, eds., *High Molecular Weight Organic Compounds* (*Frontiers in Chemistry*, Vol. VI), Interscience Publishers, New York, 1949.
3. Elias, H.-G., T. A. Ritscher, and F. Patat, *Makromol. Chem.* **27,** 1 (1958); T. A. Ritscher and H.-G. Elias, *ibid.* **30,** 48 (1959).
4. Flory, P. J., *J. Am. Chem. Soc.* **65,** 372 (1943).
5. Fuoss, R. M., and D. J. Mead, *J. Phys. Chem.* **47,** 59 (1943).
6. Gardon, J. L., and S. G. Mason, *J. Polymer Sci.* **26,** 255 (1957).
7. Gee, G., *Trans. Faraday Soc.* **40,** 261 (1944); data of A. Dobry, *J. chim. phys.* **32,** 50 (1935).
8. Glover, C. A., and R. R. Stanley, *Anal. Chem.* **33,** 447 (1961).
9. Immergut, E. H., S. Rollin, A. Salkind, and H. Mark, *J. Polymer Sci.* **12,** 439 (1954).
10. Kitson, R. E., and J. Mitchell, Jr., *Anal. Chem.* **21,** 401 (1949); R. E. Kitson, A. N. Oemler, and J. Mitchell, Jr., *ibid.*, p. 404.
11. Kokle, V., F. W. Billmeyer, Jr., L. T. Muus, and E. J. Newitt, *J. Polymer Sci.* **62,** 251 (1962).
12. Krigbaum, W. R., and P. J. Flory, *J. Polymer Sci.* **9,** 503 (1952).

13. Masson, C. R., R. F. Menzies, J. Cruickshank, and H. W. Melville, *Nature* **157,** 74 (1946).
14. Montonna, R. E., and L. T. Jilk, *J. Phys. Chem.* **45,** 1374 (1941).
15. Parrette, R. L., *J. Polymer Sci.* **15,** 447 (1955).
16. Ray, N. H., *Trans. Faraday Soc.* **48,** 809 (1952).
17. Rehner, J., Jr., *J. Polymer Sci.* **46,** 550 (1960).
18. Smith, H., *Trans. Faraday Soc.* **52,** 402 (1956).
19. Stabin, J. V., and E. H. Immergut, *J. Polymer Sci.* **14,** 209 (1954).
20. Staverman, A. J., *Rec. trav. chim.* **70,** 344 (1951); **71,** 623 (1952); A. J. Staverman, D. T. F. Pals, and Ch. A. Kruissink, *J. Polymer Sci.* **23,** 57 (1957).
21. Stockmayer, W. H., and E. F. Casassa, *J. Chem. Phys.* **20,** 1560 (1952).
22. Vofsi, D., and A. Katchalsky, *J. Polymer Sci.* **26,** 127 (1957).
23. Williamson, A. T., *Proc. Roy. Soc. (London)* **A195,** 97 (1948).
24. Zimm, B. H., and I. Myerson, *J. Am. Chem. Soc.* **68,** 911 (1946).

C. Light Scattering

The scattering of light occurs whenever a beam of light encounters matter. The nuclei and electrons undergo induced vibrations in phase with the incident light wave, and act as sources of light which is propagated in all directions, aside from a polarization effect, with the same wavelength as the exciting beam. Light scattering accounts for many natural phenomena, including the colors of the sky and the rainbow, and of most white materials.

Light scattering in gases and liquids

In 1871 Lord Rayleigh (17) applied classical electromagnetic theory to the problem of the scattering of light by the molecules of a gas. He showed that the molecules act as sources of light whose intensity per unit volume of scattering material is i_θ when observed at a distance r and at an angle θ with respect to the incident beam:

$$\frac{i_\theta r^2}{I_0} = R_\theta = \frac{2\pi^2(n-1)^2}{\lambda^4}\frac{1}{\nu}(1+\cos^2\theta) \tag{3-9}$$

where I_0 is the intensity of the primary beam, and n, ν, and λ are, respectively, the refractive index of the gas, the number of scattering particles per cubic centimeter, and the wavelength of the light.

The quantity R_θ, called the Rayleigh ratio, is often replaced by the turbidity, τ, the total scattering integrated over all angles: $\tau = \int_\theta i_\theta \, d\theta$. In the absence of absorption, τ is related to the primary beam intensities

before and after passing through the scattering medium. If the incident intensity I_0 is reduced to I in a length l of sample, then

$$I/I_0 = e^{-\tau l} \tag{3-10}$$

For particles small compared to λ, $\tau = (8\pi/3)R_{(\theta=0)}$, whence

$$\tau = \frac{32\pi^3}{3}\frac{(n-1)^2}{\lambda^4}\frac{1}{\nu} \tag{3-11}$$

The term $(1 + \cos^2\theta)$ in Eq. 3-9 arises from the fact that light can be propagated only at right angles to the direction of oscillation of the electric moment.

From the number of scattering particles per cubic centimeter, ν, either the molecular weight of the gas or Avogadro's number, N_o, may be calculated if the other is known.

In Rayleigh's treatment it was assumed that each particle scattered as a point source independent of all others. This is equivalent to assuming that the relative positions of the particles are random. In liquids this is not the case, and the scattered light intensity is reduced about fifty-fold because of destructive interference of the light scattered from different particles. In the calculation of scattered intensity in liquids (12, 19), the scattering is considered due to local thermal fluctuations in density which make the liquid optically inhomogeneous. The fluctuations in the density can be related to the compressibility κ of the liquid by comparing the thermal energy per mole RT with the work required to cause a change in density through the application of a hydrostatic pressure:

$$\tau = \frac{8\pi^3}{3\lambda^4}(n^2-1)^2 RT\kappa \tag{3-12}$$

Light scattering from solutions of particles small compared to the wavelength

In solutions and in mixtures of liquids, additional light scattering arises from irregular changes in density and refractive index due to fluctuations in composition. Debye (9) in 1944 calculated the effect of these fluctuations, relating them to the change in concentration c associated with the osmotic pressure π per mole of solute:

$$\tau = \frac{32\pi^3}{3\lambda^4}\frac{RTc}{N_o}\left(n\frac{dn}{dc}\right)^2 \Big/ \left(\frac{d\pi}{dc}\right) \tag{3-13}$$

Inserting the relation between osmotic pressure and molecular weight yields the *Debye equation*:

$$K\frac{c}{R_{90}} = H\frac{c}{\tau} = \frac{1}{M} + 2A_2c + \cdots$$

where (3-14)

$$K = \frac{2\pi^2 n^2}{N_o\lambda^4}\left(\frac{dn}{dc}\right)^2 \quad \text{and} \quad H = \frac{32\pi^3 n^2}{3N_o\lambda^4}\left(\frac{dn}{dc}\right)^2$$

Equation 3-14 forms the basis of the determination of polymer molecular weights by light scattering. Beyond the measurement of τ or R_θ, only the refractive index n and the *specific refractive increment* dn/dc require experimental determination. The latter quantity is a constant for a given polymer, solvent, and temperature, and is measured with an interferometer or a differential refractometer (3, 10) or in the analytical ultracentrifuge (Section *D*).

Equation 3-14 is correct only for vertically polarized incident light and for optically isotropic particles. The use of unpolarized light requires that τ be multiplied by $1 + \cos^2\theta$. Whether polarized or unpolarized light is used, the scattered intensity is symmetrical about the angle of observation 90°. If the scattering particles are anisotropic, a correction for depolarization is required (6, 9).

Weight-average molecular weight

In the derivation of Eq. 3-14 it is shown that the amplitude of the scattered light is proportional to the polarizability and hence to the mass of the scattering particle. Thus the intensity of scattering is proportional to the square of the particle mass. If the solute is polydisperse, the heavier molecules contribute more to the scattering than the light ones. The total scattering at zero concentration is

$$\tau = \sum_{i=1}^{\infty}\tau_i = H\sum_{i=1}^{\infty}c_iM_i = Hc\bar{M}_w \tag{3-15}$$

defining the *weight-average molecular weight* according to any of the relationships

$$\bar{M}_w = \sum_{i=1}^{\infty}c_iM_i\Big/c = \sum_{i=1}^{\infty}w_iM_i = \sum_{i=1}^{\infty}N_iM_i^2\Big/\sum_{i=1}^{\infty}N_iM_i \tag{3-16}$$

Here w_i is the weight fraction, and

$$c = \sum_{i=1}^{\infty}c_i$$

as in the corresponding definition of number-average molecular weight.

The significance of the weight- and number-average molecular weights is noteworthy. $\bar{M}_w$ is always greater than $\bar{M}_n$ except for a monodisperse

system. The ratio $\bar{M}_w/\bar{M}_n$ is a measure of the polydispersity of the system. $\bar{M}_w$ is particularly sensitive to the presence of high-molecular-weight species, whereas $\bar{M}_n$ is influenced more by species at the lower end of the molecular weight distribution: if equal weights of molecules with $M =$ 10,000 and $M = 100{,}000$ are mixed, $\bar{M}_w = 55{,}000$ and $\bar{M}_n = 18{,}200$; if equal numbers of each kind of molecule are mixed, $\bar{M}_w = 92{,}000$ and $\bar{M}_n = 55{,}000$.

Light scattering in solutions of larger particles

When the size of a scattering particle exceeds about $\lambda/20$, different parts of the particle are exposed to incident light of different amplitude and phase. The scattered light is made up of waves coming from different parts of the particle and interfering with one another. The general solution describing the distribution of scattered light for large particles is complex and has been solved only for the case of a homogeneous sphere (15). However, an approximate solution applies to most polymer systems. In this treatment, the intensity of light scattered from the elements making up the particle is shown to be proportional to

$$\sum\sum\left[\sin\left(\frac{4\pi r}{\lambda_s}\sin\frac{\theta}{2}\right)\Big/\left(\frac{4\pi r}{\lambda_s}\sin\frac{\theta}{2}\right)\right] \tag{3-17}$$

where $\lambda_s = \lambda/n$ is the wavelength of light in the medium of refractive index n, and the double summation is carried out over all pairs of scattering elements separated by a distance r.

For spherical particles, the scattering elements are symmetrically distributed, and the radial density function of Eq. 3-17 is a constant. Solution of the integral form of Eq. 3-17 leads to the following equation for the intensity of scattered light $P(\theta)$ (13, 18):

$$P(\theta) = [(3/u^3)(\sin u - u\cos u)]^2 \tag{3-18}$$

where

$$u = 2\pi(d/\lambda_s)\sin\theta/2$$

and d is the diameter of the spheres.

For a monodisperse system of randomly coiling polymers the radial density function is the Gaussian distribution, and the intensity of scattering is given (9, 20) by

$$P(\theta) = (2/v^2)[e^{-v} - (1 - v)] \tag{3-19}$$

where

$$v = (16\pi^2)(\overline{s^2}/\lambda_s^2)\sin^2\theta/2$$

and $\overline{s^2}$ is the mean-square radius of gyration of the molecule.

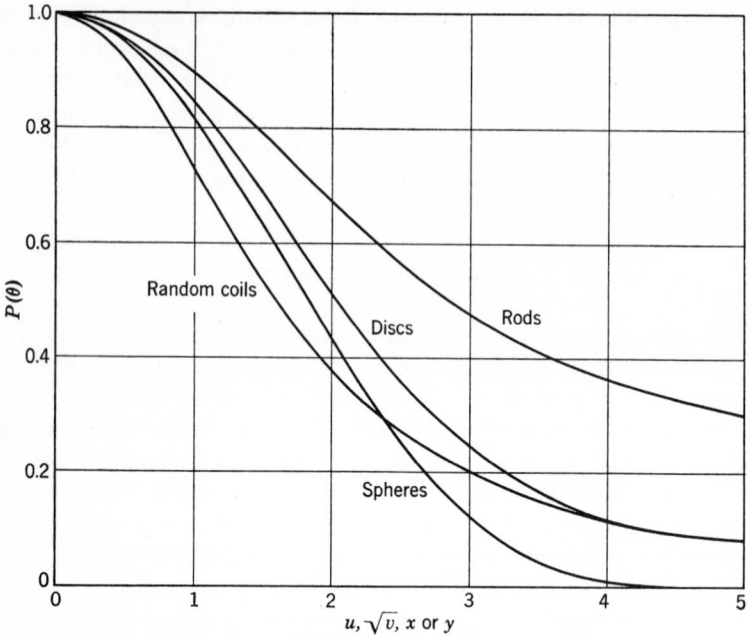

Fig. 3-4. The variation of the particle scattering function $P(\theta)$ with the size parameter u (spheres), v (monodisperse random coils), x (rods), or y (discs).

Similar functions can be derived for other models, such as rods and discs. The functions $P(\theta)$ are plotted in Fig. 3-4. It can be seen from the curves that the scattered intensity is greater in the forward than in the backward direction, falling off smoothly with angle except for rather large spheres. It is sometimes convenient to characterize the size of the scattering particles by the *dissymmetry* of light scattering, $z = i_{45°}/i_{135°}$, measured at two angles symmetrical about 90°. The dissymmetry increases rapidly with increasing particle size.

Molecular weight determination from light scattering in solutions of larger particles

Equation 3-14 is derived with the assumption that the scattering particles are small compared to λ. If this is not true, dissymmetry of scattering exists, and the turbidity of the solution is reduced by the interference between waves from different parts of the same particle. This effect is taken into account by modifying Eq. 3-14 to read

$$K\frac{c}{R_{90}} = H\frac{c}{\tau} = \frac{1}{\bar{M}_w P(\theta)} + 2A_2c + \cdots \tag{3-20}$$

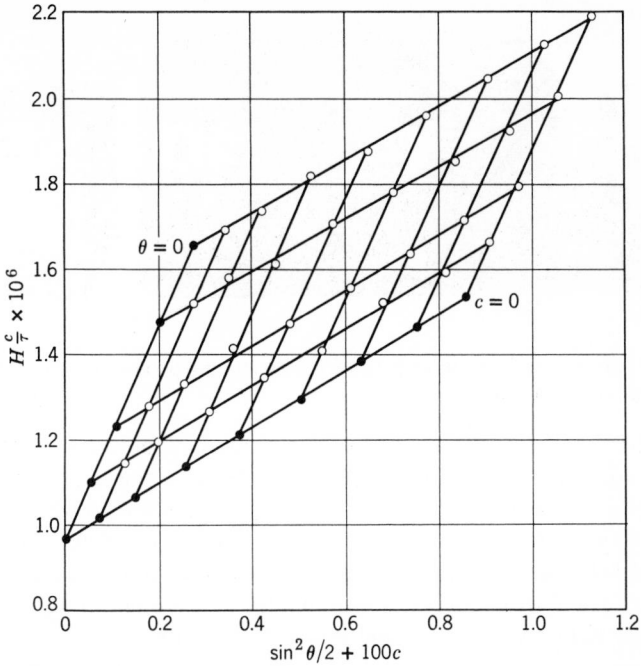

Fig. 3-5. Zimm plot showing the light scattering from a sample of polystyrene in butanone (2). Molecular parameters are $\bar{M}_w = 1{,}030{,}000$; $A_2 = 1.29 \times 10^{-4}$ ml mole g^{-2}; $(\overline{s_z^2})^{1/2} = 460$ A; $z = 1.44$.

where M is replaced by $\bar{M}_w$ for application to polydisperse systems. It has been shown (21) that a z-average dimension is obtained from $P(\theta)$ for polydisperse systems.

In practice the correction for dissymmetry can be made in two ways. In Debye's original method, the dissymmetry was measured and compared with the functions of Fig. 3-4 to evaluate particle size and thence $P(\theta)$. In more common current use is Zimm's method, in which the left-hand side of Eq. 3-20 is plotted (*Zimm plot*) against $\sin^2 \theta/2 + kc$ where k is an arbitrary constant. A rectilinear grid results (Fig. 3-5), allowing extrapolation to both $c = 0$ and $\theta = 0$, where $P(\theta) = 1$.

Determination of molecular weight by light scattering

Equipment Early light-scattering turbidimeters, measuring scattered intensity photographically or visually, have been almost entirely superseded by photoelectric instruments. Representative instruments are

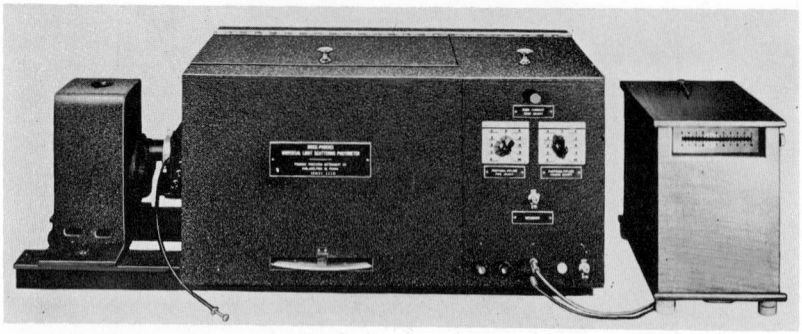

Fig. 3-6. Exterior view of the Brice-Phoenix light-scattering photometer (4). Courtesy Phoenix Precision Instrument Co.

described in the literature (1, 4, 10, 21). A brief description of a commercially available instrument (4) (Fig. 3-6) follows, the letters referring to Fig. 3-7. Light from a mercury-arc lamp L passes through a monochromatizing filter F_1, lenses (L_1, L_2), neutral filters (F_2) if desired, and a polarizer (PO). It then strikes either a calibrated reference standard or a glass cell (C) holding the polymer solution. After passing through the cell, the primary light beam is absorbed in a light-trap tube (T).

Scattered light from cell or standard is viewed by a multiplier phototube (PT) after passing through a slit system (D_4, D_5) and a second polarizer (AN). The phototube is powered by high voltage from a regulated electronic power supply. Its output signal is transmitted either to a galvanometer or to a strip chart recorder.

A typical scattering cell consists of a glass cylinder having flat entrance and exit windows. The scattering cell is centered on the axis of rotation of the receiver phototube. The receiver is rotated manually, its angular position being controlled and observed by means of a calibrated disc (D) on the outside of the scattering compartment.

Sample preparation The proper preparation of a sample for light-scattering measurement is of major importance. Polymer molecules are usually small compared to particles of dust or other extraneous material. The scattering of these extraneous substances, while usually confined to low angles, may easily outweigh the scattering from the polymer solution. Consequently, solvents and polymer solutions must be clarified by filtration or by ultracentrifugation. Proper clarification, as evidenced by obtaining a rectilinear Zimm plot, is essential for the unambiguous interpretation of light-scattering data (5, 16).

Accuracy in light-scattering measurement is favored by the proper choice of solvent. The difference in refractive index between polymer and solvent

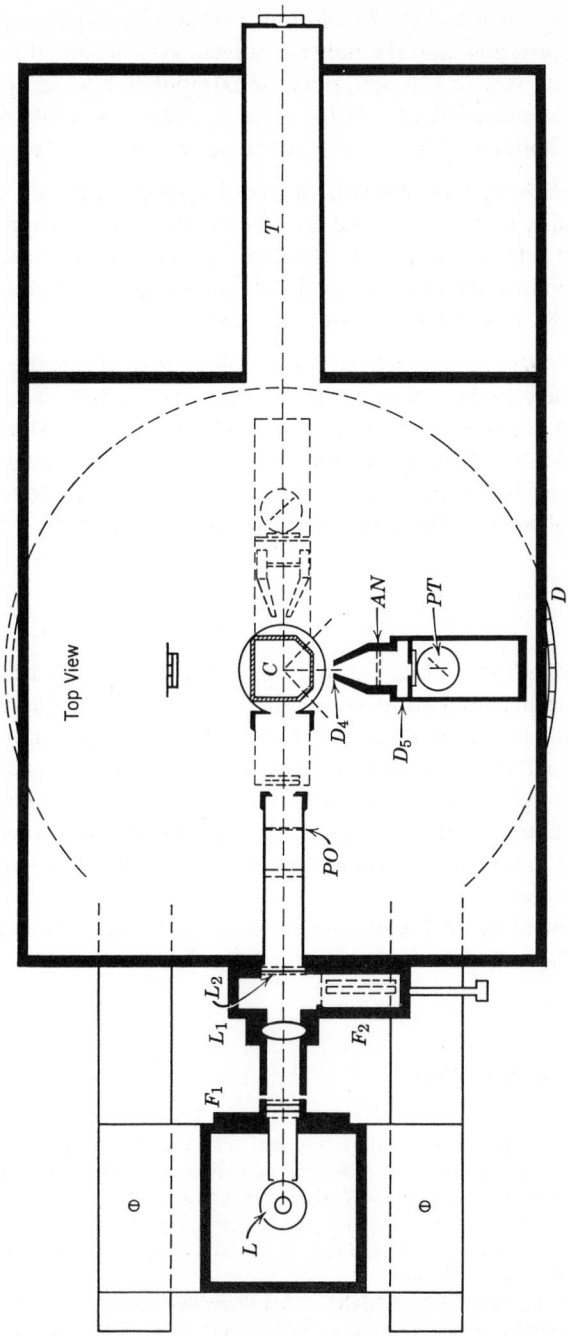

Fig. 3-7. Diagram of the essential components of the Brice-Phoenix light-scattering photometer (4). Functions of lettered parts are discussed in the text. Courtesy Phoenix Precision Instrument Co.

should be as large as possible. In addition, the solvent should itself have relatively low scattering, and the polymer-solvent system should not have too high a second virial coefficient, since the extrapolation to zero polymer concentration becomes less certain for high A_2. Mixed solvents must be avoided unless both components have the same refractive index.

Calibration Although in principle observed scattered intensity can be related to turbidity through knowledge of the geometry of the photometer, calibration with substances of known turbidity is common practice. These substances include reflecting standards (4), colloidal suspensions (14), simple liquids (8), and well-calibrated polymers (7).

Treatment of data The weight-average molecular weight is the inverse of the intercept at $c = 0$ and $\theta = 0$ in the Zimm plot (Fig. 3-5). The second virial coefficient is calculated from the slope of the lines at constant angle by Eq. 3-20, and to a first approximation is independent of angle. The radius of gyration is derived from the slope of the zero concentration line as a function of angle. The appropriate relationship, derived from Eq. 3-19, is

$$\overline{s_z^2} = \frac{3\lambda_s^2}{16\pi^2} \times \frac{\text{initial slope}}{\text{intercept}} \tag{3-21}$$

Range of applicability Light scattering from solutions has been used to measure molecular weights as low as that of sucrose (9) and as high as those of proteins (11). In practice, polymer molecular weights in the order of 10,000–10,000,000 can readily be measured, with the possibility of extending the range in either direction in favorable cases. Applications include determination of absolute values of $\bar{M}_w$, especially for proteins and other biological materials, and calibration of intrinsic viscosity-molecular weight relations (Section *E*).

The measurement by light scattering of the size of particles considerably larger than polymer molecules is discussed by Billmeyer and by van de Hulst (gen. refs.).

GENERAL REFERENCES

Billmeyer, F. W., Jr., "Principles of Light Scattering," Part I, Chap. 56, in I. M. Kolthoff and P. J. Elving, eds., *Treatise on Analytical Chemistry*, Interscience Publishers, a Division of John Wiley & Sons, New York (in preparation).

Fishman, M. M., *Light Scattering by Colloidal Systems, An Annotated Bibliography*, Technical Service Laboratories, River Edge, N.J., 1957; *Supplement* to same, 1958.

Oster, G., *Chem. Revs.* **43,** 319 (1948).

Oster, G., "Light Scattering," Chap. XXXII in A. Weissberger, ed., *op. cit.* Part 3.

Peaker F. W., "Light-Scattering Techniques," Chap. 5 in P. W. Allen, ed., *op. cit.*

Stacey, K. A., *Light-Scattering in Physical Chemistry*, Butterworth's Scientific Publications, London, 1956.

Tanford, C., *Physical Chemistry of Macromolecules*, John Wiley & Sons, New York, 1961, Secs. 17, 18.

van de Hulst, H. C., *Light Scattering by Small Particles*, John Wiley & Sons, New York, 1957.

BIBLIOGRAPHY

1. Baum, F. J., and F. W. Billmeyer, Jr., *J. Opt. Soc. Am.* **51,** 452 (1961).
2. Billmeyer, F. W., Jr., and C. B. de Than, *J. Am. Chem. Soc.* **77,** 4763 (1955).
3. Brice, B. A., and M. Halwer, *J. Opt. Soc. Am.* **41,** 1033 (1951).
4. Brice, B. A., M. Halwer, and R. Speiser, *J. Opt. Soc. Am.* **40,** 768 (1950).
5. Bryant, W. M. D., F. W. Billmeyer, Jr., L. T. Muus, J. T. Atkins, and J. E. Eldridge, *J. Am. Chem. Soc.* **81,** 3219 (1959).
6. Cabannes, J., and Y. Rocard, *La diffusion moléculaire de la lumière*, Les Presses Universitaires de France, Paris, 1929.
7. Carpenter, D. K., and W. R. Krigbaum, *J. Chem. Phys.* **24,** 1041 (1956).
8. Carr, C. I., Jr., and B. H. Zimm, *J. Chem. Phys.* **18,** 1616 (1950).
9. Debye, P., *J. Appl. Phys.* **15,** 338 (1944); *J. Phys. & Coll. Chem.* **51,** 18 (1947).
10. Debye, P. P., *J. Appl. Phys.* **17,** 392 (1946).
11. Doty, P., and J. T. Edsall, "Light Scattering in Protein Solutions," in M. L. Anson, J. T. Edsall, and K. Bailey, eds., *Advances in Protein Chemistry*, Vol. VI, Academic Press, New York, 1951, pp. 37–123.
12. Einstein, A., *Ann. Physik* **33,** 1275 (1910).
13. Gans, R., *Ann. Physik* **76,** 29 (1925).
14. Maron, S. H., and R. L. H. Lou, *J. Polymer Sci.* **14,** 20 (1954).
15. Mie, G., *Ann. Physik* **25,** 377 (1908).
16. Muus, L. T., and F. W. Billmeyer, Jr., *J. Am. Chem. Soc.* **79,** 5079 (1957).
17. Rayleigh, Lord (Strutt, J. W.), *Phil. Mag.* [4] **41,** 107, 224, 447 (1871).
18. Rayleigh, Lord (Strutt, J. W.), *Phil. Mag.* [5] **12,** 81 (1881); *Proc. Roy. Soc.* (*London*) **A90,** 219 (1914).
19. Smoluchowski, M. S., *Ann. Physik* **25,** 205 (1908); *Phil. Mag.* [6] **23,** 165 (1912).
20. Zimm, B. H., R. S. Stein, and P. Doty, *Polymer Bull.* **1,** 90 (1945).
21. Zimm, B. H., *J. Chem. Phys.* **16,** 1093, 1099 (1948).

D. Ultracentrifugation

Ultracentrifugation techniques are the most intricate of the existing methods for determining the molecular weights of high polymers. They have in the past been far more successful in application to compact protein molecules than to random coil polymers, where extended conformations increase deviations from ideality and the chances of mechanical entanglement of the polymer chains. Recent theoretical and experimental developments have combined, however, to increase the utility of the techniques for random coil polymer systems.

Experimental techniques

The ultracentrifuge consists of an aluminum rotor several inches in diameter which is rotated at high speed in an evacuated chamber. The solution being centrifuged is held in a small cell within the rotor near its periphery. The rotor may be driven electrically or by an oil or air turbine. The cell is equipped with windows, and the concentration of polymer along its length is determined by optical methods based on measurements of refractive index or absorption. A schematic view of a commercially available ultracentrifuge is shown in Fig. 3-8, and details of the rotor and drive unit are shown in Fig. 3-9. The instrument can be used over a range of speeds suitable for all types of ultracentrifugation, and at temperatures up to about 100°C.

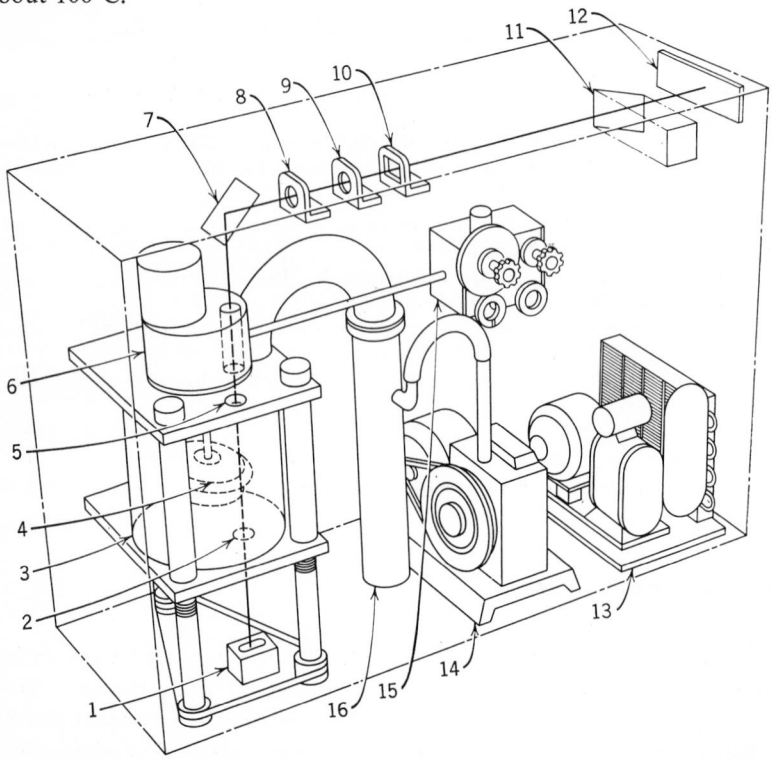

Fig. 3-8. Schematic perspective of a commercially available, electrically driven analytical ultracentrifuge. Numbered parts include: 1, lamp; 2, 5, 9, lenses; 3, vacuum chamber; 4, rotor; 6, drive unit; 7, 11, mirrors; 8, 10, diaphragm and cylindrical lens of schlieren optics; 12, photographic plate; 13, refrigeration unit; 14, 16, vacuum pumps; 15, speed-control unit. Courtesy Spinco Division, Beckman Instruments, Inc.

Fig. 3-9. Rotor compartment for the ultracentrifuge of Fig. 3-8, showing rotor (below) and drive unit (above). Courtesy Spinco Division, Beckman Instruments, Inc.

Solvents for ultracentrifugation experiments must be chosen for difference from the polymer in both density (to ensure sedimentation) and refractive index (to allow measurement). In addition, mixed solvents must usually be avoided, and low solvent viscosity is desirable. It is customary to use *schlieren optics* in the ultracentrifuge; experimental data are obtained as a (photographed) curve of dn/dr vs. r, where n is refractive index and r is the distance from the center of rotation to the point of observation in the cell. These data are converted to a curve of dc/dr vs. r through knowledge of the specific refractive increment dn/dc. A *synthetic boundary cell* (4), in which pure solvent is layered onto a polymer solution during ultracentrifugation to form a sharp boundary, can be used to calibrate the optics of the ultracentrifuge or to determine dn/dc after calibration.

Sedimentation equilibrium

In the sedimentation equilibrium experiment, the ultracentrifuge is operated at a low speed of rotation for times up to 1 or 2 weeks under constant conditions. A thermodynamic equilibrium is reached in which the polymer is distributed in the cell solely according to its molecular weight and molecular weight distribution, the force of sedimentation on each species being just balanced by its tendency to diffuse back against the concentration gradient resulting from its movement in the centrifugal field. The force on a particle of mass m at a distance r from the axis of rotation is $\omega^2 r(1 - \bar{v}\rho)m$, where ω is the angular velocity of rotation, $\bar{v}$ the partial specific volume of the polymer, and ρ the density of the solution. Writing the partial molar free energy in terms of this force and applying the conditions for equilibrium gives the result

$$\frac{M(1 - \bar{v}\rho)\omega^2 r}{RT} = \frac{1}{c}\frac{\mathrm{d}c}{\mathrm{d}r}\left(1 + c\frac{\mathrm{d}\ln\gamma}{\mathrm{d}c}\right) \tag{3-22}$$

where the second term corrects for nonideality, γ being the activity coefficient. For ideal solutions, this equation integrates to

$$M = \frac{2RT\ln(c_2/c_1)}{(1 - \bar{v}\rho)\omega^2(r_2^2 - r_1^2)} \tag{3-23}$$

where c_1 and c_2 are the concentrations at two points r_1 and r_2 in the cell. If the polymer is heterogeneous, M is the weight-average molecular weight; r_1 and r_2 must be taken at the meniscus and the cell bottom to include all the molecular species in the average. If the data are treated in terms of the refractive increment $\tilde{n}$, the difference between the refractive indices of solution and solvent, it can be shown that

$$\bar{M}_z = \frac{RT}{(1 - \bar{v}\rho)\omega^2}\frac{\left(\frac{1}{r}\frac{\mathrm{d}\tilde{n}}{\mathrm{d}r}\right)_2 - \left(\frac{1}{r}\frac{\mathrm{d}\tilde{n}}{\mathrm{d}r}\right)_1}{\tilde{n}_2 - \tilde{n}_1} \tag{3-24}$$

where $\bar{M}_z$ is the *z-average molecular weight*

$$\bar{M}_z = \sum_{i=1}^{\infty} N_i M_i^3 \Big/ \sum_{i=1}^{\infty} N_i M_i^2 = \sum_{i=1}^{\infty} w_i M_i^2 \Big/ \sum_{i=1}^{\infty} w_i M_i \tag{3-25}$$

and again positions 1 and 2 must be the ends of the cell.

The problem of nonideality is best met by working at the Θ-temperature (6), but it has been shown (3) that the apparent molecular weight is a linear function of concentration at temperatures near Θ, the slope depending primarily on the second virial coefficient.

The major disadvantage to the sedimentation equilibrium experiment lies in the fact that days or weeks may be required to reach equilibrium. Although the time can be shortened by the use of short cells (10) or a synthetic boundary cell (9) to reduce the distance the various species must travel as equilibrium is established, the experiment remains long and tedious.

In principle, the distribution of molecular weights can be obtained directly from sedimentation equilibrium measurements (11). The solution of an integral equation is involved, however, and it is usually found that the resulting distribution curves are indeterminate in the low-molecular-weight region unless $\bar{M}_n$ is known to assist in obtaining a good fit.

Equilibrium sedimentation in a density gradient In this relatively new approach (8), a mixed solvent is chosen so that the relative sedimentation of the two components gives rise to a density gradient. The solute forms a band centering at the point where its effective density equals that of the solvent mixture. The band is Gaussian in shape with respect to solute concentration, the half-width being inversely proportional to the solute molecular weight. A feature which is of major importance is the sensitivity of the method to small differences in effective density among the solute species.

Approach to equilibrium

In 1947 Archibald (1) re-examined the boundary conditions for the ultracentrifuge equation, and showed that measurement of c and dc/dr at the cell boundaries allows molecular weight to be determined at any stage in the equilibrium process. If measurements are made early enough, before the molecular species have time to redistribute in the cell, the weight-average molecular weight and second virial coefficient can be evaluated (5). In practice, measurements can be made about 10–60 min after the start of the experiment. At properly chosen times within this range, a graph of $(1/cr)(dc/dr)$ vs. r is linear near the ends of the cell, allowing simple extrapolation to the position of the meniscus or cell bottom. Redistribution of molecular species can be recognized and corrected for by extrapolating the apparent molecular weight to zero time. The method holds considerable promise for random coil polymers.

Sedimentation transport

In the *sedimentation transport* or *sedimentation velocity* experiment, the ultracentrifuge is operated at high speed so that the solute is transported to

the bottom of the cell. The rate of sedimentation is given by the *sedimentation constant*

$$s = \frac{1}{\omega^2 r}\frac{\mathrm{d}r}{\mathrm{d}t} = \frac{m(1 - \bar{v}\rho)}{f} \tag{3-26}$$

and related to particle mass through a *frictional coefficient* f. For rigid particles f is well defined (for solid spheres of diameter d, f is given by Stokes' law, $f = 3\pi\eta d$), but for random coil polymers the evaluation of f is more difficult. It can be related to the diffusion coefficient D: at infinite dilution $D = kT/f$, whence, multiplying by Avogadro's number,

$$\frac{D}{s} = \frac{RT}{M(1 - \bar{v}\rho)} \tag{3-27}$$

However, even for a single solute species the equations have not been extended to take into account simultaneously the effects of diffusion in the ultracentrifuge cell, deviations from ideality, and compressibility of solvent and solute in the high centrifugal fields required. For polydisperse solutes, additional complications arise in assigning average molecular weights, since D and s must be averaged separately. In general, simple averages such as $\bar{M}_w$ and $\bar{M}_z$ are not obtained.

The most fruitful result of the sedimentation transport experiment now appears to be examination of the distribution of sedimentation constants (2, 7). By working in a Θ-solvent, the concentration dependence of s and D is reduced enough to avoid extrapolation to $c = 0$ [although Flory (gen. ref.) has shown that it is $\mathrm{d}(s/D)/\mathrm{d}c$ which vanishes at $T = \Theta$]. The observed distribution of s over the cell must still be extrapolated to infinite time to eliminate the effects of diffusion relative to sedimentation, and correction for compression must be made. The resulting distribution of s can be converted to a distribution of molecular weight, either through detailed knowledge of the dependence of s and D on M or from the relation

$$s = kM^{1/2} \tag{3-28}$$

which is very nearly true at $T = \Theta$. The number of approximations and the complex treatment of the data required in this procedure suggest that its use be limited to qualitative observations. In this respect, however, the experiment can be extremely valuable, for it provides uniquely and directly qualitative information about the nature of the polymer species present.

Preparative separations The usefulness of the ultracentrifuge in the preparation of samples rather than in the production of analytical data should not be overlooked. Preparative ultracentrifuges have utility in fractionating polymer samples and in freeing them from easily sedimented impurities.

GENERAL REFERENCES

Baldwin, R. L., and K. E. van Holde, *Fortschr. Hochpolym.-Forsch.* (*Advances in Polymer Science*) **1,** 451 (1960).
Flory, P. J., *op. cit.*, Chap. VII-3.
Fujita, H., *Mathematical Theory of Sedimentation Analysis*, Academic Press, New York, 1962.
Nichols, J. B., and E. D. Bailey, "Determinations with the Ultracentrifuge," Chap. XVII in A. Weissberger, ed., *op. cit.*, Part 2.
Schachman, H. K., *Ultracentrifugation in Biochemistry*, Academic Press, New York, 1959.
Svedberg, T., and K. O. Pedersen, *The Ultracentrifuge*, Clarendon Press, Oxford, 1940; Johnson Reprint Corp., New York, 1959.
Tanford, C., *op. cit.*, Secs. 16, 21, 22, Appendix C.
Williams, J. W., K. E. van Holde, R. L. Baldwin, and H. Fujita, *Chem. Revs.* **58,** 715 (1958).

BIBLIOGRAPHY

1. Archibald, W. J., *J. Phys. & Coll. Chem.* **51,** 1204 (1947).
2. Cantow, H.-J., *Makromol. Chem.* **30,** 169 (1959).
3. Fujita, H., *J. Phys. Chem.* **63,** 1326 (1959).
4. Kegeles, G., *J. Am. Chem. Soc.* **74,** 5532 (1952).
5. Kegeles, G., S. M. Klainer, and W. J. Salem, *J. Phys. Chem.* **61,** 1286 (1957).
6. Mandelkern, L., L. C. Williams, and S. G. Weissberg, *J. Phys. Chem.* **61,** 271 (1957).
7. McCormick, H. W., *J. Polymer Sci.* **36,** 341 (1959).
8. Meselson, M., F. W. Stahl, and J. Vinograd, *Proc. Natl. Acad. Sci.* **43,** 581 (1957).
9. Pasternak, R. A., G. M. Nazarian, and J. R. Vinograd, *Nature* **179,** 92 (1957).
10. van Holde, K. E., and R. L. Baldwin, *J. Phys. Chem.* **62,** 734 (1958).
11. Wales, M., *J. Phys. & Coll. Chem.* **55,** 282 (1951).

E. Solution Viscosity and Molecular Size

The usefulness of solution viscosity as a measure of polymer molecular weight has been recognized ever since the early work of Staudinger (19). Solution viscosity is basically a measure of the size or extension in space of polymer molecules. It is empirically related to molecular weight for linear polymers; the simplicity of the measurement and the usefulness of the viscosity-molecular weight correlation are so great that viscosity measurement constitutes an extremely valuable tool for the molecular characterization of polymers.

Experimental methods

Measurements of solution viscosity are usually made by comparing the *efflux time t* required for a specified volume of polymer solution to flow

through a capillary tube with the corresponding efflux time t_0 for the solvent. From t, t_0, and the solute concentration are derived several quantities whose defining equations and names are given in Table 3-3. Two

TABLE 3-3. Nomenclature of solution viscosity

Common Name (3)	Recommended Name (11)	Symbol and Defining Equation
Relative viscosity	Viscosity ratio	$\eta_r = \eta/\eta_0 \simeq t/t_0$
Specific viscosity	...	$\eta_{sp} = \eta_r - 1 = (\eta - \eta_0)/\eta_0 \simeq (t - t_0)/t_0$
Reduced viscosity	Viscosity number	$\eta_{red} = \eta_{sp}/c$
Inherent viscosity	Logarithmic viscosity number	$\eta_{inh} = (\ln \eta_r/c)$
Intrinsic viscosity	Limiting viscosity number	$[\eta] = (\eta_{sp}/c)_{c=0} = [(\ln \eta_r)/c]_{c=0}$

sets of nomenclature are in use for these quantities; one (3) has had long and widespread application, the other (11) was proposed for greater clarity and precision. In this book the common nomenclature is retained. In this system, the concentration c is expressed in grams per deciliter (g/dl, g/100 ml).

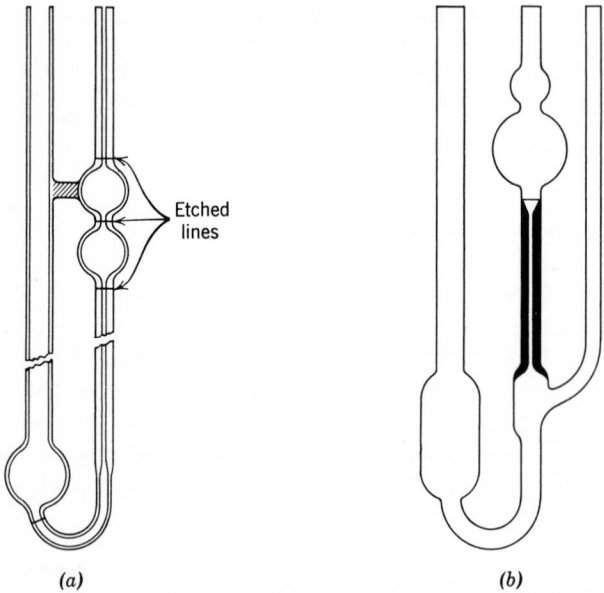

Fig. 3-10. Capillary viscometers commonly used for measurement of polymer solution viscosities (*a*) Ostwald-Fenske; (*b*) Ubbelohde.

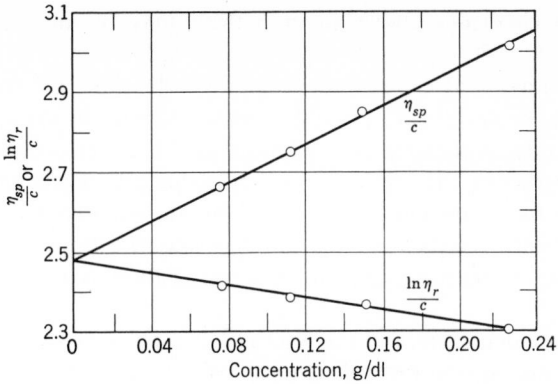

Fig. 3-11. Reduced and inherent viscosity-concentration curves for a polystyrene in benzene (5).

The intrinsic viscosity $[\eta]$ is independent of concentration by virtue of extrapolation to $c = 0$, but is a function of the solvent used. The inherent viscosity at a specified concentration, usually 0.5 g/dl, is sometimes used as an approximation to $[\eta]$.

Dilute solution viscosity is usually measured in capillary viscometers of the Ostwald-Fenske or Ubbelohde type (Fig. 3-10*a* and *b*, respectively). The latter has the advantage that the measurement is independent of the amount of solution in the viscometer; measurement at a series of concentrations can easily be made by successive dilution.

For highest precision, the following precautions are usually observed. The viscosity measurement is made in a constant-temperature bath regulated to at least $\pm 0.02°C$. The efflux time is kept long (preferably greater than 100 sec) to minimize the need for applying corrections to the observed data. For accuracy in extrapolating to $c = 0$, the solution concentration is restricted to the range which gives relative viscosities between 1.1 and 1.5.

Treatment of data Viscosity data as a function of concentration are extrapolated to infinite dilution by means of the Huggins equation (10)

$$\eta_{sp}/c = [\eta] + k'[\eta]^2 c \tag{3-29}$$

where k' is a constant for a series of polymers of different molecular weights in a given solvent. The alternate definition of the intrinsic viscosity leads to the equation (13)

$$(\ln \eta_r)/c = [\eta] - k''[\eta]^2 c \tag{3-30}$$

where $k' + k'' = \frac{1}{2}$. Typical data are shown in Fig. 3-11. The slopes of the lines vary as $[\eta]^2$ as required by Eqs. 3-29 and 3-30 for polymer species of different molecular weight in the same solvent.

At intrinsic viscosities above about 5 there may be an appreciable dependence of viscosity on the rate of shear in the viscometer. This dependence is not eliminated by extrapolation to infinite dilution. Correction by measurement at several shear rates (flow times), followed by extrapolation to zero shear rate, is required (9). Theories (2, 17, 21) of the shear rate dependence predict that the ratio of $[\eta]$ to its value $[\eta]_0$ at zero shear rate is a function of the product of shear gradient and the quantity $M\eta_0[\eta]_0/RT$, where η_0 is solvent viscosity at zero shear rate; thus M is the only property of the polymer appearing. Experimental verification is not complete, however.

The shear rate dependence of the viscosity of a polymer solution takes the form of a decrease in the apparent viscosity (shear stress/shear rate) with increase of shear rate, until at sufficiently high rates of shear (10^5 sec^{-1} compared to 10^2–10^3 sec^{-1} in the usual capillary viscometers) the viscosity again becomes constant in the *upper Newtonian region* (1). For samples with sufficiently high molecular weight, the reduced viscosity calculated from data obtained in the upper Newtonian region may be nearly independent of polymer concentration, in contrast to its behavior at low shear rate (16).

Empirical correlations between intrinsic viscosity and molecular weight for linear polymers

Staudinger's prediction in 1930 that the reduced viscosity of a polymer is proportional to its molecular weight has needed only slight modification: the intrinsic viscosity has been substituted for the reduced viscosity, and it has been recognized that the proportionality is to a power of the molecular weight somewhat less than 1. The relation is expressed in the equation

$$[\eta] = K'M^a \tag{3-31}$$

where K' and a are constants determined from a double logarithmic plot of intrinsic viscosity and molecular weight (Fig. 3-12). Such plots are usually found to be straight lines within experimental error over large ranges of the variables. For randomly coiled polymers, the exponent a varies from 0.5 in a Θ-solvent to a maximum of about 1.0. For many systems, a lies between 0.6 and 0.8. Typical values of K' range between 0.5 and 5×10^{-4}. Both K' and a are functions of the solvent as well as of the polymer type. This empirical relation between viscosity and molecular weight is valid only for linear polymers.

Intrinsic viscosity measurement leads to the *viscosity-average molecular weight* (18), defined by the equation

$$\bar{M}_v = \left[\sum_{i=1}^{\infty} w_i M_i^a\right]^{1/a} = \left[\sum_{i=1}^{\infty} N_i M_i^{1+a} \Big/ \sum_{i=1}^{\infty} N_i M_i\right]^{1/a} \tag{3-32}$$

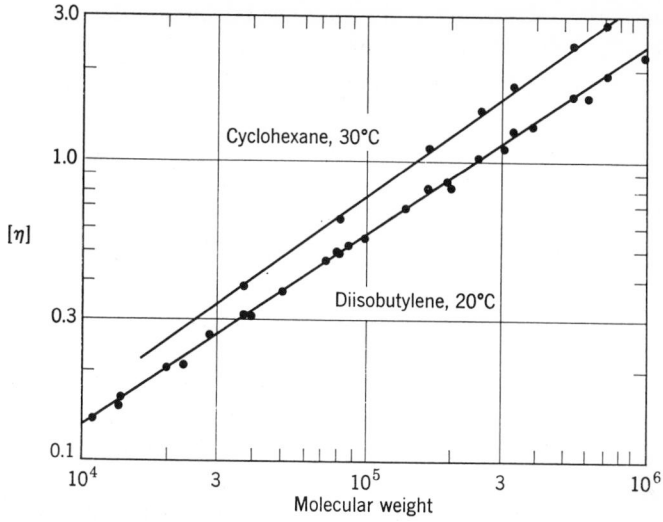

Fig. 3-12. Intrinsic viscosity-molecular weight relationships for polyisobutylene in diisobutylene and cyclohexane (6, 14).

Thus $\bar{M}_v$ depends on a as well as on the distribution of molecular species. For many polymers, $\bar{M}_v$ is 10–20% below $\bar{M}_w$ (Fig. 1-1). For $a = 1$, $\bar{M}_v = \bar{M}_w$. If fractionated polymers, for which $\bar{M}_n \simeq \bar{M}_v \simeq \bar{M}_w$, are available, any absolute molecular weight measurement may be combined with a viscosity measurement to evaluate the constants in Eq. 3-31. If the molecular weight distribution of the samples is known in sufficient detail to calculate $\bar{M}_v$ from a molecular weight average which has been measured, the constants in the equation

$$[\eta] = K'\bar{M}_v{}^a \tag{3-31a}$$

may be evaluated. Since $\bar{M}_v$ is nearer to $\bar{M}_w$ than to $\bar{M}_n$, the weight-average molecular weight is preferred for this calibration. It can be shown that polymers with fairly broad but well-known distributions are preferred to fractions for this purpose. A less precise relation,

$$[\eta] = K'\bar{M}_w{}^a \tag{3-31b}$$

is sometimes used to relate weight-average molecular weight directly to intrinsic viscosity for a limited series of samples.

Intrinsic viscosity and molecular size

Theories (4, 7, 8, 12, 24) of the frictional properties of polymer molecules in solution show that the intrinsic viscosity is proportional to the

effective hydrodynamic volume of the molecule in solution divided by its molecular weight. The effective volume is proportional to the cube of a linear dimension of the randomly coiling chain. If $(\overline{r^2})^{1/2}$ is the dimension chosen,

$$[\eta] = \Phi(\overline{r^2})^{3/2}/M \tag{3-33}$$

where Φ is a universal constant. Replacing $(\overline{r^2})^{1/2}$ by $\alpha(\overline{r_0^2})^{1/2}$ and recalling that $\overline{r_0^2}/M$ is a function of chain structure independent of its surroundings or molecular weight, it follows that

$$[\eta] = \Phi(\overline{r_0^2}/M)^{3/2}M^{1/2}\alpha^3 = KM^{1/2}\alpha^3 \tag{3-34}$$

where $K = \Phi(\overline{r_0^2}/M)^{3/2}$ is a constant for a given polymer, independent of solvent and molecular weight. The dimensions $(\overline{r^2})^{1/2}$ and $(\overline{r_0^2})^{1/2}$ and the expansion factor α used here are usually identified with those discussed in Chapter 2. The extensions of the Flory-Krigbaum treatment described at the end of Chapter 2*C* imply that Φ is not a universal constant, and that the power of α in Eq. 3-34 is somewhat less than 3 (15).

It follows from the properties of α that, at $T = \Theta$, $\alpha = 1$ and

$$[\eta]_\Theta = KM^{1/2} \tag{3-35}$$

There is ample experimental evidence for the validity of this equation. Values of K are near 1×10^{-3} for a number of polymer systems.

Determination of chain branching

The size of a dissolved polymer molecule, which can be estimated by means of its intrinsic viscosity, depends upon a number of factors. These include the molecular weight, the local chain structure as reflected in the factor $\overline{r_0^2}/M$, the perturbations reflected in α, and finally the gross chain structure, in which the degree of long-chain branching is an important variable. [The presence of short branches, such as ethyl or butyl groups in low-density polyethylene (Chapter 13*A*), has little effect on solution viscosity.] By evaluating or controlling each of the other factors, viscosity measurements may be used to estimate the degree of long-chain branching of a polymer.

The relation between molecular size and the number and type of branch points has been calculated (23), leading to a relation g between the radii of gyration for branched and linear chains of the same number of segments:

$$g = \overline{s^2}\,(\text{branched})/\overline{s^2}\,(\text{linear}) \tag{3-36}$$

Unfortunately, the several theories (7, 8, 20, 22) relating g to $[\eta]$ fail to agree on the variation of $[\eta]$ (branched)/$[\eta]$ (linear) with g: proportionality to powers of g between $\frac{1}{2}$ and $\frac{3}{2}$ is predicted. Thus at present it seems appropriate to consider the ratio $[\eta]$ (branched)/$[\eta]$ (linear) evaluated at constant molecular weight as a qualitative indication of chain branching rather than to attempt to assign numerical values to the degree of chain branching in a given polymer sample.

GENERAL REFERENCES

Bueche, F., *Physical Properties of Polymers*, Interscience Publishers, a Division of John Wiley & Sons, New York, 1962, Chap. 3.

Flory, P. J., *op. cit.*, Chaps. VII-4, XIV.

Meyerhoff, G., *Fortschr. Hochpolym.-Forsch.* (*Advances in Polymer Science*) **3**, 59 (1961).

Onyon, P. F., "Viscometry," Chap. 6 in P. W. Allen, ed., *op. cit.*

Tanford, C., *op. cit.*, Secs. 20, 23.

Tompa, H., *op. cit.*, Chap. 9.

BIBLIOGRAPHY

1. Brodnyan, J. G., F. H. Gaskins, and W. Philippoff, *Trans. Soc. Rheology* **1**, 109 (1957).
2. Bueche, F., *J. Chem. Phys.* **22**, 1570 (1954).
3. Cragg, L. H., *J. Colloid Sci.* **1**, 261 (1946).
4. Debye, P., and A. M. Bueche, *J. Chem. Phys.* **16**, 573 (1948).
5. Ewart, R. H., "Significance of Viscosity Measurements on Dilute Solutions of High Polymers," pp. 197–251 in H. Mark and G. S. Whitby, eds., *Scientific Progress in the Field of Rubber and Synthetic Elastomers* (*Advances in Colloid Science*, Vol. II), Interscience Publishers, New York, 1946.
6. Flory, P. J., *J. Am. Chem. Soc.* **65**, 372 (1943).
7. Flory, P. J., *J. Chem. Phys.* **17**, 303 (1949).
8. Flory, P. J., and T. G. Fox, Jr., *J. Polymer Sci.* **5**, 745 (1950); *J. Am. Chem. Soc.* **73**, 1904 (1951).
9. Fox, T. G., J. C. Fox, and P. J. Flory, *J. Am. Chem. Soc.* **73**, 1901 (1951).
10. Huggins, M. L., *J. Am. Chem. Soc.* **64**, 2716 (1942).
11. International Union of Pure and Applied Chemistry, *J. Polymer Sci.* **8**, 257 (1952).
12. Kirkwood, J. G., and J. Riseman, *J. Chem. Phys.* **16**, 565 (1948).
13. Kraemer, E. O., *Ind. Eng. Chem.* **30**, 1200 (1938).
14. Krigbaum, W. R., and P. J. Flory, *J. Am. Chem. Soc.* **75**, 1775 (1953).
15. Kurata, M., W. H. Stockmayer, and A. Roig, *J. Chem. Phys.* **33**, 151 (1960); M. Kurata and W. H. Stockmayer, paper presented at the International Symposium on Macromolecular Chemistry, Montreal, Can., July, 1961.
16. Merrill, E. W., *J. Polymer Sci.* **38**, 539 (1959); E. W. Merrill, H. S. Mickley, A. Ram, and G. Perkinson, *Trans. Soc. Rheology* **5**, 237 (1961); see also S. Claesson and U. Lohmander, *Makromol. Chem.* **44–46**, 461 (1961).
17. Pao, Y.-H., *J. Appl. Phys.* **28**, 591 (1957).

18. Schaefgen, J. R., and P. J. Flory, *J. Am. Chem. Soc.* **70,** 2709 (1948).
19. Staudinger, H., and W. Heuer, *Ber.* **63,** 222 (1930).
20. Stockmayer, W. H., and M. Fixman, *Ann. N. Y. Acad. Sci.* **57,** 334 (1953).
21. Takemura, T., *J. Polymer Sci.* **27,** 549 (1958).
22. Zimm, B. H., and R. W. Kilb, *J. Polymer Sci.* **37,** 19 (1959).
23. Zimm, B. H., and W. H. Stockmayer, *J. Chem. Phys.* **17,** 1301 (1949).
24. Zimm, B. H., "The Normal-Coordinate Method for Polymer Chains in Dilute Solution," Chap. 1 in F. R. Eirich, ed., *Rheology*, Academic Press, New York, Vol. 3, 1960.

F. Polyelectrolytes

Polymers with ionic groups along the chain show properties of both electrolytes and polymers. The ionic dissociation in their aqueous solutions leads to large repulsive forces among the charged groups in the chain. These forces give rise to greatly expanded conformations and very large intrinsic viscosities, as well as abnormalities in other thermodynamic properties. Examples of polyelectrolytes are sodium polyacrylate, the salt of poly(acrylic acid), an anionic polyelectrolyte:

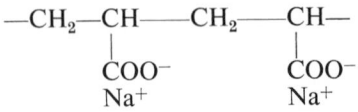

and poly(4-vinyl-N-butylpyridinium bromide), a cationic polyelectrolyte:

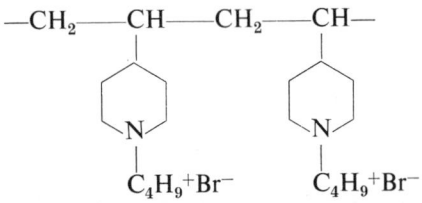

At concentrations above about 1%, the molecules in a polyelectrolyte solution are in close enough contact to overlap partially and are not appreciably expanded; the specific viscosity is approximately normal. As the solution is diluted, the molecules no longer fill all the space and some of the mobile ions leave the regions of the chains, which develop net charges and expand. The specific viscosity may actually rise with increasing dilution. At the same time, the reduced osmotic pressure π/c may rise as the ions are released to act as separate osmotic units. Related effects are shown in other physical properties, such as light scattering and conductance. The addition of a strong electrolyte suppresses the loss of the mobile ions, and

the viscosity behavior becomes normal. At constant and fairly high ionic strength, the viscosity-concentration relation of Eq. 3-29 is obeyed. The shear dependence of the viscosity of polyelectrolytes is considerably greater than that of uncharged polymers.

GENERAL REFERENCES

Conway, B. E., and A. Dobry-Duclaux, "Viscosity of Suspensions of Electrically Charged Particles and Solutions of Polyelectrolytes," Chap. 3 in F. R. Eirich, ed., *Rheology*, Vol. 3, Academic Press, New York, 1960.

Flory, P. J., *op. cit.*, Chap. XIV-5.

Fuoss, R. M., *Disc. Faraday Soc.* **11,** 125 (1951).

Rice, S. A., "Polyelectrolytes," Chap. 7 in J. L. Oncley *et al.*, eds., *Biophysical Science*, John Wiley & Sons, New York, 1959; *Revs. Modern Phys.* **31,** 69 (1959).

Rice, S. A., and M. Nagasawa, *Polyelectrolyte Solutions*, Academic Press, New York, 1961.

Tanford, C., *op. cit.*, Secs. 24, 26–28.

4

Analysis and Testing of Polymers

The purpose of this chapter is to assemble in one location information on the applicability to polymers of various standard methods of physical and chemical analysis. A background is thus established for the uninterrupted discussion in succeeding chapters of the results of such analyses and tests. In no case is the treatment in this chapter exhaustive; it is intended to provide, with minimum detail, a broad picture of the applicability of each method to polymer systems and of the type of information obtained.

A. Analytical Chemistry of Polymers

Chemical analysis

The chemical analysis of polymers is not basically different from the analysis of low-molecular-weight organic compounds, provided appropriate modification is made to ensure solubility or the availability of sites for reaction (e.g., insoluble specimens should be ground to expose a large surface area). The usual methods for functional group and elemental analysis are generally applicable, as are many other standard analytical techniques (1, 2, 6, 8). A typical scheme for analysis is given by Epprecht (gen. ref.), and another, based primarily on solubility, elemental analysis, and a few specific tests, by Gruben and Leiner (3). Schemes of this nature can be supplemented by information in compilations of the properties of plastics (7).

The analytical chemistry of the more common plastics is treated by Kline (gen. ref.). Chemical reactions of polymers (Chapter 12*D*) provide additional means of chemical analysis, as do their reactions of degradation (Chapter 12*E*).

Physical analysis of reaction products

This section, although far from comprehensive, describes two very powerful physical techniques for analyzing low-molecular-weight products of reactions (usually degradation) of polymers: mass spectrometry and gas chromatography.

Mass spectrometry (5) In the most common applications of mass spectrometry to polymer systems, the polymer is allowed to react (as in thermal degradation, Chapter 12*E*) to form low-molecular-weight fragments which are condensed at liquid-air temperature. They are then volatilized, ionized, and separated according to mass and charge by the action of electric and magnetic fields in a typical mass spectrometer analysis. From the abundances of the various ionic species found, the structures of the low-molecular-weight species can be inferred.

Gas chromatography (4) Gas chromatography is a method of separation in which gaseous or vaporized components are distributed between a moving gas phase and a fixed liquid phase or solid adsorbent. By a continuous succession of adsorption and elution steps, taking place at a specific rate for each component, separation is achieved. The components are detected by one of several methods as they emerge successively from the chromatographic column. From the detector signal, proportional to the instantaneous concentration of the dilute component in the gas stream, is obtained information about the number, nature, and amounts of the components present.

GENERAL REFERENCES

Epprecht, A. G., "The Chemical Analysis of Polymers," Chap. 3 in R. Houwink, ed., *Elastomers and Plastomers*, Vol. III, *Testing and Analysis; Tabulation of Properties*, Elsevier Publishing Co., New York, 1948.

Kline, G. M., ed., *Analytical Chemistry of Polymers*, Interscience Publishers, New York, Part I, 1959; Parts II and III, 1962.

Mitchell, J., Jr., and S. S. Lord, Jr., *Rubber Chem. Tech.* **34,** 1553 (1961).

Mass spectrometry

Stewart, D. M., "Mass Spectrometry," Chap. LI in A. Weissberger, ed., *Technique of Organic Chemistry*, Vol. 1, *Physical Methods of Organic Chemistry*, 3rd ed., Part 4, Interscience Publishers, New York, 1961.

Wall, L. A., "Mass Spectrometry," Chap. VI in G. M. Kline, ed., *op. cit.*, Part II, 1962.

Gas chromatography

Cassel, J. M., "Chromatography," Chap. X in G. M. Kline, ed., *op. cit.*, Part II, 1962.
Dal Nogare, S., and L. W. Safranski, "Gas Chromatography," pp. 91–227 in J. Mitchell, Jr., I. M. Kolthoff, E. S. Proskauer, and A. Weissberger, eds., *Organic Analysis*, Vol. IV, Interscience Publishers, New York, 1961.
Keller, R. A., *Sci. American* **205,** #4, 58 (October, 1961).

BIBLIOGRAPHY

1. Critchfield, F. E., and D. P. Johnson, *Anal. Chem.* **33,** 1834 (1961).
2. Gibson, M. E., Jr., and R. H. Heidner, *Anal. Chem.* **33,** 1825 (1961).
3. Gruben, A., and H. H. Leiner, *Modern Plastics* **37,** #11, 88 (July, 1960).
4. Guillet, J. E., W. C. Wooten, and R. L. Combs, *J. Appl. Polymer Sci.* **3,** 61 (1960).
5. Madorsky, S. L., and S. Straus, *J. Res. Natl. Bur. Standards* **40,** 417 (1948).
6. Mitchell, J., Jr., *et al.*, *J. Chem. Educ.* **35,** 76–94 (1958).
7. Plastics Properties Charts, *Modern Plastics Magazine*, Encyclopedia Issue, Breskin Publications, New York (annually).
8. Vassallo, D. A., *Anal. Chem.* **33,** 1823 (1961).

B. Infrared Spectroscopy

Emission or absorption spectra arise when molecules undergo transitions between quantum states corresponding to two different internal energies. The energy difference ΔE between the states is related to the frequency of the radiation emitted or absorbed by the quantum relation $\Delta E = h\nu$. Infrared frequencies, corresponding to wavelengths of 1–50 μ,* are associated with molecular vibration and vibration-rotation spectra.

Most infrared spectroscopy is concerned with the measurement of absorption spectra. Infrared emission spectra have been measured, but severe experimental difficulties arise from the need for a stable source and from the low intensity of the spectra. Infrared reflection spectra have been measured for polymers and are a promising source of information. Their unpopularity in the past has been largely due to difficulties in interpreting the data, which should be overcome by application of modern computational techniques.

In principle, the entire infrared absorption spectrum of any compound can be calculated by application of normal mode or equivalent methods, clarified and simplified by the application of group theory. Although the computations are complex and lengthy, this method offers great promise for the elucidation of the infrared spectra of polymers. Traditionally, however, empirical procedures have been used for polymers, based on the

* One micron (μ) is equal to 10^{-4} cm. Infrared frequencies may also be expressed in wavenumbers (1/wavelength) in cm^{-1}: $1\,\mu = 10{,}000\ cm^{-1}$; $2\,\mu = 5000\ cm^{-1}$, etc.

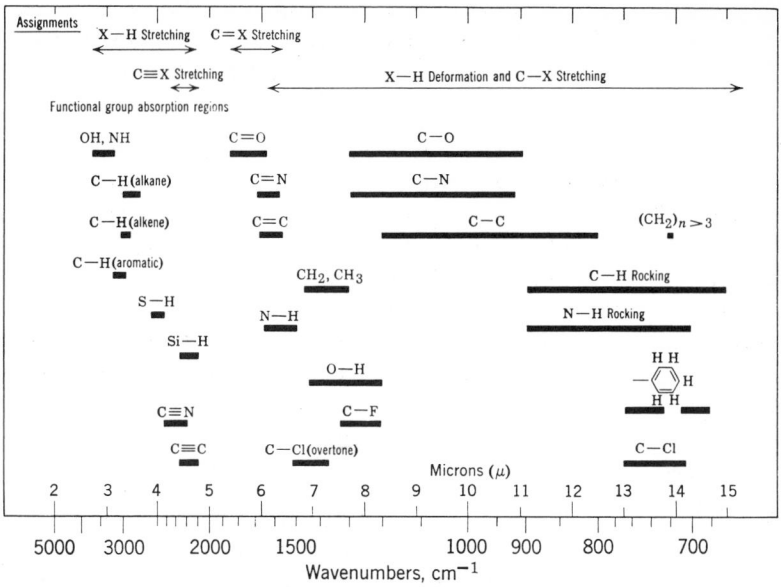

Fig. 4-1. Infrared absorption bands of interest in polymers arranged by approximate wavelength and frequency.

identification of their absorption bands with those produced by the vibrations and rotations of small groups of atoms. Infrared absorption spectra are useful in identifying chemical groups such as methyl, methylene, hydroxyl, carbonyl, vinyl, and many others. These methods are not all inclusive because some groups do not have characteristic infrared absorption. A general limitation of spectroscopic methods applies: the transition must involve a change in dipole moment for the spectrum to be observed.

The approximate wavelengths of some infrared absorption bands arising from functional group and atomic vibrations found in polymers are shown in Fig. 4-1.

Experimental methods

Sources The usual sources of radiation in infrared spectrometers are electrically heated silicon carbide or rare earth oxide rods which emit infrared radiation as black bodies. The amount of energy thus emitted varies in a well-defined way with wavelength, usually reaching a maximum at about 2 μ and falling below useful levels above 50 μ.

Dispersion methods Both diffraction gratings and prisms of inorganic salts are used to disperse the radiation in infrared spectrometers. Common prism materials are NaCl (useful below 15 μ), KBr (useful below 25 μ), and

CsBr (useful below 35 μ). For maximum dispersion, LiF is often used below 6 μ and CaF_2 below 9 μ. With reduced cost and improved quality and availability of gratings, however, the modern trend is strongly toward grating spectrometers.

Detectors The receivers of infrared energy are usually black bodies of very small mass coupled to sensitive heat detectors such as multiple junction thermocouples. Their output may be read on sensitive galvanometers, but commercial instruments usually incorporate automatic recording features. Other types of detectors include bolometers (heat-sensitive resistance elements), pneumatic detectors (depending on expansion of a gas when heated), and PbS or PbTe photocells. The first three types of detectors are sensitive to all wavelengths at which their associated black bodies absorb energy. Lead sulfide photocells are sensitive below about 2 μ, and PbTe below about 4 μ. In their useful range, the photocells are approximately 100–1000 times as sensitive as other types of detectors.

Samples and sample cells The infrared method can be applied to solid, liquid, or gaseous samples. Cells for operation at high or low temperatures are available, as are microscope attachments for studying small samples and single fibers.

To obtain transparency to infrared radiation, sample cells must be made with special window materials. The common prism materials and silver chloride are often used. With most of these materials, precautions against moisture and high humidity conditions must be observed. Infrared transmitting glasses, which are free of these difficulties, can be used at the shorter wavelengths.

One of the greatest experimental difficulties in work with substances which absorb heavily in the infrared, including many polymers, is obtaining sufficiently thin samples. Common methods of sample preparation include the following (4): compression molding, by far the most widely used preparation technique; dissolving the polymer in a solvent, such as carbon disulfide or tetrachloroethylene, whose spectrum is relatively free of intense absorption bands; preparing a thin film by microtoming or milling; casting a thin film from solution; and pressing a finely ground mixture of the sample with KBr to form a disc or wafer.

Application to polymers

Detection of chemical groups in typical spectra Figure 4-1 gives a few of the many chemical linkages or groups which can be detected in polymer spectra, together with the approximate wavelengths at which they occur. Some typical spectra are illustrated qualitatively in Fig. 4-2. That

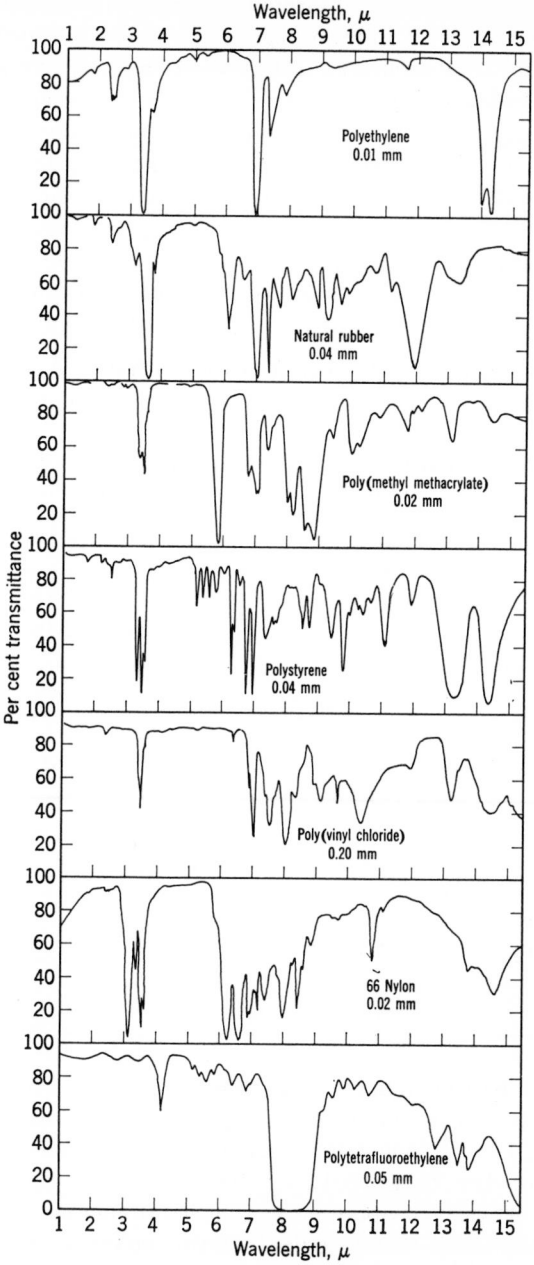

Fig. 4-2. Qualitative sketches of typical infrared spectra of polymers (4). For quantitative application, the original literature should be consulted.

of polyethylene has few intense absorptions, reflecting the simple nonpolar nature of this polymer. Absorption bands are observed in polyethylene corresponding to C—H stretching (3.4 μ), C—H bending of CH_2 groups (6.8 μ) and CH_3 groups (shoulder at 7.25 μ on an amorphous band at 7.30 μ), and CH_2 rocking of sequences of methylene groups in paraffin structures (13.9 μ). (The structure of polyethylene is discussed in more detail in Chapter 13.) Other absorption bands of interest are those due to C=C at 6.1 μ in natural rubber; carbonyl at 5.8 μ and ether at 8.9 μ in poly(methyl methacrylate); aromatic structures at 6.2, 6.7, 13.3, and 14.4 μ in polystyrene; C—Cl at 14.5 μ in poly(vinyl chloride); peptide groups at 3.0, 6.1, and 6.5 μ in nylon; and CF_2 at 8.2–8.3 μ in polytetrafluoroethylene.

Use of reference compounds in structural analysis In the identification of absorption at a given frequency with the vibrations of a specific

***TABLE 4-1. Influence of environment on the wavelength of the* C—H *valency vibration* (2)**

Group Containing C—H	Wavelength, μ
$>CH_2$	3.42 and 3.51
$=CH_2$	3.24 and 3.36
$>CH$	3.30
$-CH_3$	3.38 and 3.48

***TABLE 4-2. Absorption wavelengths of olefinic groups* (1)**

Group Containing C=C	Wavelength, μ
Vinyl, $R_1CH{=}CH_2$	10.1 and 11.0
trans-$R_1CH{=}CHR_2$	10.4
Vinylidene, $R_1R_2C{=}CH_2$	11.3
$R_1R_2C{=}CHR_3$	12.0
cis-$R_1CH{=}CHR_2$	14.2 (variable)

group of atoms, maximum use is made of the spectra of related or model compounds. This is true both in studying general band assignments, such as those in Fig. 4-1, and in investigating more specific structural influences and interactions. Thus the general assignment of absorptions in the 3.3–3.5 μ region to C—H valency vibrations was made through measurements of very simple compounds. Later studies (2) involving many low-molecular-weight hydrocarbons have shown in detail how the absorption frequencies depend upon the environment of the C—H group (Table 4-1). Similarly,

all stretching vibrations involving the C=C double bond for which there is a change in dipole moment cause absorptions in the 6.1 μ region. More detailed information can be obtained, however, by examining the C=C—H hydrogen deformation absorptions at 10–14 μ (Table 4-2) (1).

Far infrared and Raman spectra It is desirable to supplement as much as possible the observations of infrared absorption usually obtained in the 2–15 μ wavelength range. For this purpose observations at longer wavelengths (say up to 200 μ) become important. Despite serious experimental difficulties, valuable information is available in this region: for some polymers, such as polytetrafluoroethylene (6), most of the absorption bands occur above 15 μ.

Since the requirements for activity of a vibration for causing absorption in the infrared and for causing Raman scattering* are often different, information from Raman experiments in general supplements that obtainable from infrared absorption. Only the severe experimental difficulties in obtaining Raman spectra of solid polymers have prevented more widespread use of this technique.

Isotopic substitution The shift in frequency occurring when one isotope is substituted for another often provides valuable information for checking the assignment of absorption bands to specific groups. The effect is greatest when deuterium is substituted for hydrogen, causing a frequency shift of up to $\sqrt{2}$; consequently, this is the most useful case.

Polarized radiation A molecular vibration, involving a change in dipole moment, causes infrared absorption whose intensity is proportional to the cosine of the angle between the electric vector of the incident radiation and the direction of the dipole moment. In consequence, experiments with polarized infrared radiation are useful for determining the latter direction and thus some aspects of the structure of oriented polymers (3, 7, 11). The application of polarized radiation in determining degree of orientation is described in Chapter 5*G*.

Specimen variation In addition to the use of model compounds and isotopic substitution, which may be classed as specimen variation in a sense, observations as a function of temperature, thermal history, or state (crystalline, amorphous, dissolved, etc.) of the specimen being studied often yield pertinent information.

Symmetry analysis (5, 8, 12) When a molecule has symmetry, many features of its absorption spectrum can be deduced from its symmetry by

* The Raman effect results when the frequency of visible light is altered in the scattering process, by the absorption or emission of energy produced by changes in molecular vibration and vibration-rotation quantum states.

group theoretical calculations. The method predicts the number of infrared and Raman active vibrations and their polarizations. It presupposes a knowledge of the structure of the polymer, and the existence of an ordered structure, such as relatively perfect crystalline regions; thus, it can be used to test models. It yields no quantitative information, such as frequencies or force constants, however. Until recently, symmetry analysis was the most powerful theoretical tool available for the interpretation of polymer spectra.

Normal mode calculations The complete calculation of the normal vibrational frequencies of a molecule may be a formidable task even for small molecules containing only a few atoms. The computation is rewarding, however, since quantitative information results. The method was first applied to polyethylene (8, 9), where the symmetry of the atomic arrangement in the crystal is unusually high. Its application to other polymers is under way (10).

GENERAL REFERENCES

Anderson, D. H., N. B. Woodall, and W. West, "Infrared Spectroscopy," Chap. XXIX in A. Weissberger, ed., *op. cit.*, Part 3.

Bamford, C. H., A. Elliott, and W. E. Hanby, *Synthetic Polypeptides*, Academic Press, New York, 1956, Chaps. V, VI.

Elliott, A., "The Infra-Red Spectra of Polymers," pp. 214–287 in H. W. Thompson, ed., *Advances in Spectroscopy*, Vol. I, Interscience Publishers, New York, 1959.

Krimm, S., *Fortschr. Hochpolym.-Forsch.* (*Advances in Polymer Science*) **2,** 51 (1960).

Tryon, M., and E. Horowitz, "Infrared Spectrophotometry," Chap. VIII in G. M. Kline, ed., *op. cit.*, Part II, 1962.

BIBLIOGRAPHY

1. Cross, L. H., R. B. Richards, and H. A. Willis, *Disc. Faraday Soc.* **9,** 235 (1950).
2. Fox, J. J., and A. E. Martin, *Proc. Roy. Soc.* (*London*) **A175,** 208 (1940).
3. Fraser, R. D. B., *J. Chem. Phys.* **28,** 1113 (1958).
4. Hausdorff, H., paper presented at the Pittsburgh Conference on Analytical Chemistry and Applied Spectroscopy, March 7, 1951.
5. Liang, C. Y., *J. Mol. Spectroscopy* **1,** 61 (1957).
6. Liang, C. Y., and S. Krimm, *J. Chem. Phys.* **25,** 563 (1956).
7. Liang, C. Y., and S. Krimm, *J. Chem. Phys.* **27,** 327 (1957).
8. Liang, C. Y., S. Krimm, and G. B. B. M. Sutherland, *J. Chem. Phys.* **25,** 543 (1956).
9. Liang, C. Y., G. B. B. M. Sutherland, and S. Krimm, *J. Chem. Phys.* **22,** 1468 (1954).
10. Miyazawa, T., *Spectrochim. Acta* **16,** 1231, 1233 (1960).
11. Stein, R. S., *J. Appl. Polymer Sci.* **5,** 96 (1961).
12. Tobin, M. C., *J. Chem. Phys.* **23,** 891 (1955).

C. X-ray Diffraction Analysis

The *x-ray diffraction* method is a powerful tool for investigating orderly arrangements of atoms or molecules through the interaction of electromagnetic radiation to give interference effects with structures comparable in size to the wavelength of the radiation. If the structures are arranged in an orderly array or lattice, the interferences are sharpened so that the radiation is scattered or diffracted only under specific experimental conditions. Knowledge of these conditions gives information regarding the geometry of the scattering structures. The wavelengths of x-rays are comparable to interatomic distances in crystals; the information obtained from scattering at wide angles describes the spatial arrangements of the atoms. Low-angle x-ray scattering (7; Porod, gen. ref.) is useful in detecting larger periodicities, which may arise from lamellar crystallites (Chapter 5*C*) or from voids.

The results of the application of the x-ray method in determining the crystal structures of polymers are described in Chapter 5*B*. Collected crystallographic data for polymers have been tabulated (6).

Experimental methods

X-rays are usually produced by bombarding a metal target with a beam of high-voltage electrons. This is done inside a vacuum tube, the x-rays passing out through a beryllium or mica window in the tube in a well-defined beam. Choice of the target metal and the applied voltage determines the wavelength or wavelengths of x-rays produced. Experiments in which nearly monochromatic x-rays are used are of the greatest interest.

The diffracted x-rays may be detected by their action on photographic films or plates, or by means of a radiation counter and electronic amplifier feeding a recorder. Each method has its advantages. Qualitative examination of the diffraction pattern and accurate measurement of angles and distances are best made from a photographic record, whereas for precise measurement of the intensity of the diffracted beam the counting technique is preferred.

X-rays of a given wavelength are diffracted only for certain specific orientations of the sample. If the sample is a single crystal, it must be placed in all possible orientations during the experiment, usually by rotating or oscillating the sample about one of its axes to achieve the desired orientations with respect to the x-ray beam. Alternatively, a sample made up of

a powder of very small crystals may be used. If the minute powdered particles are randomly oriented, all orientations will be included within the sample. The powder method is more convenient but gives less information than the single-crystal method, since the latter allows orientations about one crystal axis at a time to be investigated.

Application to polymers

Since polymer single crystals as now prepared (Chapter 5*C*) are too small for x-ray diffraction experiments, the crystal structure of a polymer is usually determined from x-ray patterns of a fiber drawn from the polymer. Because of the alignment of the crystalline regions with the long axes of the molecules parallel to the fiber axis (Chapter 5*G*), the pattern is essentially identical to a rotation pattern from a single crystal. In such a pattern diffraction maxima occur in rows perpendicular to the fiber axis, called *layer lines* (seen, for example, in Fig. 5-39*b*).

Chain conformations As the periodicity of molecular structure of a polymer is characterized by the existence of a *repeat unit*, so the periodicity of its crystal is characterized by a *repeat distance*. The repeat distance is directly determined by measuring the distance between the layer lines: the greater the repeat distance, the closer together are the layer lines. Determination of the repeat unit is considerably more difficult. Classically, it proceeds through derivation, from the positions of the diffracted x-ray beams, of the dimensions of the *unit cell* (usually the simplest geometric volume unit which by repetition builds up the three-dimensional crystalline array). The positions of the atoms in the unit cell are then derived from the relative intensities of the diffracted beams. The diffraction patterns of polymers, however, do not provide sufficient information to allow such analyses to be carried to completion. Additional structural information is utilized from other sources, such as normal bond lengths and angles and atomic arrangements along the chain suggested by the chemical structure of the sample. For example, the repeat distance of 2.55 A in the crystals of polyethylene is readily identified with a single repeat unit in the planar zigzag conformation.

Many polymer conformations can be described in terms of atoms regularly spaced along helices (Chapter 5*B*). Methods for analyzing the diffraction patterns from helical structures (2, 3, 4) provide a relatively new and highly versatile technique for the determination of the repeat unit in such polymers (1).

Chain packing The packing of the chains is described most completely in terms of the unit cell and its contents (shown for polyethylene in Fig. 5-5). The volume of the unit cell, and hence the volume occupied in the

crystal by a single repeat unit, can be obtained from the repeat distance and the positions of the diffraction spots on the layer lines. This volume measures the density of the crystals, which is useful in determining the degree of crystallinity, as outlined in Chapter 5*F*.

The atomic arrangement within the unit cell is much more difficult to determine than the cell dimensions. Trial structures can often be deduced from these dimensions and a knowledge of the chain conformation, and can be tested in terms of calculated and observed diffraction intensities, using well-established crystallographic techniques.

Disorder in the crystal structure In the discussion so far it has been assumed that polymer crystals consist of perfect geometrical arrays of atoms. However, defects and distortions are present in crystals of all materials; and, together with grosser types of disorder, they play an important role in the crystal structures of polymers. X-ray diffraction data give a great deal of information about the qualitative aspects of disorder in the crystal structure, since disordering results in the broadening of the diffraction maxima. Since different classes of diffraction maxima are sensitive in different ways to particular types of disorder, a systematic study of the broadening yields detailed information on the types of disturbances in the crystals (1).

Quantitative measurement of the disorder is difficult because broadening of the diffraction maxima is caused both by small crystallite size and by distortions within larger crystals. As discussed in Chapter 5*D*, the former explanation was traditionally adopted in interpreting the diffraction patterns from polymers. However, the distortion of perfect crystals into *paracrystals* (5) is now considered important in many polymers.

Orientation Unless the crystalline regions in polymers are oriented, as by drawing a fiber, the diffraction maxima merge into rings, made up of the maxima from a large number of crystallites in many different orientations (Chapter 5*G*, especially Fig. 5-39*a*). If orientation is intermediate between these extremes, the rings split into arcs, and a quantitative evaluation of the crystallite orientation is made by measuring the angular spread and intensity of these arcs.

GENERAL REFERENCES

Bamford, C. H., A. Elliott, and W. E. Hanby, *op. cit.*, Chaps. VII-IX.

Bunn, C. W., *Chemical Crystallography*, 2nd ed., Oxford University Press, London, 1961.

Klug, H. P., and L. E. Alexander, *X-ray Diffraction Procedures for Polycrystalline and Amorphous Materials*, John Wiley & Sons, New York, 1954.

Lipscomb, W. N., "X-Ray Crystallography," Chap. XXV in A. Weissberger, ed., *op. cit.*, Part 2.

Porod, G., *Fortschr. Hochpolym.-Forsch.* (*Advances in Polymer Science*) **2,** 363 (1961).
Posner, A. S., "X-Ray Diffraction," Chap. II in G. M. Kline, ed., *op. cit.*, Part II, 1962.
Statton, W. O., "The Use of X-Ray Diffraction and Scattering in Characterization of Polymer Structure," pp. 246–256 in *American Society for Testing Materials Special Technical Publication No.* 247, Philadelphia, Pa., 1958.
Tanford, C., *Physical Chemistry of Macromolecules*, John Wiley & Sons, New York, 1961, Secs. 3, 4.

BIBLIOGRAPHY

1. Clark, E. S., and L. T. Muus, *Z. Krist.* **117,** 119, (1962).
2. Cochran, W., F. H. C. Crick, and V. Vand, *Acta Cryst.* **5,** 581 (1952).
3. Davies, D. R., and A. Rich, *Acta Cryst.* **12,** 97 (1959).
4. Franklin, R. E., and A. Klug, *Acta Cryst.* **8,** 777 (1955).
5. Hosemann, R., R. Bonart, and G. Schoknecht, *Z. Physik* **146,** 588 (1956); R. Hosemann and S. N. Bagchi, *Direct Analysis of Diffraction by Matter*, North-Holland Publishing Co., Amsterdam, 1962.
6. Miller, R. L., and L. E. Nielsen, *J. Polymer Sci.* **55,** 643 (1961).
7. Statton, W. O., *J. Polymer Sci.* **22,** 385 (1956); **28,** 423 (1958); **41,** 143 (1959); **58,** 205 (1962).

D. Nuclear Magnetic and Electron Paramagnetic Resonance Spectroscopy

Although the techniques of nuclear magnetic resonance (NMR) and electron paramagnetic resonance (EPR) spectroscopy detect different phenomena and, therefore, yield quite different information, they are similar in basic principles and in experimental techniques. Nuclear magnetic resonance spectroscopy detects the motion of nuclei (such as protons) having nonzero spin, while EPR spectroscopy is useful for the detection of free radical species.

NMR spectroscopy

The major application of NMR spectroscopy is in the study of molecular motion. The development of this technique has provided a new tool for studying molecular motion in solid polymers, which holds promise of at least equalling the utility of the older methods of dynamic mechanical testing (Chapter 6*C*) and dielectric studies (Section *G*). In fact, NMR has an advantage in that it allows study of the motion and positions of protons, which are not readily detected by most other means.

Experimental methods The NMR technique utilizes the property of spin (angular momentum and its associated magnetic moment) possessed

by nuclei whose atomic number and mass number are not both even. Such nuclei include the isotopes of hydrogen, C^{13}, O^{17}, and F^{19}. Application of a strong magnetic field to material containing such nuclei splits the energy levels into two, representing states with spin parallel and antiparallel to the field. Transitions between the states lead to absorption or emission of an energy

$$E = h\nu_0 = 2\mu H_0 \tag{4-1}$$

where the frequency ν_0 is in the microwave region for available fields of strength H_0 of the order of 10,000 gauss, and μ is the magnetic moment of the nucleus. The energy change is observed as a resonance peak or line in experiments where either H or ν_0 is varied with the other held constant.

With an assembly of nuclei, the field on any one is modified by the presence of the others:

$$h\nu_0 = 2\mu(H_0 + H_L) \tag{4-2}$$

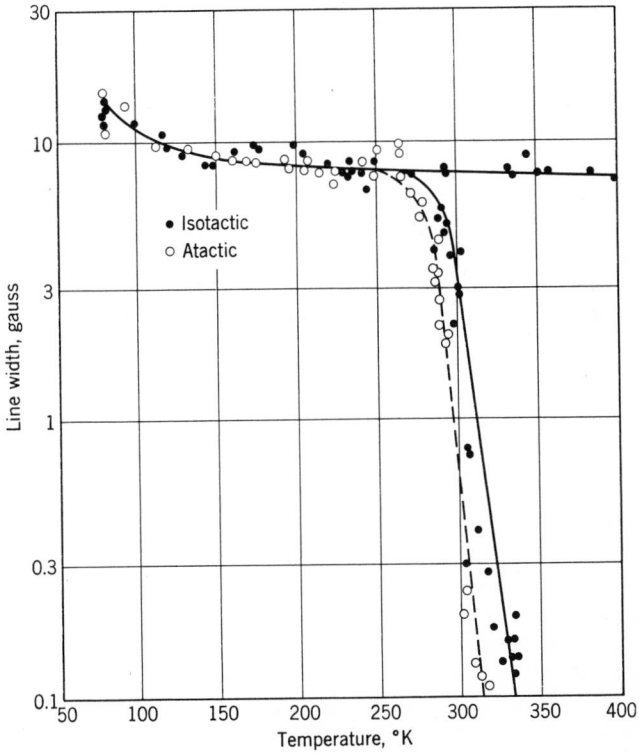

Fig. 4-3. Variation of proton NMR line width with temperature in isotactic and atactic polypropylene (3).

where H_L is the local field with a strength of 5–10 gauss. A distribution of local fields usually exists so that the resonance line becomes broadened. The variables ordinarily observed are the line width and its *second moment*, the mean-square deviation of the field from the center of the line, H_0. Both the line width (for simple systems) and the second moment can be calculated from the structure of the molecule.

Much of the value of NMR studies arises from the effect on the resonance line of the development of molecular motion. As the nucleus moves, it experiences a variety of local fields H_L, so that the average field approaches H_0 more closely and the resonance envelope narrows.

Application to polymers A study (3) of the proton resonance in atactic and isotactic polypropylene provides an example of the type of information obtainable from NMR studies on solid polymers. Figure 4-3

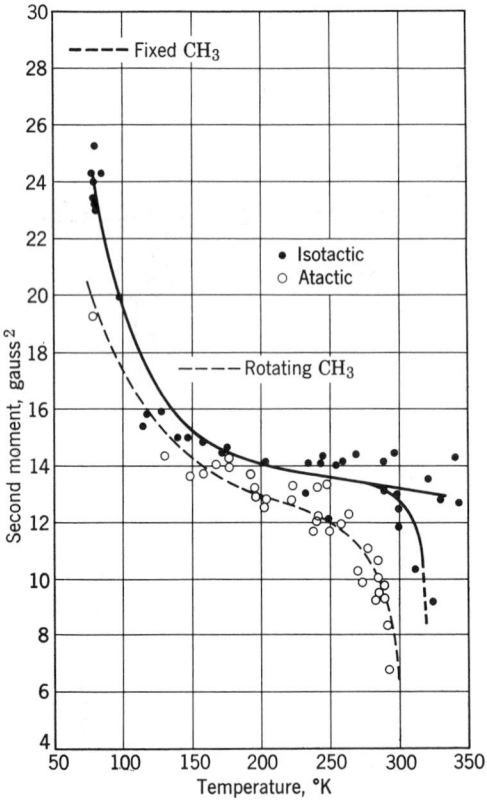

Fig. 4-4. Variation of the second moment of proton NMR with temperature in isotactic and atactic polypropylene (3).

shows the variation in line width with temperature for the two forms of polypropylene. The absorption line of the atactic form decreases rapidly in width above 250°K because of the onset of molecular motion at the glass transition (Chapter 6*D*). In isotactic polypropylene, two lines are observed above 250°K. The narrow line is attributed to the protons in the amorphous regions of the polymer, the onset of molecular motion occurring as in the atactic material but at a somewhat higher temperature. The broad line is attributed to the protons in the crystalline regions, where no large-scale motion is occurring over the temperature range covered.

The general features of the temperature dependence of line width are reflected in the behavior of the second moment (Fig. 4-4). The experimental second moment of the isotactic form is higher because of the closer packing of the polymer chains in this polymer. Also shown in the figure are values of the second moment calculated for models corresponding to a completely rigid structure and to hindered rotation of the methyl groups in an otherwise stationary molecule. From comparison of the calculated and experimental second moments, it is inferred that at some temperature below that reached experimentally (77°K) the methyl groups begin to undergo hindered rotation, resulting in the decrease in line width and second moment observed between 77 and 150°K. At about 150°K this motion is sufficiently rapid to reduce the second moment to approximately the value calculated for this model. Above this temperature the second moment and line width remain approximately constant until the large-scale motion associated with the glass transition occurs.

EPR spectroscopy

The major use of EPR spectroscopy lies in the detection of free radicals (Chapter 9*A*). These species are uniquely characterized by their magnetic moment, arising from the presence of an unpaired electron. Measurement of a magnetic property of a material containing free radicals, such as its magnetic susceptibility, gives the concentration of free radicals, but the method lacks sensitivity and cannot reveal the structure of the radicals. Electron paramagnetic resonance spectroscopy is essentially free from these limitations.

Experimental methods As in NMR spectroscopy, the action of a strong magnetic field on a material containing free radicals removes the degeneracy of their ground-state energy levels. For low radical concentrations, the new energy levels are given by two terms. The first is

$$E = h\nu_0 = g\beta\mu_0 H_0 \tag{4-3}$$

where g is a tensor relating the field direction and the symmetry directions in the radical, β is the magnetic moment of the electron spin, and μ_0 is the magnetic permeability of a vacuum. The second term represents coupling of the electron spin with the nuclear spins in the molecule. This coupling results in the splitting of the resonance line into a symmetrical group of lines whose positions and amplitudes depend on, and therefore give information about, the structure of the radical.

Applications to polymers Only a few systematic investigations have been made of the EPR spectra of polymer systems, and much theoretical and experimental work remains to be done before their results are completely understood. The investigation of free radicals produced by high-energy irradiation (Chapter 12*F*) of polytetrafluoroethylene (1, 2) serves as an example.

The EPR spectrum of irradiated polytetrafluoroethylene is interpreted as arising from radicals of the type

$$-(CF_2)_x\dot{C}F(CF_2)_y-$$

remaining trapped in the polymer after a C-F bond has been broken and the fluorine atom has diffused away. The spectrum shows fine structure due to coupling between the unpaired electron and neighboring F^{19} nuclei. The fine structure is lost below 270°K as the motion of the F^{19} nuclei is slowed down. The spectrum does not change at higher temperatures, indicating that the radical is stable to at least 550°K. Secondary products, resulting from the reaction of the primary radical with such substances as O_2 and NO, were also studied.

It may be noted that, as discussed in Chapters 1*C* and 12*F*, the strength of the C-F bond and the tendency of polytetrafluoroethylene to degrade rather than crosslink on irradiation would suggest the breaking of a C-C rather than a C-F bond as a likely source of radicals in this polymer.

GENERAL REFERENCES

Nuclear magnetic resonance

Bueche, F., *Physical Properties of Polymers*, Interscience Publishers, a Division of John Wiley & Sons, New York, 1962, Chap. 12.

Gutowsky, H. S., "Nuclear Magnetic Resonance," Chap. XLI in A. Weissberger, ed., *op. cit.*, Part 4.

Roberts, J. D., *Nuclear Magnetic Resonance*, McGraw-Hill Book Co., New York, 1959; *J. Chem. Educ.* **37,** 581 (1961).

Slichter, W. P., *Fortschr. Hochpolym.-Forsch.* (*Advances in Polymer Science*) **1,** 35 (1958); *Makromol. Chem.* **34,** 67 (1959); *Rubber Chem. Tech.* **34,** 1574 (1961).

Wall, L. A., and R. E. Florin, "Magnetic Resonance Spectroscopy," Chap. XII in G. M. Kline, ed., *op. cit.*, Part II.

Electron paramagnetic resonance

Fraenkel, G. K., "Paramagnetic Resonance Absorption," Chap. XLII in A. Weissberger, ed., *op. cit.*, Part 4.
Ingram, D. J. E., *Free Radicals as Studied by Electron Spin Resonance*, Academic Press, New York, 1958.
Pake, G. E., *Paramagnetic Resonance*, W. A. Benjamin, New York, 1962.

BIBLIOGRAPHY

1. Lebedev, Y. S., I. D. Tsvetkov, and V. V. Voevodskii, *Opt. & Spectr.* **8,** 426 (1960).
2. Rexroad, H. N., and W. Gordy, *J. Chem. Phys.* **30,** 399 (1958).
3. Slichter, W. P., and E. R. Mandell, *J. Appl. Phys.* **29,** 1438 (1958).

E. Calorimetric and Thermal Analysis

Studies of the specific heat and related thermodynamic properties of polymers are useful for predicting and explaining the energy changes associated with heating, annealing, crystallizing, or otherwise thermally treating polymers. These studies are carried out either by calorimetric measurements or by differential thermal analysis. Both methods lead to the determination, in favorable cases, of such quantities as degree of crystallinity (Chapter 5*F*), temperature ranges and heat effects in transitions, and heat effects associated with polymerization, degradation, or other chemical changes.

Calorimetry

Experimental methods Calorimetric measurements are usually carried out by placing the polymer sample in a thermally jacketed vessel (*adiabatic calorimeter*) whose temperature is raised slowly by the addition of known amounts of heat, usually from electrical resistance coils. The temperature of the jacket is adjusted to equal that of the calorimeter to maintain adiabatic conditions. Specific heat and other quantities are calculated by well-known thermodynamic methods. Alternatively, as in the measurement of a chemical change, the heat effect may be calculated from the adiabatic temperature rise in the calorimeter, without the addition of heat from an external source.

Application to polymers As an example, the specific heat-temperature curves obtained (1) on heating quenched (amorphous) specimens of poly(ethylene terephthalate) are shown in Fig. 4-5. Each curve rises linearly with temperature at low temperatures, then rises more steeply at the glass

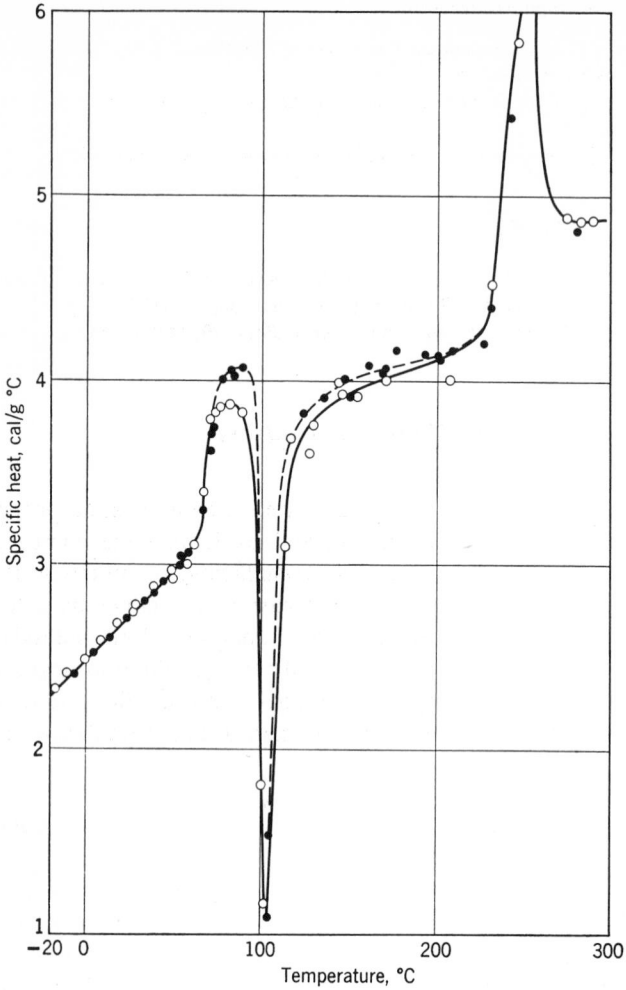

Fig. 4-5. Curves of specific heat as a function of increasing temperature for quenched (amorphous) poly(ethylene terephthalate) (1).

transition, 60–80°C. With the onset of mobility of the molecular chains above this transition, crystallization takes place, as indicated by the sharp drop in the specific heat curve. At still higher temperatures, 220–270°C, the crystals melt with a corresponding rise in the specific heat curve.

Differential thermal analysis

Experimental methods In differential thermal analysis, the sample and an inert reference substance, undergoing no thermal transitions in the

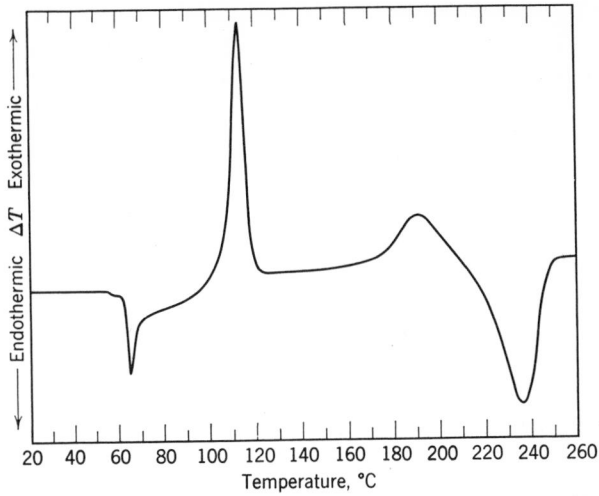

Fig. 4-6. Differential thermal analysis curve for amorphous poly(ethylene terephthalate) (Ke, gen. ref.); compare with Fig. 4-5.

temperature range of interest, are heated at exactly the same rate (1–2°C/min) in a controlled-temperature bath. The temperature difference between sample and reference is measured and plotted as a function of sample temperature. The temperature difference is finite only when heat is being evolved or absorbed because of exothermic or endothermic activity in the sample, or when the heat capacity of the sample is changing abruptly. Since the temperature difference is directly proportional to the heat capacity, the curves resemble specific heat curves, but are inverted because, by convention, heat evolution is registered as an upward peak and heat absorption as a downward peak.

Application to polymers A typical result (Ke, gen. ref.) is the differential thermal analysis curve for poly(ethylene terephthalate) shown in Fig. 4-6. Features of the curve may readily be identified by comparison with Fig. 4-5. The lower crystalline melting range in the specimen of Fig. 4-6 is attributed to impurities in the polymer.

GENERAL REFERENCES

Calorimetry

Dole, M., *Fortschr. Hochpolym.-Forsch.* (*Advances in Polymer Science*) **2**, 221 (1960).
Sturtevant, J. M., "Calorimetry," Chap. X in A. Weissberger, ed., *op. cit.*, Part 1.

Differential thermal analysis

Ke, B., "Application of Differential Thermal Analysis to High Polymers," pp. 361–392 in J. Mitchell, Jr., I. M. Kolthoff, E. S. Proskauer, and A. Weissberger, eds., *loc. cit.*

Kissinger, H. E., and S. B. Newman, "Differential Thermal Analysis," Chap. IV in G. M. Kline, ed., *op. cit.*, Part II.

Smothers, W. J., and Y. Chiang, *DTA:Theory and Practice*, Chemical Publishing Co., New York, 1959.

BIBLIOGRAPHY

1. Smith, C. W., and M. Dole, *J. Polymer Sci.* **20,** 37 (1956).

F. Light and Electron Microscopy

The techniques of polarized light microscopy and electron microscopy are well described in the literature. Their application to polymer systems requires little in the way of novel equipment, theory, or interpretation.

Light microscopy is most widely used in application to polymers in the examination of such textural features of solid polymers as spherulites (Chapter 5*D*). Observations between crossed polarizers or with phase-contrast equipment are particularly useful. Electron microscopy has become a very valuable tool for the study of polymer single crystals (Chapter 5*C*) and for more detailed observations of the morphology of solid polymers than can be resolved in the light microscope. Techniques of shadowing, replicating, and solvent etching (1, 2) are used.

GENERAL REFERENCES

Light microscopy

Hartshorne, N. H., and A. Stuart, *Crystals and the Polarizing Microscope*, 3rd ed., St. Martin's Press, New York, 1960.

Jelley, E. E., "Light Microscopy," Chap. XXI in A. Weissberger, ed., *op. cit.*, Part 2.

Electron microscopy

Hall, C. E., *Introduction to Electron Microscopy*, McGraw-Hill Publishing Co., New York, 1953.

Hamm, F. A., "Electron Microscopy," Chap. XXIV in A. Weissberger, ed., *op. cit.*, Part 2.

Keller, A., and D. C. Bassett, *J. Roy. Microscop. Soc.* **79,** 243 (1960).

Scott, R. G., "The Structure of Synthetic Fibers," pp. 121–131 in *American Society for Testing Materials Special Technical Publication No.* 257, Philadelphia, Pa., 1959.

BIBLIOGRAPHY

1. Mackie, J. S., and A. Rudin, *J. Polymer Sci.* **49,** 407 (1961).
2. Reding, F. P., and E. R. Walter, *J. Polymer Sci.* **38,** 141 (1959).

G. Physical Testing

The purpose of this section is to provide brief descriptions of the more important test methods for measuring the physical properties of polymers. The discussion is not intended to be comprehensive or detailed. Test methods related to rheology and viscoelasticity are discussed in Chapter 6.

Mechanical properties

Stress-strain properties in tension One of the most informative mechanical experiments for any material is the determination of its *stress-strain curve in tension* (6, 12, 15, 24, 28, 39). This is usually done by measuring continuously the force developed as the sample is elongated at constant rate of extension.

The generalized stress-strain curve for plastics shown in Fig. 4-7 serves to define several useful quantities, including *modulus* or *stiffness* (the slope

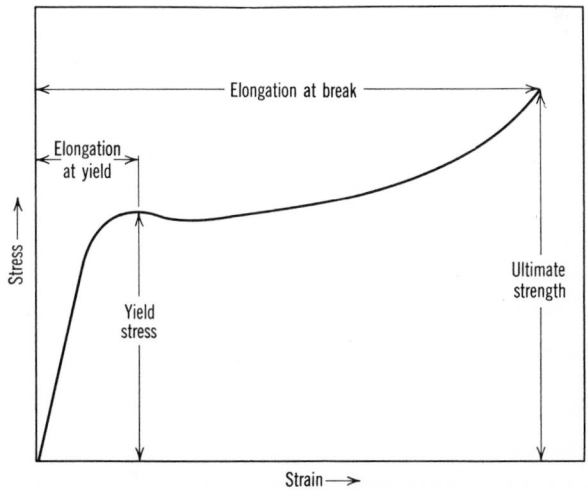

Fig. 4-7. Generalized tensile stress-strain curve for plastics (Winding and Hiatt, gen. ref.).

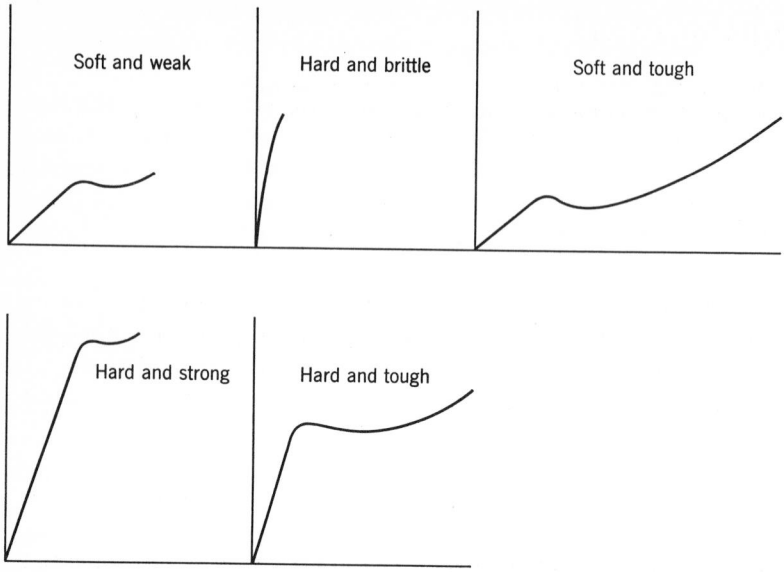

Fig. 4-8. Tensile stress-strain curves for several types of polymeric materials (Winding and Hiatt, gen. ref.).

of the curve), *yield stress*, and *strength* and *elongation at break*. This type of curve is typical of a plastic such as polyethylene. Figure 4-8 shows stress-strain curves typical of some other classes of polymeric materials. The properties of these polymer types are related to the characteristics of their stress-strain curves in Table 4-3.

Tensile properties are usually measured at rates of strain of 1–1000 %/min. At higher rates of strain (53, 59, 62, 63)—up to 10^6 %/min—tensile strength

TABLE 4-3. Characteristic features of stress-strain curves as related to polymer properties*

Description of Polymer	Characteristics of Stress-Strain Curve			
	Modulus	Yield Stress	Ultimate Strength	Elongation at Break
Soft, weak	Low	Low	Low	Moderate
Soft, tough	Low	Low	Yield Stress	High
Hard, brittle	High	None	Moderate	Low
Hard, strong	High	High	High	Moderate
Hard, tough	High	High	High	High

* Winding and Hiatt, gen. ref., after (46).

and modulus usually increase several-fold, while elongation decreases. The interpretation of these results is complicated by large temperature rises in the test specimen.

In addition to tensile measurements, tests may also be performed in *shear*, *flexure*, *compression*, or *torsion* (18, 19, 31). For materials in film form, flexural tests are often used. These may include (for stiffer materials) measurement of *flexural modulus* (21, 26) or (for less stiff materials) *flexural* or *folding endurance* tests (13) (see next paragraph).

Fatigue tests When subjected to cyclic mechanical stresses, most materials fail at a stress considerably lower than that required to cause rupture in a single stress cycle. This phenomenon is called *fatigue*. Various modes of fatigue testing in common use include alternating tensile and compressive stress (56) and cyclic flexural stress (16). Results are reported as plots of stress vs. number of cycles to fail. Many materials show a fatigue endurance limit, or a maximum stress below which fatigue failure never takes place.

Fatigue failure may arise from the absorption of energy in a material which is not perfectly elastic. This energy is manifested as heat, leading to a temperature rise, a lower modulus, and rapid failure. Energy absorption is accompanied and measured by a difference in phase between stress and strain in the cyclic test.

Impact tests In most cases, rupture in polymer samples may be divided into two classes: brittle and ductile. Brittle rupture occurs if the material behaves elastically up to the point of failure, i.e., does not yield or draw, whereas in ductile rupture the specimen is permanently distorted near the point of failure. Brittle failure is characterized by lack of distortion of the broken parts. Two aspects of brittle rupture are of interest: the temperature below which brittle failure occurs under a given set of experimental conditions, and measures of the toughness of materials at ambient temperature. The test methods used to obtain both types of information are highly empirical. Since the specimen is destroyed, multiple testing is usually employed to allow statistical evaluation of the results.

The *brittle point*, or temperature at the onset of brittleness, is usually determined by subjecting a specimen to impact in a standardized but empirical way (20). The temperature of the test is varied until that temperature is found where, statistically, half the specimens fail by brittle rupture. The brittle point is roughly related to the glass transition temperature (Chapter 6*D*).

Impact strength of plastics is commonly measured by tests in which a pendulum with a massive striking edge is allowed to hit the specimen (3, 23, 44). From the travel of the pendulum after breaking the specimen can be

calculated the energy required to cause the break. The specimen is often notched in an effort to improve the reproducibility of the mode of failure.

Other forms of impact test include a large variety in which falling objects (balls, darts, etc.) strike the specimen (43, 54). Usually, the height from which the missile must be dropped to cause failure in half the specimens is taken as a measure of the toughness of the material.

Tear resistance When plastics are used as films, particularly in packaging applications, their resistance to tearing is an important property (30). In one test of *tear strength* (17) a specimen is torn apart at a cut made by a sharp blade. Energy is provided by a falling pendulum, and the work done is measured by the residual energy of the pendulum. Tear strength and tensile strength are closely related (50).

Hardness As usually conceived, *hardness* is a composite property combining concepts of resistance to *penetration*, *scratching*, *marring*, etc. Most hardness tests for plastics are based on resistance to penetration by an indentor pressed into the plastic under a constant load (25, 37).

Abrasion resistance Abrasion resistance in plastics usually takes the form of a scratch test, in which the material is subjected to many scratches, usually from contact with an abrasive wheel (32) or a stream of falling abrasive material (33). The degree of abrasion can be determined by loss of weight for severe damage, but is more usually measured by evidence of surface marring, such as loss of gloss or development of haze in transparent specimens.

Friction, hardness, and abrasion resistance are related closely to the viscoelastic properties of polymers (49) (Chapter 6*C*).

Thermal properties

Softening temperature In addition to the rheological tests discussed in Chapter 6*A*, a thermal property of great interest is the *softening temperature* of a plastic. Various ways of measuring this property include observation of the temperatures at which (*a*) an indentor under fixed load penetrates a specified distance into the material (38) (*Vicat test*-Chapter 7*D*); (*b*) a bar, held in flexure under constant load, deforms a specified amount (14) (*deflection temperature* or *heat distortion test*); (*c*) a polymer sample becomes molten and leaves a trail when moved across a hot metal surface with moderate pressure (48) (*polymer melt* or *stick temperature test*); and (*d*) a polymer specimen fails in tension under its own weight (42) (*zero strength temperature test*).

Flammability The flammability of plastics is usually tested as the burning rate of a specified sample (9, 11, 34). The self-extinguishing tendency of the material on the removal of an external flame is also important (22).

Optical properties

Transmittance and reflectance A major determinant of the appearance of a *transparent* material (one which does not scatter light) is its *transmittance*, the ratio of the intensities of light passing through and light incident on the specimen. Similarly, the appearance of an *opaque* material (one which may reflect light but does not transmit it) is characterized by its *reflectance*, the ratio of the intensities of the reflected and the incident light. A *translucent* substance is one which transmits part and reflects part of the light incident on it.

Transmittance and reflectance may be measured as a function of the wavelength of light in a spectrophotometer (5). When adjusted to correspond to visual perception, by weighting the effects at various wavelengths according to the response of the eye, they are called *luminous transmittance* and *reflectance*. Instruments are available which measure the latter quantities directly (27).

Color Color is that aspect of the appearance of objects which depends upon the spectral composition of the light reaching the eye. The complete description of color requires specification of three variables, a common set of which are *hue*, the attribute which determines whether the color is red, green, blue, etc.; *lightness*, the attribute which permits a color to be classified equivalent to some member of a scale of grays ranging from white to black; and *saturation*, the attribute of any color possessing a hue which determines the degree of difference of the color from the gray of the same lightness. These attributes may be computed from spectrophotometric measurements (5, 27) or evaluated by visual comparison of the sample with sets of color standards (40).

Gloss Gloss is the geometrically selective reflectance of a surface responsible for its shiny or lustrous appearance. Surface reflectance is commonly at a maximum in or near the specular direction, i.e., the direction at which a mirror would reflect light. Photoelectric instruments are available for measuring gloss at a variety of angles of incidence and reflection (8, 36).

Haze For transparent materials, haze is that percentage of transmitted light which in passing through the specimen deviates from the incident

beam by forward scattering. In commercial hazemeters, only light deviating more than 2.5° from the transmitted beam direction is considered haze (29). The effect of haze is to impart a cloudy or milky appearance to the sample, but its transparency (defined in the next paragraph) need not be reduced.

Transparency For a transparent plastic material, transparency is defined as the state permitting perception of objects through or beyond the specimen. A sample of low transparency may not exhibit haze, but objects seen through it will appear blurred or distorted. Transparency can be measured as that fraction of the normally incident light which is transmitted without deviation from the primary beam direction of more than 0.1° (64).

Electrical properties

Dielectric constant and loss factor The dielectric constant of an (insulating) material is the ratio of the capacities of a parallel plate condenser measured with and without the dielectric material placed between the plates. The difference is, of course, due to the polarization of the dielectric. If the field applied to the condenser is time dependent (as in alternating current), the polarization is time dependent also. However, because of the resistance to motion of the atoms in the dielectric, there is a delay between changes in the field and changes in the polarization. This delay is often expressed as a phase difference or *loss angle*, δ. The *power factor* is then defined as $\sin \delta$ and the *dissipation factor* as $\tan \delta$. The product of the dielectric constant and the power factor is called the *loss factor*. It is proportional to the energy absorbed per cycle by the dielectric from the field. Dielectric constant and loss factor are usually measured over a frequency range from 60 cycles/sec to thousands of megacycles/sec (2).

Resistivity The resistance of most polymers to the flow of direct current is very high, and conductivity probably results from the presence of ionic impurities whose mobility is limited by the high viscosity of the medium. Both surface and volume resistivity are important properties for applications of polymers as insulating materials (4).

Dielectric strength Insulators will not sustain an indefinitely high voltage: as the applied voltage is increased, a point is reached where a catastrophic decrease in resistance takes place, accompanied by a physical breakdown of the dielectric (1). As in mechanical fatigue experiments, application of lower voltages causes eventual breakdown: curves of time to fail vs. voltage can be plotted. Such curves show two distinct regions of

failure. At short times failure is presumed (60) to occur as a result of the inability of the electrons conducting the current to dissipate rapidly enough the energy they receive from the field. Breakdown at longer times appears to be due to corona attack, which is sensitive to the atmosphere surrounding the dielectric and to the presence of mechanical strains (58).

Arc resistance The surfaces of some polymers may become carbonized and conduct current readily when exposed to an electrical discharge (7, 57). This property is important for such applications as insulation for gasoline engine ignition systems.

Chemical properties

The following properties may not involve chemical treatment or attack exclusively, but are conveniently grouped together as different from those considered in previous sections.

Resistance to solvents The effect of solvents (broadly defined as liquids in general) on polymers may take several forms: *solubility* (Chapter 2*A*); *swelling* (Chapter 2*F*), including the absorption of water (10); *environmental stress cracking*, in which the specimen fails by breaking when exposed to mechanical stress in the presence of an organic liquid or an aqueous solution of a soap or other wetting agent (45, 47, 51, 52, 55); and *crazing*, in which a specimen fails by the development of a multitude of very small cracks in the presence of an organic liquid or its vapor, with or without the presence of mechanical stress (61, 65, 66).

Vapor permeability The *permeability* of a polymer to a gas or vapor is the product of the *solubility* of the gas or vapor in the polymer and its *diffusion coefficient*. Permeability is directly measured as the rate of transfer of vapor through unit thickness of the polymer in film form, per unit area and pressure difference across the film (35).

Weathering For convenience and reproducibility, the behavior of materials on long exposure to weather is often simulated by exposure to *artificial weathering* sources, such as filtered carbon-arc lamps in controlled atmospheres (41). Such tests are usually not appreciably accelerated over natural weathering, except that they may be applied continuously.

GENERAL REFERENCES

American Society for Testing Materials, *ASTM Standards on Plastics*, Philadelphia, Pa., published at frequent intervals.

Teeple, J. H., "Methods of Testing," Chap. 2 in R. Houwink, ed., *Elastomers and Plastomers*, Vol. III, Elsevier Publishing Co., New York, 1948.

Winding, C. C., and G. D. Hiatt, *Polymeric Materials*, McGraw-Hill Book Co., New York, 1961, Chap. 3.

BIBLIOGRAPHY

1–39. American Society for Testing Materials, gen. ref., Methods of Test with ASTM Designations as follows: ref. 1, D149-55T; 2, D150-54T; 3, D256-56T; 4, D257-58T; 5, D307-44; 6, D412-51T; 7, D495-58T; 8, D523-53T; 9, D568-56T; 10, D570-57T; 11, D635-56T; 12, D638-58T; 13, D643-43; 14, D648-56; 15, D651-48; 16, D671-51T; 17, D689-44; 18, D695-54; 19, D732-46; 20, D746-57T; 21, D747-58T; 22, D757-49; 23, D758-48; 24, D759-48; 25, D785-51; 26, D790-58T; 27, D791-54; 28, D882-56T; 29, D1003-52; 30, D1004-49T; 31, D1043-51; 32, D1044-56; 33, D1242-56; 34, D1433-58; 35, D1434-58; 36, D1471-57T; 37, D1474-57T; 38, D1525-58T; 39, D1530-58T.

40. American Society for Testing Materials, *ASTM Standards* 1958, Vol. 8, Tentative Method of Test, ASTM Designation: D1535-58T.
41. American Society for Testing Materials, gen. ref., Recommended Practice, ASTM Designation: E42-57.
42. American Society for Testing Materials, gen. ref., Tentative Specification and Methods of Test, ASTM Designation: D1430-58T.
43. American Standards Association, American Standard Z26.1-1950, American Standards Association, New York.
44. Bragaw, C. G., *Modern Plastics* **33,** #10, 199 (June, 1956).
45. Carey, R. H., *SPE J.* **10,** #3, 16 (1954).
46. Carswell, T. S., and H. K. Nason, *Symposium on Plastics*, American Society for Testing Materials, Philadelphia, Pa., 1944.
47. DeCoste, J. B., F. S. Malm, and V. T. Wallder, *Ind. Eng. Chem.* **43,** 117 (1951).
48. Dennis, L. M., and R. S. Shelton, *J. Am. Chem. Soc.* **52,** 3128 (1930).
49. Flom, D. G., *Anal. Chem.* **32,** 1550 (1960); D. G. Flom and A. M. Bueche, *J. Appl. Phys.* **30,** 1725 (1959).
50. Greensmith, H. W., *J. Appl. Polymer Sci.* **3,** 183 (1960).
51. Hittmair, P., H. F. Mark, and R. Ullman, *J. Polymer Sci.* **33,** 505 (1958); P. Hittmair and R. Ullman, *J. Appl. Polymer Sci.* **6,** 1 (1962).
52. Howard, J. B., *SPE J.* **15,** #5, 395 (1959).
53. Hsaio, C. C., and J. A. Sauer, *ASTM Bull.* **172,** 29 (1951).
54. Keskkula, H., and J. W. Norton, Jr., *J. Appl. Polymer Sci.* **2,** 289 (1959).
55. Landler, L. L., *SPE J.* **16,** 1329 (1960).
56. Lazan, B. J., and A. Yorgiadis, *Symposium on Plastics*, American Society for Testing Materials, Philadelphia, Pa., 1944.
57. Martin, T. J., and R. L. Hauter, *SPE J.* **10,** #2, 13 (1954).
58. McMahon, E. J., D. E. Maloney, and J. R. Perkins, *Trans. Am. Inst. Elec. Engrs.* **78,** Part I, 654 (1959).
59. Maxwell, B., and J. P. Harrington, *Trans. Am. Soc. Mech. Engrs.* **74,** 579 (1952).
60. O'Dwyer, J. J., *Advances in Physics* **7,** 349 (1958).
61. Russell, E. W., *Nature* **165,** 91 (1950).
62. Strella, S., and L. Gilman, *Modern Plastics* **34,** #8, 158 (April, 1957).
63. Supnik, R. H., *et al.*, *J. Appl. Polymer Sci.* **4,** 257–331 (1960).
64. Webber, A. C., *J. Opt. Soc. Am.* **47,** 785 (1957).
65. Wolock, I., and D. George, *SPE J.* **12,** #2, 20 (1956).
66. Ziegler, E. E., *SPE J.* **10,** #4, 12 (1954).

5

Morphology and Order in Crystalline Polymers

A. Configurations of Polymer Chains

In this book, the word *configuration* is used to describe those arrangements of atoms which cannot be altered except by breaking and reforming primary chemical bonds. In contrast, as discussed in Chapter 2*B*, arrangements which can be altered by rotating groups of atoms around single bonds are called *conformations*. Examples involving conformations of polymer chains include *trans* vs. *gauche* arrangements of consecutive carbon-carbon single bonds and the helical arrangements found in some polymer crystal structures (Section *B*). Examples involving configurations include head-to-head, tail-to-tail, or head-to-tail arrangements in vinyl polymers (Chapter 9*A*) and the several *stereoregular* arrangements described in this section, including 1,2 and 1,4 addition, *cis* and *trans* isomers, and arrangements around asymmetric carbon atoms.

Configurations involving an asymmetric carbon atom

Staudinger (16) recognized as early as 1932 that polymers of monosubstituted olefins should contain a series of asymmetric carbon atoms along the chain, and much speculation followed over the possibility of synthesizing such polymers in stereoregular forms. As discussed in Chapter 10*D*, such syntheses have been achieved by several routes. The regular structure of the resulting polymers, notably poly(α-olefins), was recognized by Natta (7), who devised the nomenclature now accepted to describe stereoregular

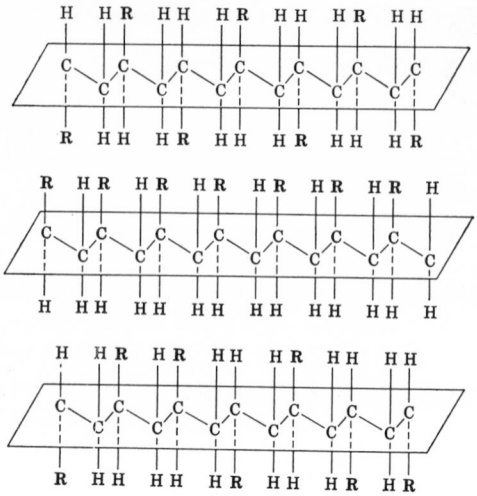

(a)

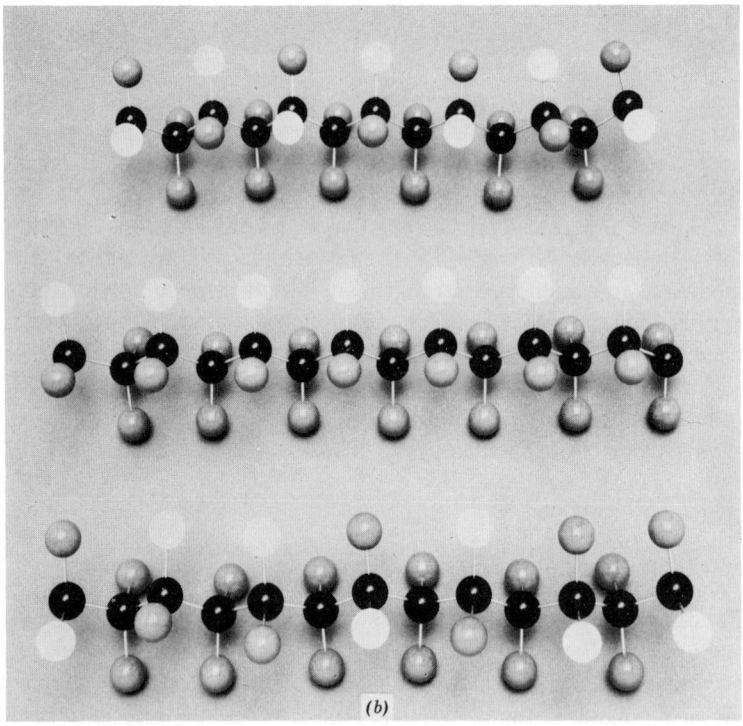

(b)

Fig. 5-1. (*a*) Diagrams and (*b*) models showing the stereoregular syndiotactic (top) and isotactic (center) and the irregular atactic (bottom) configurations in a vinyl polymer (10). The main carbon-carbon chain is depicted in the fully extended (all *trans*) planar zigzag conformation.

polymers of this type. As indicated in Fig. 5-1, where the polymer chain is depicted in the fully extended (all *trans*) planar zigzag conformation, the configuration resulting when all the substituent groups R on the vinyl polymer lie above (or below) the plane of the main chain is called *isotactic*. If substituent groups lie alternatively above and below the plane, the configuration is called *syndiotactic*, whereas a random sequence of positions is said to lead to the *atactic* configuration.

The synthesis of isotactic and syndiotactic polymers has now become almost commonplace, and many examples of each type are known. In addition to poly(α-olefins), polystyrene and poly(methyl methacrylate) are examples of polymers which can be made in the isotactic configuration; while poly(methyl methacrylate) and polybutadiene in which 1,2 addition occurs exclusively (9), giving the structure

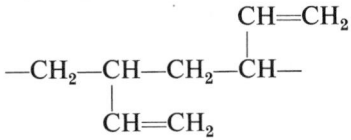

are examples of polymers which can be prepared in the syndiotactic configuration. It has been shown that some stereoregular polymers had been made earlier but were not recognized at the time, a notable example being poly(vinyl isobutyl ether) (15), which is isotactic with the same structure as poly(5-methyl-hexene-1) (8):

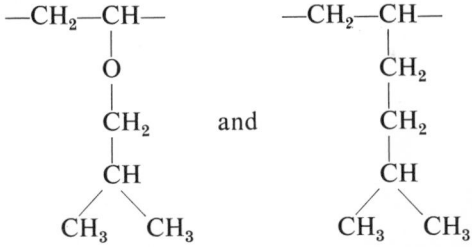

The assignment of an isotactic or syndiotactic structure can be made through determination of the crystal structure of the polymer. Several examples are treated in Section *B*. Independent confirmation of the structures has been obtained for poly(methyl methacrylate), using NMR spectroscopy (2). The method depends on differences in the environment of chain methylene groups which are part of isotactic or syndiotactic sequences. It can be applied also to polymers which are not predominantly stereoregular, yielding information about the probability σ that a monomer adds to the growing chain with the configuration established by the previous monomer: $\sigma = 0, \frac{1}{2}$, and 1 correspond to completely syndiotactic, atactic, and isotactic polymer, respectively. By this means it is found that free radical

poly(methyl methacrylate) is predominantly syndiotactic ($\sigma \simeq 0.2$) but becomes increasingly atactic when polymerized at higher temperatures.

Configurations involving two adjacent asymmetric carbon atoms

Polymers made from 1,2-disubstituted vinyl monomers of the type CHR=CHR′ can exist in several stereoregular forms. These forms are designated *ditactic* (11) (*diisotactic*, *disyndiotactic*), since each repeat unit has two asymmetric carbon atoms. Only one disyndiotactic product (Fig. 5-2*a*) is formed regardless of whether the *cis* or *trans* monomer is used, but two diisotactic forms are possible (Figs. 5-2*b* and *c*). In accordance with the convention used in naming sugars (5), these forms are designated with the prefixes *threo* when both the R and R′ substituent groups fall on the same side of the main chain, and *erythro* when this is not the case.

Configurations involving a carbon-carbon double bond

As discussed in Chapter 12*C*, polymers of 1,3 dienes, containing one residual double bond per repeat unit after polymerization, can contain sequences with several different configurations. For a monosubstituted butadiene such as isoprene, the following structures are possible:

cis-1,4

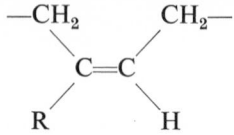

trans-1,4

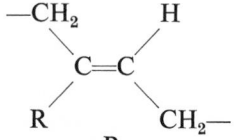

1,2

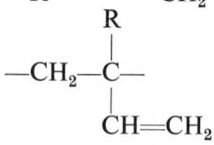

3,4

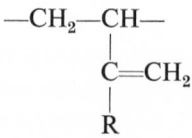

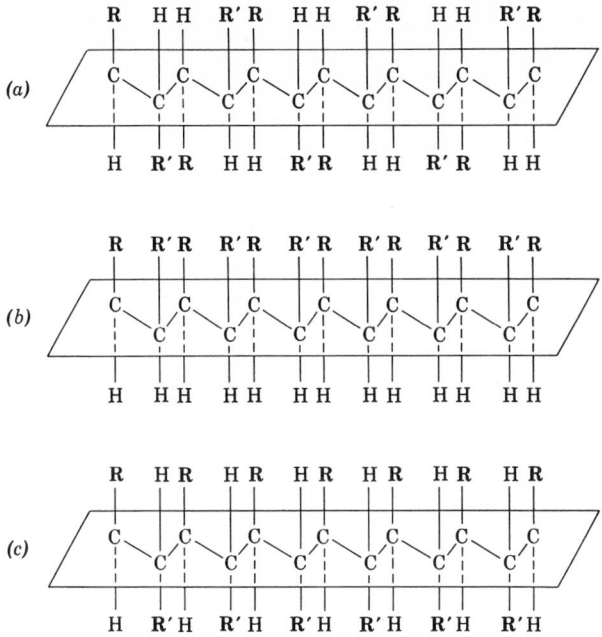

Fig. 5-2. Configurations in ditactic polymers (11): (*a*) disyndiotactic; (*b*) *threo*-diisotactic; (*c*) *erythro*-diisotactic.

Stereoregular polymers with several of these configurations are known (6, 9, 17). Their structures are confirmed by x-ray diffraction and infrared spectroscopy.

Configurations involving both asymmetric carbon atoms and double bonds

Derivatives of butadiene which are disubstituted in the 1 and 4 positions can be polymerized (13) to products having two asymmetric carbon atoms and one double bond per repeat unit. These polymers are termed *tritactic* and are optically active. In analogy with sugar nomenclature (5), the four possible configurations are named with prefixes *cis* or *trans* and *erythro* or *threo* as indicated in Fig. 5-3.

Other stereoregular configurations

There seems little doubt that stereoregularity plays an important role in the structures of proteins, nucleic acids, and other substances of biological importance. Deoxyribonucleic acid (DNA), e.g., has a highly stereoregular

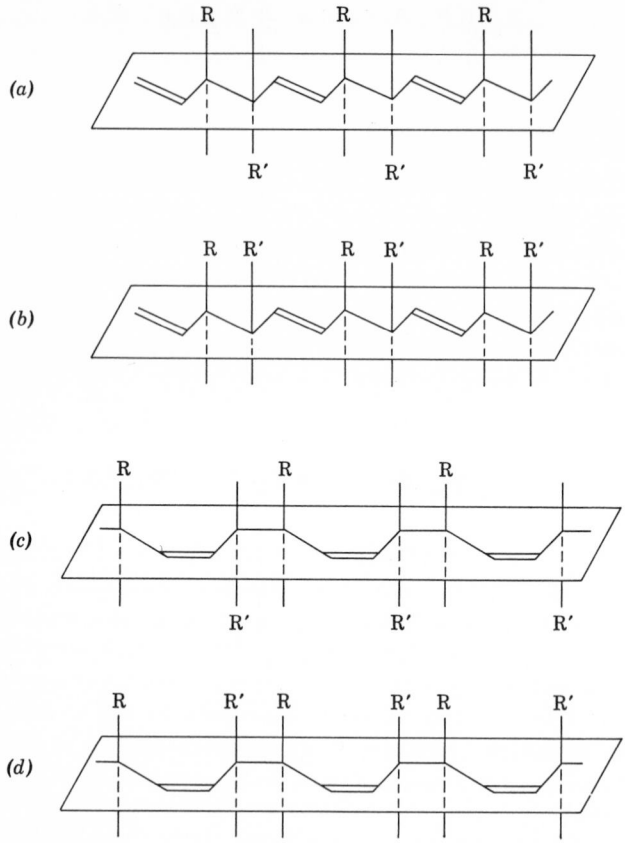

Fig. 5-3. Configurations in tritactic polymers (13): (*a*) *trans-erythro;* (*b*) *trans-threo;* (*c*) *cis-erythro;* (*d*) *cis-threo*. For clarity, carbon and hydrogen atoms have been omitted.

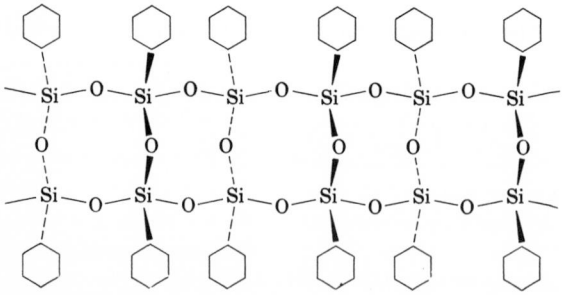

Fig. 5-4. Double-chain polymer formed by *cis* fusion of syndiotactic polyphenylsiloxane chains (3). Courtesy of J. F. Brown, Jr.

structure (4). A stereoregular double-chain structure occurs in polyphenylsilsesquioxane (3): syndiotactic polyphenylsiloxane chains are fused in the *cis* configuration by oxygen bridges at each Si atom to give a slightly pleated, ladderlike network (Fig. 5-4).

Optical activity in polymers

The requirements for obtaining optical activity in polymers have been outlined (1). In the simplest cases, the polymer is optically active if asymmetric carbon atoms, bonded to four different groups, are present in a stereoregular structure. An example is poly(propylene oxide) (14), where the starred atoms are asymmetric:

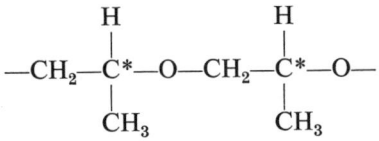

Optically active polymers can be made from optically active monomers provided that the asymmetry is not lost in the synthesis. Optically active polymers can also be made from inactive monomers if optically active catalysts are used (12). It was reported in 1962, however, that asymmetric centers make only minor direct contributions to the overall optical activity of polymers. Helical conformations developed in the stereospecific syntheses are the major source of optical rotatory power in polymer solutions. While optically active monomers are important for their stereospecific directing influence, either as catalysts or as principal ingredients, the optical activity due to asymmetric centers is masked by helical structure effects after a few additions of monomer.

GENERAL REFERENCES

Cooper, W., "Stereospecific Polymerization," pp. 279–340 in J. C. Robb and F. W. Peaker, eds., *Progress in High Polymers*, Vol. 1, Academic Press, New York, 1961.
Natta, G., *Sci. American* **205,** #2, 33 (August, 1961).

BIBLIOGRAPHY

1. Arcus, C. L., *J. Chem. Soc.* **1955,** 2801; **1957,** 1189.
2. Bovey, F. A., and G. V. D. Tiers, *J. Polymer Sci.* **44,** 173 (1960); F. A. Bovey, *ibid.* **46,** 59 (1960); see also R. L. Miller and L. E. Nielsen, *ibid.* **46,** 303 (1960).
3. Brown, J. F., Jr., and coworkers, *J. Am. Chem. Soc.* **82,** 6194 (1960).
4. Crick, F. H. C., *Sci. American* **197,** #3, 188 (September, 1957).

5. Dauben, W. G., and K. S. Pitzer, "Conformational Analysis," Chap. 1 in M. S. Newman, ed., *Steric Effects in Organic Chemistry*, John Wiley & Sons, New York, 1956, p. 10.
6. Horne, S. E., Jr., and coworkers, *Ind. Eng. Chem.* **48,** 784 (1956).
7. Natta, G., and coworkers, *J. Am. Chem. Soc.* **77,** 1708 (1955); G. Natta and F. Danusso, *J. Polymer Sci.* **34,** 3 (1959). For nomenclature see M. L. Huggins, G. Natta, V. Desreux, and H. Mark, *J. Polymer Sci.* **56,** 153 (1962).
8. Natta, G., I. Bassi, and P. Corradini, *Makromol. Chem.* **18/19,** 455 (1956).
9. Natta, G., and P. Corradini, *J. Polymer Sci.* **20,** 251 (1956).
10. Natta, G., and P. Corradini, *J. Polymer Sci.* **39,** 29 (1959).
11. Natta, G., M. Farina, and M. Peraldo, *J. Polymer Sci.* **43,** 289 (1960); G. Natta, M. Farina, M. Peraldo, and G. Bressan, *Makromol Chem.* **43,** 68 (1961).
12. Natta, G., M. Farina, M. Peraldo, and G. Bressan, *Chim. e ind.* (*Milan*) **43,** 161 (1961).
13. Natta, G., M. Farina, M. Donati, and G. Bressan, *Chim. e ind.* (*Milan*) **42,** 1363 (1960); G. Natta and coworkers, *ibid.*, p. 1361.
14. Price, C. C., and M. Osgan, *J. Am. Chem. Soc.* **78,** 4787 (1956)
15. Schildknecht, C. E., S. T. Gross, H. R. Davidson, J. M. Lambert, and A. O. Zoss, *Ind. Eng. Chem.* **40,** 2104 (1948); C. E. Schildknecht, A. O. Zoss, and F. Grosser, *ibid.* **41,** 2891 (1949).
16. Staudinger, H., *Die Hochmolekularen Organischen Verbindungen*, Springer-Verlag, Berlin, 1932.
17. Stavely, F. W., and coworkers, *Ind. Eng. Chem.* **48,** 778 (1956).

B. Crystal Structures of Polymers

Structural requirements for crystallinity

Many polymers, including most fibers, are partially crystalline. The most direct evidence of this fact is provided by x-ray diffraction studies. The x-ray patterns of crystalline polymers show both sharp features associated with regions of three-dimensional order, and more diffuse features characteristic of molecularly disordered substances like liquids. The occurrence of both types of feature is evidence that ordered and disordered regions coexist in most crystalline polymers. Additional evidence to this effect comes from other polymer properties, such as densities, which are intermediate between those calculated for completely crystalline and amorphous species.

The close relation between regularity of molecular structure and crystallizability has long been recognized. Typical crystalline polymers are those whose molecules are chemically and geometrically regular in structure. Occasional irregularities, such as chain branching in polyethylene or copolymerization of a minor amount of isoprene with polyisobutylene as in butyl rubber, limit the extent of crystallization but do not prevent its occurrence.

Typical noncrystalline polymers, on the other hand, include those in which irregularity of structure occurs: copolymers with significant amounts of two or more quite different monomer constituents, or atactic polymers.

Before the recognition of the importance of stereoregularity to crystallinity in polymers, it was thought that the bulkiness of the side groups in some atactic polymers, such as polystyrene and poly(methyl methacrylate), alone prevented their crystallization. The argument was based in part on the fact that poly(vinyl alcohol) can be made to crystallize, although it is derived by hydrolysis from poly(vinyl acetate), which has never been crystallized. The crystallinity of poly(vinyl alcohol) was once taken as evidence that it was isotactic. Therefore, poly(vinyl acetate) should have been similarly isotactic and would be expected to crystallize—perhaps not in a completely extended conformation—despite the bulkiness of the acetate groups. Poly(vinyl alcohol) is in fact atactic. Other irregular polymers, such as copolymers of ethylene and vinyl alcohol, ethylene and tetrafluoroethylene, and ethylene and carbon monoxide, are also crystalline. All these polymers have very similar structures (9). It is now clear that the CH_2, CHOH, CF_2, and C=O groups are near enough to the same size to fit into similar crystal lattices despite the stereochemical irregularity of the polymers. On the other hand, the CHCl group is apparently too large, since chlorinated polyethylene is noncrystalline. The absence of crystallinity in poly(vinyl acetate) and similar polymers is due to the combination of their atactic structure and the size of their substituent groups.

When the disordering features of atactic configuration and bulky side groups are combined with strong interchain forces, the polymer may have a *directionally ordered* structure (25) which is intermediate in kind between crystalline and amorphous. An example is polyacrylonitrile, in which steric and intramolecular dipole repulsions lead to a stiff, irregularly twisted, backbone chain conformation. The stiff molecular chains pack like rigid rods in a lattice array; lateral order between chains is present, but longitudinal order along chains is absent.

Structures based on extended chains

There is little doubt that the fully extended planar zigzag is the conformation of minimum energy for an isolated section of a hydrocarbon chain, the energy of the *trans* conformation being some 800 cal/mole less than that of the *gauche* (7). It is to be expected, therefore, that fully extended chain conformations will be favored in polymer crystal structures unless substituents on the chains cause steric hindrance relieved by the assumption of other forms. Fully extended chains are found in the crystal structures of

polyethylene, poly(vinyl alcohol), syndiotactic polymers including poly-(vinyl chloride) and poly(1,2-butadiene), most polyamides, and cellulose.

Whereas the crystal structures of syndiotactic polymers contain only regular sequences of *trans* conformations, the spatial requirements for the molten or dissolved polymer are somewhat less severe (11). Both bonds of successive pairs must be *trans* or *gauche*; also, successions of *gauche-gauche* (*gg*) pairs are excluded. The conformation is therefore of the form

$$(tt)\,(gg)\,(tt)\,(tt)\,(tt)\,(gg)\,(tt)\ \text{etc.}$$

where the average length of all-*trans* segments is estimated to be about three units (3).

Polyethylene (4) Except for end groups, the arrangement of chains in the crystal structure of polyethylene is essentially the same as that in crystals of linear paraffin hydrocarbons containing 20–40 carbon atoms (18). The unit cell of polyethylene is rectangular with dimensions $a = 7.41$ A, $b = 4.94$ A, and $c = 2.55$ A. The latter is the chain repeat distance, which is identical with the fully extended zigzag repeat distance. The arrangement of the chains with respect to one another is shown in Fig. 5-5. The positions

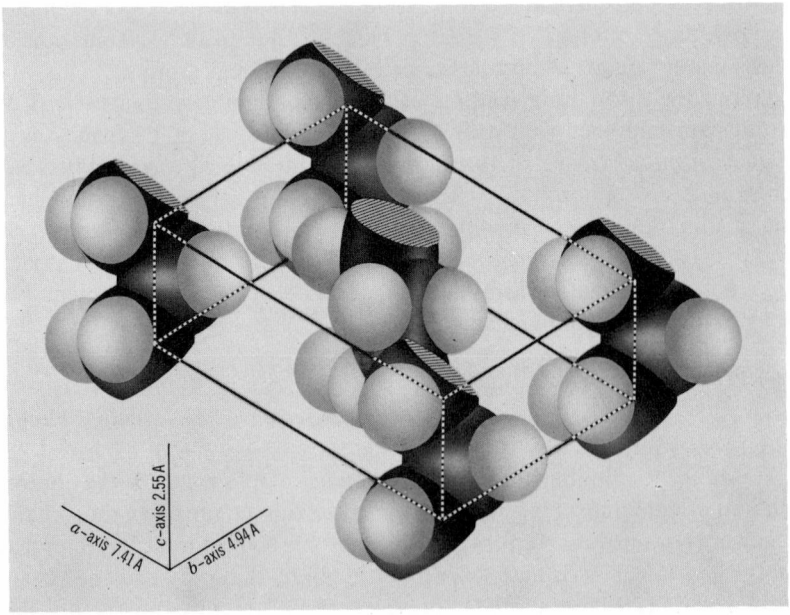

Fig. 5-5. Arrangement of chains in the unit cell of polyethylene. Courtesy of E. S. Clark.

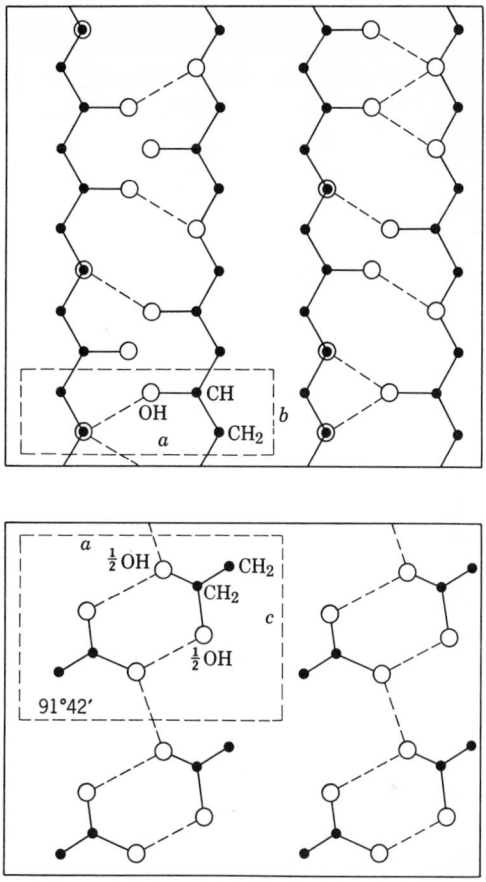

Fig. 5-6. Hydrogen bonding in the crystal structure of poly(vinyl alcohol) (5).

of the hydrogen atoms are not detectable by the x-ray method (since the scattering power of an atom is proportional to its atomic weight) but have been calculated assuming tetrahedral bonds and a C—H distance of 1.10 A.

Poly(vinyl alcohol) (5, 17) The crystal structure of poly(vinyl alcohol) is similar to that of polyethylene, as might be expected, since the CHOH group is small enough to fit into the polyethylene structure in place of a CH_2 group with little difficulty (9). The unit cell is monoclinic (i.e., distorted from rectangular by making the face perpendicular to the chain axis a nonrectangular parallelogram with one included angle designated β) with $a = 7.81$ A, $b = 2.52$ A (chain axis), $c = 5.51$ A; $\beta = 91°42'$. Pairs of chains are linked together by hydrogen bonds, and similarly linked into sheets, in so far as the stereochemical irregularity allows (Fig. 5-6).

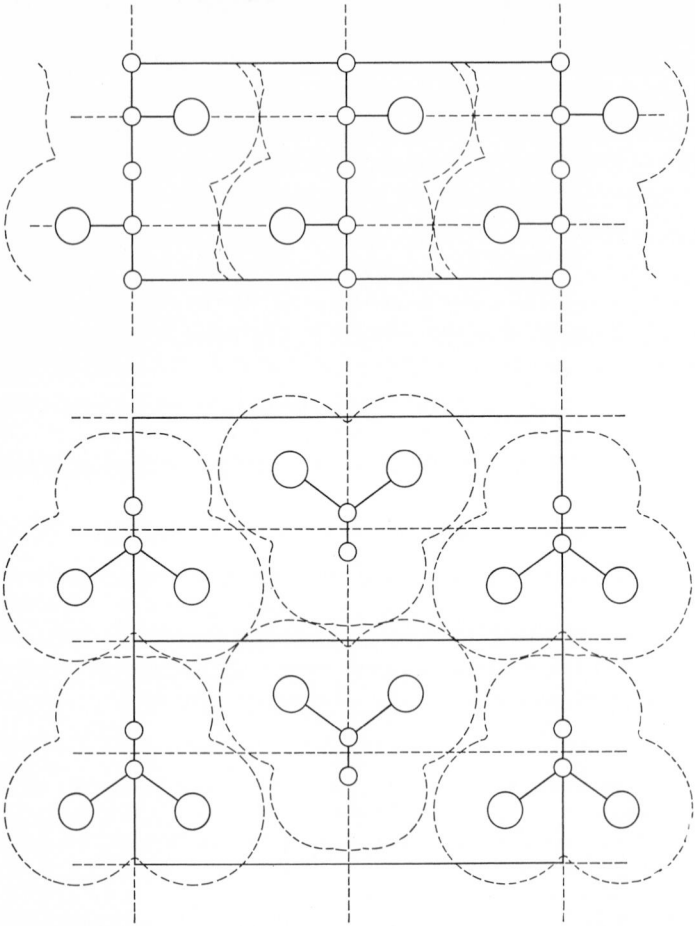

Fig. 5-7. Diagrams of the crystal structure of poly(vinyl chloride), typical of most syndiotactic polymers (19).

Poly(vinyl chloride) (19) Crystallinity in poly(vinyl chloride) is not well developed; the polymer is primarily syndiotactic but with considerable irregularity. Its crystal structure (Fig. 5-7) appears to be typical of syndiotactic polymers, with a repeat distance corresponding to four chain carbon atoms. Poly(1,2-butadiene) has a similar structure, but the carbon-carbon chains are slightly distorted from the *trans* conformation to relieve repulsion between side-chain and neighboring-chain atoms.

Polyamides The structures of poly(hexamethylene adipamide) (66 nylon), poly(hexamethylene sebacamide) (610 nylon), and polycaprolactam

(6 nylon) (6, 13) are made up of fully extended chains linked together by hydrogen bonds to form sheets (Fig. 5-8) which may be packed together in two different ways, giving two crystal modifications. Oxygen atoms of one molecule are always found opposite NH groups of a neighboring molecule, with the N—H- - -O distance of 2.8 A smaller than normal because of hydrogen bond formation. Other nylons (99, 106, 1010, and 11) contain chains slightly distorted from the planar zigzag form (1, 2).

Cellulose (16) The crystal structure of cellulose is of historical interest, since it represents the first instance in which the concepts of a crystal unit cell and a chain structure were reconciled by the concept of the chain extending through successive unit cells. The basic feature of the crystal structure of cellulose is the repeating segment of two cellobiose units, joined in 1,4 linkages, with the —CH_2OH side chains in successive units occurring on opposite sides of the chain. Many forms and derivatives of cellulose have crystal structures in which these chains are packed in slightly different ways.*

Distortions from fully extended chains

Polyesters In most aliphatic polyesters (12) and in poly(ethylene terephthalate) (Bunn, gen. ref. in R. Hill, ed.) the polymer chains are shortened by rotation about the C—O bonds to allow close packing. As a result, the main chains are no longer planar. The group 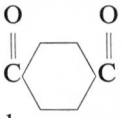 in poly(ethylene terephthalate) is planar, however, as required by resonance (Fig. 5-9).

Polyisoprenes and polychloroprene Polymers of isoprene and chloroprene in which 1,4 addition predominates can occur in all-*cis* or all-*trans* configurations; natural rubber and gutta-percha are naturally occurring *cis*-1,4-polyisoprene and *trans*-1,4-polyisoprene, respectively. All these polymers have similar crystal structures (Bunn, gen. ref. in Mark and Whitby, eds.) except that the repeat distance of the *trans* polymers corresponds to one monomer unit and that of the *cis* polymers to two.

The repeat distance of *trans*-1,4-polyisoprene (gutta-percha) (4.72–4.77 A) is smaller than that corresponding to a fully extended chain (5.04 A).

* It should be noted that some Russian workers (14) state that cellulose is amorphous. It is the consensus of workers in this country, however, that this discrepancy is semantic in origin and that, in the light of established definitions for the terms, cellulose must be considered crystalline rather than amorphous.

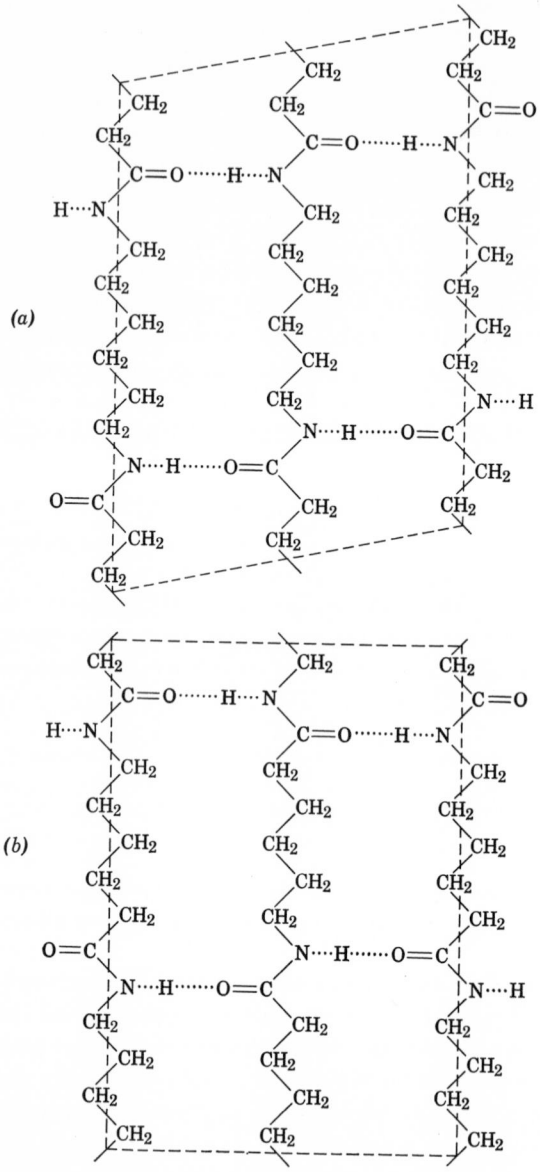

Fig. 5-8. Arrangement of chains in hydrogen-bonded sheets in the crystal structure of (*a*) 66 nylon and (*b*) 6 nylon (13).

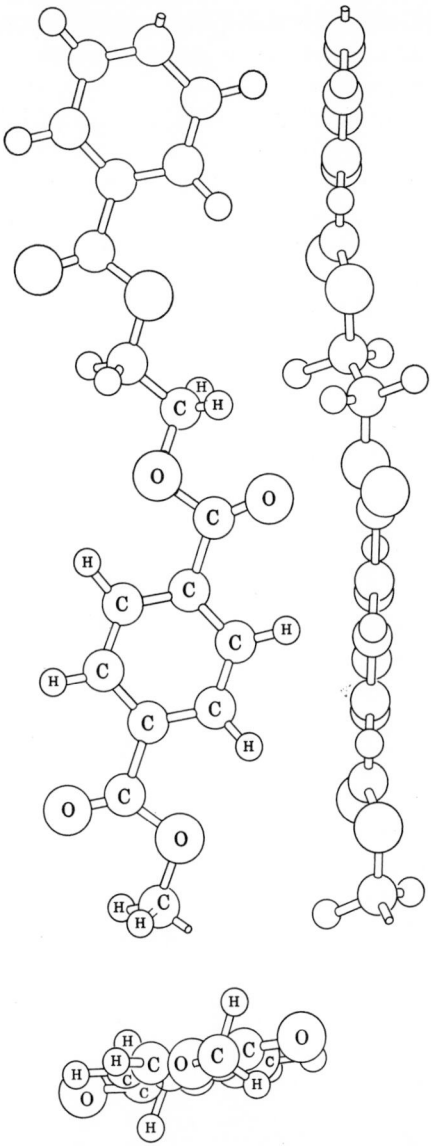
C
H
H
O
O
C
C
C
C
C
C
C
C
C
C
C
O
O
O
O
C
H
H
H

Since the atoms in the double-bonded group (Fig. 5-10, carbons 1-5) are expected to be in a single plane, the shortening must involve rotation about the bond between carbons 4 and 1*a*. This rotation can be made in either direction, giving rise to two types of molecules which are mirror images.

The molecules of gutta-percha are not symmetrical end for end: the methyl group can occur either above or below the double bond. In a polymer melt at the point of incipient crystallization there is an equal probability of finding either an "up" or a "down" molecule. The crystal structure (Fig. 5-11) must then contain equal numbers of each kind. A given crystal contains only one conformation (left- or right-handed chains) to the exclusion of the other, the orientation being determined by chance at the time of crystallization.

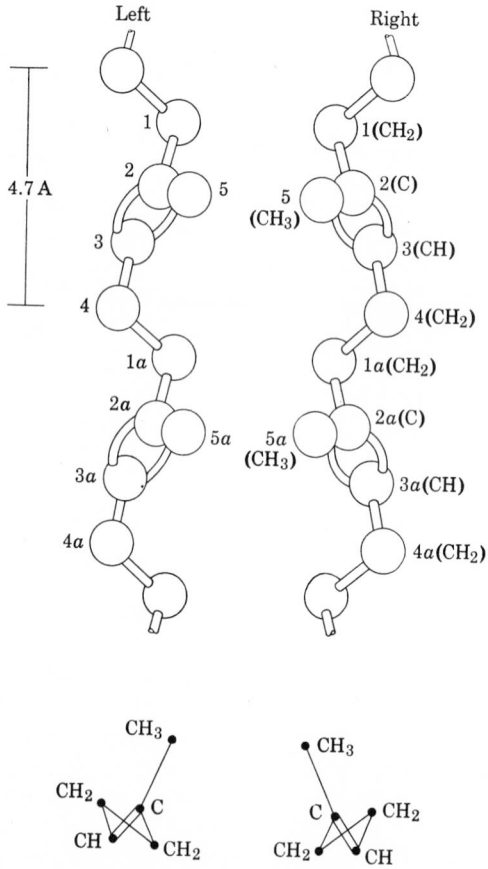

Fig. 5-10. Conformation of gutta-percha, showing left- and right-handed molecules (Bunn, gen. ref. in Mark and Whitby, eds.).

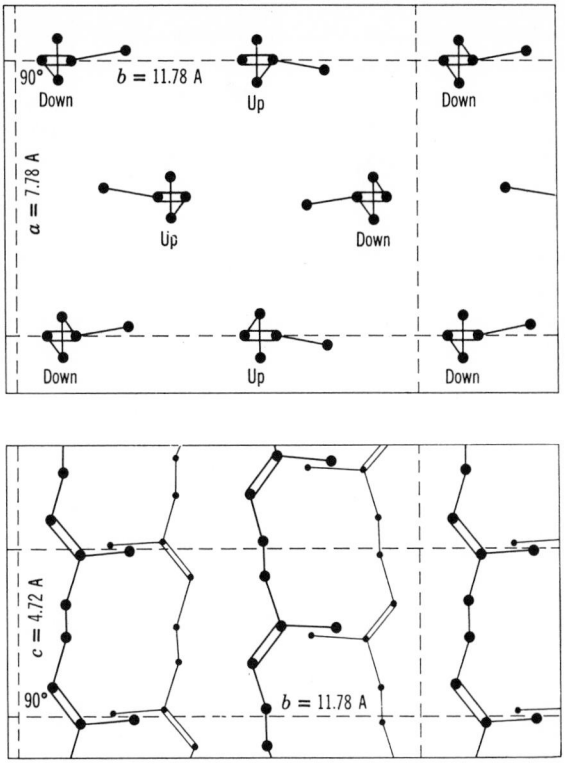

Fig. 5-11. Crystal structure of gutta-percha (Bunn, gen. ref. in Mark and Whitby, eds.).

The structure of natural rubber (*cis*-polyisoprene) is similar to that of gutta-percha except as required by the longer repeat distance due to the *cis* configuration. The chain is slightly out of plane at the single bonds and in plane around the double bond, the observed repeat distance of 8.1 A being less than the extended distance of 10.08 A. Considerations of right- and left-handed crystals and "up" and "down" chains apply to natural rubber as well as to gutta-percha.

The structure of polychloroprene is similar to that of gutta-percha except that the molecules are oriented differently with respect to one another in the crystal because of differences in polarity between the chlorine and methyl groups.

Polypeptides Proteins, including the natural fibers wool and silk, are made up of α-amino acids joined into long polypeptide chains with the

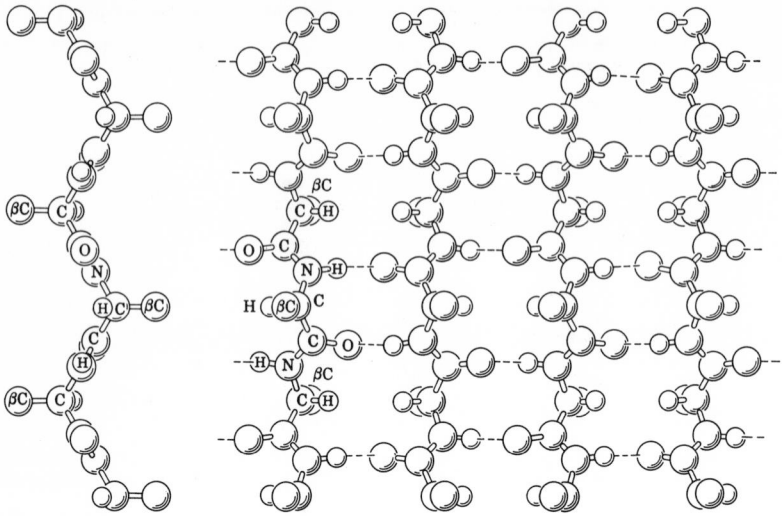

Fig. 5-12. Anti-parallel-chain pleated sheet structure of silk fibroin (21).

planar structure

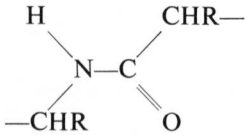

where R represents the amino acid chain. These polymer chains can exist in two quite different types of crystal structure, both of which have been under extensive investigation (21).

One of these forms, the β-keratin structure, is made up of nearly extended polypeptide chains arranged into sheets through hydrogen bonding. The sheets are not quite planar and have been described as pleated or rippled. A number of possible structures have been proposed, differing in the detailed arrangement of the chains into sheets. One of these, the anti-parallel-chain, pleated sheet structure, is shown in Fig. 5-12. It is now considered the most likely structure for silk fibroin.

The helical α-keratin structure of polypeptides is described on page 136.

Helical structures

Polymers with bulky substituents closely spaced along the chain often take on a helical conformation in the crystalline phase, since this allows the substituents to pack closely without appreciable distortion of chain bonds. Most isotactic polymers, as well as some 1,1-disubstituted ethylenes such

as isobutylene, fall in this class. Other helical structures of interest are those of polytetrafluoroethylene, discussed on page 138, and the α-keratin structure.

Isotactic polymers (7, 20) Many isotactic polymers crystallize with a helical conformation in which alternate chain bonds take *trans* and *gauche* positions. For the *gauche* positions, the rotation is always in that direction which relieves steric hindrance by placing R and H groups in juxtaposition, generating either a left-hand or a right-hand helix. Thus in the crystal

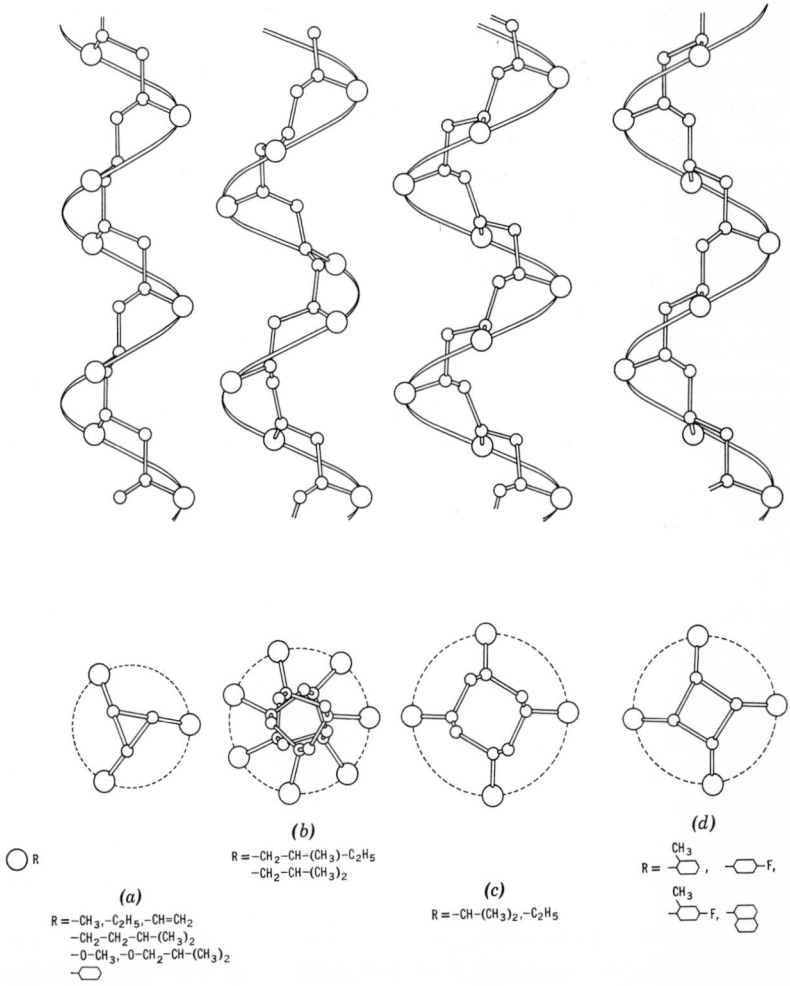

Fig. 5-13. Helical conformations of isotactic vinyl polymers (7, Gaylord and Mark, gen. ref.).

regular sequences of the type $(tg)(tg)(tg)$ or the reverse $(gt)(gt)(gt)$ occur (11). In the amorphous polymer or in solution, breaks in the helix may occur (3) where the polarity of the sequence reverses, giving segments of the type

$$(tg)(tg)(tg) \quad (gt)(gt) \quad (tg)(tg)(tg)(tg) \quad (gt) \text{ etc.}$$

The average segment length is estimated to be about four units for isotactic polystyrene, a typical example (3).

If the side group is not too bulky, the helix has exactly three units per turn and the arrangement is similar to that in Fig. 5-13*a*. This arrangement is found in isotactic polypropylene, one form of poly(1-butene), polystyrene, and others. More bulky side groups require more space, resulting in the formation of looser helices as shown in Fig. 5-13*b*, *c*, and *d*. Isotactic poly(methyl methacrylate) forms a helix with five units in two turns (Fig. 5-14) (26), while polyisobutylene forms a helix with eight units in five turns (Fig. 5-15) (15).

Polypeptides (22) Helical structures are of great importance in biological materials. In addition to helical conformations proposed for DNA and other polymers of biological interest, the α-keratin structure assumed by many polypeptides is a helix, in which about 3.6–3.7 polypeptide residues occur per turn (Fig. 5-16). Adjacent turns are held together by intramolecular hydrogen bonding. Several variations of this structure have been proposed. In addition, more complex structures made by intertwining

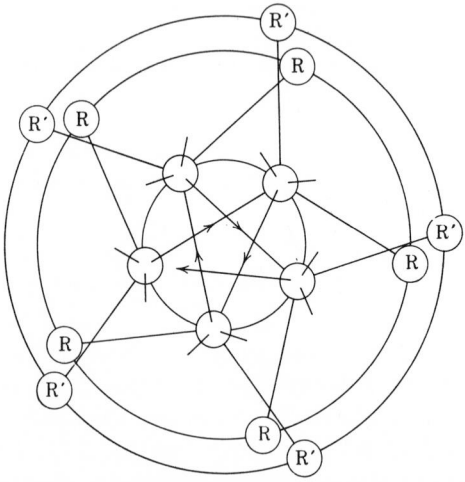

Fig. 5-14. Projection along the chain axis of the helical conformation in the crystal of isotactic poly(methyl methacrylate); five units in two turns (26).

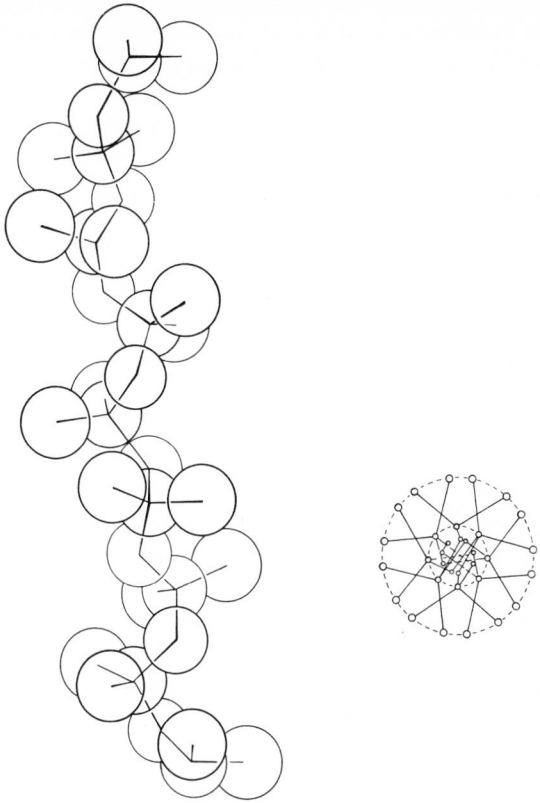

Fig. 5-15. Helical conformation in the crystal of polyisobutylene; eight units in five turns (7, 15).

the helices together in groups of three or seven have been proposed as building blocks of larger stranded structures such as hair.

Order-disorder transitions in crystalline phases

Several instances are known where a polymer exists in two or more crystal modifications. In many cases one of these forms is significantly less ordered or less closely packed than the other. Thus, poly(1-butene) crystallizes in the four-fold helical structure of Fig. 5-13*c* when first cooled from the melt, but these crystals are transformed to the more closely packed three-fold helical conformation (Fig. 5-13*a*) when the polymer is allowed to stand at temperatures between T_m and T_g or is drawn (20).

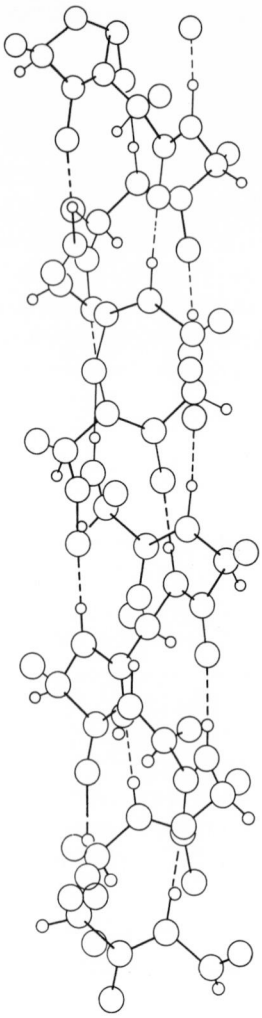

Fig. 5-16. Helical structure of α-keratin proteins (22).

Polytetrafluoroethylene This polymer exists in two helical conformations which may be described as twisted ribbons in which the fully extended planar form is distorted to have a 180° twist in thirteen CF_2 units in the more stable (low-temperature) form (8). Above 19°C this form is replaced by a slightly untwisted conformation with fifteen CF_2 units per half-twist (23). Above 19°C the x-ray diffraction pattern shows diffuse streaks which are interpreted (10) as resulting from small angular displacements of molecular segments about their long axes. Above 30°C, more diffuseness occurs in the x-ray pattern, increasing as the temperature is

raised to and above the melting point, 327°C. This additional diffuseness is attributed to random angular displacement of the molecules about their long axes.

In terms of molecular motion, it is considered that the molecules undergo slight torsional motion—i.e., twisting and untwisting of the helices—which causes the lattice to expand laterally, and at 19°C is strong enough to cause the untwisting from thirteen to fifteen CF_2 groups in each 180° twist. Between 19 and 30°C the molecules are "locked" into the three-fold helix corresponding to the fifteen CF_2 repeat distance; their motion is restricted to small-angle oscillations about their "rest positions." Above 30°C the molecules lose their preferred angular orientation, becoming randomly misaligned about their axes. Nearly cylindrical in shape, they pack like parallel rods. Above 30°C torsional motion increases, with the twisting and untwisting becoming more pronounced, until at T_m all crystalline order is lost.

Helix-coil transitions in polypeptides (11, 24, 27) The α-keratin helix structure which has been described as characteristic of polypeptides and many polymers of biological origin can undergo several phase transitions, including melting to random coils at T_m or in the presence of a solvent. A transition of particular interest results from the fact that the helix can exist as a stable conformation in solution, so that the helix-coil transition can occur without involving a solid phase.

The stability of the helix in solution results from intramolecular hydrogen bonding. Since disruption of the helix requires breaking about three such bonds per turn, some 5–10 kcal of enthalpy must be expended to realize the greater entropy of the random coil state. Once the transition is initiated, by a change in temperature or composition, the preferred conformation is perpetuated over many bonds. In the simplest case, the helical conformation occurs either not at all or in long sequences with infrequent interruption if not throughout the chain. The change is fairly abrupt and resembles a phase transition.

GENERAL REFERENCES

Bunn, C. W., "Structure of Crystalline Regions," Chap. 11 in R. Hill, ed., *Fibres from Synthetic Polymers*, Elsevier Publishing Co., New York, 1953.

Bunn, C. W., "The Study of Rubber-like Substances by X-Ray Diffraction Methods," pp. 95–145 in H. Mark and G. S. Whitby, eds., *Scientific Progress in the Field of Rubber and Synthetic Elastomers* (*Advances in Colloid Science*, Vol. II), Interscience Publishers, New York, 1946.

Gaylord, N. G., and H. Mark, *Linear and Stereoregular Addition Polymers*, Interscience Publishers, New York, 1959, Chap. V.

Miller, R. L., and L. E. Nielsen, *J. Polymer Sci.* **55**, 643 (1961).

BIBLIOGRAPHY

1. Aelion, R., *Ann. chim.* **3,** 5 (1948).
2. Baker, W. O., and C. S. Fuller, *J. Am. Chem. Soc.* **64,** 2399 (1942).
3. Birshtein, T. M., and O. B. Ptitsyn, *J. Tech. Phys.* (*U.S.S.R.*) **29,** 1048 (1959) [*Soviet Physics-Technical Physics* **4,** 954 (1960)]; T. M. Birshtein, *High Mol. Wt. Cpds.* (*U.S.S.R.*) **1,** 748 (1959); T. M. Birshtein and E. A. Sokolova, *ibid.* **1,** 1086 (1959).
4. Bunn, C. W., *Trans. Faraday Soc.* **35,** 482 (1939).
5. Bunn, C. W., *Nature* **161,** 929 (1948).
6. Bunn, C. W., and E. V. Garner, *Proc. Roy. Soc.* (*London*) **A189,** 39 (1947).
7. Bunn, C. W., and D. R. Holmes, *Disc. Faraday Soc.* **25,** 95 (1958).
8. Bunn, C. W., and E. R. Howells, *Nature* **174,** 549 (1954).
9. Bunn, C. W., and H. S. Peiser, *Nature* **159,** 161 (1947).
10. Clark, E. S., and L. T. Muus, *Z. Krist.* **117,** 119 (1962).
11. Flory, P. J., *J. Polymer Sci.* **49,** 105 (1961).
12. Fuller, C. S., *Chem. Revs.* **26,** 143 (1940).
13. Holmes, D. R., C. W. Bunn, and D. J. Smith, *J. Polymer. Sci.* **17,** 159 (1955).
14. Kargin, V. A., *J. Polymer Sci.* **30,** 247 (1958); *High Mol. Wt. Cpds.* (*U.S.S.R.*) **2,** 466 (1960).
15. Liquori, A. M., *Acta Cryst.* **8,** 345 (1955).
16. Mark, H., and K. H. Meyer, *Z. physik. Chem.* **B2,** 115 (1929).
17. Mooney, R. C. L., *J. Am. Chem. Soc.* **63,** 2828 (1941).
18. Müller, A., *Proc. Roy. Soc.* (*London*) **A120,** 437 (1928).
19. Natta, G., and P. Corradini, *J. Polymer Sci.* **20,** 251 (1956).
20. Natta, G., and P. Corradini, *J. Polymer Sci.* **39,** 29 (1959).
21. Pauling, L., and R. B. Corey, *Proc. Natl. Acad. Sci.* **37,** 729 (1951); **39,** 247, 253 (1953).
22. Pauling, L., R. B. Corey, and H. R. Branson, *Proc. Natl. Acad. Sci.* **37,** 205 (1951); L. Pauling and R. B. Corey, *Nature* **171,** 59 (1953).
23. Pierce, R. H. H., Jr., E. S. Clark, J. F. Whitney, and W. M. D. Bryant, paper presented at the 130th National Meeting of the American Chemical Society, Atlantic City, N.J., September, 1956.
24. Schellman, J. H., *J. Phys. Chem.* **62,** 1485 (1958).
25. Statton, W. O., *Ann. N. Y. Acad. Sci.* **83,** 27 (1959); C. R. Bohn, J. R. Schaefgen, and W. O. Statton, *J. Polymer Sci.* **55,** 531 (1961).
26. Stroupe, J. D., and R. E. Hughes, *J. Am. Chem. Soc.* **80,** 2341 (1958).
27. Zimm, B. H., and J. K. Bragg, *J. Chem. Phys.* **28,** 1246 (1958); **31,** 526 (1959).

C. Morphology of Polymer Single Crystals

Although the formation of single crystals of polymers was observed during polymerization many years ago (19, 22), it was long believed that such crystals could not be produced from polymer solutions because of molecular entanglement. In 1953 (20), however, and in several laboratories

in 1957 (5, 9, 23), the growth of such crystals was reported. The phenomenon has been reported for so many polymers, including gutta-percha (20), polyethylene (4, 5, 9, 23), polypropylene (16) and other poly(α-olefins) (6), polyoxymethylene (8), polyamides (7), and cellulose and its derivatives (12, 17), that it appears to be quite general and universal.

Lamellae

All the structures described as polymer single crystals have the same general appearance, being composed of thin, flat platelets about 100 A thick and often many microns in lateral dimensions. They are usually thickened by the spiral growth of additional lamellae from screw dislocations. A typical crystal of 6 nylon is shown in Fig. 5-17. The size, shape, and regularity of the crystals depend on their growth conditions, such factors as solvent, temperature, and growth rate being important (10). The thickness of the lamellae depends on crystallization temperature (10) and any subsequent annealing treatment (21).

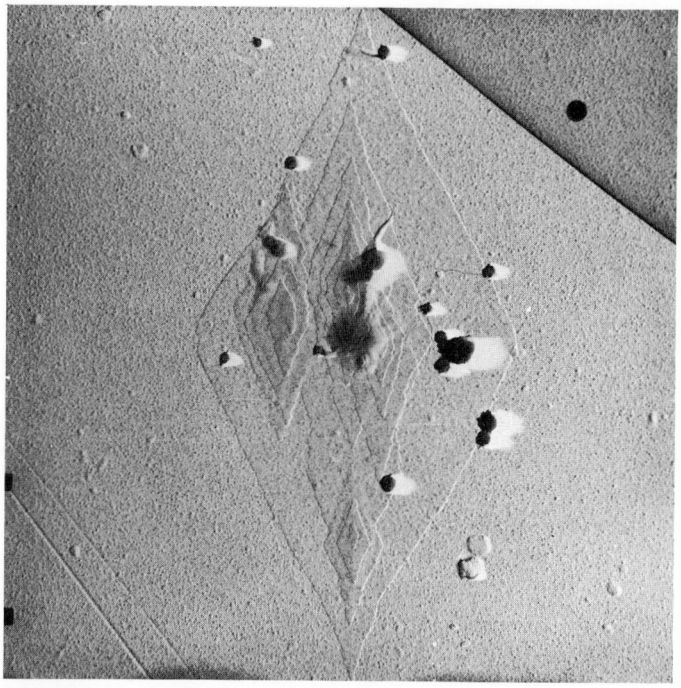

Fig. 5-17. Electron micrograph of a single crystal of 6 nylon grown by precipitation from dilute glycerine solution (7). The lamellae are about 60 A thick. Courtesy of P. H. Geil.

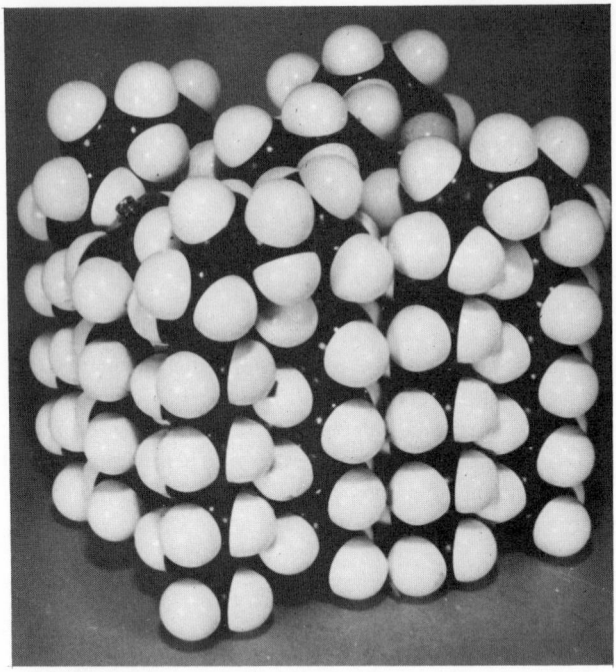

Fig. 5-18. Model of the conformation of polyethylene molecules in the folds at the surface of a single crystal (18). Courtesy of P. H. Geil.

Chain folding

Electron diffraction measurements, performed in almost all the researches cited above, indicate that the polymer chains are oriented normal or very nearly normal to the plane of the lamellae. Since the molecules in the polymers are at least 1000 A long and the lamellae are only about 100 A thick, the only plausible explanation is that the chains are folded (9). This arrangement has been shown to be sterically possible. In polyethylene, for example, the molecules can fold in such a way that only about five chain carbon atoms are involved in the fold itself (Fig. 5-18).

Pyramidal structure

Many single crystals of essentially linear polyethylene show secondary structural features, including corrugations (Fig. 5-19) and pleats (Fig. 5-20). Both these features result from the fact that many crystals of polyethylene,

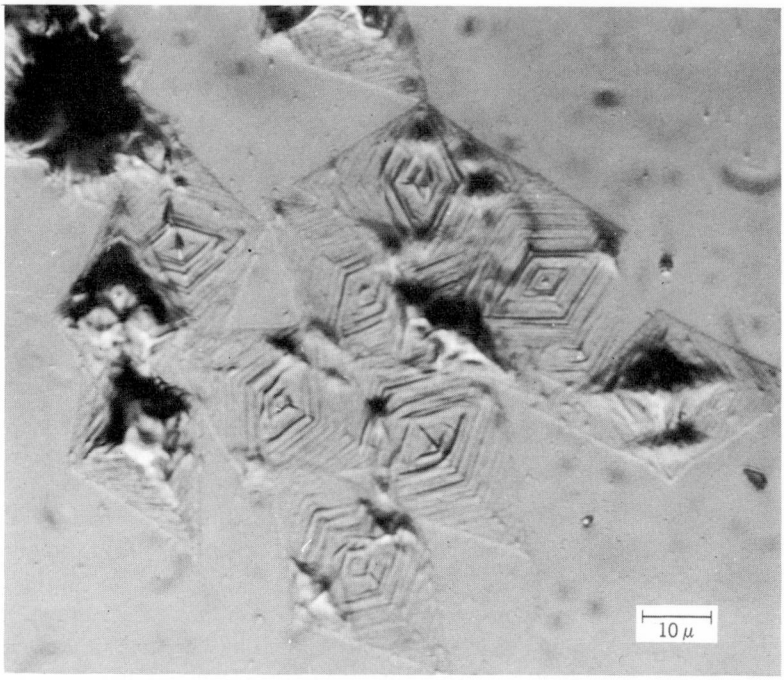

Fig. 5-19. Optical micrograph showing corrugations in a single crystal of linear polyethylene grown from a solution in perchlorethylene (18). Courtesy of P. H. Geil.

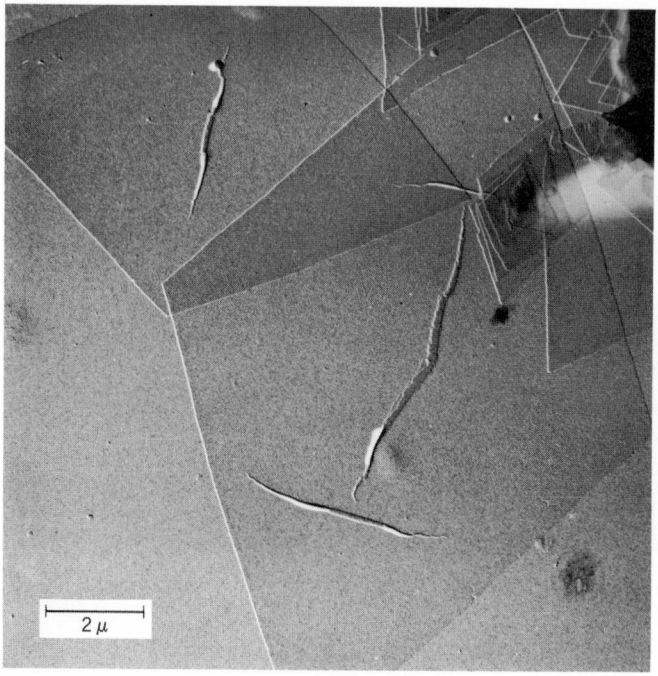

Fig. 5-20. Electron micrograph showing pleats in a crystal like that of Fig. 5-19 (18). Courtesy of P. H. Geil.

and perhaps other polymers too, grow in the form of hollow pyramids (2, 14, 15, 18). When solvent is removed during preparation of the crystals for microscopy, surface tension forces cause the pyramids to collapse.

The hollow pyramidal structure is related to the packing of the folded chains. If successive planes of folded molecules are displaced from their neighbors by an integral (or, with rotation of 180°, by a half-integral) number of repeat distances, a pyramidal structure with a slope of 32° (or 17°) results. [It should be mentioned that the orientation of the planes along which folds occur differs in the four quadrants of the crystal (1, 3), as suggested by the pattern of the corrugations in Fig. 5-19.]

More complex structures

The growth of crystal structures from polymer solutions can result in more complex structures sometimes reminiscent of features found in polymers crystallized from the melt (Section *D*). Typical of these are sheaflike arrays (Fig. 5-21) corresponding to the nuclei and initial growth habit of spherulites. Each ribbon in the array is composed of a number of lathlike lamellae with the usual 50–100 A thickness (7). Twinned crystals (11), dendritic growths (3), and clusters of hollow pyramids (13) have also been described.

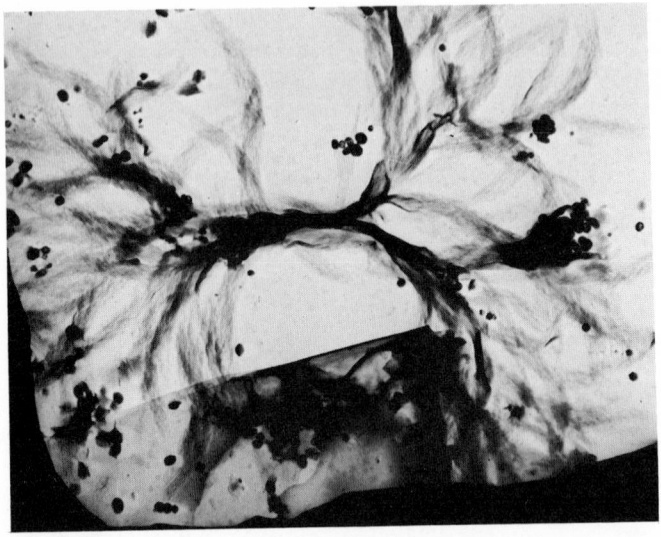

Fig. 5-21. Electron micrograph of spherulites of 6 nylon grown as in Fig. 5-17 (7). Courtesy of P. H. Geil.

GENERAL REFERENCE

Morgan, L. B., "The Fibre-Forming Properties of Polymers," pp. 233–277 in J. C. Robb and F. W. Peaker, eds., *op. cit.*

BIBLIOGRAPHY

1. Agar, A. W., F. C. Frank, and A. Keller, *Phil. Mag.* [8] **4,** 32 (1959).
2. Bassett, D. C., F. C. Frank, and A. Keller, *Nature* **184,** 810 (1959).
3. Burbank, R. D., *Bell Syst. Tech. J.* **39,** 1627 (1960); P. H. Geil and D. H. Reneker, *J. Polymer Sci.* **51,** 569 (1961).
4. Eppe, R., E. W. Fischer, and H. A. Stuart, *J. Polymer Sci.* **34,** 721 (1959).
5. Fischer, E. W., *Z. Naturforsch.* **12a,** 753 (1957).
6. Frank, F. C., A. Keller, and A. O'Conner, *Phil. Mag.* [8] **4,** 200 (1959).
7. Geil, P. H., *J. Polymer Sci.* **44,** 449 (1960).
8. Geil, P. H., Jr., N. K. J. Symons, and R. G. Scott, *J. Appl. Phys.* **30,** 1516 (1959).
9. Keller, A., *Phil. Mag.* [8] **2,** 1171 (1957).
10. Keller, A., "Morphology of Crystalline Polymers," pp. 499–532 in R. H. Doremus, B. W. Roberts, and D. Turnbull, eds., *Growth and Perfection of Crystals,* John Wiley & Sons, New York, 1958.
11. Khoury, F., and F. J. Padden, Jr., *J. Polymer Sci.* **47,** 455 (1960).
12. Manley, R. St. J., *J. Polymer Sci.* **47,** 509 (1960); *Nature* **189,** 390 (1961); *J. Polymer Sci.* **A1,** 1875 (1963).
13. Mitsuhashi, S., and A. Keller, *Polymer* **2,** 109 (1961).
14. Niegisch, W. D., *J. Polymer Sci.* **40,** 263 (1959).
15. Niegisch, W. D., and P. R. Swan, *J. Appl. Phys.* **31,** 1906 (1960).
16. Rånby, B. G., F. F. Morehead, and N. M. Walter, *J. Polymer Sci.* **44,** 349 (1960).
17. Rånby, B. G., and R. W. Noe, *J. Polymer Sci.* **51,** 337 (1961).
18. Reneker, D. H., and P. H. Geil, *J. Appl. Phys.* **31,** 1916 (1960).
19. Sauter, E., *Z. physik. Chem.* **B18,** 417 (1932); H. Staudinger, M. Staudinger, and E. Sauter, *ibid.* **B37,** 403 (1937).
20. Schlesinger, W., and H. M. Leeper, *J. Polymer Sci.* **11,** 203 (1953).
21. Statton, W. O., and P. H. Geil, *J. Appl. Polymer Sci.* **3,** 357 (1960).
22. Staudinger, H., and R. Signer, *Z. Krist.* **70,** 193 (1929).
23. Till, P. H., Jr., *J. Polymer Sci.* **24,** 301 (1957).

D. Structure of Polymers Crystallized from the Melt

The fringed micelle concept

Although x-ray diffraction studies show recognizable crystalline features in some high polymers, the Bragg reflections appear broad and diffuse compared to those from well-developed simple crystals. Diffraction theory

indicates that this broadening can arise from either small crystallite size or the presence of lattice defects (3), but the diffraction patterns from polymers are usually too weak to permit discrimination between these possibilities. Historically, the hypothesis of small crystallite size was selected as more probable.

With this assumption, rough estimates from the width of diffraction rings indicate that crystallite size rarely exceeds a few hundred angstroms. A substantial background of diffuse scattering suggests the coexistence of an appreciable amorphous fraction. It was suggested (13) that the polymer chains are precisely aligned over distances corresponding to the dimensions of the crystallites, but include more disordered segments which do not crystallize, and hence are included in amorphous regions. Since the chains are very long, they are visualized as contributing segments to several crystalline and amorphous regions, leading to a *composite single-phase* structure (2) known as the *fringed micelle* or *fringed crystallite* model (Fig. 5-22), the fringes representing transition material between the crystalline and amorphous phases.

The fringed micelle concept enjoyed popularity for many years, largely because it led to simple and appealing consequences such as the strong bonding of crystalline and amorphous regions into a composite structure of good mechanical properties, and the simple interpretation of degree of crystallinity in terms of the percentages of well-defined crystalline and amorphous regions. However, it tends to draw attention away from the details of fine structure and gives little insight into the structures of larger entities such as the spherulites described on pages 148–151.

Almost simultaneously with the discovery of polymer single crystals with a lamellar structure involving chain folding (Section *C*), it was found that spherulites are complex structures of lamellae (7, 9), and that individual lamellae in polymers crystallized from the melt are almost macroscopic in size (8). These features, incompatible with the interpretation of x-ray

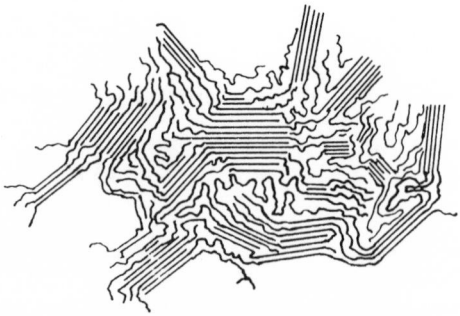

Fig. 5-22. Fringed micelle model of the crystalline-amorphous structure of polymers (2).

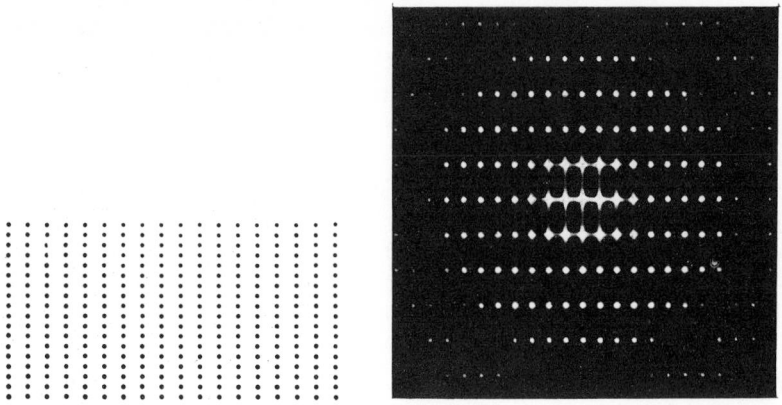

Fig. 5-23. Lattice model and optical diffraction pattern for a two-dimensional ideal crystal (15).

diffraction pattern broadening in terms of small crystallites, suggested that re-examination of the interpretation of the x-ray data was needed.

Paracrystallinity and crystal defects

In 1950, Hosemann (14) developed a theory in which lattice distortions and crystallite size are considered simultaneously in relation to x-ray line broadening. The imperfect crystallites resulting from various types of deformation are termed *paracrystals*. Two-dimensional analogs of the x-ray diffraction patterns from a perfect crystal and a paracrystal are shown in Figs. 5-23 and 5-24, respectively. The resemblance of the paracrystal pattern and others like it (15) to diffraction patterns from polymers is more than chance; there seems little doubt that the concept of paracrystallinity

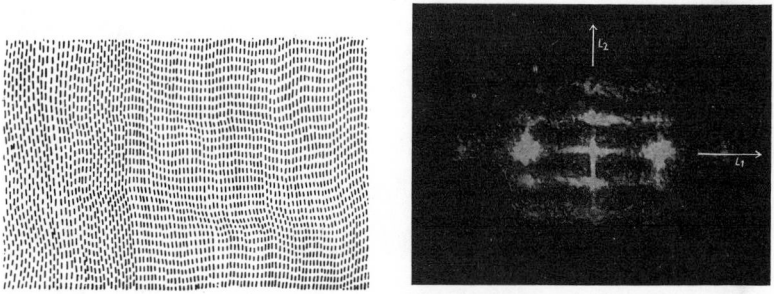

Fig. 5-24. Lattice model and optical diffraction pattern for a two-dimensional paracrystalline fiber (15).

will play an important part in explaining the structure of crystalline polymers, although the effect of crystallite size must not be ignored.

It is convenient and rewarding to consider the structure of crystalline polymers in terms of *lattice defects*, in analogy with metals. Some specific defects, such as a kink in the polymer chain resulting from the inclusion of an extra CH_2 (or, in polyoxymethylene, CH_2O) unit in the lattice, have been identified through NMR studies of oriented specimens (25). Such defects can move through the crystal from lattice point to lattice point, ultimately disappearing at its edge or being converted into other types of defect at folds or at chain ends, providing a mechanism for improving the perfection of crystals (see page 152).

Structure of spherulites and polymers crystallized from the melt

Microscopic features of spherulites The most prominent structural organization in polymers on a scale larger than lamellae is the *spherulite* (4), (ideally) a spherical aggregate ranging from submicroscopic in size to millimeters in diameter in extreme cases. Spherulites are recognized (1)

Fig. 5-25. Spherulites in a siliconelike polymer, observed in the optical microscope between crossed polarizers (23). The large and small spherulites were grown by crystallization at different temperatures. Courtesy of F. P. Price.

(Fig. 5-25) by their characteristic appearance in the polarizing microscope, where they are seen as (ideally) circular birefringent areas possessing a dark Maltese cross pattern. The birefringence effects are associated with molecular orientation resulting from the characteristic lamellar morphology now recognized in spherulites.

Relation of spherulites to crystallites Although there has not been general agreement on whether spherulites are formed by rearrangement of previously crystallized material or as products of primary crystallization, the latter interpretation is considered more probable (6, 19). It has been shown by examination of polymers which are partially spherulitic that in the usual case the spherulites are crystalline whereas the intermediate nonspherulitic material is amorphous. Spherulites thus appear to represent the crystalline portions of the sample, growing at the expense of the noncrystalline melt. The point of initiation of spherulite growth, its nucleus,

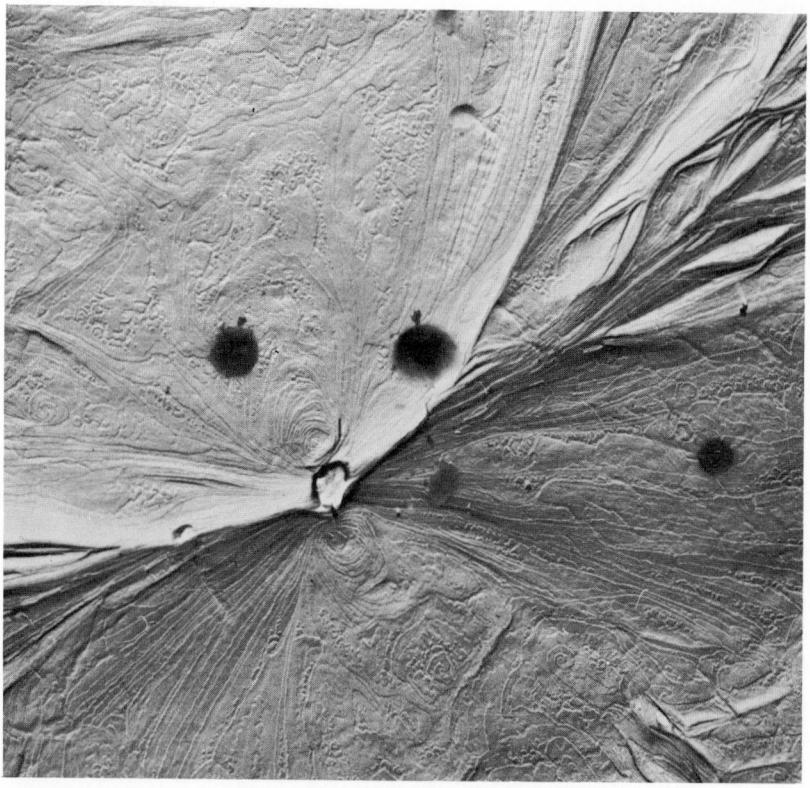

Fig. 5-26. Electron micrograph of the unrestrained surface of a portion of a spherulite grown from the melt of polyoxymethylene (12). Courtesy of P. H. Geil.

may be a foreign particle (heterogeneous nucleation) or may arise spontaneously in the melt (homogeneous nucleation).

Morphology of spherulites Electron microscopic evidence indicates that spherulites have a lamellar structure for almost all the polymers studied, including polyethylene (7, 9), polyoxymethylene (12), polycarbonates (16), and polychlorotrifluoroethylene (8), among others. The lamellar nature of the surface of a spherulite is shown in Fig. 5-26. In this instance, the spherulite was nucleated by a foreign particle seen at the lower left.

Crystallization spreads by the growth of individual lamellae. It has been suggested that growth begins both from screw dislocations at the nucleus and at spiral growths originating at crystal defects farther out in the spherulite. When two spherulites meet during crystallization, lamellae from both extend across the boundary into any uncrystallized material available. This results in interleaving, which holds the material together. Evidence that the lamellar structure persists throughout the body of spherulites, and that lamellae thus are basic structural elements of most solid polymers, is furnished by electron microscopic examination of fracture surfaces (12).

Ringed spherulites The morphology of spherulites in some polymers is complicated by the development of a ring structure (Fig. 5-27) (1, 17). In

Fig. 5-27. Ringed spherulites of poly(trimethylene glutarate) observed in the optical microscope between crossed polarizers (17). Courtesy of A. Keller.

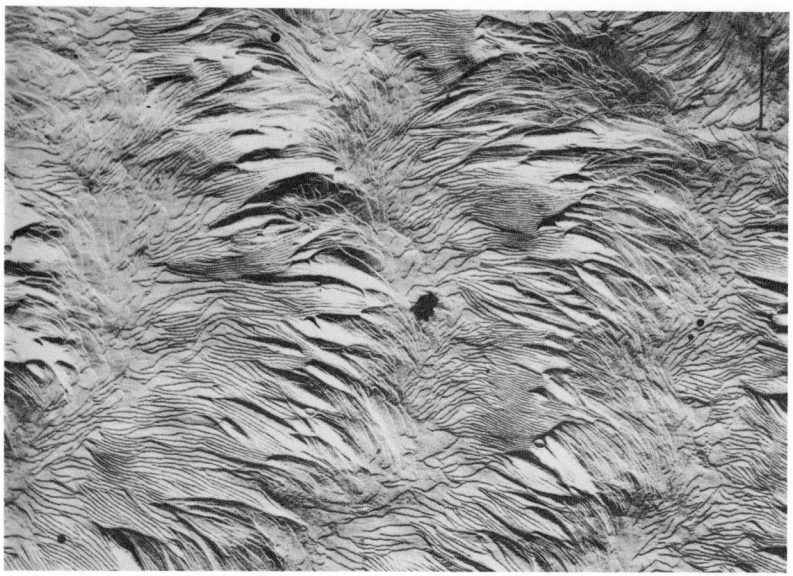

Fig. 5-28. Electron micrograph of a portion of a ringed spherulite in linear polyethylene (8).

the electron microscope it is seen (Fig. 5-28) that the ring structure is associated with a periodic twisting of the lamellae (10, 20); the cause of the twisting is unknown.

Other structural features of solid polymers Although evidence is accumulating that lamellae must be considered a basic structural element for many solid polymers, the possible existence of other features, such as fibrils (8, 18) or laminar or striated structures (5, 8) of as yet undetermined origin, must not be overlooked.

Chain folding and the crystalline-amorphous nature of polymers

Origins of chain folding The causes of chain folding and, in particular, of the regularity with which the fold period is maintained, are still obscure. Both thermodynamic (22) and kinetic (11, 20, 21, 24) theories of chain folding have been devised. In the thermodynamic theory, folds or other defects are shown to stabilize the crystals by their influence on torsional and vibrational modes of molecular motion. The kinetic theories relate the fold period to the stability of nuclei rather than fully grown crystals:

nuclei with too short fold periods are energetically unstable and those with too long periods are less likely to occur. The fold period which persists is that which forms and grows fastest. The theories disagree on the surface free energy of the crystals: values required for agreement with the experimental temperature dependence of the fold period (27) are 2–3 times higher for the kinetic than for the thermodynamic theories. It is possible, however, that kinetic considerations control initial crystallization, but the crystals then improve in perfection (through motion of defects) with a reduction in surface free energy as they approach the perfection required by the thermodynamic theory.

Origin of amorphous x-ray scattering Electron microscopic examination of highly crystalline polymers crystallized from the melt appears to reveal no evidence for the existence of the discrete amorphous regions expected on the basis of the fringed micelle concept. In addition, the interpretation of the broadening of x-ray diffraction patterns in terms of paracrystallinity appears to be gaining in favor. It is suggested (12, 26) that imperfections or defects within the crystal lattice, in which some atoms depart from positions associated with perfect arrangements, give rise to the amorphous halo present in x-ray diffraction patterns from highly crystalline polymers. Such imperfections include thermal displacements and those associated with folds, branch points, slight irregularities in the fold period, terminal groups, etc. Quantitative correlation has not yet been made between the numbers of atoms associated with such imperfections and the resulting x-ray patterns, but the order of magnitude of the amorphous scattering appears correct. For polymers of intermediate degree of crystallinity, a structure involving paracrystals plus discrete amorphous regions is considered probable; only for polymers of quite low crystallinity do structures resembling the classical fringed micelle appear appropriate. Although acceptance of the paracrystallinity concept is by no means universal, re-evaluation of many concepts of polymer properties involving the nature of both the crystalline and the amorphous regions seems indicated.

Molecular motion and chain folding The concept of the lamellar structure of polymers implies large-scale molecular motion in both the melt and the solid crystalline polymer. Either prefolding and some local orientation exist in the crystallizing polymer, or the molecules must undergo a considerable degree of motion during crystallization. Evidence also exists (27) that extensive melting and crystallization with an extremely rapid, several-fold increase in fold period occur during annealing of polymer single crystals and melt-crystallized lamellae; the extent to which the crystal morphology is disrupted, even at temperatures well below T_m, is illustrated in Fig. 5-29. The development of spherulites in polymers

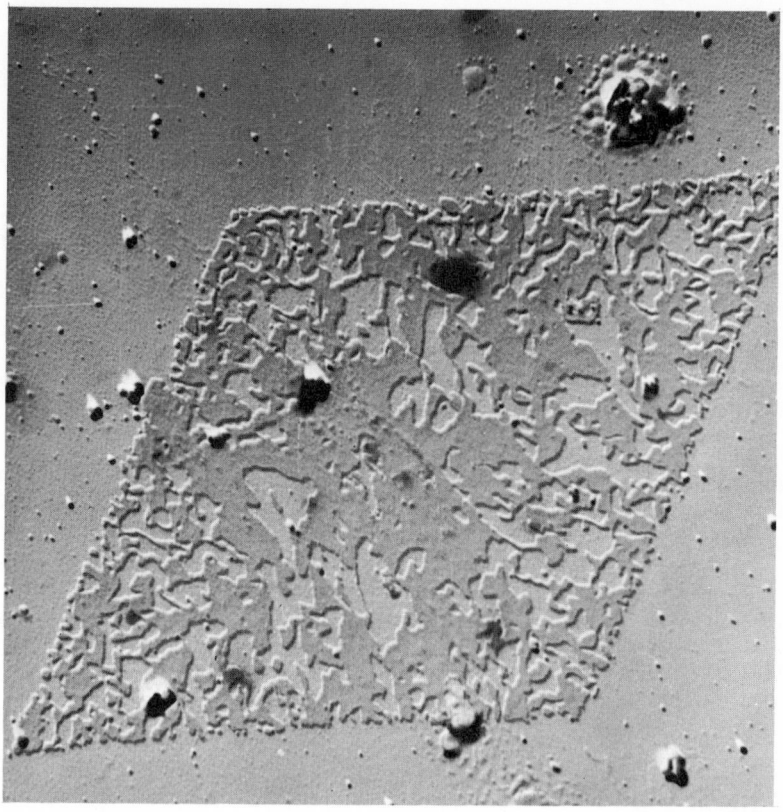

Fig. 5-29. A single crystal of linear polyethylene crystallized from perchlorethylene solution, then annealed for 30 min at 125°C, at least 10° below T_m (27). The fold period increased from about 100 A to almost 200 A during annealing. Compare with Fig. 5-20. Courtesy of P. H. Geil.

quenched from the melt and then held at temperatures between T_m and T_g is additional evidence of high chain mobility below T_m (28). The existence of mobile lattice defects (page 148) provides a plausible mechanism for this mobility in crystalline material.

GENERAL REFERENCES

Flory, P. J., *J. Am. Chem. Soc.* **84,** 2857 (1962).

Keller, A., "Morphology of Crystalline Polymers," pp. 499–532 in R. H. Doremus, B. W. Roberts, and D. Turnbull, eds., *Growth and Perfection of Crystals*, John Wiley & Sons, New York, 1958; see also A. Keller, *Makromol. Chem.* **34,** 1 (1959).

Morgan, L. B., *loc. cit.*

BIBLIOGRAPHY

1. Billmeyer, F. W., Jr., P. H. Geil, and K. R. van der Weg, *J. Chem. Educ.* **37,** 460 (1960).
2. Bryant, W. M. D., *J. Polymer Sci.* **2,** 547 (1947).
3. Bunn, C. W., *Chemical Crystallography*, Oxford University Press, London, 2nd ed., 1961, Chap. XI.
4. Bunn, C. W., and T. C. Alcock, *Trans. Faraday Soc.* **41,** 317 (1945).
5. Bunn, C. W., A. J. Cobbold, and R. P. Palmer, *J. Polymer Sci.* **28,** 365 (1958).
6. Burnett, B. B., and W. F. McDevit, *J. Appl. Phys.* **28,** 1101 (1957).
7. Claver, G. C., Jr., R. Buchdahl, and R. L. Miller, *J. Polymer Sci.* **20,** 202 (1956).
8. Eppe, R., E. W. Fischer, and H. A. Stuart, *J. Polymer Sci.* **34,** 721 (1959).
9. Fischer, E. W., *Z. Naturforsch.* **12a,** 753 (1957).
10. Fischer, E. W., *Kolloid-Z.* **159,** 108 (1958).
11. Frank, F. C., *Disc. Faraday Soc.* **25,** 208 (1958); F. C. Frank and M. Tosi, *Proc. Roy. Soc. (London)* **A263,** 323 (1961).
12. Geil, P. H., *J. Polymer Sci.* **47,** 65 (1960).
13. Herrmann, K., O. Gerngross, and W. Abitz, *Z. physik. Chem.* **B10,** 371 (1930).
14. Hosemann, R., *Z. Physik* **128,** 1, 465 (1930); R. Hosemann, R. Bonart, and G. Schoknecht, *ibid.* **146,** 588 (1956); R. Hosemann and S. N. Bagchi, *Direct Analysis of Diffraction by Matter*, North-Holland Publishing Co., Amsterdam, 1962.
15. Hosemann, R., and R. Bonart, *Kolloid-Z.* **152,** 53 (1957).
16. Kämpf, G., *Kolloid-Z.* **172,** 50 (1960).
17. Keller, A., *J. Polymer Sci.* **17,** 291 (1955); **39,** 151 (1959).
18. Keller, A., *J. Polymer Sci.* **36,** 361 (1959).
19. Keller, A., G. R. Lester, and L. B. Morgan, *Phil. Trans.* **A247,** 1 (1954); L. B. Morgan, *ibid.*, p. 13; F. D. Hartley, F. W. Lord, and L. B. Morgan, *ibid.*, p. 23.
20. Keller, A., and A. O'Conner, *Disc. Faraday Soc.* **25,** 114 (1958).
21. Lauritzen, J. I., Jr., and J. D. Hoffman, *J. Research Natl. Bur. Standards* **64a,** 73 (1960); *J. Chem. Phys.* **31,** 1680 (1959); J. D. Hoffman and J. I. Lauritzen, Jr., *J. Research Natl. Bur. Standards* **65a,** 297 (1961).
22. Peterlin, A., *J. Appl. Phys.* **31,** 1934 (1960); A. Peterlin and E. W. Fischer, *Z. Physik*, **159,** 272 (1962).
23. Price, F. P., p. 466 in R. H. Doremus, B. W. Roberts, and D. Turnbull, eds., *Growth and Perfection of Crystals*, John Wiley and Sons, New York, 1958.
24. Price, F. P., *J. Polymer Sci.* **42,** 49 (1960).
25. Reneker, D. H., *J. Polymer Sci.* **59,** S39 (1962); R. W. Kedzie, *Bull. Am. Phys. Soc.* II, **7,** 240 (1962) (abstract of paper).
26. Sella, C., and J.-J. Trillat, *Compt. rend.* **248,** 410 (1959).
27. Statton, W. O., and P. H. Geil, *J. Appl. Polymer Sci.* **3,** 357 (1960); W. O. Statton, *J. Appl. Phys.* **32,** 2332 (1961).
28. Stuart, H. A., paper presented at the XVIIIth Congress of Pure and Applied Chemistry, Montreal, Can., August, 1961.

E. Crystallization Kinetics

Experimental observations of the development of crystallinity in polymers are of two types. In one case (2; Wood, gen. ref.) a property varying

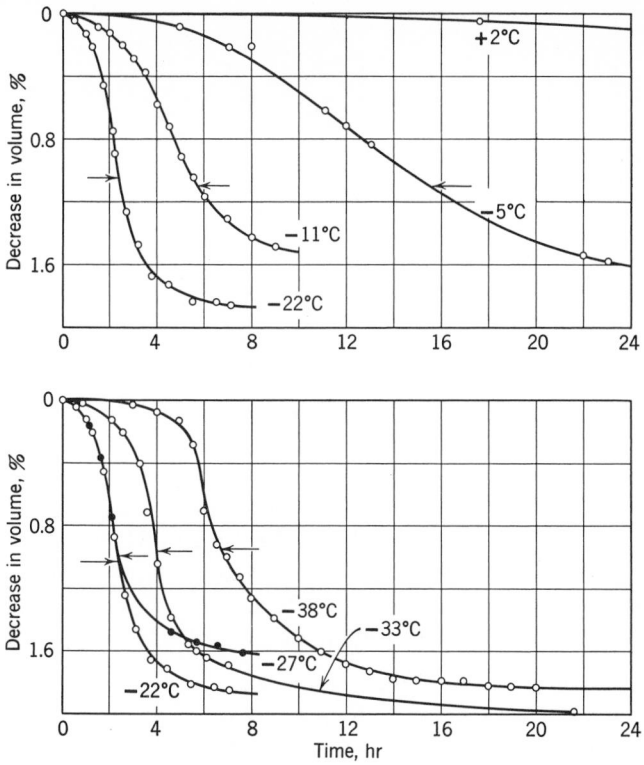

Fig. 5-30. Volume-time relations during the crystallization of natural rubber (Wood, gen. ref.). Arrows indicate points where crystallization is half complete.

with total amount of crystallinity, such as specific volume, is observed as a function of time at constant temperature. Alternatively (3, 5) the rate of formation and growth of spherulites is observed directly with the microscope. The two techniques lead to essentially the same results.

The development of crystallinity in polymers is not instantaneous. Curves of specific volume as a function of time at temperatures below the crystalline melting point (Fig. 5-30) show that crystallization sometimes cannot be considered complete for long periods. Since the time for complete crystallization is somewhat indefinite, it is customary to define the rate of crystallization at a given temperature as the inverse of the time needed to attain one-half the total volume change.

The rate so defined is a characteristic function of temperature (Fig. 5-31). As the temperature is lowered, the rate increases, goes through a maximum, and then decreases as the mobility of the molecules decreases and crystallization becomes diffusion controlled. At temperatures at which the rate is

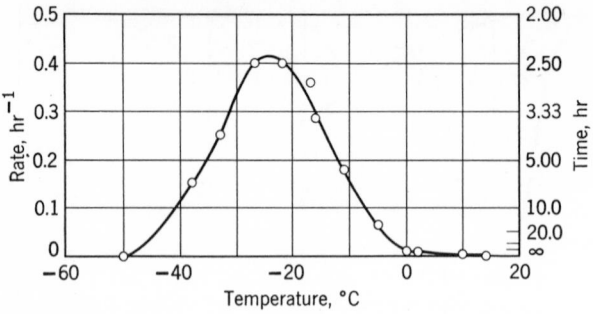

Fig. 5-31. Rate of crystallization of rubber as a function of temperature (Wood, gen. ref.).

very low, the polymer may be supercooled and maintained in the amorphous state.

If data for specific volume as a function of time during crystallization, such as those for natural rubber in Fig. 5-30, are plotted against log time, all the curves have the same shape and can be superposed by a shift along the time axis, as shown in Fig. 5-32. This behavior is also typical of the crystallization of low-molecular-weight substances, and curves of the type shown in Fig. 5-32 can be fitted with an equation due to Avrami (1):

$$\ln\left[(V_\infty - V_t)/(V_\infty - V_0)\right] = -(1/w_c)kt^n \qquad (5\text{-}1)$$

where V_∞, V_t, and V_0 are specific volumes at the times indicated in the subscripts, w_c is the weight fraction of material crystallized, k is a constant describing the rate of crystallization, and n is an exponent varying with the type of nucleation and growth process.

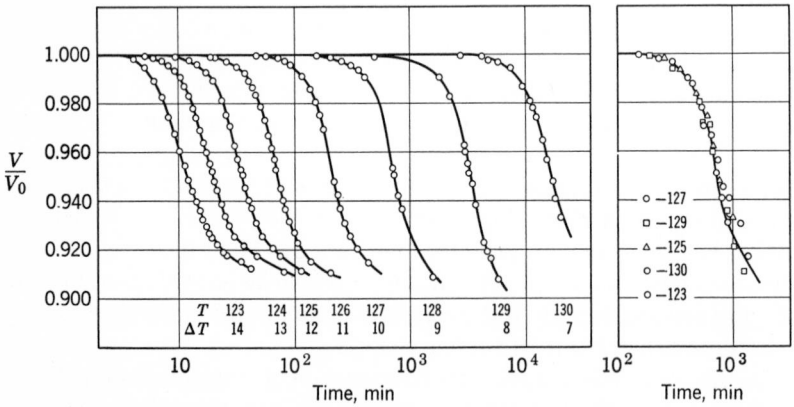

Fig. 5-32. Specific volume as a function of time in the crystallization of polyethylene (Mandelkern, gen. ref.). At right, the data for various temperatures are shifted to superimpose on those obtained at 128°C.

Although this analysis provides a satisfying analogy between crystallization in polymers and that in low-molecular-weight substances, quantitative aspects of the agreement with theory are not completely satisfactory. The shape of the curves, as in Fig. 5-32, is not very sensitive to values of the exponent n in the expected range of 2-4, and it is not always possible to determine n or indeed to ascertain whether n is constant throughout the crystallization process. Much of this difficulty can be resolved by appropriate choice of zero time in the Avrami equation, values of k and n being very sensitive to this parameter (7).

In addition, it is generally found that crystallization continues in polymer systems for much longer times than Eq. 5-1 predicts (4). This may result from the fact that crystallization in polymers involves the steps of (primary) nucleation and relatively rapid spherulite growth, followed by a slow, kinetically difficult improvement in spherulite order. "Amorphous" chains remaining between the spherulite lamellae after the first stage also crystallize in the slow second step (6).

GENERAL REFERENCES

Mandelkern, L., "Crystallization Kinetics in Polymeric Systems," pp. 467–497 in R. H. Doremus, B. W. Roberts, and D. Turnbull, eds., *op. cit.* See also L. Mandelkern, *Chem. Revs.* **56,** 903 (1956).

Morgan, L. B., *loc. cit.*

Wood, L. A., "Crystallization Phenomena in Natural and Synthetic Rubbers," pp. 57–95 in H. Mark and G. S. Whitby, eds., *op. cit.*

BIBLIOGRAPHY

1. Avrami, M., *J. Chem. Phys.* **7,** 1103 (1939); **8,** 212 (1940).
2. Bekkedahl, N., *J. Research Natl. Bur. Standards* **13,** 411 (1934).
3. Hartley, F. D., F. W. Lord, and L. B. Morgan, *Phil. Trans.* **A247,** 23 (1954).
4. Kovacs, A. J., *Ricerca sci.* (*Suppl. A*) **25,** 668 (1955).
5. Price, F. P., *J. Am. Chem. Soc.* **74,** 311 (1952).
6. Stuart, H. A., paper presented at the XVIIIth Congress of Pure and Applied Chemistry, Montreal, Can., August, 1961.
7. Ueberreiter, K., private communication.

F. Thermodynamics of Crystallization and Degree of Crystallinity

Determination of T_m

The disappearance of a polymer crystalline phase at the melting point is accompanied by changes in physical properties: the material becomes a

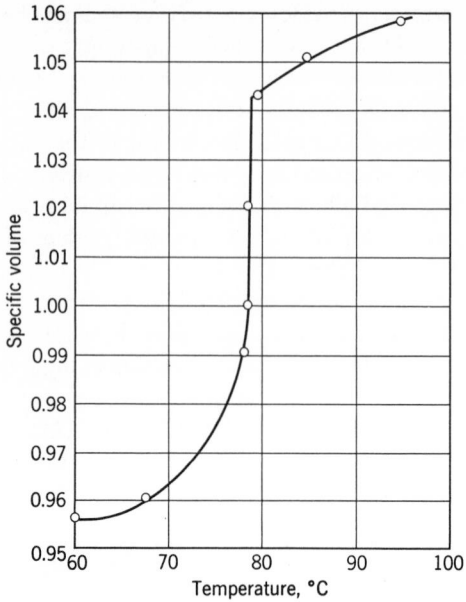

Fig. 5-33. Change of specific volume on crystalline melting in poly(decamethylene sebacate) (4).

(viscous) liquid, with discontinuous changes in density, refractive index, heat capacity, transparency, and other properties. Measurement of any of these properties may be used to detect the *crystalline melting point*,* T_m. Since melting occurs over a temperature range (see next page), the melting point determined by a particular method may depend to some extent on the sensitivity of the method. For example, crystalline melting points may be determined by noting the temperature of disappearance of the last traces of crystallinity as evidenced by birefringence observed between crossed polarizers on a hot-stage microscope (18). This melting point is usually several degrees higher than those obtained from methods depending on gross changes in properties such as specific volume and specific heat. Because of retention of order above T_m and the existence of a melting range in polymers, properties sensitive to the amount of crystalline material present undergo change over a relatively wide range in temperature, as evidenced by experimental data for specific volume (Fig. 5-33) or specific heat (Fig. 5-34).

Other methods which could be used to detect T_m are x-ray diffraction and infrared and NMR spectroscopy, each of these being suited to the detection

* Note that, while T_m is often reported in °C, its value in °K is required in equations such as those found in this section.

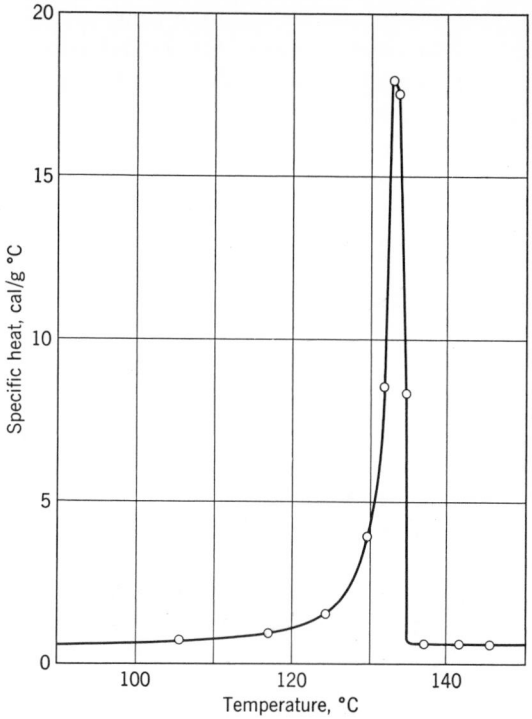

Fig. 5-34. Specific heat as a function of temperature over the crystalline melting range in linear polyethylene [Mandelkern, gen. ref., from (23)].

of crystalline material as described in Chapter 4. As the experiments are usually performed, they are not adapted to detecting T_m but rather are carried out at constant temperature. Differential thermal analysis, however, affords direct and sensitive detection of T_m (Chapter 4*E*).

Thermodynamics of crystalline melting

For a polymer homogeneous in molecular weight, a statistical thermodynamic analysis based on a lattice model yields the relation (5)

$$\frac{1}{T} - \frac{1}{T_m} = \frac{R}{\Delta H_m}\left(\frac{1}{xw_a} + \frac{1}{x - \zeta + 1}\right) \tag{5-2}$$

where w_a is the weight fraction of material that is amorphous at temperature T. The equilibrium melting point of the pure polymer at infinite chain length is $T_m = \Delta H_m/\Delta S_m$. The degree of polymerization is x, and ζ is a parameter characterizing the crystallite size. Equation 5-2 indicates that,

even for a polymer of uniform chain length, melting occurs over a finite temperature range. The physical reason for this behavior is the higher surface free energy of smaller crystallites. Equation 5-2 also shows that the temperature interval for fusion decreases with increasing molecular weight, and further analysis indicates that it broadens with increasing breadth of molecular weight distribution. On the other hand, as w_a approaches unity its derivative $\partial w_a/\partial T$ does not vanish; in consequence, the last traces of crystallinity disappear at a well-defined temperature, as confirmed by experiment.

Considerations of lamellar crystallites lead to similar results (Section *D*) in which the parameter ζ is replaced by the fold period: lamellae with longer fold periods have higher melting points. In addition, annealing of both single crystals and bulk polymers above the temperature at which they were crystallized leads to increases in the fold period, indicating a recrystallization process in which chain refolding occurs. The new crystalline phase which develops on continued annealing is more highly ordered and has a higher melting point which is clearly not the result of an increase in the degree of crystallinity (21).

For polymer-diluent systems, use of the Flory-Huggins expression for the free energy of mixing (Chapter 2*C*) leads to the expression

$$\frac{1}{T_m} - \frac{1}{T_m^0} = \frac{R}{\Delta H_m}\frac{V_2}{V_1}(v_1 - \chi_1 v_1^2) \tag{5-3}$$

where T_m and T_m^0 are the melting points of the polymer-diluent mixture and the pure polymer, respectively, V is molar volume and v volume fraction, and χ_1 is the polymer-diluent interaction parameter defined in Chapter 2*C*. Thus the melting point depression depends on the volume fraction of diluent added and its interaction with the polymer. For thermodynamically good solvents, large depressions (40–50°C) are sometimes observed in dilute solutions. Since the melting temperature is only slightly dependent on molecular weight, crystallization does not provide an efficient method of molecular weight fractionation.

Heats and entropies of fusion

Measurement of the heat and entropy of fusion yields important experimental data for assessing the effect of polymer structure on T_m, as discussed in Chapter 7*C*.

Heats of fusion are readily derived from several experimental sources (Dole, gen. ref.). An observed heat of fusion ΔH^* is calculable from specific heat measurements. This is not ΔH_m, the heat of fusion per mole of

repeating unit, but is related to it by the degree of crystallinity w_c: $\Delta H^* = w_c \Delta H_m$. It is ΔH_m which appears directly in Eq. 5-3, however, and in the related equation for the melting points of copolymers,

$$\frac{1}{T_m} - \frac{1}{T_m^0} = -\frac{R}{\Delta H_m} \ln n \tag{5-4}$$

where n is the mole fraction of crystallizing units. Thus ΔH_m can be determined from melting point data for these two types of systems even if w_c is not known.

In some cases ΔH_m can be inferred from the heats of fusion of low-molecular-weight homologs. It is found, for example, that the increment per CH_2 group in the heat of fusion of n-paraffins becomes constant for chain length greater than six carbon atoms (1, 3, 13). With certain assumptions (1), this increment may be taken as the heat of fusion of crystalline polyethylene. The method can be extended to the calculation of the lattice or sublimation energy at 0°K, a measure of the intermolecular forces in the crystal (1, 11, 22).

A comparison is made in Table 5-1 of the values of ΔH_m for polyethylene determined from specific heat measurements, the melting point depressions of polymer-diluent mixtures, and extrapolation of data on paraffins. Heats of fusion for several other polymers are included.

The total entropy change on fusion can be calculated from T_m and the heat of fusion, since $\Delta F_m = \Delta H_m - T\,\Delta S_m = 0$ at the melting point. The total change in entropy is made up of the change in entropy on fusion at constant volume $(\Delta S_m)_v$ plus the change in entropy due to expansion to the volume of the liquid phase. The latter can be evaluated by a simple thermodynamic relation to yield the equation

$$(\Delta S_m)_v = \Delta S_m - (V_L - V_S)\left(\frac{\partial P}{\partial T}\right)_{v,n} \tag{5-5}$$

TABLE 5-1. Heat and entropy of fusion of crystalline polymers

Polymer	ΔH_m, cal/g	Reference	ΔS_m, cal/g °K	$(\Delta S_m)_v$, cal/g °K	Reference
Polyethylene			0.162	0.126	19
From specific heat	66.2	23			
From diluents	67.0	15			
From extrapolation	65.9	1			
Polyoxymethylene	59.6	19	0.131	0.079	19
Polytetrafluoroethylene	13.7	19	0.023	0.015	19
Natural rubber	15.3	16	0.051	0.025	8
Gutta-percha	45.3	8	0.130	0.075	8

where V_S and V_L are the volumes of the solid and liquid phases, respectively. Values of $(\Delta S_m)_v$ so calculated, as well as of ΔS_m, are included in Table 5-1.

Degree of crystallinity

The determination of the degree of crystallinity of a polymer is affected by the conceptual difficulties discussed in Section *D*. The interpretation of measurements of crystallinity depends on the model used in designing the experiments: each of the methods to be described involves a simplifying assumption, usually concerned with the degree to which defects and disordered regions in paracrystalline material are weighted in the determination of the amorphous fraction. Fortunately, independent techniques lead to unexpectedly good experimental agreement, implying that differences in their resolution, specifically the difference in the lateral extent of lattice order detected, are relatively minor.

The three major methods of determining crystallinity, discussed in the following paragraphs, are based on specific volume, x-ray diffraction, and infrared spectroscopy. Polyethylene is used throughout as an example, although each method is applicable, at least in principle, to other polymers as well.

Another well-established method, based on measurement of heat content as a function of temperature through the fusion range (Dole, gen. ref.), is used less frequently because of the exacting and time-consuming nature of precision calorimetric measurements. Other methods are based on increased swelling (6) or increased chemical reactivity of the amorphous regions. Nuclear magnetic resonance spectroscopy (Chapter 4*D*) has been correlated with crystallinity, but this technique places more emphasis on molecular motion than directly on differences between crystalline and amorphous character.

Specific volume The crystallinity of a material is given in terms of the specific volumes of the specimen (V), the pure crystals (V_c), and the completely amorphous material (V_a), as

$$w_c = (V_a - V)/(V_a - V_c) \tag{5-6}$$

This relation assumes additivity of the specific volumes, but this requirement is fulfilled for polymers of homogeneous structure, as far as is known. It is implicit that the sample be free of voids. Since specific volume can be determined to 1 or 2 parts in 10^4, the method can attain high precision. Its accuracy depends on the uncertainty in V_a and V_c, which are known to about 1 part in 10^3 for polyethylene. The crystalline specific volume is

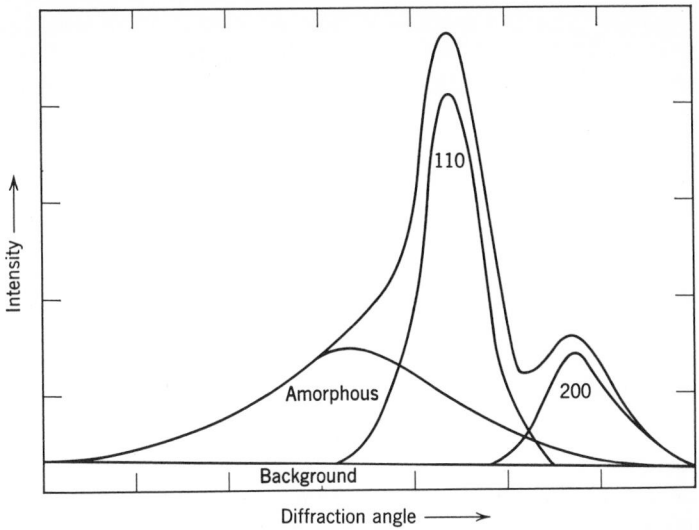

Fig. 5-35. Resolution of the x-ray scattering curve of a polyethylene into contributions from two crystalline peaks (indices 110 and 200), the amorphous peak, and background.

determined from x-ray unit cell dimensions, while V_a is usually obtained by extrapolation of the specific volumes of polymer melts (7).

X-ray diffraction (12) The x-ray method allows calculation of the relative amounts of crystalline and amorphous material in a sample if it is possible to resolve the contributions of the two types of structure to the x-ray diffraction pattern. A favorable situation is depicted in Fig. 5-35, where the scattering envelope of a polyethylene sample (outer line) has been resolved into contributions of two crystalline peaks (with indices 110 and 200) and a broad amorphous peak (2, 9, 14). The estimation of amount of crystallinity is usually based on comparison of the areas under the peaks, but other measures, such as peak height (20), may be used. With proper attention to experimental detail, this method provides one of the fundamental measures of crystallinity in polymers.

Infrared absorption Suitable infrared absorption bands can be used as primary measures of crystallinity in polymers. The application requires that the band bear a simple and unambiguous relationship to the crystalline or amorphous character of the polymer, and that absorption data can be obtained or inferred for the pure crystalline and amorphous polymers. These conditions are satisfied, e.g., for an infrared band at 7.67 μ in polyethylene (10) identified with a methylene wagging motion in the amorphous regions, and a band at 10.6 μ in 610 nylon (17) whose specific absorbance varies linearly with specific volume.

GENERAL REFERENCES

Dole, M., *Fortschr. Hochpolym.-Forsch.* (*Advances in Polymer Science*) **2,** 221 (1960)
Mandelkern, L., *Rubber Chem. Tech.* **32,** 1392 (1959); see also *Chem. Revs.* **56,** 903 (1956).

BIBLIOGRAPHY

1. Billmeyer, F. W., Jr., *J. Appl. Phys.* **28,** 1114 (1957).
2. Bryant, W. M. D., J. P. Tordella, and R. H. H. Pierce, Jr., paper presented at the 118th National Meeting of the American Chemical Society, Chicago, Ill., Sept. 3–8, 1950.
3. Dole, M., W. P. Hettinger, Jr., N. R. Larson, and J. A. Wethington, Jr., *J. Chem. Phys.* **20,** 781 (1952).
4. Evans, R. D., H. R. Mighton, and P. J. Flory, *J. Am. Chem. Soc.* **72,** 2018 (1950).
5. Flory, P. J., *J. Chem. Phys.* **17,** 223 (1949).
6. Kast, W., O. Kratky, G. Porod, and H. A. Stuart, pp. 247–253 in H. A. Stuart, ed., *Die Physik der Hochpolymeren*, Bd. III, Springer-Verlag, Berlin, 1955.
7. Krimm, S., and A. V. Tobolsky, *J. Polymer Sci.* **7,** 57 (1951); S. Krimm, *J. Phys. Chem.* **57,** 22 (1953).
8. Mandelkern, L., F. A. Quinn, Jr., and D. E. Roberts, *J. Am. Chem. Soc.* **78,** 926 (1956).
9. Matthews, J. L., H. S. Peiser, and R. B. Richards, *Acta Cryst.* **2,** 85 (1949).
10. Miller, R. G. J., and H. A. Willis, *J. Polymer Sci.* **19,** 485 (1956).
11. Müller, A., *Proc. Roy. Soc.* (*London*) **A154,** 624 (1936).
12. Nichols, J. B., *J. Appl. Phys.* **25,** 840 (1954).
13. Parks, G. S., and J. R. Mosely, *J. Chem. Phys.* **17,** 691 (1949).
14. Pierce, R. H. H., Jr., J. Holmes, F. C. Wilson, and W. M. D. Bryant, paper presented at the 132nd National Meeting of the American Chemical Society, New York, September, 1957.
15. Quinn, F. A., Jr., and L. Mandelkern, *J. Am. Chem. Soc.* **80,** 3178 (1958).
16. Roberts, D. E., and L. Mandelkern, *J. Am. Chem. Soc.* **77,** 781 (1955).
17. Sandeman, I., and A. Keller, *J. Polymer Sci.* **19,** 401 (1956).
18. Sperati, C. A., W. A. Franta, and H. W. Starkweather, Jr., *J. Am. Chem. Soc.* **75,** 6127 (1953).
19. Starkweather, H. W., Jr., and R. H. Boyd, *J. Phys. Chem.* **64,** 410 (1960).
20. Statton, W. O., *J. Appl. Polymer Sci.* **7,** 803 (1963).
21. Stuart, H. A., paper presented at the XVIIIth Congress of Pure and Applied Chemistry, Montreal, Can., August, 1961.
22. Ubbelohde, A. R., *Trans. Faraday Soc.* **34,** 282 (1938).
23. Wunderlich, B., and M. Dole, *J. Polymer Sci.* **24,** 201 (1957).

G. Orientation and Drawing

When a polymer mass is crystallized in the absence of external forces, there is no preferred direction in the specimen along which the polymer chains lie. If such an unoriented crystalline polymer is subjected to an

external stress, it undergoes a rearrangement of the crystalline material. Changes in the x-ray diffraction pattern suggest that the polymer chains align in the direction of the applied stress. At the same time the physical properties of the sample change markedly (see page 224).

Cold drawing

When the orientation process is carried out below T_m but above the glass transition temperature T_g, as when an unoriented fiber is stretched rapidly, the sample does not become gradually thinner but suddenly becomes thinner at one point, in a process known as "necking down" (Fig. 5-36). As the stretching is continued, the thin or drawn section increases in length at the expense of the undrawn portion of the sample. The diameters of the drawn and undrawn portions remain about the same throughout the

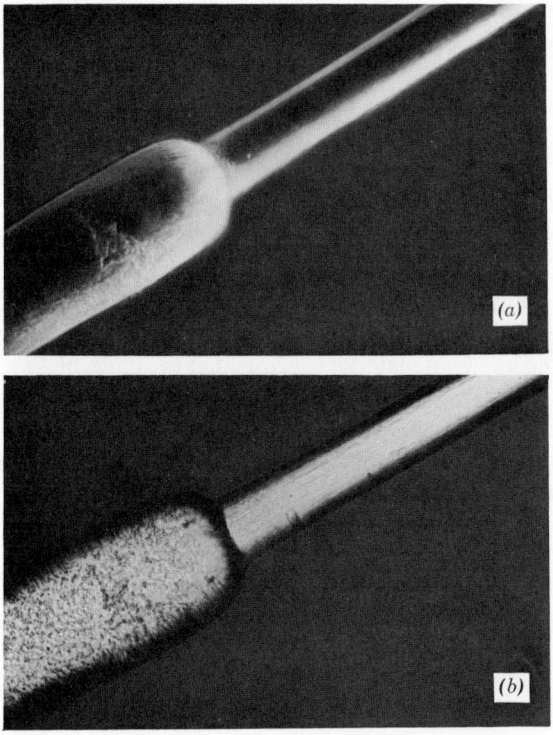

Fig. 5-36. (*a*) Cold drawn fiber of polyethylene, showing sections of drawn and undrawn material (Bunn, gen. ref.). (*b*) Same, observed between crossed polarizers, showing evidence of random orientation in the undrawn section and orientation parallel to the direction of stress in the drawn section. Courtesy of C. W. Bunn.

process. The *draw ratio*, or ratio of the length of the drawn fiber to that of the undrawn, is about 4 or 5 to 1 for a number of polymers, including branched polyethylene, polyesters, and polyamides, but is much higher (10 to 1 or more) in linear polyethylene.

In general, the degree of crystallinity in the specimen does not change greatly during drawing if crystallinity was previously well developed. If the undrawn polymer was amorphous or only partially crystallized, crystallinity is likely to increase during cold drawing.

Necking down during cold drawing is postulated (10) to occur because the process is carried out adiabatically. The energy expended in orientation appears as heat, causing a rise in temperature with a consequent reduction in viscosity at the point of draw. The neck forms as a result of the increased tendency to flow. An alternate explanation (17) holds that the polymer specimen is mechanically unstable at the point of draw because of its decreased modulus, and that heat considerations are not involved. The remarkably definite draw ratio and decrease in thickness for a given polymer have not been explained quantitatively.

Morphology changes during orientation

Observations in the electron microscope indicate that, when polymer single crystals are stressed, fibrils 50–100 A in diameter are drawn across the break (9) (Fig. 5-37). These fibrils must contain many molecules; smaller fibrils have not been resolved in the microscope. The fibrils appear to come from individual lamellae, and the sharp discontinuity between the drawn and the undrawn material suggests that the molecules are unfolding. In some cases the fibrils appear to have a periodicity of 100–400 A along their length; a similar period is observed by low-angle x-ray diffraction (14).

Examination of melt or solution cast specimens shows that the lamellae tend to rotate toward the draw direction (11) and on further elongation break up (12) into microlamellae and finally into submicroscopic units. Since the elongation at break in cold drawn specimens is less than could be obtained by complete unfolding, it has been suggested (12) that the smallest units are microcrystallites in which the chains remain folded. Periodic spacing of these crystallites within the fibrils would account for their observed periodicity.

Spherulites tend to remain intact during the first stages of drawing, often elongating to markedly ellipsoidal shapes (13). Rupture of the sample usually occurs at spherulite boundaries. In some cases it is accompanied by the production of fibrous material (8) (Fig. 5-38).

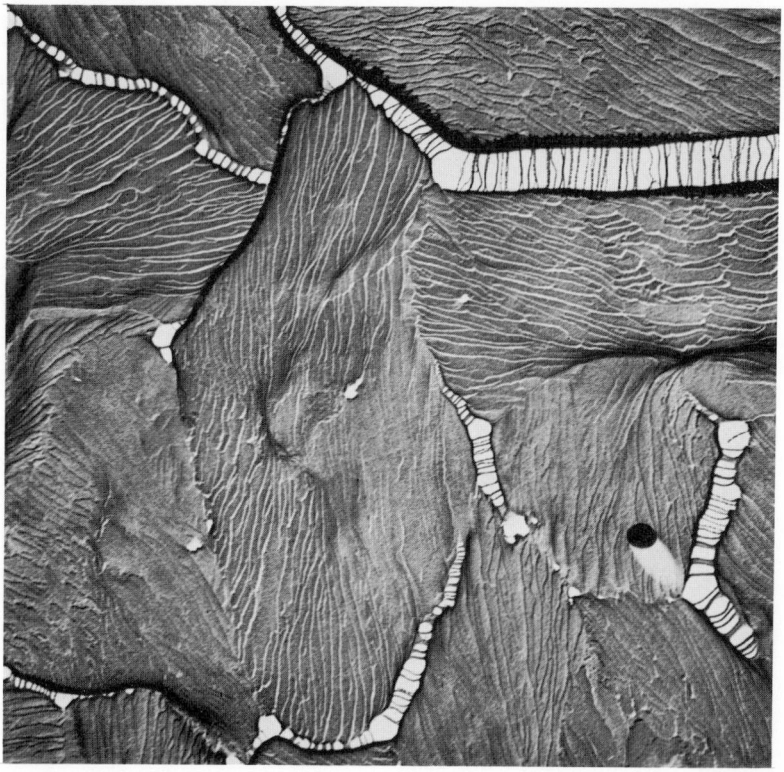

Fig. 5-37. Electron micrograph of a film of linear polyethylene crystallized by evaporation of the solvent (9). Fibrils drawn across the breaks are 50-100 A in diameter and appear to have a 100–300 A periodicity along their length. Courtesy of A. Keller.

The structure of fibers, whether formed by drawing or crystallized from an oriented melt, is still unsettled. X-ray evidence indicates that the molecules are aligned and shows that a long periodicity is present. Some sort of folded structure is regarded as possible, although the available evidence can also be explained (15) in terms of alternating crystalline-amorphous regions within elementary fibrils consisting of aggregates of several hundred molecules. A nonfolded paracrystalline structure has also been postulated (18) in which loops or isolated folds and chain ends form dislocations and lattice vacancies analogous to those found in metals.

Annealing of both drawn and undrawn material (Sections *D* and *F*) indicates a recrystallization process in which chain refolding takes place to give more ordered structures with higher melting points. In undrawn material, recrystallization is preceded by premelting (substantial loss of

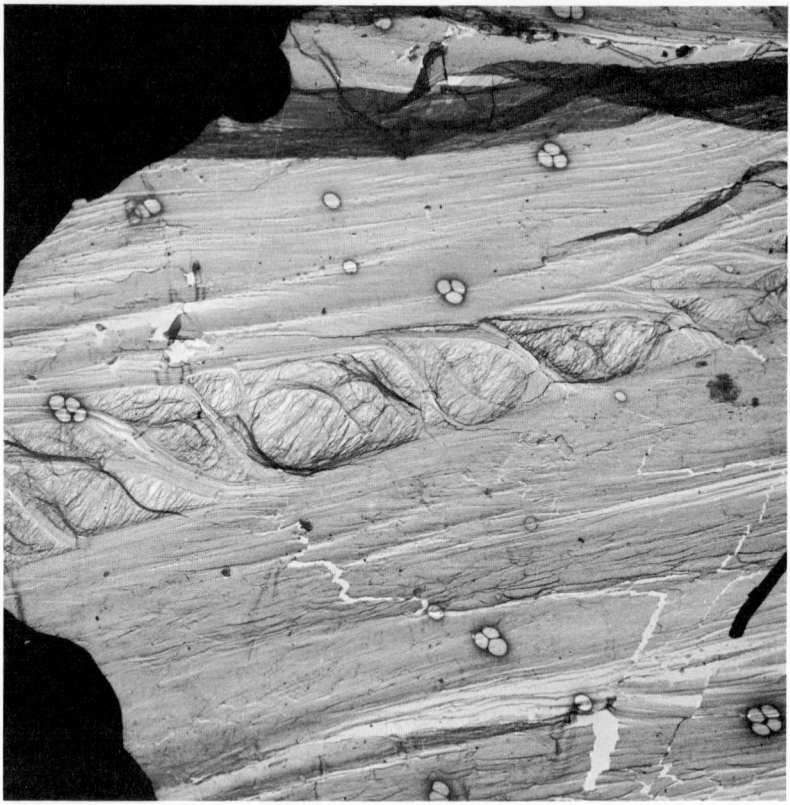

Fig. 5-38. Electron micrograph of a drawn polyethylene film, showing the production of fibrils during drawing. Courtesy of P. H. Geil.

crystalline order), whereas in drawn material premelting is not observed (16). Hence it is concluded that melting is not a prerequisite to refolding of chains.

Degree of orientation

As in the measurement of degree of crystallinity (Section *F*), the determination of degree of orientation is complicated both by the complexity of the types of orientation which can occur (5, 7) and by the conceptual difficulties involved in the current picture of the structure of an oriented specimen. The interpretation of the measurements depends on the model and simplifying assumptions in each method, but these differences are relatively minor, and the agreement among results by different methods is in general satisfactory.

X-ray diffraction (3) As indicated in Chapter 4*C*, x-ray diffraction patterns of unoriented polymers resemble small-molecule powder photographs, characterized by rings rather than diffraction spots. As the specimen is oriented, these rings break into arcs, and at high degrees of orientation approach the relatively sharp patterns characteristic of paracrystals (Fig. 5-39). From the distribution of scattered intensity in the

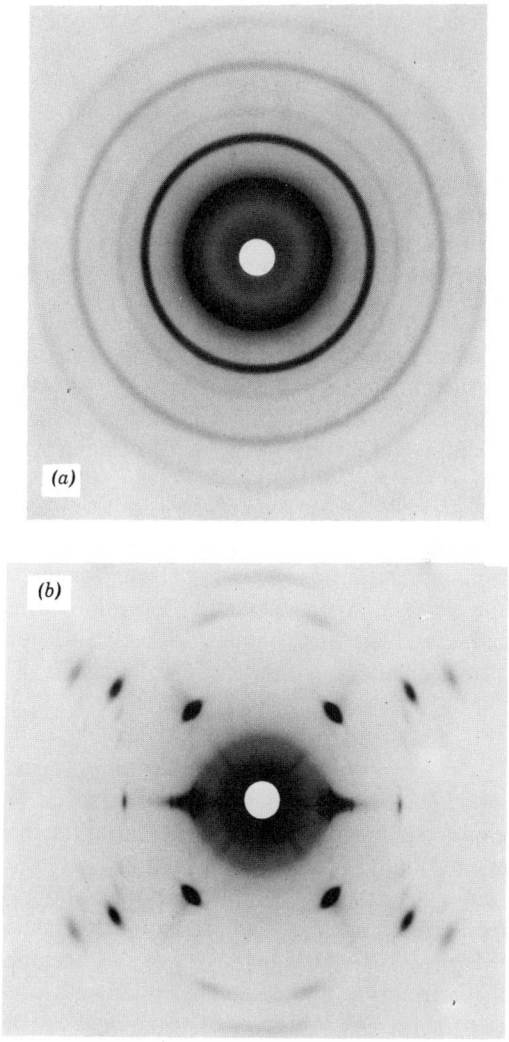

Fig. 5-39. X-ray diffraction patterns for unoriented (*a*) and oriented (*b*) polyoxymethylene. Courtesy of E. S. Clark.

arcs or spots, the degree of orientation can be measured in terms of distribution functions. These functions (Stein, gen. ref.) describe the amount of material occurring in crystalline regions having angular coordinates describing their orientation which fall within certain limits. For practical purposes, only average values of the functions are important.

Birefringence Birefringence, or change in index of refraction with direction, is evidenced by the ability of a material to rotate the plane of polarized light. The birefringence of a crystalline polymer is made up of contributions from the crystalline and amorphous regions plus a contribution, *form birefringence*, resulting from the shape of the crystals or the presence of voids. For completely unoriented material, the contributions of both the crystalline and the amorphous regions are zero. The increase in birefringence on orientation is due primarily to the crystalline regions and is proportional to the degree of crystallinity, the intrinsic birefringence of the crystals (which can be calculated), and an orientation factor which can be identified with that determined by the x-ray technique.

Infrared dichroism Infrared absorption is dependent upon change in dipole moment. Such a change is a vector quantity, i.e., one which is confined to definite directions for certain groups. In oriented samples the amount of absorption of plane polarized infrared radiation may vary markedly with change in direction of the plane of polarization. Thus the stretching vibrations of the C=O and N—H groups in polyamides involve changes in dipole moment which are almost exactly perpendicular to the chain axis; the corresponding absorption bands are weak for polarized radiation vibrating along the chain axis and strong for that vibrating perpendicular to the axis (1). When separate absorptions can be found for both the crystalline and the amorphous regions, the dichroism of these bands gives information about orientation in both regions of the specimen.

Orientation in noncrystalline polymers

The presence of crystalline material is not essential for the development of molecular orientation. Even polymers such as polystyrene, which do not crystallize, undergo considerable molecular orientation when subjected to stress at temperatures above T_g. This orientation (which is often seen in injection molded pieces) is stable below T_g and is accompanied by the development of birefringence and of enhanced strength properties in the direction of the applied stress. This type of orientation can be measured by birefringence, by a property change associated with the orientation, or by a change on subsequent heating above T_g. The force with which the

specimen tends to retract on heating above T_g is a suitable and readily measured property characterizing orientation in amorphous polymers (2, 4, 6).

GENERAL REFERENCES

Bunn, C. W., "Polymer Texture: Orientation of Molecules and Crystals in Polymer Specimens," Chap. 10-2 in R. Hill, ed., *op. cit.*

Morgan, L. B., *loc. cit.*

Stein, R. S., *The Optical and Mechanical Properties of High Polymers*, J. D. Loconti, N. S. Schneider, and H. W. Coles, eds., *Research Report No.* 14, High Polymer Series, U.S. Army Quartermaster Research and Engineering Center, Natick, Mass., 1960. Chaps. VIII, IX-3.

BIBLIOGRAPHY

1. Ambrose, E. J., A. Elliott, and R. B. Temple, *Proc. Roy. Soc.* (*London*) **A199,** 183 (1949).
2. Bailey, J., *India Rubber World* **118,** 225 (1948).
3. Bunn, C. W., "The Structure of Polythene," Chap. 5 in A. Renfrew and P. Morgan, eds., *Polythene*, 2nd ed., Interscience Publishers, New York, 1960.
4. Fortner, C. P., *SPE J.* **9,** #5, 21 (1953).
5. Gailey, J. A., *Anal. Chem.* **33,** 1831 (1961).
6. Hathaway, C. T., *SPE J.* **17,** 567 (1961).
7. Heffelfinger, C. J., and R. L. Burton, *J. Polymer Sci.* **47,** 289 (1960).
8. Kargin, V. A., *J. Polymer Sci.* **30,** 247 (1958).
9. Keller, A., *Phil. Mag.* [*8*] **2,** 1171 (1957).
10. Marshall, I., and A. B. Thompson, *Proc. Roy. Soc.* (*London*) **A221,** 541 (1954).
11. Peck, V., and W. Kaye, *J. Appl. Phys.* **25,** 1465 (1954) (abstract of paper).
12. Sella, C., and J.-J. Trillat, *Compt. rend.* **248,** 410 (1959).
13. Starkweather, H. W., Jr., G. E. Moore, J. E. Hansen, T. M. Roder, and R. E. Brooks, *J. Polymer Sci.* **21,** 189 (1956).
14. Statton, W. O., *J. Polymer Sci.* **28,** 423 (1958).
15. Statton, W. O., *J. Polymer Sci.* **41,** 143 (1959).
16. Stuart, H. A., paper presented at the XVIIIth Congress of Pure and Applied Chemistry, Montreal, Can., August, 1961.
17. Vincent, P. I., *Polymer* **1,** 7 (1960).
18. Zaukelies, D. A., reported in *Chem. Eng. News* **40,** #16, 48 (April 16, 1962).

6

Rheology and the Mechanical Properties of Polymers

Rheology is, by definition, the science of deformation and flow of matter. The rheological behavior of polymers involves several widely different phenomena, which can be related to some extent to different molecular mechanisms. These phenomena and their associated major mechanisms are as follows:

a. Viscous flow, the irreversible bulk deformation of polymeric material, associated with irreversible slippage of molecular chains past one another.

b. Rubberlike elasticity, where the local freedom of motion associated with small-scale movement of chain segments is retained, but large-scale movement (flow) is prevented by the restraint of a diffuse network structure.

c. Viscoelasticity, where the deformation of the polymer specimen is reversible but time dependent, and is associated (as in rubber elasticity) with the distortion of polymer chains from their equilibrium conformations through activated segment motion involving rotation about chemical bonds.

d. Hookean elasticity, where the motion of chain segments is drastically restricted, and probably involves only bond stretching and bond angle deformation: the material behaves like a glass.

These four phenomena are discussed in Sections *A–D*, respectively. Together they form the basis for a description of the mechanical properties of amorphous polymers. The mechanical properties of semicrystalline polymers, however, depend intimately on the restraining nature of their crystalline regions and can be inferred only in part from the rheological behavior of amorphous polymers. The mechanical properties of crystalline polymers are therefore discussed separately in Section *E*.

A. Viscous Flow

Phenomena of viscous flow

If a force per unit area s causes a layer of liquid at a distance x from a fixed boundary wall to move with a velocity v, the viscosity η is defined as the ratio between the shear stress s and the velocity gradient $\partial v/\partial x$ or rate of shear $\dot{\gamma}$:

$$s = \eta \frac{\partial v}{\partial x} = \eta \dot{\gamma} \tag{6-1}$$

If η is independent of the rate of shear, the liquid is said to be *Newtonian* or to exhibit ideal flow behavior (Fig. 6-1*a*). Two types of deviation from Newtonian flow are commonly observed in polymer solutions and melts. One is *shear thinning*, a reversible decrease in viscosity with increasing shear rate (Fig. 6-1*b*). Shear thinning results from the tendency of the applied force to disturb the long chains from their favored equilibrium conformation (Chapter 2*B*), causing elongation in the direction of shear. An opposite effect, *shear thickening*, in which viscosity increases with increasing shear rate, is rarely observed in polymers.

A second deviation from Newtonian flow is the exhibition of a *yield value*, a critical stress below which no flow occurs. Above the yield value, flow may be either Newtonian (as indicated in Fig. 6-1*c*) or non-Newtonian. For most polymer melts, only an apparent yield value is observed.

For low-molecular-weight liquids, the temperature dependence of viscosity is found to follow the simple exponential relationship

$$\eta = A e^{-(E/RT)} \tag{6-2}$$

where E is an *activation energy for viscous flow*, and A is a constant.

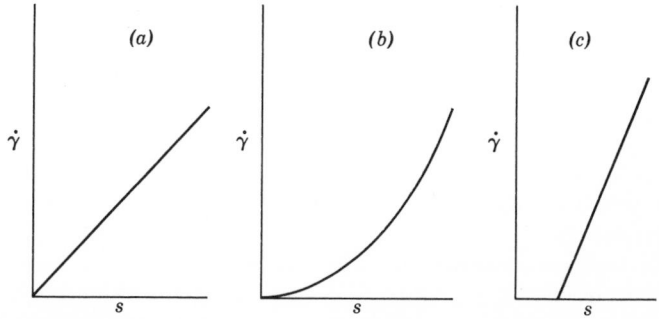

Fig. 6-1. Dependence of shear rate $\dot{\gamma}$ on shearing force s for three types of liquid viscosity: (*a*) Newtonian, (*b*) non-Newtonian, (*c*) yield stress followed by Newtonian flow.

These features of the flow of liquids can be explained in terms of several molecular theories. That of Eyring (8) is based upon a lattice structure for the liquid, containing some unoccupied sites or holes. These sites move at random throughout the liquid as they are filled and created anew by molecules jumping from one site to another. Under an applied stress the probability of such jumps is higher in the direction which relieves the stress. If each jump is made by overcoming an energy barrier of height E, the theory leads to Eq. 6-2. The energy of activation E is expected to be related to the latent heat of vaporization of the liquid, since the removal of a molecule from the surroundings of its neighbors forms a part of both processes. Such a relation is indeed found and is taken as evidence that the particle which moves from site to site is probably a single molecule.

As molecular weight is increased in a homologous series of liquids up to the polymer range, the activation energy of flow E does not increase proportionally with the heat of vaporization but levels off at a value independent of molecular weight. This is taken to mean that in long chains the unit of flow is considerably smaller than the complete molecule. It is rather a segment of the molecule whose size is of the order of 5–50 carbon atoms. Viscous flow takes place by successive jumps of segments (with of course some degree of coordination) until the whole chain has shifted.

Flow measurement

Methods commonly used for measuring the viscosity of polymer solutions and melts are listed in Table 6-1 (see Fox *et al.*, and Oka, gen. refs.). The most important of these methods involve rotational and capillary devices.

Rotational viscometry Rotational viscometers are available with several different geometries, including concentric cylinders, two cones of

TABLE 6-1. Summary of methods for measuring viscosity

Method	Approximate Useful Viscosity Range, Poises
Capillary pipette	10^{-2}–10^{3}
Falling sphere	1–10^{5}
Capillary extrusion	1–10^{8}
Parallel plate	10^{4}–10^{9}
Falling coaxial cylinder	10^{5}–10^{11}
Stress relaxation	10^{3}–10^{10}
Rotating cylinder	1–10^{12}
Tensile creep	10^{5}–$>10^{12}$

different angles, a cone and a plate, or combinations of these. Measurements with rotational devices become difficult to interpret at very high shear stresses, owing to the generation of heat in the specimen because of dissipation of energy, and to the tendency of the specimen to migrate out of the region of high shear.

A simple rotational instrument used in the rubber industry is the *Mooney viscosimeter* (10). This empirical instrument measures the torque required to revolve a rotor at constant speed in a sample of the polymer at constant temperature. It is used to study changes in the flow characteristics of rubber during milling or mastication (Chapter 22).

Capillary viscometry Capillary rheometers, usually made of metal and operated either by dead weight or by gas pressure, or at constant displacement rate, have advantages of good precision, ruggedness, and ease of operation. They may be built to cover the range of shear stresses found in commercial fabrication operations. However, they have the disadvantage that the shear stress in the capillary varies from zero at the center to a maximum at the wall.

An elementary capillary rheometer [extrusion plastometer (15)] is used to determine the flow rate of polyethylene in terms of *melt index*, defined as the mass rate of flow of polymer through a specified capillary under controlled conditions of temperature and pressure.

Experimental results

Molecular weight and shear dependence As discussed further in Chapter 7, the most important structural variable determining the flow properties of polymers is molecular weight or, alternatively, chain length, Z (the number of atoms in the chain). Although early data (6) suggested that $\log \eta$ was proportional to $Z^{1/2}$, it has now been observed (7; Fox *et al.*, gen. ref.) for essentially all polymers studied that, for values of Z above a critical value Z_c,

$$\log \eta = 3.4 \log \bar{Z}_w + k \tag{6-3}$$

where k is temperature dependent. This equation is valid only for shear stress sufficiently low (10^2–10^3 dynes/cm^2) that the viscosity is Newtonian.

The weight-average chain length $\bar{Z}_w$ is usually assumed to be the appropriate average for the above conditions. Studies of the Newtonian melt viscosity of polyethylene (4) show, however, that, for $Z > Z_c$,

$$\log \eta = 3.4 \log \bar{Z}_v + (3.4/a) \log w_2 + k \tag{6-3a}$$

where $\bar{Z}_v$ is the viscosity average, defined for the exponent a as in Chapter 3E, and w_2 is the weight fraction of polyethylene in mixtures of that polymer

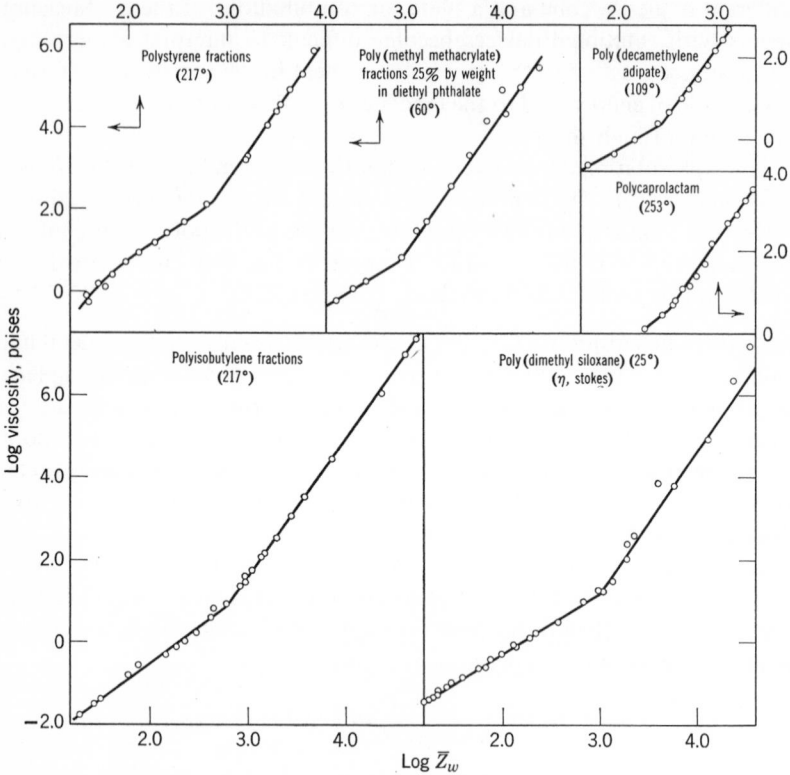

Fig. 6-2. Dependence of melt viscosity η on chain length $\bar{Z}_w$ for low shear rate, showing the regions below $Z_c \simeq 600$, where η is approximately proportional to $\bar{Z}_w^{1.75}$, and above Z_c, where $\eta \sim \bar{Z}_w^{3.4}$ (Fox *et al.*, gen. ref.).

with a low-molecular-weight homologous diluent, paraffin wax. Equation 6-3*a* holds for both linear and branched polyethylene, over the entire concentration range, and for molecular weights from that corresponding to Z_c up to 10^6. The exponent a ranges from about 0.25 for branched polymers up to about 0.8 for linear polyethylene, and is roughly identified with the exponent in the modified Staudinger equation (3-31) for the latter case.

For chain lengths below Z_c, which is about 600 for many polymers, the viscosity is found to depend upon a power of $\bar{Z}_w$ (and hence $\bar{M}_w$) in the range 1.75–2.0. In this range, shear rate has little effect on viscosity. Typical experimental data for the two regions described above are shown in Fig. 6-2.

At higher shear stress (10^4–10^7 dynes/cm^2), well within the non-Newtonian

region, it is now recognized (9, 14) that melt viscosity varies as a somewhat lower power of chain length, which is near 2.5 but varies with shear stress. The exponent appears to be independent of molecular weight above that corresponding to Z_c, on the basis of rather meager evidence.

Flow instabilities (16) At shear stresses in the neighborhood of 2×10^6 dynes/cm^2 for many polymers, instabilities in the flow appear, with the result that the upper Newtonian region (Chapter 3*E*) is rarely realized in bulk polymers. These instabilities are manifested as a striking and abrupt change in the shape of the polymer stream emerging from the capillary of the rheometer. At and above a critical stress, the shape of the emerging stream changes from that of a regular cylinder to a rough or distorted one.

Emergent streams of polymers extruded under unstable flow conditions have one of two rather different characteristic shapes: either a gross distortion or a fine-scale surface roughness without gross distortion. The gross distortion (sometimes called *melt fracture*) is illustrated in Fig. 6-3 for a poly(methyl methacrylate). The shape change is small at the critical stress and increases in magnitude with increasing stress. Qualitatively, the

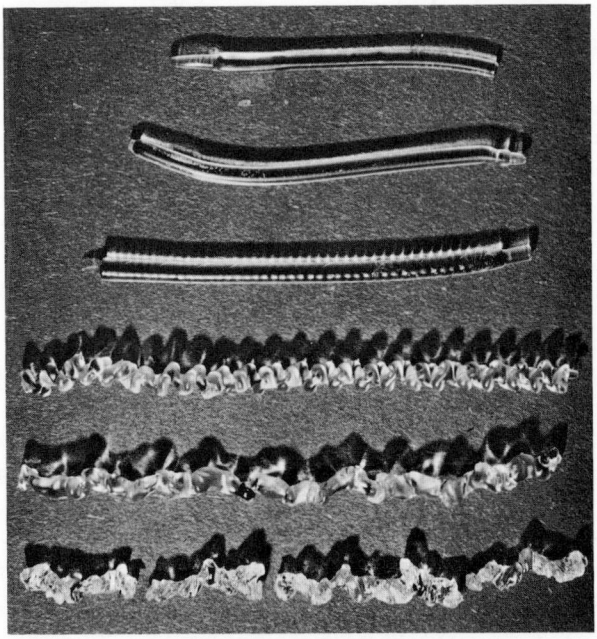

Fig. 6-3. Specimens of poly(methyl methacrylate) extruded at 170°C at various stresses, showing the effect of unstable flow on the shape of the emerging stream (16). Courtesy of J. P. Tordella.

"wavelength" of the disturbance is comparable in size to the diameter of the filament or stream. Associated with the change in shape is a slight decrease in effective viscosity, suggesting that some type of yielding occurs at the critical stress. This yielding results in unsteady axially unsymmetric flow at the inlet to the capillary, and appears to be elastic or viscoelastic in nature.

The second type of stream irregularity is much smaller than the first, appearing as a ripple on the surface of the polymer with a "wavelength" about $\frac{1}{10}$ to $\frac{1}{100}$ the diameter of the emerging stream. This irregularity is probably associated with unstable flow at the wall of the capillary, which appears to result in slip at the wall and plug flow.

Temperature dependence of viscosity On close examination for polymer systems, the activation energy for viscous flow of Eq. 6-2 is found to be constant only for small ranges of temperature. For pure liquids, it was found many years ago (1) that most of the change in viscosity with temperature is associated with the concurrent change in volume. This observation led to theories of viscosity based on the concept of free volume, whose application to polymers is discussed in Section *D*. A major result of these theories is the WLF equation (Eqs. 6-29 and 6-30), in which the temperature dependence of melt viscosity is expressed in terms of the glass transition temperature T_g (or another reference temperature) and universal constants. Since the terms describing the variation of melt viscosity with temperature and with molecular weight are independent, the WLF equation can be combined with Eq. 6-3 to yield the relation, for low shear rates,

$$\log \eta = 3.4 \log \bar{Z}_w - \frac{17.44(T - T_g)}{51.6 + T - T_g} + k' \tag{6-4}$$

where k' is a constant depending only on polymer type. This equation holds over the temperature range from T_g to about $T_g + 100°K$.

Theories of polymer flow

Several theories (3, 11, 12, 13) relate the steady-state viscosity of molten polymers, along with other viscoelastic parameters, to molecular structure through the distribution of relaxation times, as discussed in Section *C*. The dependence of melt viscosity on molecular weight for low shear rates was derived (2) by considering the motion of a polymer chain in a "solvent" composed of similar polymer molecules in an extension of theories (5) of dilute solution viscosity (Chapter 3*E*). As chain length increases, an abrupt

change in the molecular weight dependence occurs at the point where entanglement of the chains begins. The viscosity is predicted to vary with the 3.5 power of chain length above the critical value, in agreement with the experimentally observed exponent of 3.4. No theoretical treatment has been published for the high shear region, where melt viscosity is proportional to a lower power of Z.

GENERAL REFERENCES

Bueche, F., *Physical Properties of Polymers*, Interscience Publishers, a Division of John Wiley & Sons, New York, 1962, Chap. 3.

Fox, T. G., S. Gratch, and S. Loshaek, "Viscosity Relationships for Polymers in Bulk and in Concentrated Solution," Chap. 12 in F. R. Eirich, ed., *Rheology*, Academic Press, New York, Vol. 1, 1956.

Oka, S., "The Principles of Rheometry," Chap. 2 in F. R. Eirich, ed., *op. cit.*, Vol. 3, 1960.

Tanford, C., *Physical Chemistry of Macromolecules*, John Wiley & Sons, New York, 1961, Sec. 19.

BIBLIOGRAPHY

1. Batchinski, A. J., *Z. physik. Chem.* **84,** 643 (1913).
2. Bueche, F., *J. Chem. Phys.* **20,** 1959 (1952); **25,** 599 (1956).
3. Bueche, F., *J. Chem. Phys.* **22,** 603 (1954); F. Bueche and S. W. Harding, *J. Polymer Sci.* **32,** 177 (1958).
4. Busse, W. F., and R. Longworth, *J. Polymer Sci.* **58,** 49 (1962).
5. Debye, P., *J. Chem. Phys.* **14,** 636 (1946).
6. Flory, P. J., *J. Am. Chem. Soc.* **62,** 1057 (1940).
7. Fox, T. G., Jr., and P. J. Flory, *J. Am. Chem. Soc.* **70,** 2384 (1948); *J. Appl. Phys.* **21,** 581 (1950); *J. Polymer Sci.* **14,** 315 (1954); T. G. Fox and S. Loshaek, *J. Appl. Phys.* **26,** 1080 (1955).
8. Glasstone, S., K. J. Laidler, and H. Eyring, *The Theory of Rate Processes*, McGraw-Hill Book Co., New York, 1941, Chap. 9.
9. Merrill, E. W., Private communication.
10. Mooney, M., *Ind. Eng. Chem.* (*Anal. Ed.*) **6,** 147 (1934).
11. Pao, Y.-H., *J. Appl. Phys.* **28,** 591 (1957); see also R. H. Boyd, *ibid.* **29,** 953 (1958).
12. Ree, T., and H. Eyring, "The Relaxation Theory of Transport Phenomena," Chap. 2 in F. R. Eirich, ed., *op. cit.*, Vol. 2, 1958.
13. Rouse, P. E., *J. Chem. Phys.* **21,** 1272 (1953).
14. Rudd, J. F., *J. Polymer Sci.* **44,** 459 (1960).
15. Tordella, J. P., and R. E. Jolly, *Modern Plastics* **31,** #2, 146 (October 1953); see also American Society for Testing Materials, *ASTM Standards on Plastics*, Tentative Method of Test, ASTM Designation: D1238-57T.
16. Tordella, J. P., *J. Appl. Phys.* **27,** 454 (1956); *Trans. Soc. Rheology* **1,** 203 (1957); *Rheologica Acta* **2/3,** 216 (1961); *J. Appl. Polymer Sci.* **7,** 215, (1963).

B. Kinetic Theory of Rubber Elasticity

Rubberlike elasticity is in many respects a unique phenomenon, involving properties markedly different from those of low-molecular-weight solids, liquids, or gases. The properties of typical elastomers are defined by the following requirements:

a. They must stretch rapidly and considerably under tension, reaching high elongations (500–1000%) with low damping, i.e., little loss of energy as heat.

b. They must exhibit high tensile strength and high modulus (stiffness) when fully stretched.

c. They must retract rapidly, exhibiting the phenomenon of *snap* or *rebound.*

d. They must recover their original dimensions fully on the release of stress, exhibiting the phenomena of *resilience* and *low permanent set.*

Although the thermodynamics associated with rubber elasticity was developed in the middle of the nineteenth century, the molecular requirements for the exhibition of rubbery behavior were not recognized until 1932 (1). Theories of the mechanism relating these molecular structure requirements to the phenomena of rubber elasticity were developed soon after.

As discussed further in Chapter 7*F*, the molecular requirements of elastomers may be summarized as follows:

a. The material must be a high polymer.

b. It must be above its glass transition temperature T_g to obtain high local segment mobility.

c. It must be amorphous in its stable (unstressed) state for the same reason.

d. It must contain a network of crosslinks to restrain gross mobility of its chains.

Thermodynamics of rubber elasticity

A discussion of rubber elasticity in thermodynamic terms can be based on considerations of either free energy or work content. Since the constant-pressure experiment is more usual, free energy is used in the present treatment.

The ideal elastomer In discussions of the mechanical properties of a polymer, parameters related to its distortion must be included, as well as

the variables defining the state of the system, such as pressure and temperature. In particular, when a sample of rubber is stretched, work is done on it and its free energy is changed. By restricting the development to stretching in one direction, for simplicity, the elastic work W_{el} equals $f\,\mathrm{d}l$, where f is the retractive force and $\mathrm{d}l$ the change in length. If the change in free energy is F,

$$f = \left(\frac{\partial F}{\partial l}\right)_{T,p} = \left(\frac{\partial H}{\partial l}\right)_{T,p} - T\left(\frac{\partial S}{\partial l}\right)_{T,p} \tag{6-5}$$

In analogy with an ideal gas, where

$$\text{Ideal gas:} \qquad \left(\frac{\partial E}{\partial V}\right)_T = 0; \quad p = T\left(\frac{\partial S}{\partial V}\right)_T \tag{6-6}$$

an *ideal elastomer* is defined by the condition that

$$\text{Ideal elastomer:} \quad \left(\frac{\partial H}{\partial l}\right)_{T,p} = 0; \quad f = -T\left(\frac{\partial S}{\partial l}\right)_{T,p} \tag{6-7}$$

The negative sign results since work is done on the specimen to increase its length.

Entropy elasticity Equation 6-7 shows that the retractive force in an ideal elastomer is due to its decrease in entropy on extension. The molecular origin of this *entropy elasticity* is the distortion of the polymer chains from their most probable conformations in the unstretched sample. As described in Chapter 2*B*, the distribution of these conformations is Gaussian, the probability of finding a chain end in a unit volume of the space coordinates x, y, z at a distance r from the other end being

$$W(x, y, z) = (b/\sqrt{\pi})^3 \mathrm{e}^{-b^2r^2} \tag{6-8}$$

where $b^2 = \frac{3}{2}xl^2$, x being the number of links, with length l. Since the entropy of the system is proportional to the logarithm of the number of configurations it can have,

$$S = (\text{const.}) - kb^2r^2 \tag{6-9}$$

where k is Boltzmann's constant. It follows that, for a single polymer chain, the retractive force f' for an extension of magnitude $\mathrm{d}r$ is

$$f' = -T\frac{\mathrm{d}S}{\mathrm{d}r} = 2kTb^2r \tag{6-10}$$

It is customarily assumed that the retractive force f (Eq. 6-7) for a bulk polymer sample can be identified as the sum of the forces f' for all the chains in the specimen. This assumption, which is inaccurate in detail but justified in most cases (3), implies that individual chains contribute additively and

without interaction to the elasticity of the macroscopic sample, and that the distribution of end-to-end distances undergoes transformation identically with the sample dimensions (affine transformation).

Temperature coefficient of retractive force Equation 6-10 states that, unlike the Hookean restoring force of a spring, the retractive force of a rubber increases with increasing temperature. It was shown by Lord Kelvin in 1857 that

$$\left(\frac{\partial f}{\partial T}\right)_{p,l} = \frac{C_p}{T}\left(\frac{\partial T}{\partial l}\right)_{p,\,\text{adiabatic}} \tag{6-11}$$

where C_p is the specific heat. Thus, as is readily checked, rubbers heat on adiabatic stretching. It follows also that

$$\left(\frac{\partial T}{\partial f}\right)_{p,\,\text{adiabatic}} = -\frac{T}{C_p}\left(\frac{\partial l}{\partial T}\right)_{p,f} \tag{6-12}$$

which states that rubbers contract, rather than expand, on heating. These facts were demonstrated as long ago as 1806 (Flory, gen. ref.).

Deviations from ideal elastomeric behavior Since stretching a rubber may involve overcoming energy barriers in real polymer chains (4), the

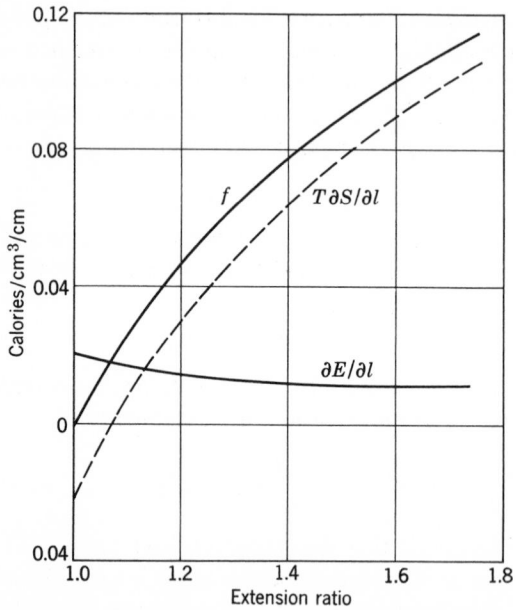

Fig. 6-4. Contributions of the energy and entropy terms to the retractive force in natural rubber [Treloar, gen. ref., after (5)].

internal energy of the sample may be expected to depend slightly on the elongation. Since

$$\left(\frac{\partial H}{\partial l}\right)_{T,p} = \left(\frac{\partial E}{\partial l}\right)_{T,p} + p\left(\frac{\partial V}{\partial l}\right)_{T,p} \tag{6-13}$$

and the second term on the right is negligible for normal pressures, this represents a departure from the ideal behavior defined in Eq. 6-7. As indicated in Fig. 6-4, the contribution of internal energy to the retractive force

$$f = \left(\frac{\partial E}{\partial l}\right)_{T,p} - T\left(\frac{\partial S}{\partial l}\right)_{T,p} \tag{6-14}$$

is small except for very low elongations (5). At much greater elongations, however, the energy term becomes very large negatively as crystallization takes place (8).

Stress-strain behavior of elastomers

The stress-strain curves of typical elastomers (Fig. 6-5) show marked deviations from the straight line required by Hooke's law. The relatively

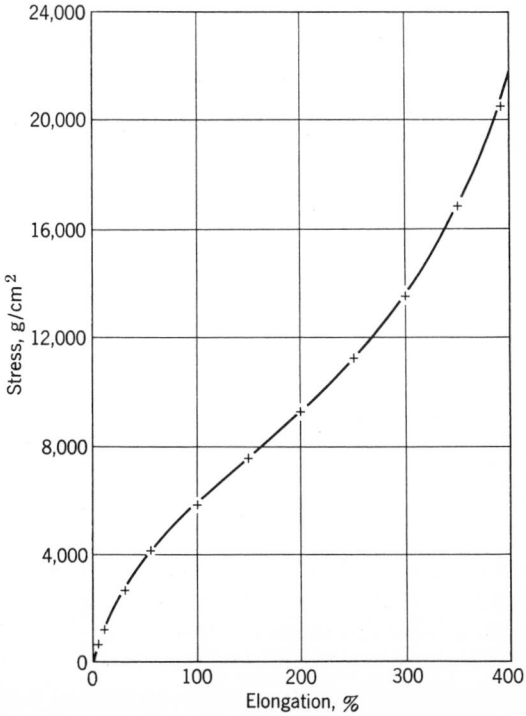

Fig. 6-5. Stress-strain curve for a typical elastomer (Guth *et al.*, gen. ref.).

low slope of the curve (defined near the origin as the modulus) decreases to about one-third its original value over the first hundred per cent elongation, and later increases, often to quite high values at high elongations. To explain this behavior, it is convenient first to examine the stress-strain properties of a simple model, and then to consider the relation of the model to actual elastomers.

A simple model of an elastomeric network (6; Guth, James, and Mark, gen. ref.) It is assumed that the actual tangled mass of polymer chains may be represented by an idealized network of flexible chains, irregular in detail but homogeneous and isotropic, extending throughout the sample. The network consists of m chains of average length $(\overline{r^2})^{1/2}$ per unit volume directed along each of three perpendicular axes. To obtain the proper space-filling properties of the model it is assumed that the space between the chains is filled with an incompressible fluid exerting a hydrostatic pressure p outward against the elastic tension of the chains.

If a cube of this material of unit dimensions is stretched by a force s along the z-axis, it will be converted into a parallelepiped of reduced dimensions α, β, and γ, where $\alpha = \beta$. Since it is incompressible, its volume is $\alpha\beta\gamma = 1$. The condition for equilibrium on a face parallel to the stress is obtained by equating the inward and outward forces. By the conventions of the model the outward force is $p\alpha\gamma$. The inward force is that developed by the m chains of the network seeking to find their equilibrium length and is given by Eq. 6-10. Setting the outward and inward forces equal yields

$$p\alpha\gamma = 2mkTb^2\beta \tag{6-15}$$

On an end face the stretching force must be considered also:

$$s + p\alpha\beta = 2mkTb^2\gamma \tag{6-16}$$

Rearranging and eliminating p,

$$s = 2mkTb^2(\gamma - 1/\gamma^2) \tag{6-17}$$

The modulus is given by

$$G = \mathrm{d}s/\mathrm{d}\gamma = 2mkTb^2(1 + 2/\gamma^3) \tag{6-18}$$

As the strain increases from zero ($\gamma = 1$) to large values of γ, the slope decreases to one-third its initial value as required by the experimental facts. Equation 6-18 predicts the stress-strain curve of actual elastomers very well up to elongations of 300% or more (Fig. 6-6).

It can be shown (Treloar, gen. ref.) that this oversimplified model yields results in accord with more general cases.

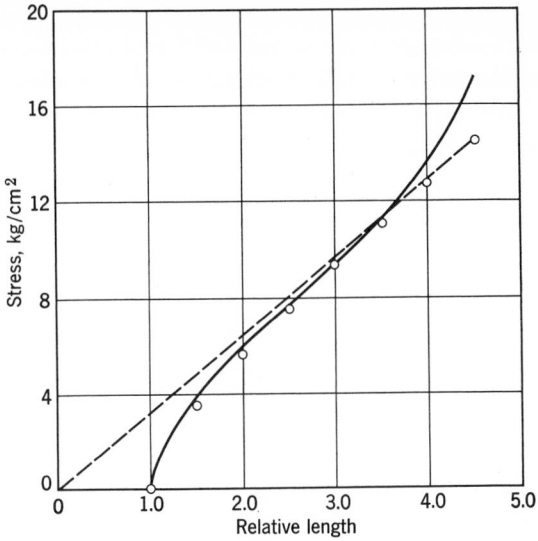

Fig. 6-6. Experimental and theoretical stress-strain curves (Guth *et al.*, gen. ref.). Solid line, experimental; circles, computed from Eq. 6-18; dashed line, asymptote of Eq. 6-18 for high elongations.

Behavior at high extensions Figure 6-6 suggests that, at elongations greater than about 300%, the stress-strain curves of actual elastomers have a higher slope than that predicted by Eq. 6-18. This is far below the ultimate elongation of 1000% for a good elastomer. One reason for the failure of the theory is that the Gaussian distribution of chain lengths which holds at low elongations fails in the region of interest for elastomers, as the actual lengths approach those of fully extended chains. A better approximation than the Gaussian is available (7). With its use the theory predicts the actual stress-strain curves rather well for elastomers which do not crystallize on stretching. In the cases where crystallinity does develop, the slope still increases faster than predicted. As is true in general for the problem of the mechanical behavior of crystalline polymers (Section *E*), a quantitative theory of the effect of crystallinity on the modulus of elastomers has not been developed.

Network structure and elasticity

In any real polymer network, all the chains cannot respond to external stress. Some of them are attached to other chains at one end only and contribute nothing to the elasticity. These are designated *terminal sections* in

contrast to the *principal sections* attached at both ends. Other types of flaws, in which one chain loops around another, can also be visualized. For a constant number of crosslinks per unit volume or mass of polymer, the number of terminal sections increases as the average molecular weight of the polymer decreases, and the number of chain ends correspondingly increases. At some critical molecular weight just equal to the average molecular weight of a primary segment, the network structure can no longer exist.

If the network is assumed to consist of N chains crosslinked together at $\nu/2$ points (i.e., with ν mers chemically linked together in $\nu/2$ pairs), there are $2N$ terminal segments and $\nu + N$ segments in all, hence $\nu - N$ principal segments. If the number of crosslink points were just $N - 1$ ($\simeq N$), all the chains would be joined together, but with no closed loops an elastic network would not exist. Each additional crosslink point, however, adds one loop useful in generating elasticity. Thus the number of crosslink points effective in producing elasticity (the number of loops) is $(\nu_e/2) = (\nu/2) - N$, defining the number of effective crosslinked mers as $\nu_e = \nu - 2N = \nu(1 - 2N/\nu)$.

It can be shown (Flory, gen. ref.) that the entropy of the network is given, in analogy with Eq. 6-9, by

$$S = (\text{const.}) - k(\nu_e/2)r^2 \tag{6-19}$$

If the volume of the network containing ν_e effective crosslinked mers is V, it follows from a derivation similar to that on page 184 that

$$s = \frac{kT\nu_e}{V}\left(\gamma - \frac{1}{\gamma^2}\right) = \frac{kT\nu}{V}\left(1 - \frac{2N}{\nu}\right)\left(\gamma - \frac{1}{\gamma^2}\right) \tag{6-20}$$

$$G = \frac{kT\nu_e}{V}\left(1 + \frac{2}{\gamma^3}\right) = \frac{kT\nu}{V}\left(1 - \frac{2N}{\nu}\right)\left(1 + \frac{2}{\gamma^3}\right) \tag{6-21}$$

The first term on the right shows that, for the usual case of $\nu \gg 2N$, the stress and modulus are proportional to the density of crosslinks ν/V. The second term introduces the dependence on molecular weight: at $\nu = 2N$, a critical molecular weight is reached below which (higher N) the network cannot exist. This conclusion suggested a method (2) for determining the critical molecular weight: a series of butyl rubber fractions was vulcanized to a constant density of crosslinks. It was found that network structures were obtained only for molecular weights above a limiting value. For higher molecular weights, G varied with M as predicted by Eq. 6-21 and shown in Fig. 6-7.

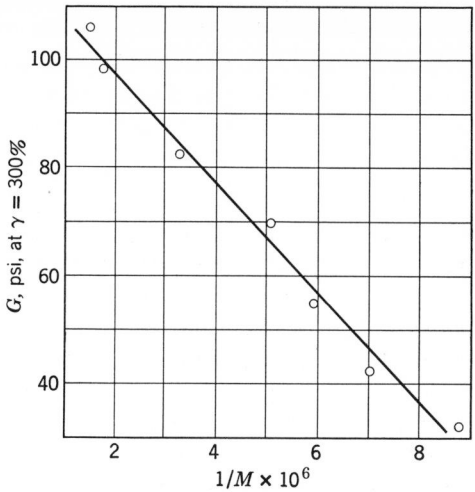

Fig. 6-7. Dependence of the modulus of vulcanized butyl rubber on its molecular weight before vulcanization (2).

GENERAL REFERENCES

Bueche, F., *op. cit.*, Chap. 2.

Ciferri, A., *J. Polymer Sci.* **54,** 149 (1961).

Flory, P. J., *Principles of Polymer Chemistry*, Cornell University Press, Ithaca, N.Y., 1953, Chap. XI.

Guth, E., H. M. James, and H. Mark, "The Kinetic Theory of Rubber Elasticity," pp. 253–299 in H. Mark and G. S. Whitby, eds., *Scientific Progress in the Field of Rubber and Synthetic Elastomers* (*Advances in Colloid Science*, Vol. II), Interscience Publishers, New York, 1946.

Stein, R. S., *The Optical and Mechanical Properties of High Polymers*, J. D. Loconti, N. S. Schneider, and H. W. Coles, eds., *Research Report No.* 14, High Polymer Series, U.S. Army Quartermaster Research and Engineering Center, Natick, Mass., 1960, Chap. II.

Treloar, L. R. G., *The Physics of Rubber Elasticity*, 2nd ed., Clarendon Press, Oxford, 1958.

BIBLIOGRAPHY

1. Busse, W. F., *J. Phys. Chem.* **36,** 2862 (1932).
2. Flory, P. J., *Chem. Revs.* **35,** 51 (1944); *Ind. Eng. Chem.* **38,** 417 (1946).
3. Flory, P. J., paper presented at the 139th National Meeting of the American Chemical Society, St. Louis, Mo., March, 1961.
4. Flory, P. J., C. A. J. Hoeve, and A. Ciferri, *J. Polymer Sci.* **34,** 337 (1959).
5. Gee, G., *Trans. Faraday Soc.* **42,** 585 (1946).

6. James, H. M., and E. Guth, *J. Chem. Phys.* **11,** 455 (1943).
7. Kuhn, W., and F. Grün, *Kolloid-Z.* **101,** 248 (1942).
8. Wood, L. A., and F. L. Roth, *J. Appl. Phys.* **15,** 781 (1944).

C. Viscoelasticity

Sections *A* and *B* have dealt with the equilibrium response of linear and network polymer structures to external stress. In this section are considered the time-dependent mechanical properties of amorphous polymers. As in studies of rate phenomena, theories of dynamic mechanical phenomena in polymers are less thoroughly developed than those referring to the equilibrium states. They are also more dependent on the details of models. The present problem is, first, to find suitable models or mathematical functions with which to describe the several types of molecular motion postulated on page 172 to be associated with the processes of elasticity, viscoelasticity, and viscous flow in polymers; and, second, to utilize these models in relating the behavior of the polymers to their molecular structure.

Experimental methods

Stress relaxation If elongation is stopped during the determination of the stress-strain curve of a polymer (Chapter 4*G*), the force or stress decreases with time as the specimen approaches equilibrium or quasi equilibrium under the imposed strain. The direct observation and measurement of this phenomenon constitutes the *stress relaxation* experiment. Usually the sample is deformed rapidly to a specified strain, and stress at this strain is observed for periods ranging from several minutes to several days or longer.

It has been shown (4) both experimentally and theoretically that the stress relaxation behavior of rubbery polymers can be factored into independent functions of strain and time. At small strains, the stress-strain function is almost linear and can be represented by a time-dependent modulus of elasticity $G(t)$. Typical stress relaxation data (8) for a well-studied sample of polyisobutylene are shown in Fig. 6-8, where $G(t)$ is plotted against time for experiments performed at several temperatures.

By means of time-temperature superposition (see page 196) these data may be shifted to produce a "master curve," as shown in Fig. 6-9. The curve indicates that any specified modulus can be observed after a period that depends, through the factor a_T, on temperature. As temperature is increased, a given modulus is observed at shorter times.

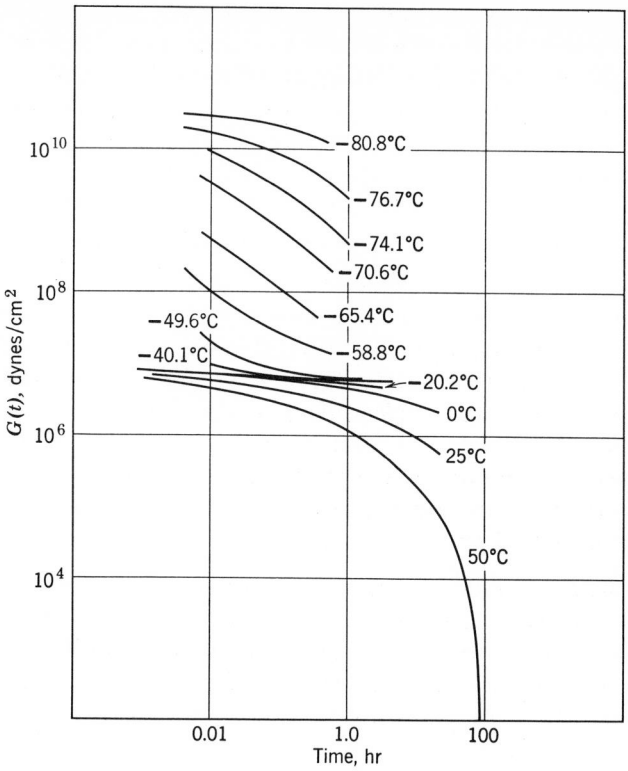

Fig. 6-8. Stress relaxation of polyisobutylene (8).

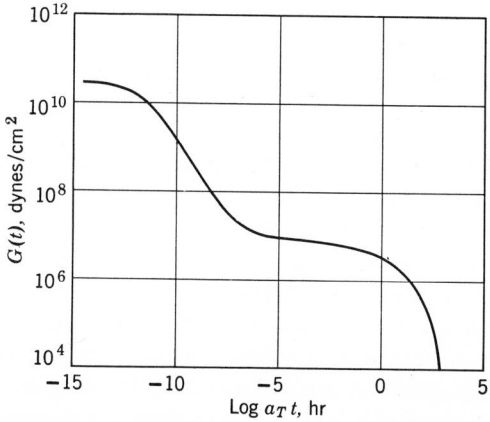

Fig. 6-9. Stress relaxation master curve for polyisobutylene (8).

Master curves such as that of Fig. 6-9 show the different types of viscoelastic behavior usually observed with amorphous materials. At small times the high modulus and low slope are characteristic of glassy behavior. The following region of rapidly decreasing modulus represents the glass transition, described further in Section *D*. This is followed by a flat region of rubbery behavior produced either through entanglements among relatively long molecules or through permanent crosslinks. The ultimate slope at long times represents the region of viscous flow in uncrosslinked polymers.

Creep Creep is studied by subjecting a sample rapidly to a constant stress and observing the resulting time-dependent strain for relatively long periods of time, frequently for a week or more or even for a year or more. Creep and stress relaxation are complementary aspects of plastics behavior and in many cases may provide equivalent information for studies of both fundamental viscoelastic properties and performance in practical applications. Creep experiments are usually easier, more economical, and more feasible for long periods of time.

Dynamic methods The delayed reaction of a polymer to stress and strain also affects its dynamic properties. If a simple harmonic stress of angular frequency ω is applied to the sample, the strain lags behind the stress by a phase angle δ, whose tangent measures the *internal friction* $\Delta E/E$,

$$\Delta E/E = 2\pi \tan \delta \tag{6-22}$$

where ΔE is the energy dissipated in taking the sample through a stress cycle, and E is the energy stored in the sample when the strain is a maximum. The internal friction is a maximum when the dynamic modulus of the material (the ratio of the stress to that part of the strain which is in phase with the stress) is in a region of relaxation. In terms of frequency the relaxation behavior of the dynamic modulus and the internal friction is illustrated in Fig. 6-10.

At low frequencies (about 1 sec^{-1}) it is convenient to measure the internal friction of polymers by observing the free decay of torsional oscillations of a thin fiber or film loaded with a large moment of inertia (*torsion pendulum*, Fig. 6-11) (5). The internal friction is the logarithm of the ratio of the amplitudes of successive free oscillations, while the modulus is calculated from the frequency of the oscillations (7).

A natural extension of this technique is to force oscillation with an external driving force. In such techniques the specimen is driven in flexural, torsional, or longitudinal oscillations by an oscillating force of constant amplitude but variable frequency, and the displacement amplitude of the resulting oscillation is observed. These techniques are particularly useful

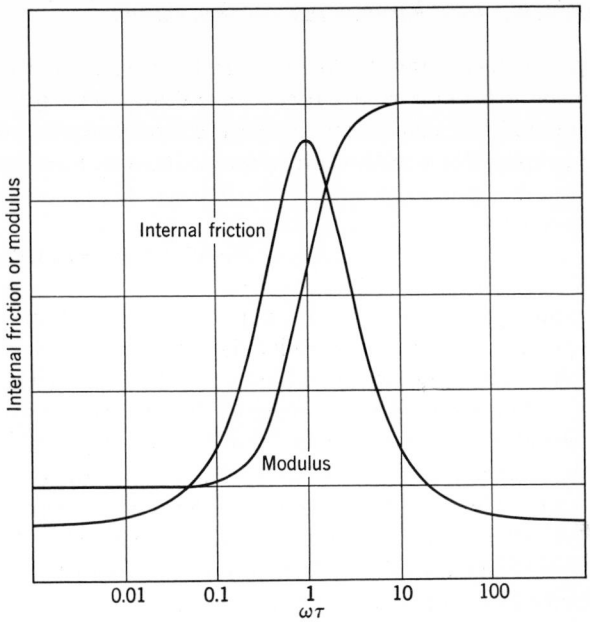

Fig. 6-10. Variation of dynamic modulus and internal friction with frequency.

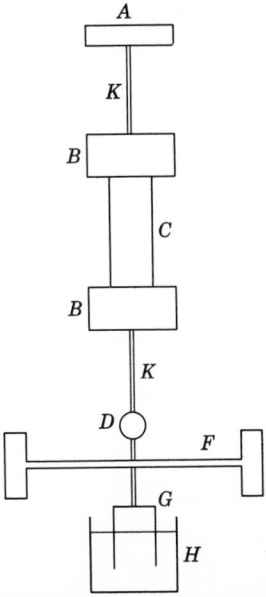

Fig. 6-11. Schematic diagram of a torsion pendulum apparatus for measuring torsion modulus and internal friction (5). *A*, fixed support; *B*, specimen clamps; *C*, specimen; *D*, observation mirror; *F*, inertia arm; *G*, *H*, cylinder in oil bath to prevent swinging motion; *K*, support rods.

in the kilocycle frequency range. At very high frequencies (above 1 megacycle), the dimensions of the sample become awkwardly small for mechanical techniques, and it is convenient to measure the attenuation of sound waves in the sample. These methods are discussed fully by Ferry (gen. ref.).

Models of viscoelastic behavior

Before attempting to devise a model to duplicate the viscoelastic behavior of an actual polymer, it is well to examine the response γ of ideal systems to a stress s. An ideal elastic element is represented by a spring which obeys Hooke's law, with a modulus of elasticity G. The elastic deformation is instantaneous and independent of time: $\gamma = (1/G)s$ (Fig. 6-12*a*). A completely viscous response is that of a Newtonian fluid, whose deformation is linear with time while the stress is applied, and is completely irrecoverable: $d\gamma/dt = (1/\eta)s$, where η is the viscosity of the fluid (Fig. 6-12*b*). A simple mechanical analogy of a Newtonian fluid is a dashpot.

The two elements of spring and dashpot can be combined in two ways. If they are placed in series, the resulting *Maxwell element* (Fig. 6-13) exhibits flow plus elasticity on the application of stress (Fig. 6-12*c*): when the stress is applied, the spring elongates while the dashpot slowly yields. On the removal of the stress the spring recovers but the dashpot does not. The strain is given by the equation

$$\frac{d\gamma}{dt} = \frac{1}{\eta}s + \frac{1}{G}\frac{ds}{dt} \tag{6-23}$$

The relation of creep to stress relaxation may be seen by considering the experiment in which a strain is obtained and then held by fixing the ends of the system: $d\gamma/dt = 0$ in Eq. 6-23. The equation can then be solved:

$$s = s_0 e^{-(G/\eta)t} = s_0 e^{-t/\tau} \tag{6-24}$$

where the stress s relaxes from its initial value s_0 exponentially as a function of time. The time η/G after which the stress reaches $1/e$ of its initial value is called the *relaxation time* τ.

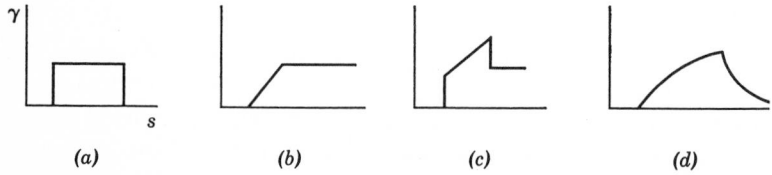

Fig. 6-12. Strain-time relationships at constant stress for simple models (Alfrey, gen. ref.): (*a*) ideal elastic spring; (*b*) Newtonian fluid (dashpot); (*c*) Maxwell element; (*d*) Voigt element.

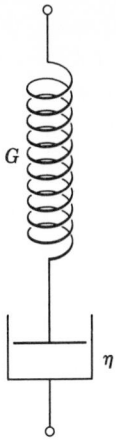

Fig. 6-13. A Maxwell element.

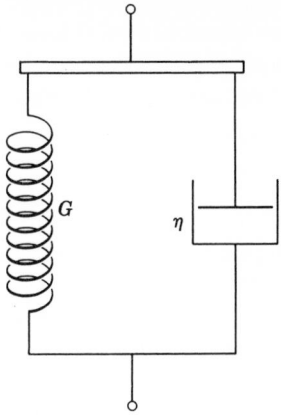

Fig. 6-14. A Voigt element.

The response of a tangled mass of polymer chains to a stress is better represented by a parallel combination of spring and dashpot in a *Kelvin* or *Voigt element* (Fig. 6-14). This element shows a *retarded elastic* or *visco-elastic* response (Fig. 6-12*d*). The dashpot acts as a damping resistance to the establishment of the equilibrium of the spring. The equation for the strain is

$$\eta \frac{d\gamma}{dt} + G\gamma = s \tag{6-25}$$

If a stress is applied and after a time removed (Fig. 6-12*d*), the deformation-time curve is given by:

$$\gamma = \frac{s}{G}(1 - e^{-(G/\eta)t}) = \frac{s}{G}(1 - e^{-t/\tau}) \tag{6-26}$$

where τ is a *retardation time*. When the stress is removed the sample returns to its original shape along the exponential curve:

$$\gamma = \gamma_0 e^{-t/\tau} \tag{6-27}$$

General mechanical models for an amorphous polymer The three components which make up the simplest behavior of an actual polymer sample in creep can be represented by a mechanical model of the type shown in Fig. 6-15, which combines a Maxwell and a Voigt element in series. If a stress is suddenly applied to this model, the strain changes with time as shown in Fig. 6-16; the corresponding behavior of the model is shown in Fig. 6-17.

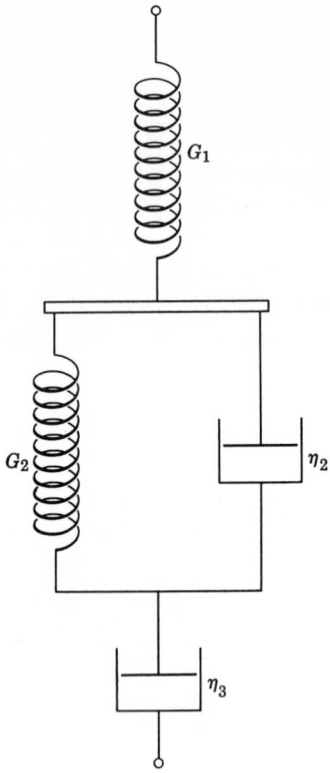

Fig. 6-15. A generalized mechanical model for creep combining elasticity, viscoelasticity, and flow.

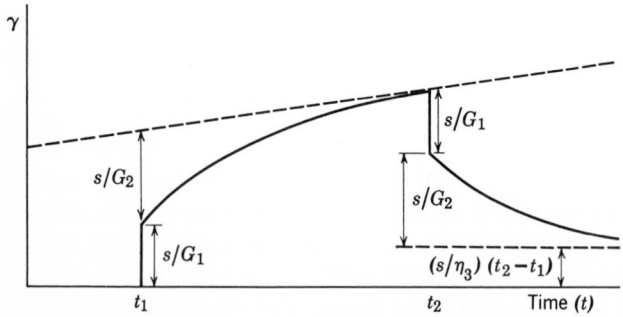

Fig. 6-16. Strain-time relationship for the mechanical model of Fig. 6-15 (Alfrey, gen. ref.).

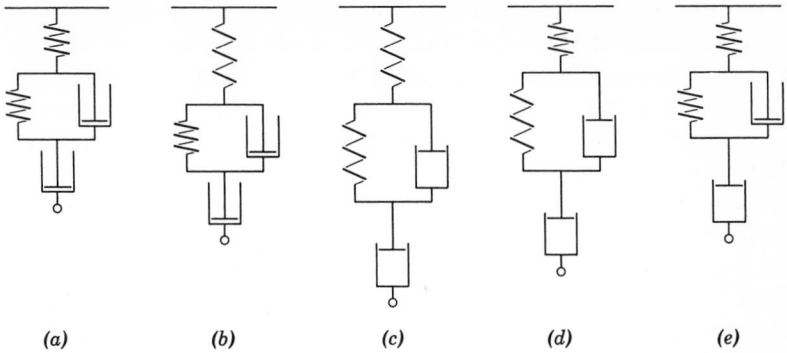

Fig. 6-17. Interpretation of strain-time curve of Fig. 6-16 in terms of the mechanical model.

The model departs from the initial conditions (*a*) at time t_1 by an elastic deformation s/G_1 (*b*). A viscoelastic response approaching s/G_2 as an equilibrium value and a viscous flow at the rate s/η_3 follow (*c*). On the removal of the stress at time t_2, the elastic element relaxes immediately (*d*) and the viscoelastic one slowly (*e*), but the viscous flow is never recovered.

Alternatively, the creep and stress relaxation experiments can be described with a generalized model consisting of a Maxwell and a Voigt element arranged in parallel. The two generalized models are entirely equivalent.

Although these models exhibit the chief characteristics of the viscoelastic behavior of polymers, they are nevertheless very much oversimplified. The flow of the polymer is probably not Newtonian, and its elastic response may not be Hookean. Moreover, the behavior of a real polymer cannot be characterized by a single relaxation time, but requires a spectrum of relaxation times to account for all phases of its behavior.

Treatment of experimental data

Distribution of relaxation times Equation 6-26 indicates that most of the relaxation associated with a single element takes place within one cycle of log time. In contrast, experiments indicate (as in Fig. 6-9) that relaxation phenomena in polymers extend over much wider ranges of time. Thus, the actual behavior of polymers can be described only in terms of a distribution of model elements and an associated distribution of relaxation or retardation times:

$$\gamma(t) = s \int_{-\infty}^{\infty} \bar{J}(\log \tau)(1 - e^{-t/\tau})\, d \log \tau \qquad (6\text{-}28)$$

where $J = 1/G$ is the *elastic compliance*, and $\bar{J}(\log \tau)$ the distribution of retardation times.

In principle, knowledge of $\gamma(t)$ over the entire range of time allows $\bar{J}(\log \tau)$ to be evaluated explicitly. This is so difficult in practice, however, that it is necessary for mathematical simplicity to adopt simple empirical forms of the distribution function $\bar{J}(\log \tau)$ and evaluate their parameters from the more limited experimental data at hand (Tobolsky, gen. ref.). The distribution may be assumed to be continuous, having the shape of a "box" (i.e., a distribution having an equal population of relaxation times between specified minimum and maximum values) or a wedge (in which the population decreases linearly over the range between the ends of the box). Alternatively, the experimental data may be fitted graphically by summing the theoretical curves corresponding to a small number (three or four) of discrete relaxation times.

The usefulness of the distribution of relaxation times arises from its identification with certain molecular parameters of the specimen, as described below. For many polymers, the experimental data are fitted well by the sum of a wedge distribution with relaxation times corresponding to the region of the glass transition, plus a box distribution with relaxation times corresponding to the region of rubbery flow.

Time-temperature superposition An increase in temperature accelerates molecular and segmental motion, bringing the system more rapidly to equilibrium or apparent equilibrium and accelerating all types of viscoelastic processes. A convenient way of formulating this effect of temperature is in terms of the ratio $a_T(\tau, T)$ of a relaxation time τ at temperature T to its value τ_0 at a convenient reference temperature T^0. For many cases, including most nonpolar amorphous polymers, a_T does not vary with τ, so that changes in temperature shift the distribution of relaxation times to smaller or greater values of τ but do not otherwise alter it. Time and temperature affect viscoelasticity only through the product of a_T and actual time, and a_T is the *shift factor* which must be applied to experimental data to produce master curves of the type of Fig. 6-9.

Despite the marked dependence on molecular structure of the relation between a_T and absolute temperature, nearly general empirical relations have been derived by expressing the temperature for each material in terms of its glass transition temperature or some nearly equivalent reference temperature. Among the most successful of these relations is the Williams-Landel-Ferry (WLF) equation (9):

$$\log a_T = \frac{-17.44(T - T_g)}{51.6 + T - T_g} \tag{6-29}$$

This equation holds over the temperature range from T_g to about $T_g + 100°K$. The constants are related to the free volume as described in Section *D*.

Molecular structure and viscoelasticity

The phenomena of relaxation processes can be thought of in terms of the effect of thermal motion on the orientation of polymer molecules. When a mechanical stress is applied to a polymer, introducing deformations of the chains, the entropy of the system decreases as less probable conformations are taken up. The free energy correspondingly increases. If the sample is kept in the deformed state, stress relaxation takes place as a result of the thermal motions of the chains, the molecular deformations are obliterated, and the excess free energy is dissipated as heat. The details of the stress relaxation process depend upon the multiplicity of ways in which the polymer molecules can regain their most probable conformations through thermal motion. These complex motions of a polymer molecule can be expressed as a series of characteristic modes requiring various degrees of long-range cooperation among the segments of the chain. Thus (Fig. 6-18) the first mode corresponds to translation of the entire molecule, requiring maximum cooperation, the second corresponds to motion of the ends of the chain in opposite directions, requiring somewhat less cooperation, and so on. With each of these modes is associated a characteristic relaxation time; there are so many modes that over most of the time scale the discrete spectrum of relaxation times can be approximated by a continuous distribution.

Theories based on the above considerations (1, 6, 10), derived for dilute solutions of polymer molecules, have been combined (Ferry, gen. ref., Chapter 10) with the concepts of entanglement of long molecules (2) successful in predicting the molecular weight dependence of viscous flow (Section *A*). Although exact numerical agreement is not achieved, e.g., in calculating the dynamic modulus of polyisobutylene over a wide range of frequency and molecular weight (3), the qualitative situation seems extremely satisfactory.

In the rubbery region, where the viscoelastic data are well fitted by a box distribution, the maximum relaxation time is strongly dependent on molecular weight. In this region, the motions of the molecules are long range in

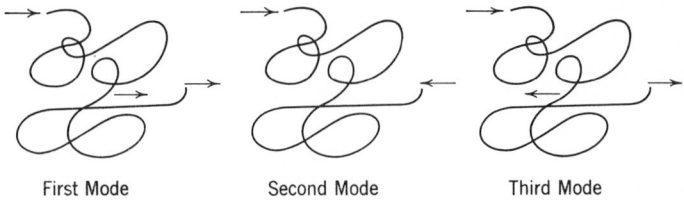

Fig. 6-18. Schematic diagram of cooperative modes of motion of a polymer molecule (Ferry, gen. ref.).

nature, involving motions of units of the order of the length of the molecule itself. This is the region where entanglements are important.

In the region around the glass transition, however, where the wedge distribution is found, only vibrations of the parts of the molecule between entanglements are important. So long as the molecules are long enough for entanglements to exist, this portion of the curve is expected, and found, to be essentially independent of the molecular weight and to depend primarily upon the local structure of the polymer.

GENERAL REFERENCES

Alfrey, T., Jr., *Mechanical Behavior of High Polymers*, Interscience Publishers, New York, 1948; see also T. Alfrey, Jr., and E. F. Gurnee, "Dynamics of Viscoelastic Behavior," Chap. 11 in F. R. Eirich, ed., *op. cit.*, Vol. 1, 1956.

Bueche, F., *op. cit.*, Chaps. 6–9.

Ferry, J. D., *Viscoelastic Properties of Polymers*, John Wiley & Sons, New York, 1961; see also J. D. Ferry, "Experimental Techniques for Rheological Measurements on Viscoelastic Bodies," Chap. 11 in F. R. Eirich, ed., *op. cit.*, Vol. 2, 1958.

Leaderman, H., "Viscoelasticity Phenomena in Amorphous High Polymeric Systems," Chap. 1 in F. R. Eirich, ed., *op. cit.*, Vol. 2, 1958.

Stein, R. S., *op. cit.*, Chaps. IV, V.

Tobolsky, A. V., *Properties and Structure of Polymers*, John Wiley and Sons, New York, 1960; see also A. V. Tobolsky, "Stress Relaxation Studies of the Viscoelastic Properties of Polymers," Chap. 2 in F. R. Eirich, ed., *op. cit.*, Vol. 2, 1958.

BIBLIOGRAPHY

1. Bueche, F., *J. Chem. Phys.* **20,** 1959 (1952); **25,** 599 (1956).
2. Bueche, F., *J. Chem. Phys.* **22,** 603 (1954).
3. Ferry, J. D., R. F. Landel, and M. L. Williams, *J. Appl. Phys.* **26,** 359 (1955).
4. Guth, E., P. E. Wack, and R. L. Anthony, *J. Appl. Phys.* **17,** 347 (1946).
5. McCrum, N. G., *J. Polymer Sci.* **34,** 355 (1959).
6. Rouse, P. E., Jr., *J. Chem. Phys.* **21,** 1272 (1953).
7. Schmieder, K., and K. Wolf, *Kolloid-Z.* **127,** 65 (1952).
8. Tobolsky, A. V., and E. Catsiff, *J. Polymer Sci.* **19,** 111 (1956).
9. Williams, M. L., R. F. Landel, and J. D. Ferry, *J. Am. Chem. Soc.* **77,** 3701 (1955).
10. Zimm, B. H., *J. Chem. Phys.* **24,** 269 (1956).

D. The Glassy State and the Glass Transition

As anticipated in the previous sections, and in particular in Fig. 6-9, all amorphous polymers assume at sufficiently low temperatures the characteristics of glasses (13), including hardness, stiffness, and brittleness. One property associated with the glassy state is a low volume coefficient of

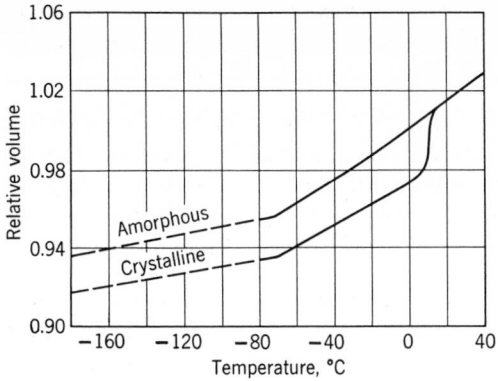

Fig. 6-19. Volume behavior of rubber near the glass transition temperature (1, 2).

expansion. This low coefficient occurs as the result of a change in slope of the curve of volume vs. temperature at the point called the *glass transition temperature* T_g. This behavior is shown for natural rubber in Fig. 6-19 (1, 2). In the high-temperature region, the slope of the curve (expansion coefficient) is characteristic of a rubber; below T_g at about $-70°C$, it is characteristic of a glass.

Figure 6-19 illustrates another general phenomenon: the amorphous regions in partially crystalline polymers also assume a glassy state, T_g being independent of degree of crystallinity to a first approximation. The magnitude of the phenomena associated with T_g decreases with decreasing amorphous content, however. As a result, T_g is sometimes difficult to detect in highly crystalline polymers. In terms of the single-phase lamellar or paracrystalline model (Chapter 5*D*) the glass transition is considered to involve defect regions within or at the boundaries of the lamellae.

In contrast to crystalline melting at a temperature T_m (about $+10°C$ in Fig. 6-19), there is not an abrupt change in volume at T_g, but only a change in the slope of the volume-temperature curve. In analogy to thermodynamic first- and second-order transitions (see footnote, p. 10), T_g is sometimes referred to as a second-order or apparent second-order transition. This nomenclature is considered poor, however, since it implies more thermodynamic significance than the nature of the transition warrants (9).

Measurement of T_g

The glass transition temperature can be detected in a variety of experiments, which can be roughly classified into those dealing with bulk properties of the polymer, and those measuring the nature and extent of

molecular motion. The classification is to some extent arbitrary, since, as indicated below, T_g is in fact the temperature of onset of extensive molecular motion.

Bulk properties Perhaps the most common way of estimating T_g is by means of the volume expansion coefficient, as indicated above. Other bulk properties whose temperature coefficients undergo marked changes at T_g, and which can therefore be used for its determination, include heat content (Chapter 4*E*), refractive index, stiffness, and hardness.

Molecular motion Experiments which are sensitive to the onset of molecular motion in polymer chains may be used to detect the glass transition. Such methods include the measurement of internal friction (Section *C*), dielectric loss in polar polymers, and NMR spectroscopy (Chapter 4*D*).

Phenomena related to T_g The onset of brittleness, as measured in impact tests, and the softening of amorphous polymers, as measured in thermal tests of various sorts, take place at temperatures near T_g (see Chapter 4*G*).

Viscoelastic phenomena near T_g

As indicated in Section *C*, the effects of time and temperature in determining the viscoelastic behavior of polymers may be formulated in terms of the shift factor a_T leading to time-temperature superposition. If the molecular phenomena responsible for viscoelasticity are the result of activated processes (as is, e.g., viscous flow), the temperature dependence of a_T may be analyzed by the Arrhenius treatment to obtain an apparent activation energy. At temperatures well above T_g, this activation energy has a value (5–40 kcal/mole) characteristic of viscous flow; below, it is characteristic of the glassy state. In the vicinity of T_g, however, the apparent activation energy goes through a high maximum, of the order of 50–300 kcal/mole. This value has no significance in terms of an actual activation energy. Rather, it evidences the fact that different molecular mechanisms are dominant above and below the transition, and is a measure of the rate at which the change from one to the other occurs.

The temperature dependence of a_T in the region just above T_g can be expressed in the form of a general curve applicable to all polymers if $\log a_T$ is plotted against $T - T_s$, where T_s is a reference temperature which is a function of polymer type. The empirical equation fitting this universal curve is

$$\log a_T = \frac{-8.86(T - T_s)}{101.6 + T - T_s} \tag{6-30}$$

which is another form of the WLF equation (21). Since T_s is equal to $T_g + 50°C$ with fair precision, Eq. 6-30 is equivalent to (but somewhat more precise than) 6-29.

The sharp increase in the apparent activation energy of a_T near T_g can be explained qualitatively in terms of the *free volume* v_f of the polymer. Although poorly defined theoretically, the free volume is a useful semi-quantitative concept representing the volume present as holes of the order of molecular (monomeric) dimensions or smaller (5, 7). The relative amount of free volume decreases sharply with decreasing temperature above T_g, and is assumed (9) to be constant at T_g. It can be shown (3, 5, 6, 21) that

$$\log a_T = B\left(\frac{1}{f} - \frac{1}{f_g}\right) \tag{6-31}$$

where $f = v_f/v$ is the volume fraction of free volume, and the subscript g indicates measurement at T_g. If f increases linearly with temperature, e.g., $f = f_g + \alpha_f(T - T_g)$, the WLF equation results.

Time effects near T_g

If a polymer sample is cooled rapidly to a temperature just below T_g, its volume continues to decrease for many hours. In consequence, the value observed as T_g in a volume-temperature experiment depends on the time scale of the measurements (14). As indicated in Fig. 6-20, the results of volume measurements after 0.02 and 100 hr gave values of T_g some 9° apart for poly(vinyl acetate). Thus, it is convenient to define T_g in terms of an arbitrary convenient (but not highly critical) time interval, such as 10 min to 1 hr (13).

For similar reasons, other tests for T_g give results somewhat dependent on the time scale of the experiments, with tests requiring shorter times yielding higher values for T_g. The brittle temperature as determined in an impact test is, e.g., normally somewhat higher than T_g as otherwise measured.

Molecular interpretation of T_g

In the glassy state, large-scale molecular motion does not take place. Rather, atoms and small groups of atoms move against the local restraints of secondary bond forces, much as atoms vibrate around their equilibrium positions in a crystal lattice, except that the glassy state does not have the regularity of the crystalline state. The glass transition corresponds to the onset of liquidlike motion of much longer segments of molecules, characteristic of the rubbery state. This motion requires more free volume than the short-range excursions of atoms in the glassy state. The rise in the

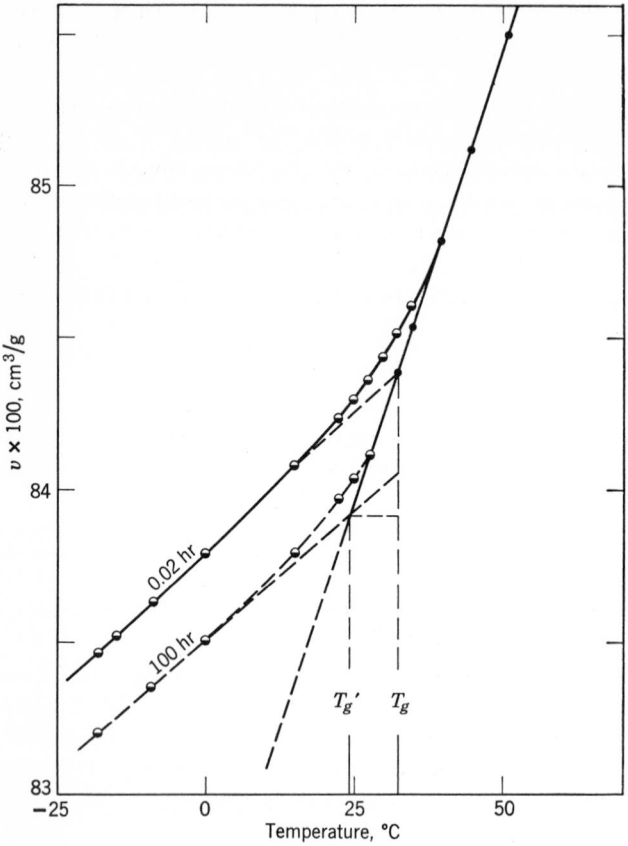

Fig. 6-20. Curves of specific volume vs. temperature for poly(vinyl acetate), measured after cooling the specimen quickly from well above T_g[Ferry, gen. ref., after (14)]. ●, equilibrium values; ◒, measured 0.02 hour and 100 hours after cooling, as indicated.

relative free volume with increasing temperature above T_g leads to the higher observed volume expansion coefficient in this region.

A theoretical examination (8, 10) of the temperature dependence of the number of conformations available for the polymer molecule leads to the following line of reasoning. Since the fully extended chain is the conformation of minimum energy (4; Chapter 5*B*), it tends to be assumed more frequently as the temperature is lowered. As the molecules thus straighten out, the free volume decreases. In consequence, flow becomes more difficult and a_T changes more rapidly with T. The glass transition (observed at infinite time)—or, alternatively, the onset of crystallization where possible—is taken as the point where the number of possible conformations of the amorphous phase decreases sharply toward one.

Molecular motion below T_g

The foregoing discussion is concerned with the transition involving the motion of long segments of the polymer chain. At lower temperatures, other transitions may occur, produced by the motion of short sections of the main chain or of side chains. Although some characteristics of the glassy state, such as brittleness, may occasionally occur only below one of

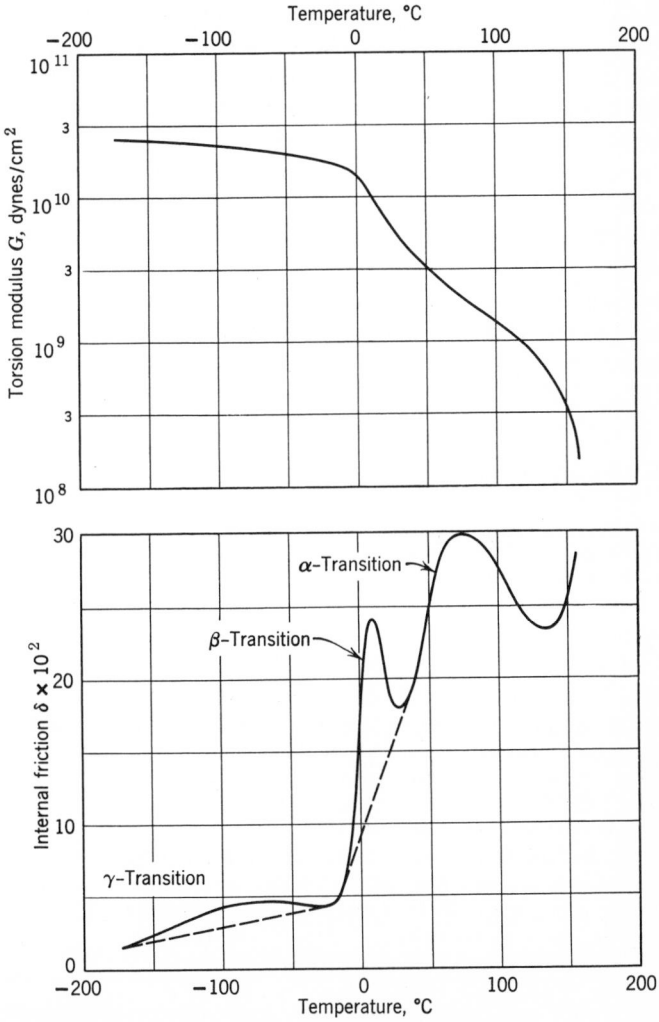

Fig. 6-21. Dependence on temperature of the internal friction and torsion modulus of isotactic polypropylene (16).

these lower transitions (16), it is proposed (20) that the transition of highest temperature be called T_g. Alternatively (17), the transitions may be denoted α, β, γ, etc., in order of descending temperature.

Transitions due to the motion of short segments of the main polymer chain occur most prominently in crystalline polymers such as polyethylene (16), polypropylene (16), and polytetrafluoroethylene (15). Such polymers also typically exhibit an α-transition.

The internal friction results for isotactic polypropylene (Fig. 6-21) are typical. The onset of brittleness in this polymer is accounted for by the β-transition occurring at 0°C at the torsion pendulum frequency of 1 cycle/sec but shifted to room temperature or higher at the frequencies prevalent in impact tests. In linear polyethylene, however, the β-transition appears as a shoulder on the large α-transition; the associated decrease in modulus is spread out over a wider temperature range because of the combined effects of the two phenomena. The onset of brittleness occurs in the temperature range of the γ-transition, at about −120°C in the torsion pendulum experiment.

Side-chain transitions occur in methacrylate polymers (11, 12, 18, 19) as a result of the relaxation of the carbomethoxy side chain at about 20°C (torsion pendulum) and the relaxation of the aliphatic ester group below −150°C.

GENERAL REFERENCES

Bueche, F., *op. cit.*, Chaps. 4, 5, 11.
Ferry, J. D., *op. cit.*, Chaps. 10–12, 14.
Stein, R. S., *op. cit.*, Chap. VI.

BIBLIOGRAPHY

1. Bekkedahl, N., *J. Research Natl. Bur. Standards* **13,** 411 (1934).
2. Boyer, R. F., and R. S. Spencer, "Second-Order Transition Effects in Rubber and Other High Polymers," pp. 1–57 in H. Mark and G. S. Whitby, eds., *Scientific Progress in the Field of Rubber and Synthetic Elastomers* (*Advances in Colloid Science*, Vol. II), Interscience Publishers, New York, 1946.
3. Bueche, F., *J. Chem. Phys.* **24,** 418 (1956).
4. Bunn, C. W., and D. R. Holmes, *Disc. Faraday Soc.* **25,** 95 (1958).
5. Doolittle, A. K., *J. Appl. Phys.* **22,** 1471 (1951).
6. Doolittle, A. K., and D. B. Doolittle, *J. Appl. Phys.* **28,** 901 (1957).
7. Eyring, H., and J. O. Hirschfelder, *J. Phys. Chem.* **41,** 249 (1937).
8. Flory, P. J., *Proc. Roy. Soc.* (*London*) **A234,** 60 (1956).
9. Fox, T. G., Jr., and P. J. Flory, *J. Appl. Phys.* **21,** 581 (1950).
10. Gibbs, J. H., *J. Chem. Phys.* **25,** 185 (1956); J. H. Gibbs and E. A. DiMarzio, *ibid.* **28,** 373 (1958); E. A. DiMarzio and J. H. Gibbs, *ibid.* **28,** 807 (1958); *J. Polymer Sci.* **40,** 121 (1959).

11. Hoff, E. A. W., D. W. Robinson, and A. H. Willbourn, *J. Polymer Sci.* **18,** 161 (1955).
12. Iwayanagi, S., *J. Sci. Research Inst.* (*Tokyo*) **49,** 23 (1955).
13. Kauzmann, W., *Chem. Revs.* **43,** 219 (1948).
14. Kovacs, A. J., *J. Polymer Sci.* **30,** 131 (1958); abstracts, 48th International Colloquium on Rheology, Paris, June 29–July 4, 1960.
15. McCrum, N. G., *J. Polymer Sci.* **34,** 355 (1959).
16. Muus, L. T., N. G. McCrum, and F. C. McGrew, *SPE J.* **15,** 368 (1959).
17 Oakes, W. G., and D. W. Robinson, *J. Polymer Sci.* **14,** 505 (1954).
18. Sinnott, K. M., *J. Polymer Sci.* **42,** 3 (1960).
19. Tuckett, R. F., *Trans. Faraday Soc.* **40,** 448 (1944).
20. Willbourn, A. H., *Trans. Faraday Soc.* **54,** 717 (1958).
21. Williams, M. L., R. F. Landel, and J. D. Ferry, *J. Am. Chem. Soc.* **77,** 3701 (1955).

E. The Mechanical Properties of Crystalline Polymers

The models developed in the preceding sections represent well the rheological and mechanical properties of amorphous polymers. The viscoelastic properties of crystalline polymers are much more complex, however, and are not amenable to adequate theoretical explanation for three reasons.

First, an amorphous polymer is isotropic. This means that models suitable for describing shear stress, e.g., are adequate to describe tensile stress or other types. Since crystalline polymers are not isotropic, this universality does not hold and the range of application of any model is severely limited.

Second, the homogeneous nature of amorphous polymers ensures that an applied stress is distributed uniformly throughout the system, at least down to very small dimensions. In crystalline polymers the relatively large crystallites are bound together in such a way that large stress concentrations inevitably develop.

Finally, a crystalline polymer is a mixture of regions of different degrees of order ranging all the way from completely ordered crystallites to completely amorphous regions. As the stress on the sample changes, the amounts of these regions change continuously as the crystallites melt or grow. This change of composition with respect to ordering is the most difficult obstacle to overcome in formulating a theory of the mechanical behavior of crystalline polymers. Even in the simplest cases the necessity of having the mechanical model change continuously with the applied stress has led to insurmountable difficulties.

An orderly, if qualitative, discussion of the mechanical properties of crystalline polymers requires their classification into several categories, as indicated in Table 6-2. It is in the range of intermediate degree of crystallinity that the properties, unique to polymers, of most importance for mechanical and engineering applications are found.

Polymers with low crystallinity include plasticized poly(vinyl chloride) and so-called elastic polyamides (Chapter 17*A*). These materials behave like lightly crosslinked amorphous polymers, their crystalline regions acting like crosslinks which are stable with respect to time but unstable with respect to temperature. The viscoelastic properties of these polymers are much like those of amorphous polymers except that the transition region between glassy and rubbery or liquid behavior is very much broadened on the temperature scale.

TABLE 6-2. Classification of crystalline polymers

Predominant Properties in Temperature Range	Degree of Crystallinity		
	Low (5–10%)	Intermediate (20–60%)	High (70–90%)
Above T_g	Rubbery	Leathery, tough	Stiff, hard (brittle)
Below T_g	Glassy, brittle	Hornlike, tough	Stiff, hard, brittle

At very low extensions (<1%), at temperatures well below T_m, and at not too long times, polymers with intermediate degrees of crystallinity (such as low-density polyethylene) behave much like those with very low crystallinity. Time-temperature superposition is probably not applicable to such polymers (1); it should be applied with strong reservations and only after an extensive comparison between creep or stress relaxation and dynamic modulus data. The transition regions of these polymers in modulus and the corresponding distributions of relaxation times are exceptionally broad, as indicated by the master curves of Fig. 6-22. (It should be pointed out that these curves were themselves derived by application of time-temperature superposition, and must be considered only as idealizations.)

At higher extensions, these polymers exhibit the phenomena of a yield stress and cold drawing, with the accompanying changes in crystalline morphology described in Chapter 5*G*. No adequate theory relating rheological and mechanical behavior in this region to structure has yet been formulated. Polymers of intermediate degree of crystallinity are characteristically leathery or horny in texture, and exhibit good impact resistance which in many cases is retained even below T_g. The exact structural features responsible for this toughness have not been well defined.

The major effects of a further increase in crystallinity to very high values include (*a*) a further increase in modulus, as the high modulus characteristic of the crystalline regions is approached; and (*b*) the onset of a tendency

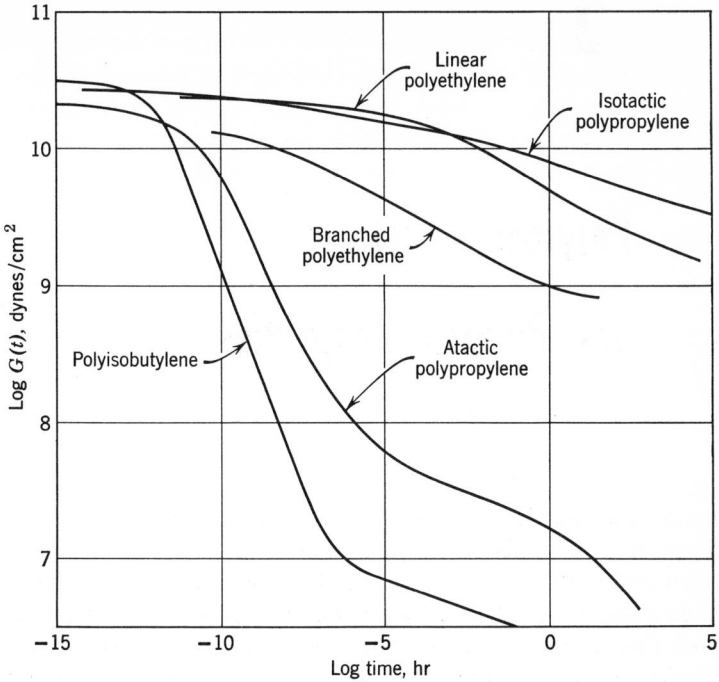

Fig. 6-22. Stress relaxation master curves for several crystalline and amorphous polymers [Tobolsky, gen. ref., after (1)].

toward brittleness, in the sense of failure at low strains. Such polymers, of which high-density polyethylene is typical, can be cold drawn only with difficulty. Tensile failure usually occurs at or slightly beyond the yield stress, accompanied by distortion or deformation which appears to occur at slip boundaries or dislocations, reminiscent of the viscoelastic behavior of metals.

GENERAL REFERENCES

Ferry, J. D., *op. cit.*, Chap. 15.
Hopkins, I. L., and W. O. Baker, "The Deformation of Crystalline and Crosslinked Polymers," Chap. 10 in F. R. Eirich, ed., *op. cit.*, Vol. 3, 1960.
Stein, R. S., *op. cit.*, Chaps. X, XI.
Tobolsky, A. V., *op. cit.*, Chap. IV-11.

BIBLIOGRAPHY

1. Faucher, J. A., *Trans. Soc. Rheology* **3,** 81 (1959).

7

Polymer Structure and Physical Properties

Although interrelations between the molecular structure of polymers and their properties are inferred throughout this book, they are emphasized in this chapter. In Section *A* are classified the structural features of polymers most directly responsible for determining their properties, and succeeding sections describe how these structures influence various classes of physical properties. Finally, the property requirements associated with the uses of various polymers are reviewed in Section *F*.

A. Basic Determinants of Polymer Properties

Since the acceptance of the macromolecular hypothesis in the 1920's it has been recognized that the unique properties of polymers—e.g., the elasticity and abrasion resistance of rubbers, the strength and toughness of fibers, and the flexibility and clarity of films—must be attributed to their long-chain structure. In the more recent examination of structure-property relationships it has been advantageous to classify properties into those involving large and small deformations. The former class includes such properties as tensile strength and phenomena observed in the melt, while properties involving only small deformations include electrical and optical behavior, such mechanical properties as stiffness and yield point, and the glass and crystalline melting transitions.

Properties involving large deformations depend primarily on the long-chain nature of polymers and the gross configuration of their chains. Important factors for this group of properties include molecular weight

and its distribution, chain branching and the related category of side-chain substitution, and crosslinking.

Physical properties associated with small deformations are influenced most by factors determining the manner in which chain atoms interact at small distances. The ability of polymers to crystallize, set by considerations of symmetry and steric effects, has major importance here, as do the flexibility of the chain bonds and the number, nature, and spacing of polar groups. To the extent that they influence the achievement of local order, gross configurational properties are also important. Similar considerations apply to amorphous polymers below the glass transition.

In crystalline polymers, the nature of the crystalline state introduces another set of variables influencing mechanical properties. These variables include nature of the crystal structure, degree of crystallinity, size and number of spherulites, and orientation. Some of these phenomena are in turn influenced by the conditions of fabrication of the polymer.

Finally, the properties of polymers can be varied importantly by the addition of other materials, such as plasticizers or reinforcing fillers. Properties involving both large and small deformations may be influenced in this way.

GENERAL REFERENCES

Bawn, C. E. H., *The Chemistry of High Polymers*, Interscience Publishers, New York, 1948, Chap. 7.

D'Alelio, G. F., *Fundamental Principles of Polymerization*, John Wiley & Sons, New York, 1952, pp. 100–166.

B. Properties Involving Large Deformations

Melt properties

Melt viscosity As discussed in Chapter 6*A*, the viscosity of a polymer melt is a strong function of weight-average (or more properly viscosity-average) molecular weight. Melt viscosity is also influenced by chain branching: in polyethylene (7) and in silicone polymers (3) melt viscosity decreases with increasing degree of long-chain branching at constant weight-average molecular weight, whereas in poly(vinyl acetate) (5) melt viscosity increases under the same circumstances. The reason for this difference appears to lie in branch length: in experiments with poly(vinyl acetate) branched by graft polymerization (5), melt viscosity decreased at constant molecular weight when the added branches were shorter than the critical chain length Z_c at which the melt viscosity power law changes

(Chapter 6*A*). Only when branches longer than Z_c were added did the viscosity increase.

Crosslinking has a pronounced effect on melt viscosity in that the latter becomes essentially infinite along with $\bar{M}_w$ at the onset of gelation. However, small, tightly crosslinked network particles (1) may behave like rigid spheres and have little effect on melt viscosity.

The addition of low-molecular-weight species, as in plasticization, reduces melt viscosity by lowering average molecular weight. Bulky side groups may have a similar effect (2).

Other melt properties For polymers with very high melt viscosity such as polytetrafluoroethylene, the tensile strength of the melt becomes a property of importance (11). Like melt viscosity, it increases with increasing molecular weight. The viscoelastic or elastic properties of polymer melts decrease in magnitude with increasing molecular weight and with increasing chain branching (10).

Tensile strength and related properties

Many polymer properties, including tensile strength, can be described by an equation of the type

$$\text{property} = a - b/\bar{M}_n \tag{7-1}$$

This is the type of relation predicted for properties depending on the number of ends of polymer chains. Many such properties, including density and refractive index, attain constant values at molecular weights well below the polymer range. Tensile strength, however, varies significantly with molecular weight in the range of interest for polymers (9), although the variation may in fact be with an average between $\bar{M}_n$ and $\bar{M}_w$ (6). It has been shown (4) that dependence on $\bar{M}_n$ implies that the tensile strength $\overline{TS}$ of a mixture of components with tensile strengths $(TS)_i$ is the weight average

$$\overline{TS} = \sum_i w_i(TS)_i \tag{7-2}$$

If a polymer exhibits a yield point and then undergoes extensive elongation before tensile failure, its ultimate tensile strength increases with increasing molecular weight (10, 12). Typical data for branched polyethylene (10) are shown in Fig. 7-1. Important structure variables in this polymer were found to be essentially those described in Fig. 1-3; in Fig. 7-1 crystallinity is replaced by its equivalent, density, and molecular weight by the logarithm of melt viscosity.

Although Fig. 7-1 shows that tensile strength is independent of degree of crystallinity for branched polyethylene, this independence does not carry

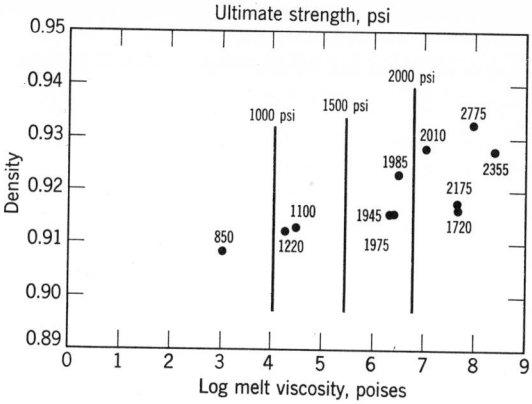

Fig. 7-1. Dependence of ultimate tensile strength of polyethylene on molecular structure variables (10).

over to linear polyethylenes, with densities in the range 0.94–0.96: the latter polymers have higher tensile strengths than branched polyethylenes of the same molecular weight (see also Chapter 6*E*). The difference is attributed to changes in the crystal morphology of the high-density polymers during tensile elongation (Chapter 5*G*).

Effect of molecular weight distribution

For many polymers an upper limit to $\bar{M}_w$ is set by fabrication requirements on the melt viscosity. At the same time, a lower limit on $\bar{M}_n$ may be set by requirements involving tensile strength, brittleness, or other mechanical properties. In such cases, which include polypropylene (13) and probably other polyolefins, the best balance of properties is achieved when the distribution of molecular weights is made as narrow as possible.

However, the anticipated gain in ease of fabrication on decreasing $\bar{M}_w$ at constant $\bar{M}_n$ may not be fully realized, since fabricability depends upon melt viscosity at high shear stress. In contrast to low shear (Newtonian) viscosity, which depends upon $\bar{M}_w$, high shear viscosity depends on a molecular weight average between $\bar{M}_w$ and $\bar{M}_n$ (8).

BIBLIOGRAPHY

1. Bryant, W. M. D., F. W. Billmeyer, Jr., L. T. Muus, J. T. Atkins, and J. E. Eldridge, *J. Am. Chem. Soc.* **81,** 3219 (1959).
2. Bueche, F., *J. Appl. Phys.* **26,** 738 (1955).

3. Charlesby, A., *J. Polymer Sci.* **17,** 379 (1955).
4. Flory, P. J., *J. Am. Chem. Soc.* **67,** 2048 (1945).
5. Hobbs, L. M., G. C. Berry, and V. C. Long, paper presented at the International Symposium on Macromolecular Chemistry, Montreal, Can., July, 1961.
6. McCormick, H. W., F. M. Brower, and L. Kin, *J. Polymer Sci.* **39,** 87 (1959).
7. Moore, L. D., Jr., *J. Polymer Sci.* **36,** 155 (1959).
8. Rudd, J. F., *J. Polymer Sci.* **44,** 459 (1960).
9. Sookne, A. M., and M. Harris, *Ind. Eng. Chem.* **37,** 478 (1945).
10. Sperati, C. A., W. A. Franta, and H. W. Starkweather, Jr., *J. Am. Chem. Soc.* **75,** 6127 (1953).
11. Sperati, C. A., and J. L. McPherson, paper presented at the 130th National Meeting of the American Chemical Society, Atlantic City, N.J., Sept. 18, 1956.
12. Thomas, P. E., J. F. Lontz, C. A. Sperati, and J. L. McPherson, *SPE J.* **12,** #5, 89 (June, 1956).
13. Wijga, P. W. O., "Structure and Properties of Polypropylene," pp. 35–45 in *The Physical Properties of Polymers*, S.C.I. Monograph No. 5, The Macmillan Co., New York, 1959.

C. Transitions

Relation between T_m and T_g

With few exceptions, polymer structure affects the glass transition T_g and the crystalline melting point T_m similarly. This is not unexpected, since similar considerations of cohesive energy and molecular packing apply to the amorphous and crystalline or paracrystalline regions, respectively, in accounting for the temperature levels at which the transitions occur. In consequence, T_m and T_g are rather simply related for many polymers (2, 5): depending on symmetry, T_g °K is approximately one-half or two-thirds of T_m °K (Fig. 7-2). Some exceptions to this general rule are discussed later in this section.

Effect of molecular weight

The crystalline melting point T_m is independent of molecular weight in the polymer range. From the criterion that the free energy of melting is zero at T_m,

$$T_m = \frac{\Delta H_m}{\Delta S_m} = \frac{H_0 + xH_1}{S_0 + xS_1} \rightarrow \frac{H_1}{S_1} \text{ as } x \rightarrow \infty \tag{7-3}$$

where the heat and entropy of fusion, ΔH_m and ΔS_m, are made up of molecular weight-independent terms H_0 and S_0 plus increments H_1 and S_1 for each chain unit, and x is the degree of polymerization. The glass transition temperature, however, retains in the molecular weight range of interest a molecular weight dependence of the form

$$T_g = T_g^{\infty} - k/\bar{M}_n \tag{7-4}$$

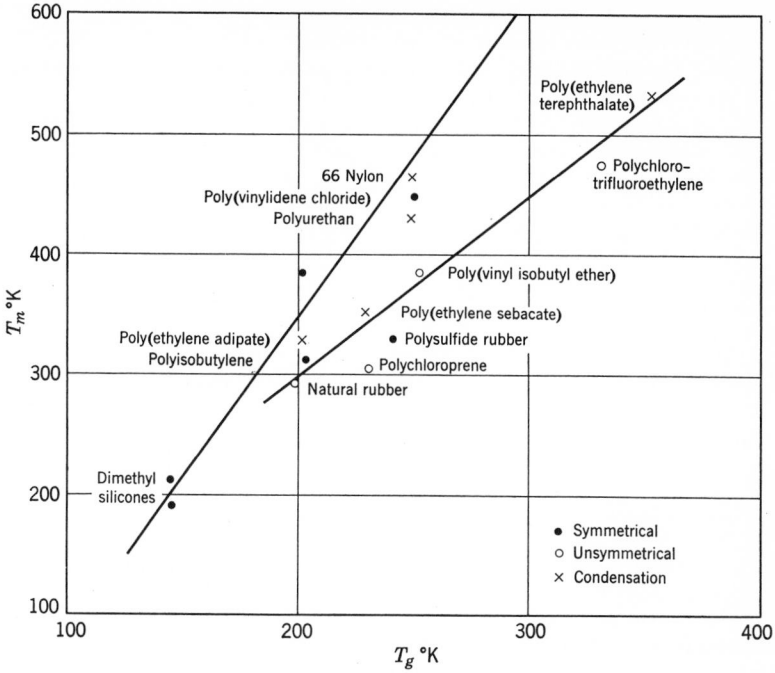

Fig. 7-2. Relation between T_m and T_g for various polymers (5).

derived from temperature-volume considerations (10), where T_g^{∞} is the glass transition temperature at infinite molecular weight, and k is about 2×10^5 for polystyrene (9) and poly(methyl methacrylate) (3).

From the considerations of Chapter 5, it is probable that chain ends, which lead to the terms H_0 and S_0 for low-molecular-weight substances, are usually associated with defects in the crystalline regions of polymers. Hence, they are unlikely to make a significant contribution to T_m, which is the melting point of the most perfect crystalline regions. In amorphous polymers, however, the effect of chain ends on free volume (Chapter 6*D*) should retain importance in proportion to the concentration of ends; thus it is not surprising that T_g is found to vary with $\bar{M}_n$ in the polymer range.

Effect of copolymerization

Crystalline melting point When copolymers are made from monomers which form crystalline homopolymers, degree of crystallinity and crystalline melting point decrease as the second constituent is added to either

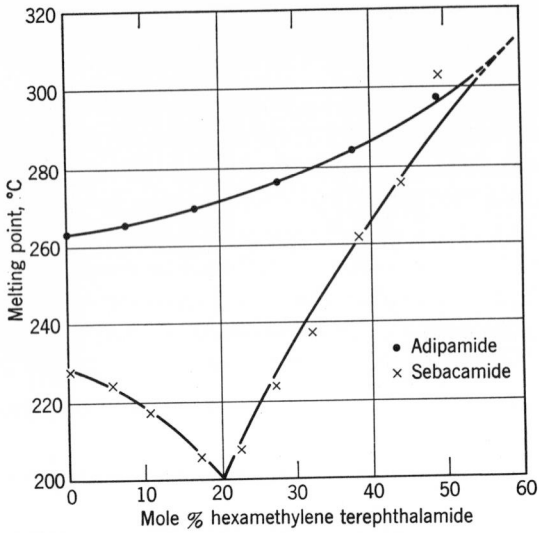

Fig. 7-3. Melting points of copolymers of hexamethylene adipamide and terephthalamide, and of hexamethylene sebacamide and terephthalamide (7).

homopolymer. The melting point depends on the mole fraction n of the crystallizing constituent by the relation of Eq. 5-4 (8):

$$\frac{1}{T_m} - \frac{1}{T_m^0} = -\frac{R}{\Delta H_m} \ln n \tag{7-5}$$

where T_m^0 is the melting point of the homopolymer and ΔH_m is its heat of fusion. A typical case is the copolymer of hexamethylene terephthalamide and hexamethylene sebacamide (Fig. 7-3).

If, however, the comonomers are isomorphous, i.e., capable of replacing each other in the crystals, the melting point may vary smoothly over the composition range. An example is the copolymer of hexamethylene terephthalamide and hexamethylene adipamide, also shown in Fig. 7-3. Other variations may occur, including the formation of an alternating copolymer with a crystal structure and melting point far different from those of either homopolymer. Block and graft copolymers may exhibit two crystalline melting points, one for each type of chain segment.

Glass transition The glass transition temperatures of random copolymers usually fall between those of the corresponding homopolymers (11, 13, 18), T_g for the copolymer being a weighted average given by

$$a_1c_1(T_g - T_{g1}) + a_2c_2(T_g - T_{g2}) = 0 \tag{7-6}$$

where T_{g1} and T_{g2} refer to the homopolymers, c_1 and c_2 are the weight fractions of monomers 1 and 2 in the copolymer, and a_1 and a_2 depend on

monomer type (19). However, random copolymers of methyl methacrylate and acrylonitrile were found to have glass transitions lower than that of either homopolymer (4).

In the case just cited (4), block copolymers had glass transitions between those of the two homopolymers. It is sometimes found, however, that block and graft copolymers may have long enough homogeneous chain segments to exhibit the properties of both homopolymers, rather than intermediate values. Thus, a block copolymer in which one homopolymer has high, and the other low, softening and brittle temperatures may exhibit both a high softening point and a low brittleness temperature.

Effect of heat and entropy of fusion

Although an explanation of the effect of polymer structure on crystalline melting point in terms of intermolecular forces has been accepted for some years (Bunn, gen. ref.), examination of heats and entropies of fusion as a function of polymer structure (Dole and Wunderlich, gen. ref.) suggests that these quantities are of primary importance in determining T_m, as indicated by Eq. 7-3.

Even if the effect of entropy is neglected, it is dangerous to substitute the cohesive energy for the heat of fusion in examining the effect of ΔH on T_m. The cohesive energy, E_v, is the energy of evaporation (corrected for expansion work), while $\Delta H_m = E_s - E_v$, where E_s is the energy of sublimation similarly corrected. Only if $E_s - E_v$ is proportional to E_v can the substitution be made.

The melting points of homologous series of various types of polymers are plotted in Fig. 7-4. The effect of the spacing of the polar groups on T_m is much as predicted by considering the polymers as derived from polyethylene by replacing methylenes with polar groups, but the over-all level of T_m for each series remains to be considered.

Polyesters Although the low melting points of linear aliphatic polyesters are commonly attributed to unusual flexibility of the C-O chain bond, implying high entropies of fusion, both ΔS_m and ΔH_m are significantly lower in these polyesters than in polyethylene. Thus low values of T_m must be attributed to low heats of fusion, for reasons not yet clarified.

Polyamides For polyamides it is observed that ΔH_m is lower than for polyethylene; therefore, molar cohesion cannot account for high values of T_m. It is believed that these result from low liquid-state entropies, leading to low values of ΔS_m. These may arise from partial retention of hydrogen

bonding in the melt and from chain stiffening due to the tendency for resonance of the type

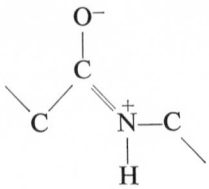

within the amide group.

Similar considerations may apply to the melting points of polyurethans and polyureas, but values of ΔH_m and ΔS_m are not available.

As the spacing between the polar groups is increased, the melting points approach that of polyethylene. When the homologous series are examined in more detail, the melting points are found to vary in a more complex way with the spacing of the polar groups than is suggested in Fig. 7-4. An alternation of T_m with spacing is typical (Fig. 7-5). It results from differences in the crystal structure, which alternates in type for chain repeat units with odd and even numbers of carbon atoms.

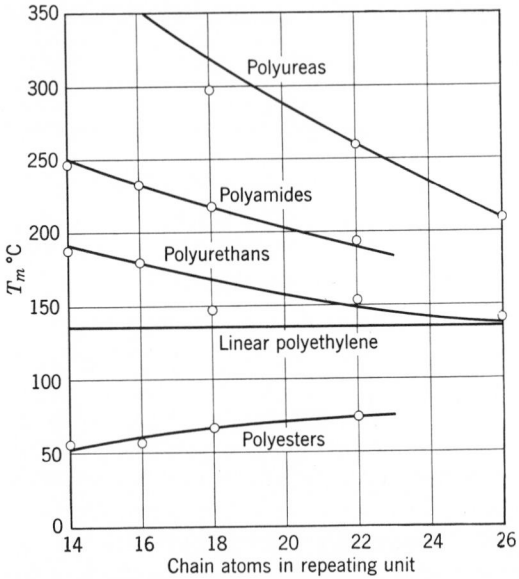

Fig. 7-4. Trend of crystalline melting points in homologous series of aliphatic polymers [after (12)].

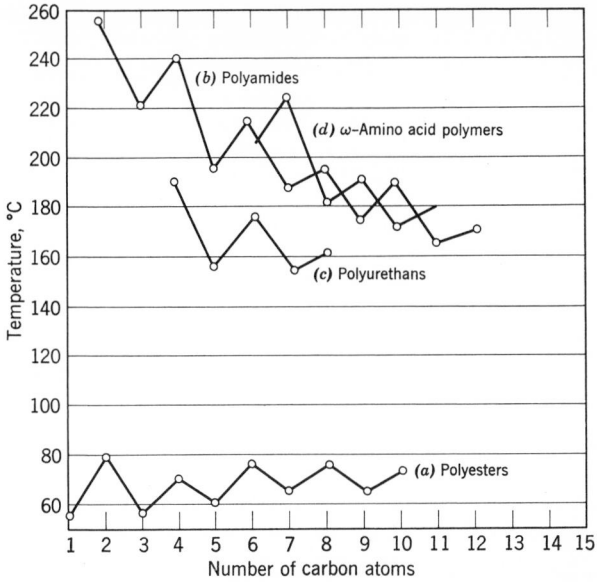

Fig. 7-5. Dependence of crystalline melting point on spacing of polar groups. Number of carbon atoms refers to (*a*) acid for polyesters made with decamethylene glycol; (*b*) diamine for polyamides made with sebacic acid; (*c*) diamine for polyurethans made with tetramethylene glycol; and (*d*) ω-amino acid polymers (Bannerman and Magat, gen. ref.).

Effect of chain flexibility and other steric factors

Chain flexibility The flexibility of chain molecules arises from rotation around saturated chain bonds. The potential energy barriers hindering this rotation range from 1 to 5 kcal/mole, the same order of magnitude as molecular cohesion forces. It is not surprising therefore that the flexibility of polymer chains is an important factor in determining their melting points. Thus polytetrafluoroethylene ($T_m = 327°C$) melts much higher than polyethylene because of its low entropy of fusion (17), which results from the high stiffness of the polymer chains. Similarly, the high melting point of isotactic polypropylene ($T_m = 165°C$) is attributed to low entropy of fusion arising from stiffening of the chain in the melt because of the higher energy barrier for rotation about C-C bonds than in polyethylene. In neither case can a high heat of fusion account for the high value of T_m, since for both polytetrafluoroethylene and polypropylene ΔH_m is well below that of polyethylene as indicated in Table 5-1 (Dole and Wunderlich, gen. ref.).

The substitution of an inflexible group like the *p*-phenylene group,

TABLE 7-1. Effect of a p-phenylene group on the melting point of condensation polymers (7)

Repeating Unit	T_m, °C
$—O(CH_2)_2OCO(CH_2)_6CO—$	45
$—O(CH_2)_2OCO$  $CO—$	265
$—NH(CH_2)_6NHCO(CH_2)_6CO—$	235
$—NH(CH_2)_6NHCO$  $CO—$	350*
$—O(CH_2)_8OCO(CH_2)_8CO—$	75†
$—OCH_2$ 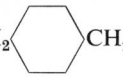 CH_2OCOCH_2 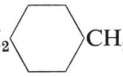 $CH_2CO—$	146
$—CH_2CH_2—$	135
$—CH_2$  $CH_2—$	380

* Decomposes.
† Estimated.

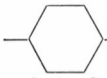 , for six chain CH_2 groups causes a marked rise in the melting point of the polymer (7). Some examples of this change are shown in Table 7-1. The rise is considerable when the *p*-phenylene group is connected to CH_2 groups, but greater still when it is connected to carbonyl, 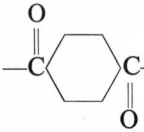 , since the group can resonate as a unit. Among other chain-stiffening groups are *p,p'*-diphenyl, 1,5- or 2,6-naphthyl, diketopiperazine: 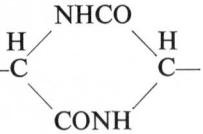 , triazole: 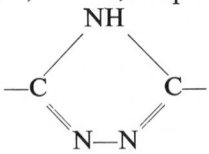 , acetal, and thioketal.

Side-chain substitution In most cases, the substitution of nonpolar groups for hydrogens of a polymer chain leads to a reduction in T_m (or T_g)

or possibly complete loss of crystallinity. If the substitution is random, as in branched polyethylene, the primary effect is a reduction in the size and perfection of the crystalline regions, usually accompanied by a decrease in the degree of crystallinity. The crystalline melting point of polyethylene is lowered 20–25° on going from the linear to the branched material.

Replacement of an amide hydrogen with an alkyl group has a much larger effect, since hydrogen bonding is destroyed. In general, N-methyl nylons melt at least 100° lower than their unsubstituted counterparts.

TABLE 7-2. Effect of side-chain structure on the crystalline melting point of isotactic poly (α-olefins) (1,6)

Side Chain	T_m, °C
$-CH_3$	165
$-CH_2CH_3$	125
$-CH_2CH_2CH_3$	75
$-CH_2CH_2CH_2CH_3$	−55
$-CH_2CH(CH_3)CH_2CH_3$	196
$-CH_2C(CH_3)_2CH_2CH_3$	350

When an alkyl group is regularly substituted into a methylene chain, with the retention of stereoregularity so that crystallization is possible, two effects compete in setting T_m. As indicated for isotactic poly(α-olefins) (1, 6) in Table 7-2 (with polyethylene omitted because of its widely different crystal structure), an increase in the length of the side chain results in a looser crystal structure with an increasingly lower melting point. On the other hand, an increase in the bulkiness of the side chain increases T_m, since rotation in the side chain is hindered in the liquid state, with consequent decrease in ΔS_m.

The glass transition is similarly affected by side-chain substitution, e.g., in poly(*n*-alkyl methacrylates) (14, 16), where increasing chain length from one to twelve carbon atoms reduces T_g from 105 to −65°C. Again, an increase in the bulkiness of the side chain causes an increase in T_g (15). When the side chains become quite long, they can themselves crystallize even though the backbone chain cannot in this case. For example, the

predominant transition observed in poly(octadecyl methacrylate) is crystalline melting at 37.5°C.

Plasticization and crosslinking

The introduction of crosslinks or plasticizer molecules primarily affects the glass transition, except where the effects are extensive enough to cause loss of crystallinity. Plasticization lowers T_g by reducing the interchain forces and increasing the mobility of the structure. The effect of crosslinking is opposite, and T_g is raised.

GENERAL REFERENCES

Bannerman, D. G., and E. E. Magat, "Polyamides and Polyesters," Chap. VII in C. E. Schildknecht, ed., *Polymer Processes*, Interscience Publishers, New York, 1956.

Bueche, F., *Physical Properties of Polymers*, Interscience Publishers, a Division of John Wiley & Sons, New York, 1962, Chap. 13.

Bunn, C. W., "Melting and Second Order Transition Phenomena," Chap. 12 in R. Hill, ed., *Fibres from Synthetic Polymers*, Elsevier Publishing Co., New York, 1953. See also C. W. Bunn, *J. Polymer Sci.* **16,** 323 (1955).

Dole, M., and B. Wunderlich, *Makromol. Chem.* **34,** 29 (1959).

Wilfong, R. E., *J. Polymer Sci.* **54,** 385 (1961).

BIBLIOGRAPHY

1. Bawn, C. E. H., *Chem. & Ind.* **1960,** 388.
2. Beaman, R. G., *J. Polymer Sci.* **9,** 470 (1942).
3. Beevers, R. B., and E. F. T. White, *Trans. Faraday Soc.* **56,** 744 (1960).
4. Beevers, R. B., and E. F. T. White, *Trans. Faraday Soc.* **56,** 1529 (1960).
5. Boyer, R. F., *J. Appl. Phys.* **25,** 825 (1954).
6. Campbell, T. W., and A. C. Haven, Jr., *J. Appl. Polymer Sci.* **1,** 73 (1959).
7. Edgar, O. B., and R. Hill, *J. Polymer Sci.* **8,** 1 (1952).
8. Flory, P. J., *J. Chem. Phys.* **17,** 223 (1949).
9. Fox, T. G., Jr., and P. J. Flory, *J. Appl. Phys.* **21,** 581 (1950).
10. Fox, T. G., and S. Loshaek, *J. Polymer Sci.* **15,** 371 (1955).
11. Grieveson, B. M., *Polymer* **1,** 499 (1960).
12. Hill, R., and E. E. Walker, *J. Polymer Sci.* **3,** 609 (1948).
13. Jenckel, E., *Z. physik. Chem.* **A190,** 24 (1942).
14. Mark, H., *Ann. N.Y. Acad. Sci.* **57,** 445 (1953).
15. Rehberg, C. E., W. A. Faucette, and C. H. Fisher, *J. Am. Chem. Soc.* **66,** 1723 (1944).
16. Rogers, S. S., and L. Mandelkern, *J. Phys. Chem.* **61,** 985 (1957).
17. Starkweather, H. W., Jr., and R. H. Boyd, *J. Phys. Chem.* **64,** 410 (1960).
18. Ueberreiter, K., *Angew. Chem.* **53,** 247 (1940); *Kolloid-Z.* **102,** 272 (1943).
19. Wood, L. A., *J. Polymer Sci.* **28,** 319 (1958).

D. Mechanical Properties of Solid Polymers

This section is primarily concerned with structural determinants of such properties as stiffness, yield stress, elongation, and impact strength. Related to these properties are hardness, abrasion resistance, and flexural fatigue life, among others. The influence of structure on tensile strength where plastic flow is important is treated in Section *B*.

A major determinant of the behavior of a polymer on impact is the relation between the yield stress and the tensile strength in brittle failure, which may be designated the brittle strength (9). If the yield stress or strength is lower than the brittle strength, plastic flow begins (at the yield point in a tensile experiment) and the polymer is tough. If the brittle strength is lower, brittle failure takes place on impact.

The properties of crystalline polymers are emphasized in this section for two reasons: it is these polymers which are most widely utilized because of their mechanical properties, and structural features related to crystallinity may have profound effects on these properties.

Effect of crystallinity

The degree of crystallinity alone is effective in determining the stiffness and yield point for most crystalline plastics (5, 7, 8, 10). As indicated for branched polyethylene in Fig. 7-6, these properties are independent of molecular weight. As a result, they can be expressed as single-valued functions of degree of crystallinity. This is illustrated in Fig. 7-7. As crystallinity decreases, both stiffness and yield stress decrease. As a result of the latter change, the chance of brittle failure is reduced.

As the degree of crystallinity decreases with temperature during the approach to T_m (Chapter 5*F*), stiffness and yield stress decrease correspondingly. These factors often set limits on the temperature at which a plastic is useful for mechanical purposes.

Combined effects of crystallinity and molecular weight

A number of mechanical properties (5, 8, 9), including hardness, flexural fatigue resistance or flex life, softening temperature, elongation at tensile break (where plastic flow occurs), and sometimes impact strength (7, 9), are influenced by both degree of crystallinity and molecular weight. Typical examples are the softening temperature of branched polyethylene (5) as measured in the Vicat test (Chapter 4*G*), which increases with increasing

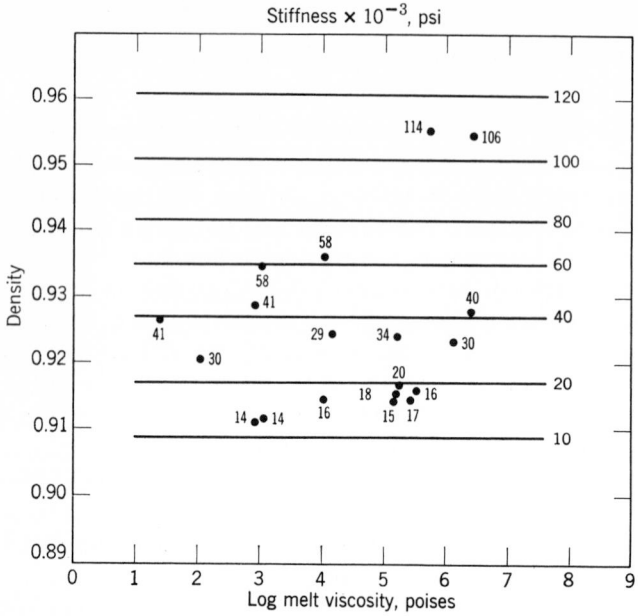

Fig. 7-6. Dependence of stiffness of branched polyethylene on molecular structure variables (5). (See discussion for Fig. 7-1.)

molecular weight and increasing crystallinity (Fig. 7-8), and the flex life of polytetrafluoroethylene (8), which increases with increasing molecular weight and decreasing crystallinity (Fig. 7-9; "standard specific gravity" is a measure of molecular weight, as discussed in Chapter 16*A*). Flex life is sometimes associated with brittleness, samples having low resistance to flexing being brittle (10). The chance of brittle failure is decreased by raising molecular weight, which increases brittle strength, and by reducing crystallinity as indicated previously. For amorphous polymers (2), impact strength is found to depend on the weight-average molecular weight.

Effects of spherulites and orientation

Spherulite size and number A decrease in spherulite size, usually accompanied by an increase in the number of spherulites, has a marked effect on some mechanical properties of polymers (1, 6, 10), but the direction of the effect may differ with the type of polymer. As indicated in Table 7-3, impact strength (or flex life) always increases as spherulite size decreases [although this may in part be limited to experiments with thin films, where the spherulite diameter is an appreciable fraction of the film thickness (4, 7)].

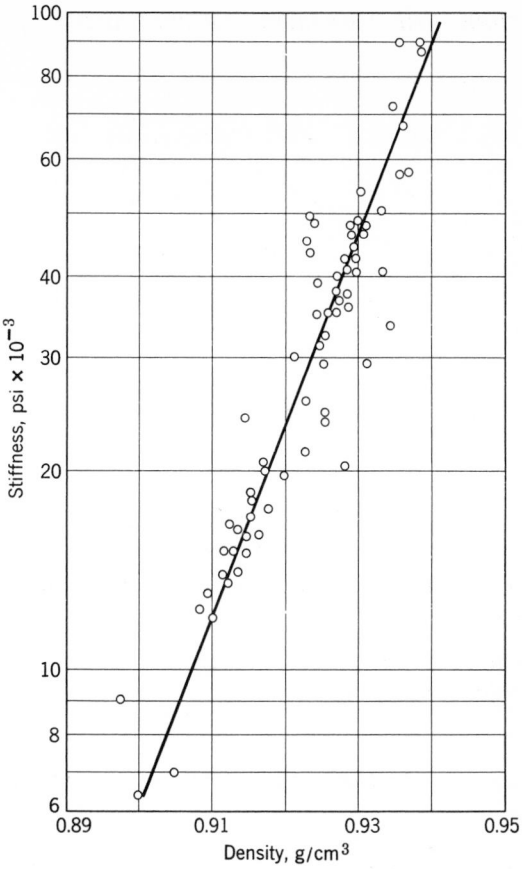

Fig. 7-7. Dependence of stiffness on crystallinity in polyethylene (5).

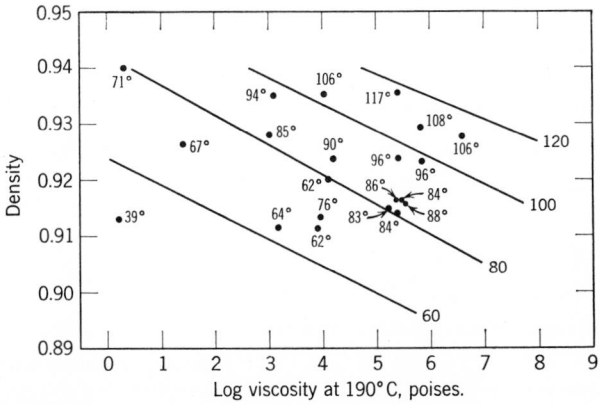

Fig. 7-8. Dependence of softening temperature, as measured by the Vicat test, of branched polyethylene on molecular structure parameters (5). (See discussion for Fig. 7-1.)

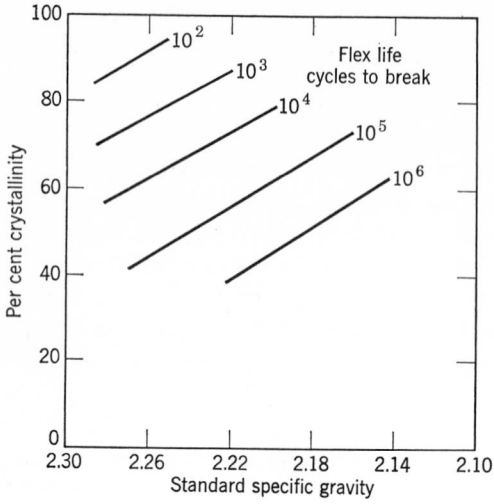

Fig. 7-9. Lines of equal flex life on a crystallinity-molecular weight "map" for polytetrafluoroethylene (8).

TABLE 7-3. Effect of a decrease in spherulite size on the mechanical properties of polymers

Property	Polymer		
	Polypropylene (10)	66 Nylon (6)	Polyoxymethylene (1)
Impact strength	Increased	Increased*	Increased
Yield stress	Decreased	Increased	Decreased
Ultimate elongation	Decreased	Decreased	Increased

* No effect except in thin films; also true for linear polyethylene (4).

On the other hand, the change in yield stress and ultimate elongation (or ductility) with spherulite size is not consistent among the polymers studied.

Orientation The effects of orientation on the physical properties of polymers are profound. They lie in the generally expected direction of increased tensile strength and stiffness with increasing orientation. As orientation is increased, anisotropy of the mechanical properties appears. Oriented (drawn) fibers are much stronger along their length than perpendicular to it.

Effect of copolymerization

The addition of a comonomer to a crystalline polymer usually causes a marked loss in crystallinity, unless the second monomer crystallizes

isomorphous with the first (see Section *C*). Crystallinity typically decreases very rapidly, accompanied by reductions in stiffness, hardness, and softening point, as relatively small amounts (10–20 mole per cent) of the second monomer are added. In many cases, a rigid, fiber-forming polymer is converted to a highly elastic, rubbery product by such minor modification.

The dependence of mechanical properties on copolymer composition in systems which do not crystallize results primarily from changes in intermolecular forces as measured by cohesive energy. Higher cohesive energy results in higher stiffness and hardness and generally improved mechanical properties.

The additional variable introduced by block or graft copolymerization can be used to alter the properties of such copolymers by changing their method of preparation. As described in Section *E*, a block or graft copolymer consisting of long-chain segments of widely differing polarity can exist in solution with one or the other type of segment extended, the second relatively contracted, depending on solvent type. When isolated from solution, the copolymer has properties resembling those of the homopolymer corresponding to the extended segments (3). Thus, natural rubber with chains of poly(methyl methacrylate) grafted to it was hard and stiff with a nontacky surface when isolated from a solution where the rubber chains were collapsed and the poly(methyl methacrylate) chains were extended. When prepared under the opposite conditions, it was limp, flabby, and self-adherent like rubber. A third form with intermediate properties was isolated from solvents in which both chain segments were relatively extended.

Effect of plasticization, reinforcement, or crosslinking

Plasticization The addition of a plasticizer usually reduces stiffness, hardness, and brittleness, and has a similar effect on other mechanical properties, since interchain forces are effectively reduced. These changes are accompanied by a reduction in T_g, as noted in Section *C*. As indicated in Chapter 20*D*, plasticization is usually restricted to amorphous polymers or polymers with a low degree of crystallinity because of the limited compatibility of plasticizers with highly crystalline polymers.

Reinforcement and crosslinking Whether carried out by chemical crosslinking in an unmodified amorphous polymer system or by the addition of a reinforcing filler such as carbon black in rubber (a process involving chemical bonding between polymer and filler), the addition of crosslinks leads to stiffer, stronger, tougher products, usually (in the case of rubbers) with enhanced tear and abrasion resistance as well. However, extensive crosslinking in a crystalline polymer may cause loss of crystallinity, with

attendant deterioration of the mechanical properties depending on this factor. When this occurs, the initial trend of properties may be toward either enhancement or deterioration, depending on the degree of crystallinity of the unmodified polymer and the method of formation and location (crystalline or amorphous regions) of the crosslinks.

BIBLIOGRAPHY

1. Hammer, C. F., T. A. Koch, and J. F. Whitney, *J. Appl. Polymer Sci.* **1,** 169 (1959).
2. McCormick, H. W., F. M. Brower, and L. Kin, *J. Polymer Sci.* **39,** 87 (1959).
3. Merrett, F. M., *J. Polymer Sci.* **24,** 467 (1957).
4. Ohlberg, S. M., J. Roth, and R. A. V. Raff, *J. Appl. Polymer Sci.* **1,** 114 (1959).
5. Sperati, C. A., W. A. Franta, and H. W. Starkweather, Jr., *J. Am. Chem. Soc.* **75,** 6127 (1953).
6. Starkweather, H. W., Jr., and R. E. Brooks, *J. Appl. Polymer Sci.* **1,** 236 (1959).
7. Starkweather, H. W., Jr., G. E. Moore, J. E. Hansen, T. M. Roder, and R. E. Brooks, *J. Polymer Sci.* **21,** 189 (1956).
8. Thomas, P. E., J. F. Lontz, C. A. Sperati, and J. L. McPherson, *SPE J.* **12,** #5, 89 (June, 1956).
9. Vincent, P. I., *Polymer* **1,** 425 (1960).
10. Wijga, P. W. O., "Structure and Properties of Polypropylene," pp. 35–45 in *The Physical Properties of Polymers*, S. C. I. Monograph No. 5, The Macmillan Co. New York, 1959.

E. Other Properties

This section is primarily concerned with structural determinants of two classes of properties. The first includes solubility and related phenomena, such as swelling, cloud points, sorption of liquids, permeability to gases, and compatibility of plasticizers. The second class under discussion includes the effects of electromagnetic radiation, in such optical properties as refractive index and transparency and such electrical properties as dielectric constant, dielectric loss, and dielectric strength.

Effect of molecular weight

Solubility Where the presence of a crystalline phase is not involved, solubility and related phenomena are inverse functions of molecular weight. This fact is reflected in the equations for the thermodynamic properties of polymer solutions, and forms the basis of fractionation methods, as discussed in Chapter 2.

Electrical and optical properties The interactions of electromagnetic radiation with polymers involve, at the most, the cooperative movement of small groups of atoms. The molecular weight dependence of these properties, obeying relations of the type of Eq. 7-1, vanishes at molecular weights far below the polymer range. Except as molecular weight influences some more direct structural determinant of these properties, they are independent of this variable.

Effect of crystallinity

Solubility and related properties As pointed out in Chapter 2, crystallinity decreases the solubility of polymers markedly, since the process of solution involves overcoming the heat and entropy factors associated with crystallization as well as those of the intermolecular interactions in the amorphous regions. Properties related to solubility, such as the cloud point of dilute solutions, are often functions of crystallinity relatively independent of molecular weight (4).

The solubility of liquids and gases in polymers is also strongly dependent on crystallinity, since solubility is usually confined to the amorphous regions. Permeability, the product of solubility and diffusivity, behaves similarly.

Plasticization is closely related to solubility, and the selection of an efficient and compatible plasticizer (see Chapter 20) involves considerations similar to those for the selection of solvents. Plasticization usually results in loss of crystallinity; however, if crystallinity is well developed it may not be possible to find a plasticizer sufficiently compatible with (soluble in) the polymer to have a significant effect on its properties.

Electrical and optical properties The primary effect of crystallinity on the electrical and optical properties is associated with the changes in dielectric constant and refractive index arising from the difference in density between the crystalline and amorphous regions. In the case of visible light, this difference leads to scattering, which may be large if the regions responsible (crystallites or lamellae, and spherulites) are significant in size compared to the wavelength of the light. Thus crystalline plastics usually appear translucent or opaque, their transparency increasing with decreasing spherulite size.

Effect of polar groups

Solubility and related properties In general, the introduction of polar groups into polymers tends to decrease solubility, since strong polymer-polymer bonds usually develop. The situation is complicated, however, by

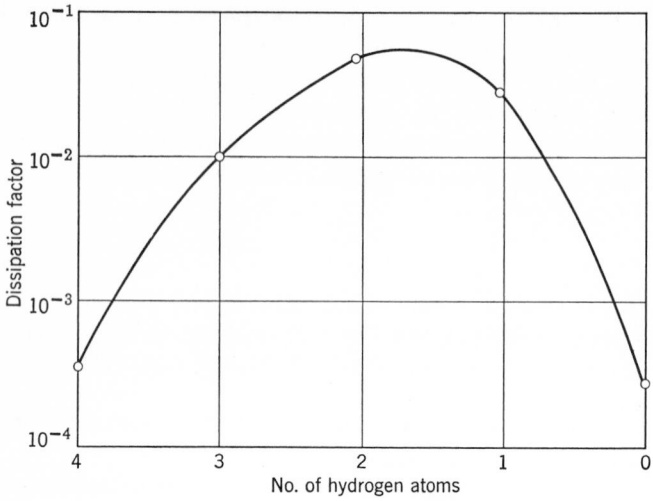

Fig. 7-10. Dielectric loss as a function of composition for the series polyethylene to polytetrafluoroethylene (3).

factors such as the arrangement and bulkiness of the groups, which in turn influence crystallinity.

The solubility behavior of cellulose and its derivatives (2) provides examples of the effect of polarity and substitution on solubility. Cellulose itself is insoluble because of strong hydrogen bonding and stiff chains which prevent hydrating molecules from penetrating its crystalline* regions. Substitution of less polar groups for the hydroxyls on cellulose leads first to solubility in alkalis, which only swell the parent polymer, and then to water solubility when 0.5–1 hydroxyl per glucose unit is replaced. Crystallinity has usually disappeared at this stage. At higher degrees of substitution, water solubility is replaced by alcohol solubility, and then by solubility in organic solvents whose type depends on the nature of the substituent group. When all the substituents are alike, complete substitution leads to crystallization and more limited solubility.

The permeability of polymers to gases and liquids decreases with increasing polarity, since a more polar polymer has a higher activation energy for diffusion. Even methyl groups contribute sufficiently to high cohesion and high activation energy in rubbers to impart low gas permeability. Thus, butyl, nitrile, and chloroprene rubbers all have lower permeability than natural or butadiene-based rubbers.

Electrical properties The electrical properties of polymers depend more on an unbalance or asymmetry of dipoles than on the presence of polar

* See footnote on page 129.

groups *per se*. Thus, both polyethylene and polytetrafluoroethylene have low dielectric constant and dielectric loss (dissipation factor), but these quantities are much larger for polymers containing both hydrogen and fluorine or chlorine (Fig. 7-10). Similarly, the dielectric loss of poly(2,5-dichlorostyrene) is much less than that of poly(3,4-dichlorostyrene) because of cancellation of the dipoles of the *para* chlorines.

Effect of copolymerization

Solubility In essentially random copolymers of monomers whose homopolymers are noncrystalline, a property such as solubility varies more or less regularly from that of one homopolymer to that of the other as the relative proportions of the components are varied. The solubility of copolymers of this type is frequently low in solvents for either homopolymer, but high in mixtures of these solvents.

The solubility of graft and block copolymers is often unusually high, especially if the two components have widely different polarities. Block copolymers of polystyrene with poly(vinyl alcohol) or poly(acrylic acid), e.g., are soluble in benzene, acetone, and water (1). In water the hydrophilic blocks are solubilized and extended, holding the tightly coiled hydrocarbon segments in solution much as a detergent solubilizes a hydrocarbon by micelle formation. In benzene the opposite situation occurs, but in an intermediate solvent such as acetone both blocks are relatively extended, as indicated by higher solution viscosity in intermediate solvents. These polymers act as efficient detergents and emulsifying and compatibilizing agents, but their virtually universal solubility makes it difficult to isolate or purify them.

GENERAL REFERENCES

Burlant, W. J., and A. S. Hoffman, *Block and Graft Copolymers*, Reinhold Publishing Corp., New York, 1960.

Colborne, R. S., "Plasticizers," Chap. 10 in R. Houwink, ed., *Elastomers and Plastomers*, Vol. I, Elsevier Publishing Co., New York, 1950.

Van Amerongen, G. J., "Physics and Structure," Chap. 6 in R. Houwink, ed., *op. cit.*, especially Sec. 5, "Permeability to Gases and Liquids."

Walker, E. E., "Solubility," Chap. 13 in R. Hill, ed., *op. cit.*

BIBLIOGRAPHY

1. Mark, H., *Rec. Chem. Progr.* **12,** 139 (1951).
2. Rånby, B. G., and S. A. Rydholm, pp. 417–428 in "Cellulose and Cellulose Derivatives," Chap. IX in C. E. Schildknecht, ed., *Polymer Processes*, Interscience Publishers, New York, 1956.

3. Sperati, C. A., private communication.
4. Sperati, C. A., W. A. Franta, and H. W. Starkweather, Jr., *J. Am. Chem. Soc.* **75**, 6127 (1953).

F. Property Requirements and Polymer Utilization

Combinations of properties unique to polymers are evidenced in each of the major uses, including elastomers, fibers, and plastics, to which macromolecules are put. In this section the property requirements of these end uses are described and related to polymer structure.

Elastomers

It is a matter of experience that all substances exhibiting a high degree of rubberlike elasticity contain long-chain structures. The restoring force leading to elastic behavior results directly from the decrease in entropy associated with the distortion of a chain macromolecule from its most probable conformation. Two additional property requirements are imposed by the condition that there be sufficient freedom of molecular motion to allow the distortions to take place rapidly: first, the polymer must at its use temperature be above T_g; and, second, it must be amorphous, at least in the undistorted state.

In contrast to the high local mobility of chain segments implied by these requirements, the gross mobility of chains in elastomers must be low. The motions of chains past one another must be restricted in order that the material can regain its original shape when the stress is released. This restriction of gross mobility is usually obtained by the introduction of a network of primary bond crosslinks in the material. (It cannot be obtained through secondary bond forces; these must be kept low in order to gain local segment mobility.) The crosslinks must be relatively few and widely separated, however, so that stretching to large extensions can take place without rupture of primary bonds.

The requirement of low cohesive energy limits the family of elastomers to polymers which are largely hydrocarbon (or fluorocarbon or silicone) in nature, with polar groups distributed at random and in not too great number, so that crystallinity is absent and T_g is sufficiently low, ideally in the range -50 to $-80°C$. The requirement of sites for crosslinking often leads to the choice of a diene as a monomer or comonomer for an elastomer.

In contrast to its equilibrium properties, a stretched elastomer should

have the high tensile strength and modulus usually associated with crystalline plastics. Thus, rubbers in which crystallinity can develop on stretching, such as natural rubber and its stereoregular synthetic counterparts, usually have more desirable properties than those with less regular structures. However, a reinforcing filler can sometimes impart to a rubber properties similar to those obtained on the development of crystallinity. Thus, reinforced styrene-butadiene rubber has properties nearly equivalent to those of reinforced natural rubber, whereas its properties when crosslinked but not reinforced are much poorer.

Fibers

In contrast to elastomers, the requirements of high tensile strength and modulus characteristic of fibers are almost always obtained by utilizing the combination of molecular symmetry and high cohesive energy associated with a high degree of crystallinity. Usually the fiber is oriented to provide optimum properties in the direction of the fiber axis.

The end use requirements of fibers, particularly those involving textiles, lead to restrictions on several properties. The crystalline melting point, T_m, must be above a certain minimum, say 200°C, if the resulting fabric is to be subjected to ironing. On the other hand, spinning the polymer into a fiber requires either that T_m be below, say, 300°C and well below the temperature of decomposition of the polymer, or that the polymer be soluble in a solvent from which it can be spun. Other requirements limit solubility, e.g., in solvents useful for dry cleaning.

The requirement of orientation in the fiber usually implies that T_g be not too high (since orientation by cold drawing and ironing are typically carried out at or above this temperature) or too low (since orientation and related characteristics, such as crease retention after ironing, must be maintained at room temperature).

Thus the selection of a polymer for use as a fiber involves a number of compromises, usually met by choosing a linear polymer with high symmetry and high intermolecular forces resulting from the presence of polar groups, high enough in molecular weight so that tensile strength and related properties are fully developed. Branching in the polymer chain is in general detrimental to fiber properties because branch points disrupt the crystalline lattice, lower the crystalline melting point, and decrease stiffness. Crosslinking, on the other hand, offers the possibility of obtaining strong interchain bonding. If crosslinks are formed after the polymer is spun into fiber, and are relatively few in number, improvement in fiber properties may result. Thus poly(vinyl alcohol), polyurethans, and protein fibers may be crosslinked with formaldehyde to give higher melting point, lower

solubility and moisture regain, and improved hand, while wool, a natural protein fiber, is crosslinked with cystine links.

Plastics

The wide range of end uses of plastics requires a variety of property combinations; correspondingly, a wider variety of structures is important. In general, the properties of plastics are intermediate between those of fibers and elastomers, with much overlapping on either end. Thus, typical plastics may have cohesive energies higher than those of elastomers but lower than those of fibers (Table 1-4). However, a polymer useful as a fiber when oriented in this form may also be useful as a plastic, where orientation is not readily achieved in the massive pieces used; an example is nylon.

Optical applications The requirement of good optical properties, especially in massive pieces, imposes severe limitations on the structure of polymers. In general, crystallinity must be absent, and most amorphous polymers exhibit softness and brittleness which exclude them from many applications. In thin films, structural requirements are not as severe, since crystallinity can often be tolerated if spherulite growth is inhibited and the material can be processed to give sufficiently smooth surfaces. Thus, the normally incompatible requirements of clarity and toughness can both be met.

Electrical applications To achieve low dielectric loss over a wide frequency range, the structure of a polymer must be selected on the basis of low polarity. All other requirements are of considerably less importance. It follows that polyethylene and polytetrafluoroethylene are the best materials for low-loss applications, particularly at high frequencies. At low frequencies, however, other plastics, such as poly(vinyl chloride), are useful.

Mechanical applications Perhaps the most important property requirement for the use of polymers in mechanical applications is toughness. This property is usually achieved by selection of a polymer with a moderate, but not too high, degree of crystallinity (see the discussion of the mechanical properties of crystalline polymers, Chapter 6*E*). Often, a delicate balance of structural features is needed to achieve the desired combination of properties. Composite structures (as in polyblends, Chapter 15*A*) are sometimes useful.

Economic aspects Though the preceding discussion could be extended considerably, it should be pointed out that a major factor in the selection of the appropriate plastic for a given use is economic. In many (some

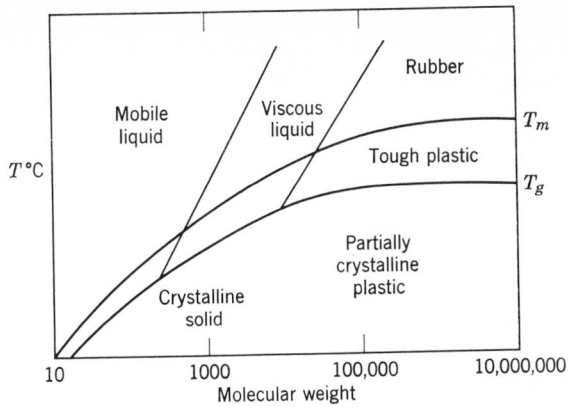

Fig. 7-11. Approximate relations among molecular weight, T_g, T_m, and polymer properties.

observers would say far too many) cases, a mass market is achieved by sacrificing properties for price. The plastics manufacturer must, therefore, consider what structures give the optimum combination of melt and solid-state properties. He would like, e.g., to lower molecular weight to achieve low melt viscosity and more rapid fabrication; but he must maintain levels of molecular weight consistent with the development of good mechanical properties. The selection of raw materials is also important: the rise in prominence of olefin polymers is closely related to the low cost of the monomers. The cost of polymerization itself appears less important in most cases, but it is clear that monomers which are expensive to polymerize have less opportunity to yield large-volume plastics.

Summary

The variables necessary to define the mechanical and physical properties of polymers have now been discussed (Fig. 7-11). The increase of T_m with molecular weight, leveling off as polymer molecular weights are reached, the related approximate behavior of T_g, and the continual increase of viscosity with molecular weight serve to define, in terms of the variables molecular weight and temperature, regions in which the properties of typical plastics, rubbers, viscous liquids, etc., may be found.

III

Polymerization

8

Step-Reaction (Condensation) Polymerization

A. Classification of Polymers and Polymerization Mechanisms

In 1929 W. H. Carothers suggested a classification of polymers into two groups, *condensation* and *addition* polymers. Condensation polymers are those in which the molecular formula of the repeat unit of the polymer chain lacks certain atoms present in the monomer from which it is formed (or to which it can be degraded). For example, a polyester is formed by typical condensation reactions between bifunctional monomers, with the elimination of water:

$$x\text{HO—R—OH} + x\text{HOCO—R}'\text{—COOH} \longrightarrow \text{HO[—R—OCO—R}'\text{—COO—]}_x\text{H} + (2x - 1)\text{H}_2\text{O}$$

Addition polymers (Chapter 9) are those in which this loss of a small molecule does not take place. The most important group of addition polymers includes those derived from unsaturated vinyl monomers:

$$\underset{\displaystyle X}{\underset{|}{\text{CH}_2{=}\text{CH}}} \longrightarrow \text{—CH}_2\text{—}\underset{\displaystyle X}{\underset{|}{\text{CH}}}\text{—CH}_2\text{—}\underset{\displaystyle X}{\underset{|}{\text{CH}}}\text{—, etc.}$$

Carother's original distinction between addition and condensation polymers was amended by Flory, who placed emphasis on the *mechanisms* by which the two types of polymer are formed. Condensation polymers are usually formed by the stepwise intermolecular condensation of reactive groups; addition polymers ordinarily result from chain reactions involving some sort of active center. Different chain structures may result from these

mechanisms: the structural units of condensation polymers are usually joined by interunit functional groups, whereas most addition polymers do not have such functional groups in their backbone chain. However, this distinction is not always observed, leading to the breakdown of the classification as applied to polymerization products, shown in the following examples:

$$CH_2{=}CH_2 \xrightarrow{\text{chain reaction}} \underset{\text{typical addition polymer}}{-CH_2-CH_2-} \xleftarrow{\text{step reaction}} Br(CH_2)_{10}Br + 2Na$$

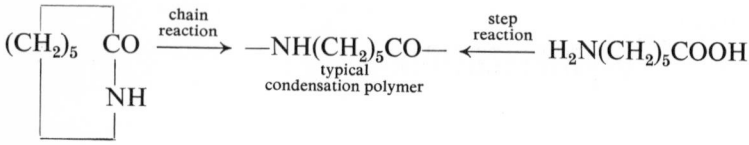

Polymerization mechanisms can be classified in two ways. If a stoichiometric basis is used, condensation polymerization may be said to involve the loss of a small molecule at each reaction step, whereas in addition polymerization this feature is absent. This scheme, however, classes the formation of polyurethans with vinyl chain polymerization, which is kinetically quite different, rather than with the formation of polyesters, polyamides, etc., proceeding by entirely similar processes.

The classification adopted in this book is based on the reaction mechanism, dividing polymerization into *step reactions*, commonly producing

TABLE 8-1. Distinguishing features of chain- and step-polymerization mechanisms

Chain Polymerization	Step Polymerization
Only growth reaction adds repeating units one at a time to the chain.	Any two molecular species present can react.
Monomer concentration decreases steadily throughout reaction.	Monomer disappears early in reaction: at DP* 10, less than 1% monomer remains.
High polymer is formed at once; polymer molecular weight changes little throughout reaction.	Polymer molecular weight rises steadily throughout reaction.
Long reaction times give high yields but affect molecular weight little.	Long reaction times are essential to obtain high molecular weights.
Reaction mixture contains only monomer, high polymer, and about 10^{-8} part of growing chains.	At any stage all molecular species are present in a calculable distribution.

* Degree of polymerization.

step-reaction or condensation polymers, and *chain reactions*, commonly producing chain-reaction or addition polymers (1). Polymerizations are classified without regard to loss of a small molecule (e.g., polyurethans are formed by step-reaction polymerization) or type of interunit linkage (e.g., phenol-formaldehyde resins result from stepwise polymerization even though they lack interunit functional groups). Wherever precise differentiation on the basis of mechanism is required, the terms *step reaction* and *chain reaction* are used; in deference to well-established tradition, the common terms *condensation* and *addition* are permissible where no confusion can result.

Some of the consequences of the differences between the mechanisms of chain and stepwise polymerization are shown in Table 8-1.

GENERAL REFERENCES

Flory, P. J., *Principles of Polymer Chemistry*, Cornell University Press, Ithaca, N.Y., 1953, Chaps. II-1, II-2.

Mark, H., and G. S. Whitby, eds., *Collected Papers of Wallace Hume Carothers on High Polymeric Substances*, Interscience Publishers, New York, 1940, Part One; W. H. Carothers, *J. Am. Chem. Soc.* **51,** 2548 (1929).

BIBLIOGRAPHY

1. Mark, H., and A. V. Tobolsky, *Physical Chemistry of High Polymeric Systems*, Interscience Publishers, New York, 1950, Chaps. XI, XII.

B. Chemistry of Stepwise Polymerization

Types of condensation polymers

Table 8-2 lists some representative products of stepwise polymerization. Most of them are, stoichiometrically, condensation polymers. Proteins and cellulose are included on the basis that they can be degraded hydrolytically to monomers differing from their repeating units by the addition of the elements of a molecule of water.

The type of product formed in a condensation reaction is determined by the *functionality* of the monomers, i.e., by the average number of reactive functional groups per monomer molecule. Monofunctional monomers give only low-molecular-weight products. Bifunctional monomers give linear polymers, as illustrated earlier in this chapter. Polyfunctional monomers, with more than two functional groups per molecule, give branched

TABLE 8-2. Typical step-reaction polymers

Type	Interunit Linkage	Examples
Polyester	$-\overset{O}{\overset{\|\|}{C}}-O-$	$HO(CH_2)_xCOOH \longrightarrow HO[-(CH_2)_xCOO-]_yH + H_2O$
		$HO(CH_2)_xOH + HOOC(CH_2)_{x'}COOH \longrightarrow HO[-(CH_2)_xO\overset{O}{\overset{\|\|}{C}}(CH_2)_{x'}\overset{O}{\overset{\|\|}{C}}O-]_yH + H_2O$
		$\begin{array}{l} CH_2OH \\ \| \\ CHOH \\ \| \\ CH_2OH \end{array} + HOOC(CH_2)_xCOOH \longrightarrow$ three-dimensional network $+ H_2O$
Polyanhydride	$-\overset{O}{\overset{\|\|}{C}}-O-\overset{O}{\overset{\|\|}{C}}-$	$HOOC(CH_2)_xCOOH \longrightarrow HO[-CO(CH_2)_xCOO-]_yH + H_2O$
Polyacetal	$-O-\overset{H}{\underset{R}{\overset{\|}{\underset{\|}{C}}}}-O-$	$HO(CH_2)_xOH + CH_2(OR)_2 \longrightarrow HO[-(CH_2)_xOCH_2O-]_y(CH_2)_xOH + ROH$
Polyamide	$-\overset{O}{\overset{\|\|}{C}}-NH-$	$NH_2(CH_2)_xCOOH \longrightarrow H[-NH(CH_2)_xCO-]_yOH + H_2O$
		$NH_2(CH_2)_xNH_2 + HOOC(CH_2)_{x'}COOH \longrightarrow H[-NH(CH_2)_xNHCO(CH_2)_{x'}CO-]_yOH + H_2O$
Polyurethan	$-O-\overset{O}{\overset{\|\|}{C}}-NH-$	$HO(CH_2)_xOH + OCN(CH_2)_{x'}CNO \longrightarrow [-O(CH_2)_xOCONH(CH_2)_{x'}NHCO-]_y$
Polyurea	$-NH-\overset{O}{\overset{\|\|}{C}}-NH-$	$NH_2(CH_2)_xNH_2 + OCN(CH_2)_{x'}CNO \longrightarrow [-NH(CH_2)_xNHCONH(CH_2)_{x'}NHCO-]_y$
Silk fibroin	$-\overset{O}{\overset{\|\|}{C}}-NH-$	$NH_2CH_2COOH + NH_2CHRCOOH \longleftarrow H[-NHCH_2CONHCHRCO-]_yOH + H_2O$

Cellulose	—C—O—C—	$C_6H_{12}O_6 \longleftarrow —[C_6H_{10}O_4]—O—[C_6H_{10}O_4]— + H_2O$
Phenol-aldehyde	—CH_2— —CH_2OCH_2—	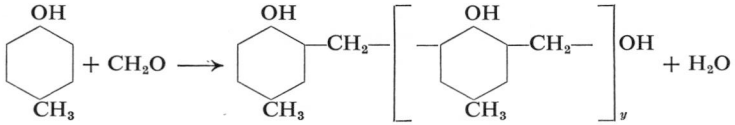  + $CH_2O \longrightarrow$ three-dimensional network with —CH_2— and —CH_2OCH_2— bridges between positions on rings *o*- and *p*- to hydroxyls + H_2O
Urea-aldehyde	—NH—CHR—NH—	$NH_2CONH_2 + CH_2O$ (1 to 1 ratio) $\longrightarrow$ —NHCONH—CH_2—NHCONH—CH_2— + H_2O
	—CHR—N(—CHR—)—CHR—	With excess CH_2O, three-dimensional network + H_2O
Polysulfide	—S—	$Cl(CH_2)_xCl + Na_2S \longrightarrow Cl[—(CH_2)_xS—]_yNa + NaCl$
	—S—S—	$HS(CH_2)_xSH$ + oxidation $\longrightarrow HS[—(CH_2)_xSS—]_y(CH_2)_xSH + H_2O$ or H_2S
	S S ‖ ‖ —S—S—	$Cl(CH_2)_xCl + Na_2S_4 \longrightarrow Cl[—(CH_2)_xS(=S)—S(=S)—]_yNa + NaCl$
Polysiloxane	R R —Si—O—Si— R R	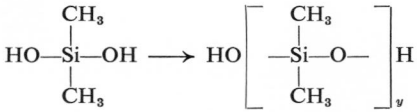
	R R —Si—O—Si— R O R—Si—R	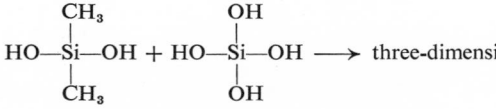 three-dimensional network + H_2O

or crosslinked (three-dimensional) polymers. The properties of the linear and the three-dimensional polymers differ widely.

The chemistry of step-reaction polymerization is discussed in this section, with particular reference to polyesters and polyamides. Further information on the organic chemistry of the polymer types listed in Table 8-2 is found in Part IV.

Types of stepwise reactions

Direct reaction The direct reaction of a dibasic acid and a glycol to form a polyester, or a dibasic acid and a diamine to form a polyamide, works well and is widely used in practice. In esterification, a strong acid or acidic salt often serves as a catalyst. The reaction may be carried out by heating the reactants together and removing water, usually applying vacuum in the later stages.

An important modification of the direct reaction is the use of a salt, as in the preparation of poly(hexamethylene adipamide) (66 nylon) by heating the hexamethylene diamine salt of adipic acid above the melting point in an inert atmosphere. The stringent requirement of stoichiometric equivalence to obtain high molecular weight is easily met by purifying the salt by recrystallization.

Interchange The reaction between a glycol and an ester,

$$x\text{HO—R—OH} + x\text{R}''\text{OCO—R}'\text{—COOR}'' \longrightarrow$$
$$\text{R}''\text{O(—CO—R}'\text{—COO—R—O)}_x\text{H} + (2x - 1)\text{R}''\text{—OH}$$

is often used to produce polyesters, especially where the dibasic acid has low solubility. Frequently the methyl ester is used, as in the production of poly(ethylene terephthalate) from ethylene glycol and dimethyl terephthalate. The reaction between a carboxyl and an ester link is much slower, but other interchange reactions, such as amine-amide, amine-ester, and acetal-alcohol, are well known.

Acid chloride or anhydride Either of these species can be reacted with a glycol or an amine to give a polymer. The anhydride reaction is widely used to form an alkyd resin from phthalic anhydride and a glycol:

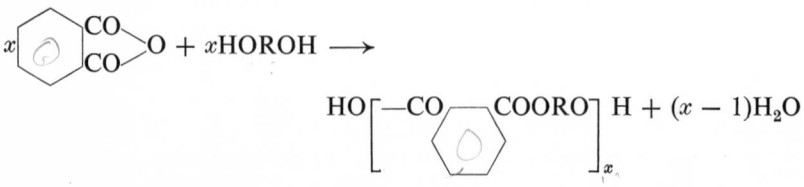

The condensation in bulk of an acid chloride with a glycol is not useful because of side reactions leading to low-molecular-weight products, but the reaction of an acid chloride with a diamine is a valuable means of preparing polyamides.

Interfacial condensation

The reaction of an acid halide with a glycol or a diamine proceeds rapidly to high-molecular-weight polymer if carried out at the interface between two liquid phases, each containing one of the reactants (1). Very-high-molecular-weight polymer can be formed. Typically, an aqueous phase containing the diamine or glycol and an acid acceptor is layered at room temperature over an organic phase containing the acid chloride. The polymer formed at the interface can be pulled off as a continuous film or filament (Fig. 8-1). The method has been applied to the formation of polyamides, polyurethans, polyureas, polysulfonamides, and polyphenyl esters. It is particularly useful for preparing polymers which are unstable at the higher temperatures usual in step-reaction polymerization.

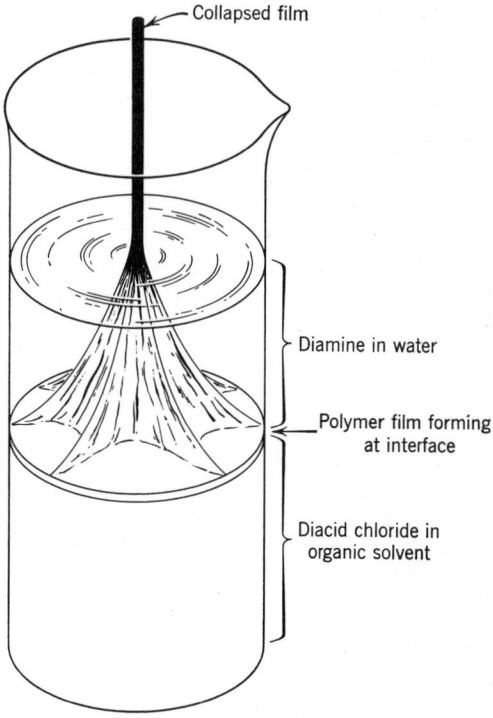

Fig. 8-1. Formation of a polyamide by interfacial polymerization (1).

Ring versus chain formation

In addition to polymer formation, bifunctional monomers may react intramolecularly to produce a cyclic product. Thus, hydroxy acids may give either lactones or polymers on heating,

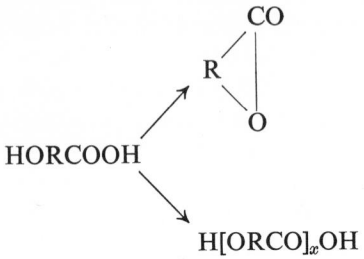

amino acids may give lactams or linear polyamides, etc.

The chief factor governing the type of product is the size of the ring which can be formed. If the ring contains less than five atoms or more than seven, the product is usually linear polymer. If a ring of five atoms can form, it will do so to the exclusion of linear chains. If six- or seven-membered rings can form, either type of product can result. Larger rings can sometimes be formed under special conditions.

The lack of formation of rings with less than five atoms is explained by the strain imposed by the valence angles of the ring atoms. Five-membered rings are virtually strain free, and all larger rings can be strain free if non-planar forms are possible. As the ring size increases, the statistical probability of forming rings becomes smaller. The formation of five-membered rings to the complete exclusion of linear polymer is not thoroughly understood.

GENERAL REFERENCES

Mark, H., and G. S. Whitby, eds., *loc. cit.*

Marvel, C. S., *Introduction to the Organic Chemistry of High Polymers*, John Wiley & Sons, New York, 1959, pp. 11–34.

Sorenson, W. R., and T. W. Campbell, *Preparative Methods of Polymer Chemistry*, Interscience Publishers, New York, 1961, Chap. III.

BIBLIOGRAPHY

1. Morgan, P. W., and S. L. Kwolek, *J. Chem. Educ.* **36,** 182 (1959); P. W. Morgan, *SPE J.* **15,** 485 (1959); E. L. Wittbecker, P. W. Morgan, and coworkers, *J. Polymer Sci.* **40,** 289–407 (1959).

C. Kinetics and Statistics of Linear Stepwise Polymerization

Reactivity and molecular size

The similarity between monofunctional and polyfunctional step reactions is illustrated in the chemical equations in this chapter. Other similarities exist in the influence of temperature and catalysts on the rates of the reactions. It is tempting to substitute the simple concept of reactions between functional groups for that of the myriad separate reactions of each molecular species with all the others. With this substitution, the kinetics of stepwise polymerization is essentially identical with that of simple condensation; without it, analysis of the kinetics would be almost hopelessly difficult.

If the simplification is to be made, the rate of the reaction of a group must be independent of the size of the molecule to which it is attached. This assumption is amply justified by experimental evidence. The rate constants of condensation reactions in a homologous series reach asymptotic values independent of chain length quite rapidly and show no tendency to drop off with increase in molecular size. The diluting effect of the large chain must, of course, be taken into account by adjusting to constant molar composition of reactive groups. In addition, the rate constants for monofunctional and bifunctional reagents are identical for sufficiently long chains separating the reactive groups of the bifunctional compound.

In explanation of these results, it may be recalled that chemical reactions in condensed phases usually take place during a period of many collisions between reactive groups before they diffuse apart. Although a polymer chain as a whole diffuses very slowly, the mobility of the terminal functional group on the chain is much greater than that of the entire chain. Such a group diffuses readily over a considerable region through rearrangements in the conformations of nearby chain segments. Its collision rate with its neighbors is at least comparable to that prevailing in liquids. A somewhat lower diffusion rate not only prolongs the time before two groups diffuse into the same neighborhood, but proportionally prolongs the time during which the groups are close together and colliding. Stepwise polymerizations take place at rates such that only about one collision in 10^{13} leads to chemical reaction; within the length of time necessary for this number of collisions there is ample opportunity for diffusion to maintain the concentration of reactive pairs essentially at equilibrium.

In very dilute solution, the reactivity of a functional group may be reduced because the segments of its own molecule shield it from functional

groups on other molecules. In concentrated systems, however, the molecules are extensively intertwined and show no preference for units of their own chain over those of other molecules. Here it is sufficient to dispose of the shielding problem by taking the diluent effect of nonreactive segments into account by computing the concentrations of functional groups per unit volume.

Kinetics of stepwise polymerization

With the concept of functional group reactivity independent of molecular weight, the kinetics of stepwise polymerization becomes quite simple. The formation of a polyester from a glycol and a dibasic acid may be taken as an example. It is well known that this reaction is catalyzed by acids. In the absence of added strong acid, a second molecule of the acid being esterified acts as a catalyst. The reaction is followed by measuring the rate of disappearance of carboxyl groups:*

$$-\mathrm{d}[\mathrm{COOH}]/\mathrm{d}t = k[\mathrm{COOH}]^2[\mathrm{OH}] \tag{8-1}$$

If the concentrations c of carboxyl and hydroxyl groups are equal,

$$-\mathrm{d}c/\mathrm{d}t = kc^3$$

and

$$2kt = 1/c^2 - \text{const.} \tag{8-2}$$

It is convenient to introduce the *extent of reaction*, p, defined as the fraction of the functional groups that has reacted at time t. Then

$$c = c_0(1 - p) \tag{8-3}$$

and

$$2c_0^2kt = 1/(1 - p)^2 + \text{const.} \tag{8-4}$$

or a plot of $1/(1 - p)^2$ should be linear in time. Typical experimental data (Fig. 8-2) bear this out over a wide range of reaction times.

If only bifunctional reactants are present and no side reactions occur,

* Stepwise polymerization typically involves equilibrium reactions of the type

$$\mathrm{A} + \mathrm{B} \underset{k_r}{\overset{k_f}{\rightleftharpoons}} \mathrm{C} + \mathrm{D}$$

where the rates of the forward and reverse reactions are $k_f[\mathrm{A}][\mathrm{B}]$ and $k_r[\mathrm{C}][\mathrm{D}]$, respectively. At equilibrium these rates are equal, whence $K = k_r/k_f = [\mathrm{A}][\mathrm{B}]/[\mathrm{C}][\mathrm{D}]$. If the system is far from equilibrium, as in the initial stages of polymerization, the reverse reaction is negligibly slow, and changes in the concentrations of the reactants may be considered to result from the forward reaction alone, as in Eq. 8-1 and those following.

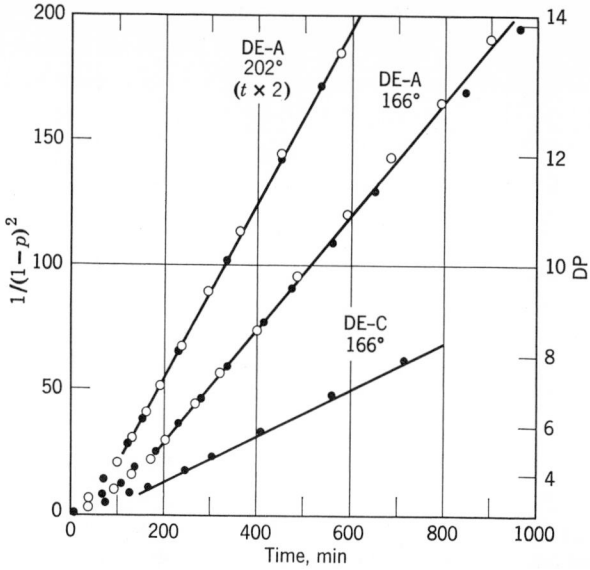

Fig. 8-2. Reactions of diethylene glycol (DE) with adipic acid (A) and caproic acid (C) (Flory, in Burk and Grummitt, eds., gen. ref.). Time values at 202°C have been multiplied by 2.

the number of unreacted carboxyl groups equals the total number of molecules in the system N. If acid or glycol groups separately (not in pairs) are defined as structural units, the initial number of carboxyls present is equal to the total number of structural units present, N_0. The number-average degree of polymerization $\bar{x}_n$ is simply

$$\bar{x}_n = N_0/N = c_0/c = 1/(1-p) \tag{8-5}$$

The scale of D.P. in Fig. 8-2 shows that uncatalyzed esterifications require quite long times to reach high degrees of polymerization. Greater success is achieved by adding to the system a small amount of catalyst, whose concentration is constant throughout the reaction. In this case the concentration of the catalyst may be included in the rate constant:

$$\begin{aligned} -\mathrm{d}[\mathrm{COOH}]/\mathrm{d}t &= k'[\mathrm{COOH}][\mathrm{OH}] \\ -\mathrm{d}c/\mathrm{d}t &= k'c^2 \\ c_0k't &= 1/(1-p) + \text{const.} \end{aligned} \tag{8-6}$$

Here $\bar{x}_n$ increases linearly with reaction time (Fig. 8-3). Similar reactions have been shown to be linear up to at least $\bar{x}_n = 90$, corresponding to molecular weights of 10,000.

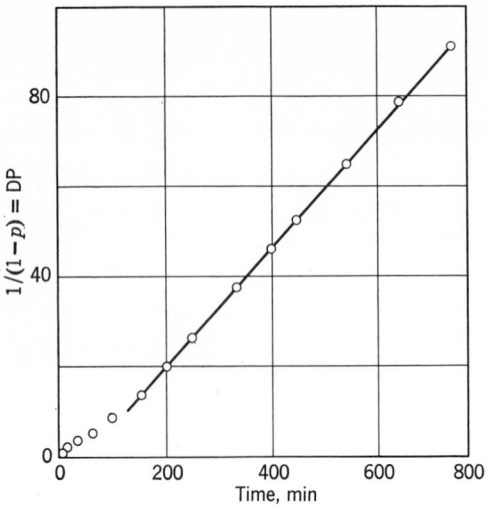

Fig. 8-3. Reaction of diethylene glycol with adipic acid, catalyzed by 0.4 mole per cent of p-toluenesulfonic acid (Flory, in Burk and Grummitt, eds., gen. ref.).

Statistics of linear step-reaction polymerization

Rate equations similar to Eq. 8-6 may be written for each molecular species in the reaction mixture. From these it is possible to derive the distribution of molecular weights in a step-reaction polymer. The same relation is, however, more readily derived from statistical considerations.

The statistical analysis of stepwise polymerization yields results which are equivalent to and in some cases more easily obtained than those of kinetic considerations. The method assumes independence of reaction rate and molecular size. The extent of reaction p is defined as the probability that a functional group has reacted at time t. This definition is entirely equivalent to the previous definition of p as the fraction of the functional groups which has reacted. It follows that the probability of finding a functional group unreacted is $1 - p$.

It is now necessary to find the probability that a given molecule selected at random is an x-mer, i.e., contains x repeating units. Such a molecule contains $x - 1$ reacted functional groups (e.g., carboxyls) and, on the end, one unreacted group of this type. The probability of finding a single reacted carboxyl group in the molecule is p, and that of finding $x - 1$ of them in the same molecule is p^{x-1}. The presence of one unreacted group has a probability of $1 - p$. Hence the probability of finding the complete molecule is $p^{x-1}(1 - p)$. This is also equal to the fraction of all the

molecules which are x-mers. If there are N molecules in all, the total number of x-mers is

$$N_x = Np^{x-1}(1 - p) \tag{8-7}$$

If the total number of units present is N_0, $N = N_0(1 - p)$, and

$$N_x = N_0(1 - p)^2 p^{x-1} \tag{8-8}$$

This is the number-distribution function for a linear stepwise polymerization at extent of reaction p.

The weight fraction w_x of x-mers is given by

$$\begin{aligned} w_x &= xN_x/N_0 \\ &= x(1 - p)^2 p^{x-1} \end{aligned} \tag{8-9}$$

This is the weight-distribution function for a linear stepwise polymerization at extent of reaction p. Equations 8-8 and 8-9 are illustrated in Figs. 8-4 and 8-5, respectively. On a number basis, monomers are more plentiful than any other molecular species at all stages of the reaction. On a weight basis, however, the proportion of low-molecular-weight material is very small and decreases as the average molecular weight increases. The maximum in the weight-distribution curve occurs near the number-average molecular weight.

Various average degrees of polymerization may be derived from the weight- or number-distribution functions according to the definitions of

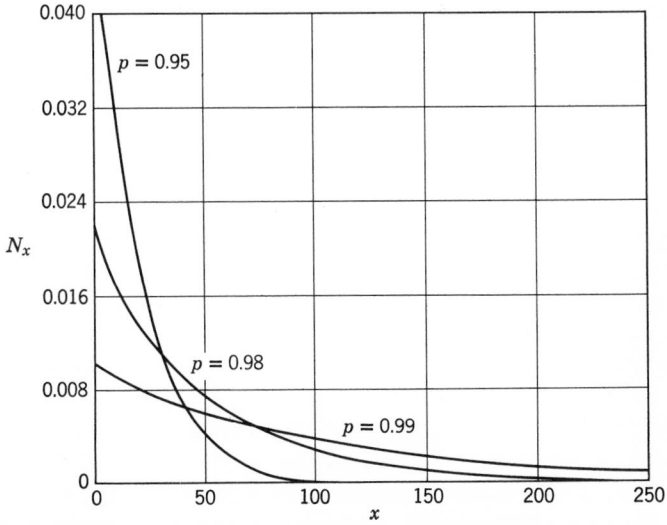

Fig. 8-4. Number or mole-fraction distribution of chain molecules in a linear step-reaction polymer for several extents of reaction p (Flory, in Burk and Grummitt, eds., gen. ref.).

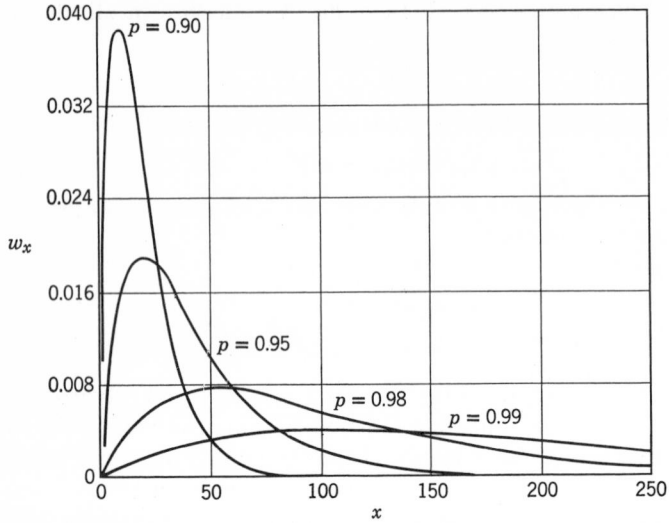

Fig. 8-5. Weight-fraction distribution of chain molecules in a linear step-reaction polymer for several extents of reaction p (Flory, in Burk and Grummitt, eds., gen. ref.).

the corresponding molecular weight averages in Chapter 3. The number-average degree of polymerization $\bar{x}_n$ calculated in this way is that given in Eq. 8-5. The weight-average degree of polymerization is

$$\bar{x}_w = (1 + p)/(1 - p) \tag{8-10}$$

The breadth of the molecular weight-distribution curve, $\bar{x}_w/\bar{x}_n = 1 + p$; at large extents of reaction $\bar{x}_w/\bar{x}_n \rightarrow 2.0$. This has been confirmed experimentally for nylon (2).

Molecular weight control

Although initial molecular weight control in a stepwise polymerization can be achieved by stopping the reaction (e.g., by cooling) at the desired point, the stability of the product to subsequent heating may not be adequate with respect to changes in molecular weight.

The easiest way to avoid this situation is to adjust the composition of the reaction mixture slightly away from stoichiometric equivalence, by adding either a slight excess of one bifunctional reactant or (as in molecular weight stabilization of nylons by acetic acid) a small amount of a monofunctional reagent. Eventually the functional group deficient in amount is completely used up, and all chain ends consist of the group present in excess. If only bifunctional reactants are present, the two types of groups being designated A and B and initially present in numbers $N_A < N_B$ such

that the ratio $r = N_A/N_B$, the total number of monomers present is $(N_A + N_B)/2 = N_A(1 + 1/r)/2$. At extent of reaction p (defined for A groups; for B groups, extent of reaction $= rp$), the total number of chain ends is $N_A(1 - p) + N_B(1 - rp) = N_A[1 - p + (1 - rp)/r]$. Since this is twice the number of molecules present,

$$\bar{x}_n = \frac{N_A(1 + 1/r)/2}{N_A[1 - p + (1 - rp)/r]/2} = \frac{1 + r}{1 + r - 2rp} \tag{8-11}$$

As $p \to 1$,

$$\bar{x}_n = (1 + r)/(1 - r) \tag{8-12}$$

Thus if 1 mole per cent of stabilizing groups is added,

$$\bar{x}_n = [1 + (100/101)]/[1 - (100/101)] = 201.$$

This illustrates the precision needed in maintaining stoichiometric balance in order to obtain high degrees of polymerization. Loss of one ingredient, side reactions, or the presence of monofunctional impurities may limit severely the degree of polymerization which can be achieved.

The analysis can be applied to the case of an added monofunctional reagent, retaining Eqs. 8-11 and 8-12 unchanged if r is appropriately defined in terms of the numbers of functional groups present.

Interchange reactions

In interchange reactions, which may occur freely at elevated temperature, as in polymer melts, the molecular weights of the reacting molecules can change, since, e.g., two molecules of average length may react to give one longer and one shorter than average. However, the total number of molecules and hence $\bar{x}_n$ do not change.

It can be shown that, if free interchange takes place, the final molecular weight distribution is always the most probable distribution defined by Eqs. 8-8 and 8-9. The change can be demonstrated by mixing two polymers of different $\bar{x}_n$ and following a weight-average property (such as melt viscosity) as a function of time at elevated temperature.

Multichain polymer

Another type of step-reaction polymer of interest is produced by polymerizing a bifunctional monomer of type A—B with a small amount of monomer of functionality f of the type R—A_f. If the reaction is carried nearly to completion, the resulting polymer consists of f chains growing out from a central unit R. Network structures cannot occur, since no units

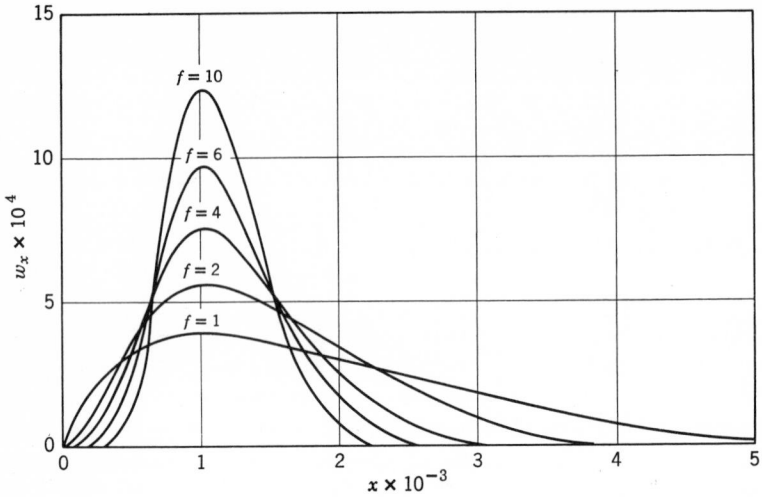

Fig. 8-6. Molecular weight distributions for multichain polymers with functionality f as shown (1).

of the type B—B are present. The case $f = 1$ corresponds to the most probable distribution of Eq. 8-9. The polymer formed when $f = 2$ is also linear; those with higher functionalities are branched.

Since the length of each of the branches is statistically determined independent of all the others, the probability of having a molecule with several branches all longer or shorter than average becomes smaller as the number of branches increases. The molecular weight distribution thus becomes narrower with increasing functionality, as shown in Fig. 8-6 (1). The distribution breadth is given by

$$\bar{x}_w/\bar{x}_n = 1 + 1/f \tag{8-13}$$

For $f = 2$, $\bar{x}_w/\bar{x}_n = 1.5$.

GENERAL REFERENCES

Burnett, G. M., *Mechanism of Polymer Reactions*, Interscience Publishers, New York, 1954, Chap. XIIA, B.

Flory, P. J., *op. cit.*, Chaps. II-3, III, and VIII-1.

Flory, P. J., "Condensation Polymerization and Constitution of Condensation Polymers," pp. 211–283 in R. E. Burk and O. Grummitt, eds., *High Molecular Weight Organic Compounds* (*Frontiers in Chemistry*, Vol. VI), Interscience Publishers, New York, 1949; see also P. J. Flory, *Chem. Revs.* **39,** 137 (1946).

Howard, G. J., "The Molecular Weight Distribution of Condensation Polymers" pp. 185–231 in J. C. Robb and F. W. Peaker, eds., *Progress in High Polymers*, Vol. 1, Academic Press, New York, 1961.

BIBLIOGRAPHY

1. Schulz, G. V., *Z. physik. Chem.* **B43,** 25 (1939).
2. Taylor, G. B., *J. Am. Chem. Soc.* **69,** 638 (1947).

D. Polyfunctional Step-Reaction Polymerization

Three-dimensional step-reaction polymers are produced from the polymerization of reactants with more than two functional groups per molecule. The structures of these polymers are more complex than those of linear step-reaction polymers. Three-dimensional polymerization is complicated experimentally by the occurrence of *gelation*, or the formation of essentially infinitely large polymer networks in the reaction mixture. The sudden onset of gelation marks the division of the mixture into two parts: the *gel*, which is insoluble in all nondegrading solvents; and the *sol*, which remains soluble and can be extracted from the gel. As the polymerization proceeds beyond the *gel point* the amount of gel increases at the expense of sol, and the mixture rapidly transforms from a viscous liquid to an elastic material of infinite viscosity. An important feature of the onset of gelation is the low number-average molecular weight of the mixture at the gel point, where its weight-average molecular weight becomes infinite.

In the statistical consideration of three-dimensional step polymerization, it is assumed that all functional groups are equally reactive, independent of molecular weight or viscosity. This assumption is not correct, e.g., for glycerol, where the secondary hydroxyl group is known to be less reactive than the two primary hydroxyls; but this complication does not affect the general conclusions of the theory.

It is also assumed that all the reactions occur between functional groups on different molecules. This is known to be somewhat in error, for it can be shown that the number of interunit links formed is always somewhat greater than the corresponding decrease in the number of molecules present. The error again affects only the numerical agreement between theory and experiment and not the over-all considerations of the theory.

Gelation

Prediction of the gel point In order to calculate the point in the reaction at which gelation takes place, a *branching coefficient* α is defined as the

probability that a given functional group on a *branch unit* (i.e., a unit of functionality greater than 2) is connected to another branch unit.

The value of α at which gelation becomes possible can be deduced as follows. Consider the case where bifunctional A—A and B—B units are present, as well as polyfunctional units A_f with functionality f. The structures resulting in this system consist of chain segments of the type

$$A_{f-1}\text{—}A[B\text{—}BA\text{—}A]_iB\text{—}BA\text{—}A_{f-1}$$

where i may have any value, plus other segments which lack branch units on one or both ends. The criterion for gel formation is that at least one of the $f - 1$ segments radiating from the end of a segment of the type shown is in turn connected to another branch unit. The probability of this occurring is just $1/(f - 1)$; hence the critical value of α for gelation is

$$\alpha_c = 1/(f - 1) \tag{8-14}$$

Here f is the functionality of the branch units; if more than one type of branch unit is present, an average f over all types of branch units may be used in Eq. 8-14.

To relate α to the extent of reaction, consider the probability of obtaining a segment of the type shown. If the extents of reaction for A and B groups are p_A and p_B, and the ratio of A groups on branch units to all A groups in the mixture is ρ, the probability that a B group has reacted with a branch unit is $p_B\rho$; with a bifunctional A, $p_B(1 - \rho)$. The probability that a segment of the type shown is obtained is given by

$$p_A[p_B(1 - \rho)p_A]^i p_B\rho$$

Summing over all values of i gives

$$\alpha = \frac{p_A p_B \rho}{1 - p_A p_B(1 - \rho)} \tag{8-15}$$

Either p_A or p_B can be eliminated from this expression by defining $r = N_A/N_B$, whence $p_B = rp_A$. Then

$$\alpha = \frac{rp_A{}^2\rho}{1 - rp_A{}^2(1 - \rho)} = \frac{p_B{}^2\rho}{r - p_B{}^2(1 - \rho)} \tag{8-16}$$

Simpler relations can be derived for several special cases. When equal numbers of A and B groups are present, $r = 1$ and $p_A = p_B = p$:

$$(r = 1) \quad \alpha = \frac{p^2\rho}{1 - p^2(1 - \rho)} \tag{8-17}$$

When there are no A—A units, $\rho = 1$ and

$$(\rho = 1) \quad \alpha = rp_A{}^2 = \frac{p_B{}^2}{r} \tag{8-18}$$

If both conditions apply,

$$(r = \rho = 1) \quad \alpha = p^2 \tag{8-19}$$

Finally, with only branch units present the probability that a functional group on a branch unit leads to another branch unit is just the probability that it has reacted:

$$\text{(branch units only)} \quad \alpha = p \tag{8-20}$$

The equations hold for all branch unit functionalities with the definitions given for r and ρ.

Experimental observations of the gel point The gel point can be observed precisely as the time when the polymerizing mixture suddenly loses fluidity, e.g., when bubbles no longer rise in it. If the extent of reaction has been followed as a function of time, say by removing aliquots of the solution and titrating for the number of functional groups present, the value of p at the gel point can be determined.

In a series of experiments (1) with glycerol and dibasic acids, gelation occurred at $p = 0.765$. Here Eq. 8-19 was appropriate; hence $\alpha_c = p^2 = 0.58$. Correction for the lower reactivity of the secondary hydroxyl of glycerol would lower this value slightly. From Eq. 8-14 the theoretical value of α_c is 0.50.

In several other cases investigated by Flory, observed values of α_c were always slightly higher than those theoretically required. This discrepancy is attributed to the reaction of some functional groups to form intramolecular links, which do not contribute to network structures. The reactions must, therefore, be carried slightly farther to reach the critical point.

Molecular weight distributions in three-dimensional step-reaction polymers

The distribution functions for three-dimensional polymers are derived with somewhat more difficulty than those for the linear case, Eqs. 8-8 and 8-9. They depend upon the functionality and relative amounts of all the units involved. Only one example is discussed, for simplicity that of the reaction of equivalent quantities of two trifunctional monomers, all three of the functional groups of each monomer being equally reactive. In this case the weight-distribution function is

$$w_x = [(fx - x)!f/(x - 1)!(fx - 2x + 2)!]p^{x-1}(1 - p)^{fx-2x+2} \tag{8-21}$$

This equation is analogous to Eq. 8-9. The first factor (in brackets) arises because of the numerous geometric isomers of a polyfunctional x-mer. (In

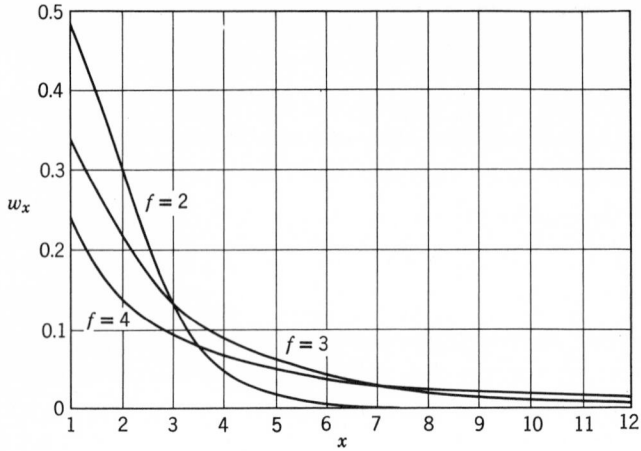

Fig. 8-7. Molecular size distribution in stepwise polymerizations for monomers with functionality f, at $p = 0.3$ (2).

the linear case this factor is just x.) The second term is the probability of finding $x - 1$ links in an x-mer, and the last term is the probability of finding $(f - 2)x + 2 = fx - 2x + 2$ unreacted links or ends.

Equation 8-21 is illustrated in Figs. 8-7, 8-8, and 8-9. In comparison with the linear case, the weight distributions of branched step-reaction polymers of increasing functionality are progressively broader at equivalent extents of reaction (Fig. 8-7). Figure 8-8 shows that the distributions broaden out with increasing extent of reaction p (or, alternatively, α, in the trifunctional case).

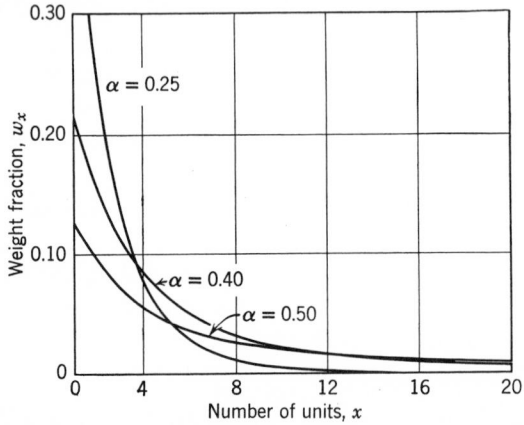

Fig. 8-8. Molecular size distribution in step-reaction polymer formed from trifunctional units at various stages in the reaction, denoted by $\alpha = p$ (Flory, in Burk and Grummitt, eds., gen. ref.).

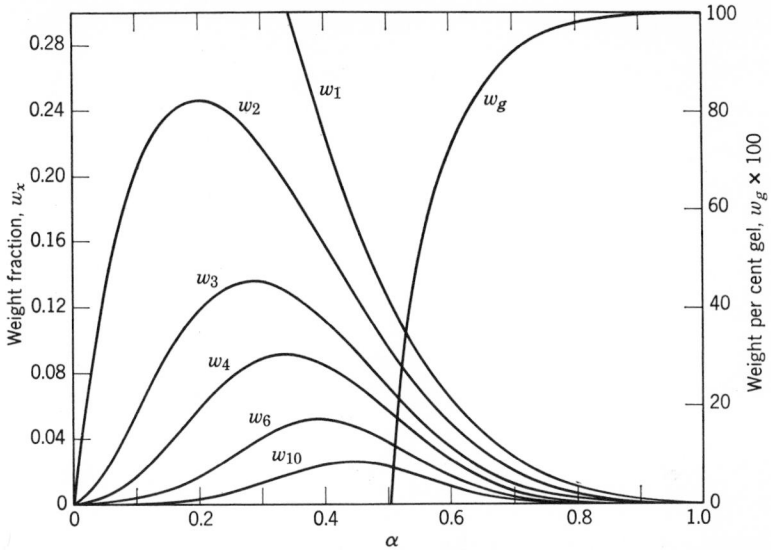

Fig. 8-9. Weight fractions of various molecular species in a trifunctional stepwise polymerization as a function of $\alpha = p$ (Flory, in Burk and Grummitt, eds., gen. ref.). The weight fractions of the finite species are calculated from Eq. 8-21, that of the gel from Eq. 8-24.

In Fig. 8-9 the weight fraction of the various molecular species is plotted against $\alpha = p$. In contrast to the linear case (Figs. 8-8 and 8-9 vs. Fig. 8-5), the weight fraction of monomer is always greater than the amount of any one of the other species, the weight fractions of higher species being successively lower. The extent of reaction at which the weight fraction of any species reaches its maximum shifts continuously to higher values for higher molecular weights. In no case, however, does the maximum occur beyond the gel point ($\alpha = \frac{1}{2}$).

Up to the gel point the sum of the weight fractions of all the species present must equal unity; beyond this point the sum of the weight fractions of all finite species drops below unity as the weight fraction of gel increases. At $\alpha = 1$ only gel is present.

The number-average degree of polymerization is given by

$$\bar{x}_n = \frac{1}{1 - fp/2} \tag{8-22}$$

and the weight-average degree by

$$\bar{x}_w = xw_x = \frac{1 + p}{1 - (f - 1)p} \tag{8-23}$$

At the gel point the weight-average degree of polymerization becomes

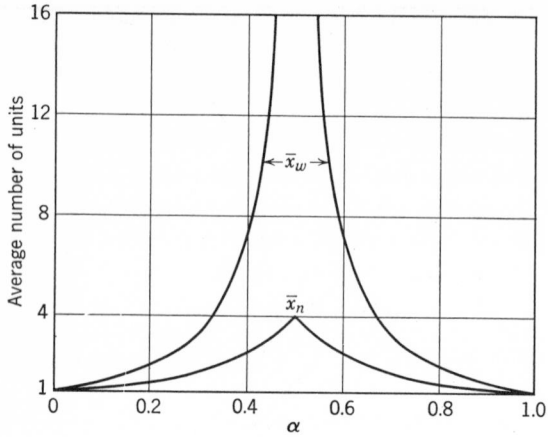

Fig. 8-10. Weight- and number-average degree of polymerization as a function of α for a trifunctional step-reaction polymer (Flory, in Burk and Grummitt, eds., gen. ref.).

infinite. As may be seen in Fig. 8-10, where both averages are plotted against $\alpha = p$, $\bar{x}_n$ attains only a value of $1/(1 - \frac{3}{4}) = 4$ at this point. The very large values of $\bar{x}_w/\bar{x}_n$ near the gel point illustrate the extreme breadth of the distributions. (The portions of the curves in Fig. 8-10 above the gel point refer only to the sol fraction.)

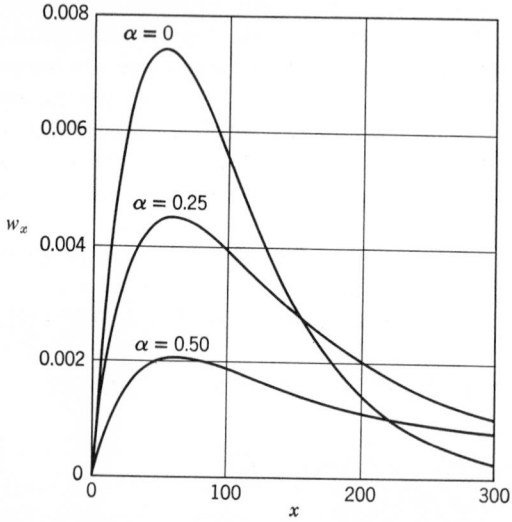

Fig. 8-11. Weight distributions for step-reaction polymers made to $\bar{x}_n = 50$ with various degrees of branching α induced by the addition of trifunctional units: $\alpha = 0$ indicates no branching, and $\alpha = 0.50$ is the critical point for gelation (Flory, in Burk and Grummitt, eds., gen. ref.).

The increase in distribution breadth with increasing degree of branching is illustrated in Fig. 8-11. These curves represent the calculated distributions for bifunctional step reactions in which the branching factor α was varied by varying the small amount of added trifunctional units. The curves were calculated for $\bar{x}_n = 50$ in each case. The curves are drawn so that the total area under each curve extended to infinite x is the same.

Equation 8-21 is valid for finite species beyond the gel point. If it is summed over all values of x and subtracted from unity, the weight fraction of gel results. For trifunctional units this is

$$w_g = 1 - [(1 - p)^3/p^3] \tag{8-24}$$

The distribution functions for $\alpha > \frac{1}{2}$ are identical with those at the corresponding lower value $1 - \alpha$ except that they are reduced by a factor $w_s = 1 - w_g$, the total weight fraction of sol present at that value of α. Thus the complexity of the distribution of the sol fraction above the gel point is presumed to become less until, at $\alpha = 1$, only monomer is left in a vanishingly small amount. The reduction of $\bar{x}_w/\bar{x}_n$ shown for values of α increasing above $\frac{1}{2}$ in Fig. 8-10 is another way of expressing the same phenomenon.

GENERAL REFERENCES

Burnett, G. M., *op. cit.*, Chap. XII C-F.
Flory, P. J., *op. cit.*, Chap. IX-1 and IX-2.
Flory, P. J., "Condensation Polymerization and Constitution of Condensation Polymers," pp. 211–283 in R. E. Burk and O. Grummitt, eds., *op. cit.;* see also P. J. Flory, *Chem. Revs.* **39,** 137 (1946).
Howard, G. J., *loc. cit.*

BIBLIOGRAPHY

1. Kienle, R. H., P. A. van der Meulen, and F. E. Petke, *J. Am. Chem. Soc.* **61,** 2258, 2268 (1939); R. H. Kienle and F. E. Petke, *ibid.* **62,** 1053 (1940); **63,** 481 (1941).
2. Mark, H., and A. V. Tobolsky, *Physical Chemistry of High Polymeric Systems*, 2nd ed., Interscience Publishers, New York, 1950, Chap. XI-H.

E. Ring-Scission Polymerization

In Section *A* it was mentioned that some step-reaction polymers may alternatively be prepared by routes, involving ring opening, which must be classed stoichiometrically as addition, since no small molecule is split off

TABLE 8-3. Typical ring-scission polymers

Type	Example
Lactone	$O(CH_2)_xCO \longrightarrow -[O(CH_2)_xCO-]_y$
Lactam	$HN(CH_2)_xCO \longrightarrow -[HN(CH_2)_xCO-]_y$
Cyclic ether	$(CH_2)_xO \longrightarrow -[(CH_2)_xO-]_y$
Cyclic anhydride	$CO(CH_2)_xCO \longrightarrow -[CO(CH_2)_xCOO-]_y$
N-carboxyanhydride	$COCHRNHCO \longrightarrow -[COCHRNH-]_y + CO_2$

in the reaction. Some typical examples are given in Table 8-3. The polymerization of these compounds usually proceeds by a ring-opening step induced either catalytically or by the presence of a small amount of end-group-forming substance, followed by an interchange reaction. Thus the polymerization of caprolactam in the presence of small amounts of water or of uncyclized monomer presumably proceeds by the mechanism

$$HN(CH_2)_5CO + H_2O \longrightarrow H_2N(CH_2)_5COOH \xrightarrow{+\text{lactam}} H[HN(CH_2)_5CO]_xOH$$

Similar mechanisms can be written for the polymerization of lactones, cyclic ethers such as ethylene oxide, and cyclic anhydrides, including N-carboxyanhydrides. The latter polymerize to form polypeptides with the elimination of CO_2:

$$COCH_2NHCO \longrightarrow -[COCH_2NH-]_x + CO_2$$

Most of these cyclic compounds also polymerize by ionic chain mechanisms in the presence of strong acids or bases when water and alcohols are excluded. These polymerizations are often very rapid. They are of commercial interest in the polymerization of caprolactam, ethylene oxide, and 3,3-bis(chloromethyl)oxetane.

The stepwise polymerization of these compounds resembles chain polymerization in that it proceeds by the addition of monomer (but never of larger units) to growing chain molecules. However, the chain-initiating and subsequent addition steps are similar and proceed at similar rates; these are not chain reactions in the kinetic sense. As in stepwise polymerization, the polymer molecules continue to increase in molecular weight throughout the reaction. On the other hand, ionic ring-scission polymerization is a true chain reaction involving initiation, an extremely rapid propagation step, and termination, as discussed in Chapter 10.

In stepwise ring-scission polymerization, all the chains are initiated simultaneously and grow for the same length of time. In consequence, the distribution of molecular weights in the resulting polymer should be extremely narrow (1). If the number of monomers added to each chain is n,

$$\bar{x}_w/\bar{x}_n = 1 + n/(n + 1)^2 \tag{8-25}$$

which approaches unity as n becomes large.

GENERAL REFERENCES

Eastham, A. M., *Fortschr. Hochpolym.-Forsch.* (*Advances in Polymer Science*) **2,** 18 (1960).

Flory, P. J., *op. cit.*, Chap. II-5.

Furukawa, J., and T. Saegusa, *Polymerization of Aldehydes and Oxides*, Interscience Publishers, a Division of John Wiley & Sons, New York, 1962.

Sorenson, W. R., and T. W. Campbell, *loc. cit.*

BIBLIOGRAPHY

1. Flory, P. J., *J. Am. Chem. Soc.* **62,** 1561 (1940).

9

Radical Chain (Addition) Polymerization

A. Chemistry of Vinyl Polymerization

The polymerization of unsaturated monomers typically involves a chain reaction. It can be initiated by methods typical for simple gas-phase chain reactions, including the action of ultraviolet light. It is susceptible to retardation and inhibition. In a typical chain polymerization, one act of initiation may lead to the polymerization of thousands of monomer molecules.

The characteristics of chain polymerization listed in Table 8-1 suggest that the active center responsible for the growth of the chain is associated with a single polymer molecule through the addition of many monomer units. Thus polymer molecules are formed from the beginning, and almost no species intermediate between monomer and high-molecular-weight polymer are found. Of several postulated types of active center, three have been found experimentally: cation, anion, and free radical. Free radical polymerization is discussed in this chapter, and the related cases of ionic and coordination polymerization are described in Chapter 10.

The concept of vinyl polymerization as a chain mechanism is not new, dating back to Staudinger's work in 1920 (8). However, an alternative mechanism of a stepwise reaction involving hydrogen transfer (10) was seriously considered as late as 1936. The controversy was largely settled by Flory's analysis of the kinetics of vinyl polymerization in 1937 (2). He showed conclusively that radical polymerization proceeds by and requires the steps of initiation, propagation, and termination typical of chain reactions in low-molecular-weight species.

Vinyl monomers

Electronic structures of organic compounds As noted in Chapter 1*C*, most organic compounds involve covalent or shared electron bonds. Single covalent bonds have bond electrons which are distributed within a considerable volume of space between the two nuclei and symmetrically about the line connecting them. In a double bond, the presence of the second pair of electrons destroys the axial symmetry, modifying the volume within which the electrons are found to a ribbonlike band between the nuclei. As a result there is a high energy barrier preventing free rotation, and *cis-trans* isomers become possible. The fact that the second electron pair is held only about 70% as firmly as the first leads to high polarizability and chemical reactivity in unsaturated compounds.

The positions and reactivities of the electrons in unsaturated molecules are subject to several geometrical and electrical influences. *Mesomerism* and *resonance* are used to explain certain facts, such as the equivalence of the reactivity of all the hydrogen atoms in benzene. The presence of strongly electropositive or electronegative groups in a molecule may cause an *inductive shift* of electrons. In chloroacetic acid, e.g., the shift of electrons toward the chlorine atom effectively increases the strength of the acid:

$$\mathrm{Cl} \longleftarrow \mathrm{CH_2} \longleftarrow \overset{\overset{\displaystyle O}{\|}}{C} \longleftarrow \mathrm{O} \longleftarrow \mathrm{H}$$

A further effect is purely *electrostatic* in nature: if, e.g., an atom or group of atoms is substituted for a hydrogen atom adjacent to a double bond, the electrostatic force exerted by the electrons of the substituent on those of the double bond affects the reactivity of the molecule.

The end result of these influences is that the reaction of the double bond with a free radical proceeds well for compounds of the type $CH_2{=}CHX$ and $CH_2{=}CXY$, called *vinyl monomers.* (Monomers in which fluorine is substituted for hydrogen may be included in this class.) The polymerization of monomers with more than one double bond is considered in Chapter 12.

Not all vinyl monomers yield high polymer as a result of radical polymerization. Aliphatic hydrocarbons other than ethylene polymerize only to oils; 1,2-disubstituted ethylenes not at all. Among compounds of the type $CH_2{=}CXY$, those in which both groups are larger than CH_3 polymerize slowly if at all.

Mechanism of vinyl radical polymerization

Generation of free radicals Many organic reactions take place through intermediates having an odd number of electrons and, consequently, an unpaired electron. Such intermediates are known as *free radicals.* They can be generated in a number of ways, including thermal or photochemical decomposition of organic peroxides or hydroperoxides, or azo or diazo compounds. Two reactions commonly used to produce radicals for polymerization are the thermal or photochemical decomposition of benzoyl peroxide:

$$(C_6H_5COO)_2 \longrightarrow 2C_6H_5COO\cdot \longrightarrow 2C_6H_5\cdot + 2CO_2$$

and of azobisisobutyronitrile:

$$\underset{\underset{CN}{|}}{(CH_3)_2C}N{=}N\underset{\underset{CN}{|}}{C}(CH_3)_2 \longrightarrow 2\underset{\underset{CN}{|}}{(CH_3)_2C}\cdot + N_2$$

The stability of radicals varies widely. Primary radicals are less stable and more reactive than secondary radicals, which are in turn less stable than tertiary ones. (Some tertiary radicals, such as the triphenylmethyl, can be isolated in the solid state without decomposition.) The phenyl radical is more reactive than the benzyl radical, the allyl radical is quite unreactive, and so on.

Initiation When free radicals are generated in the presence of a vinyl monomer, the radical adds to the double bond with the regeneration of another radical. If the radical formed by decomposition of the *initiator* I is designated R·,

$$I \longrightarrow 2R\cdot$$

$$R\cdot + CH_2{=}CHX \longrightarrow RCH_2\overset{\overset{H}{|}}{\underset{\underset{X}{|}}{C}}\cdot$$

The regeneration of the radical is characteristic of chain reactions.

Evidence for the radical mechanism of addition polymerization comes not only from the capability of radicals to accelerate vinyl polymerization, but also from the demonstration that the polymers so formed contain fragments of the radicals. The presence of heavy atoms, such as bromine or iodine, or radioactive atoms, in the initiator has been shown many times to lead to polymers from which these atoms cannot be removed.

The efficiency with which radicals initiate chains can be estimated by comparing the amount of initiator decomposed with the number of polymer chains formed. The decomposition of the initiator can usually be followed

by analytical methods. The most direct method of finding the initiator efficiency then depends upon analyzing the polymer for initiator fragments. This is not difficult in cases where the initiator leaves a reactive end group on the polymer or is radioactively tagged. In other polymerizations it is more useful to determine the number of polymer molecules formed from the number-average molecular weight of the polymer. Still another alternative, sometimes of doubtful validity, is to react the chain radical stoichiometrically with an inhibitor (see page 268). Most initiators in typical vinyl polymerizations have efficiencies between 0.6 and 1.0; i.e., between 60 and 100% of all the radicals formed ultimately initiate polymer chains.

Propagation The chain radical formed in the initiation step is capable of adding successive monomers to propagate the chain:

$$\mathrm{R{-}(CH_2CHX{-})_x CH_2\overset{\overset{\displaystyle H}{|}}{\underset{\underset{\displaystyle X}{|}}{C}}\cdot} + \mathrm{CH_2{=}CHX} \longrightarrow \mathrm{R{-}(CH_2CHX{-})_{x+1} CH_2\overset{\overset{\displaystyle H}{|}}{\underset{\underset{\displaystyle X}{|}}{C}}\cdot}$$

Termination Propagation would continue until the supply of monomer was exhausted were it not for the strong tendency of radicals to react in pairs to form a paired-electron covalent bond with loss of radical activity. This tendency is compensated for in radical polymerization by the small concentration of radical species compared to monomers.

The termination step can take place in two ways:

Combination or *coupling:*

$$\mathrm{{-}CH_2\overset{\overset{\displaystyle H}{|}}{\underset{\underset{\displaystyle X}{|}}{C}}\cdot} + \mathrm{\cdot\overset{\overset{\displaystyle H}{|}}{\underset{\underset{\displaystyle X}{|}}{C}}CH_2{-}} \longrightarrow \mathrm{{-}CH_2\overset{\overset{\displaystyle H}{|}}{\underset{\underset{\displaystyle X}{|}}{C}}{-}\overset{\overset{\displaystyle H}{|}}{\underset{\underset{\displaystyle X}{|}}{C}}CH_2{-}}$$

Disproportionation:

$$\mathrm{{-}CH_2\overset{\overset{\displaystyle H}{|}}{\underset{\underset{\displaystyle X}{|}}{C}}\cdot} + \mathrm{\cdot\overset{\overset{\displaystyle H}{|}}{\underset{\underset{\displaystyle X}{|}}{C}}CH_2{-}} \longrightarrow \mathrm{{-}CH_2\overset{\overset{\displaystyle H}{|}}{\underset{\underset{\displaystyle X}{|}}{C}}{-}H} + \mathrm{\overset{\overset{\displaystyle H}{|}}{\underset{\underset{\displaystyle X}{|}}{C}}{=}CH{-}}$$

in which hydrogen transfer results in the formation of two molecules with one saturated and one unsaturated end group. Each type of termination is known. Studies (1) of the number of initiator fragments per molecule show that polystyrene terminates predominantly by combination, whereas poly(methyl methacrylate) terminates entirely by disproportionation at polymerization temperatures above 60°C, and partly by each mechanism at lower temperatures.

Radical-molecule reactions

Although the three steps of initiation, propagation, and termination are both necessary and sufficient for chain polymerization, other steps can take place during polymerization. As these often involve the reaction between a radical and a molecule, they are conveniently so classified.

Chain transfer It was recognized by Flory (2) that the reactivity of a radical could be transferred to another species, which would usually be capable of continuing the chain reaction. The reaction involves the transfer of an atom between the radical and the molecule. If the molecule is saturated, like a solvent or other additive, the atom must be transferred to the radical:

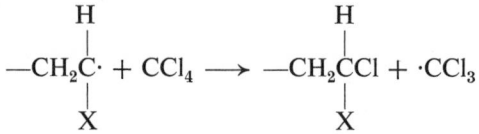

If the molecule is unsaturated, like a monomer, the atom transferred (usually hydrogen) can go in either direction:

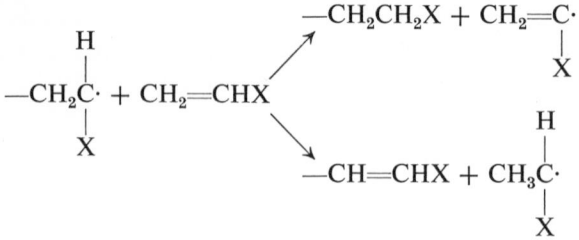

The major effect of chain transfer to a saturated small molecule (solvent, initiator, or deliberately added *chain transfer agent*) is the formation of additional polymer molecules for each radical chain initiated. Transfer to polymer, and transfer to monomer with subsequent polymerization of the double bond, lead to the formation of branched molecules. The latter reaction has a pronounced effect on molecular weight distribution (Section *E*) and is important in the production of graft copolymers (Chapter 11*D*).

The efficiency of compounds as chain transfer agents varies widely with molecular structure. Aromatic hydrocarbons are rather unreactive unless they have benzylic hydrogens. Aliphatic hydrocarbons become more reactive when substituted with halogens. Carbon tetrachloride and carbon tetrabromide are quite reactive. The reactivity of various polymer radicals to transfer varies widely. Transfer reactions offer a valuable means for comparing radical reactivities.

Retardation A *retarder* is defined as a substance which can react with a radical to form products incapable of adding monomer. If the retarder is very effective, no polymer may be formed; this condition is sometimes called *inhibition* and the substance an *inhibitor*. The distinction is merely one of degree. The two phenomena are illustrated in Fig. 9-1, where the rate of polymerization is displayed. Benzoquinone is an inhibitor for the polymerization of styrene; nitrobenzene is a retarder. Nitrosobenzene is

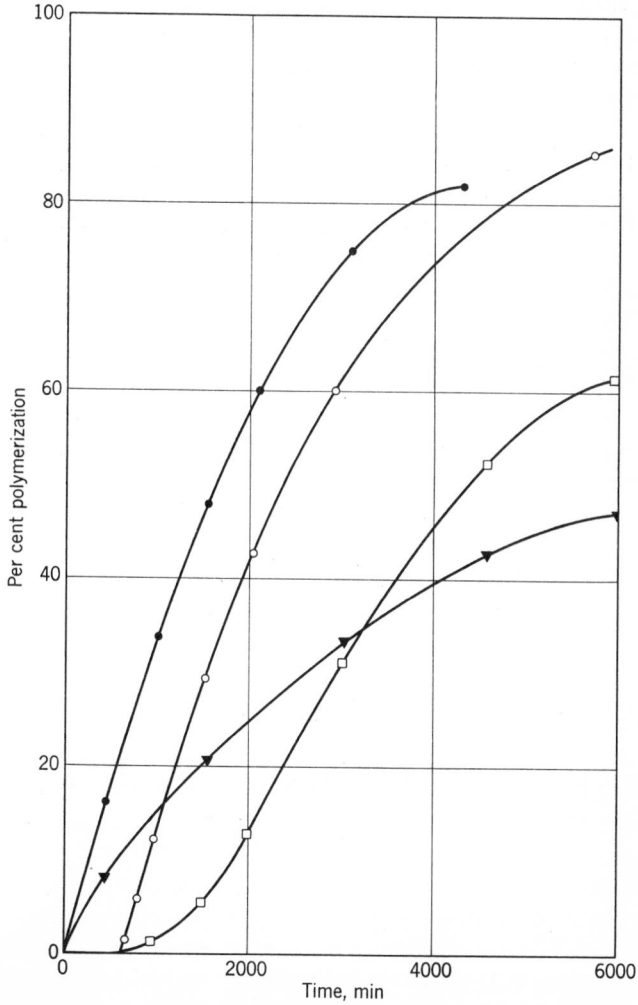

Fig. 9-1. Thermal polymerization of styrene at 100°C (7). ●, no additive; ○, with 0.1% benzoquinone; ▼, with 0.5% nitrobenzene; □, with 0.2% nitrosobenzene.

an inhibitor which reacts to form a substance which is a retarder. The action of a retarder is two-fold: it both reduces the concentration of radicals and shortens their average lifetime and thus the length of the polymer chains.

In the simplest case, the retarder may be a free radical, such as triphenylmethyl or diphenylpicrylhydrazyl, which is too unreactive to initiate a polymer chain. The mechanism of retardation is simply the combination or disproportionation of radicals. If the retarder is a molecule, the chemistry of retardation is more complex, but the product of the reaction must be a radical too unreactive to initiate a chain.

Inhibitors are useful in determining initiation rates, since their reaction with radicals is so rapid that the decomposition of inhibitor is independent of its concentration but gives directly the rate of generation of radicals. As a result the length of the *induction period* before polymerization starts is directly proportional to the number of inhibitor molecules initially present. This number then represents the number of radicals produced during the time of the induction period. For this analysis to be valid it must be known that one molecule of inhibitor is exactly equivalent to one radical. There are a number of reasons why this may not be so. Molecular inhibitors generate free radicals in their action; if these free radicals disappear by disproportionating, an inhibitor molecule reappears, so that eventually one inhibitor molecule stops two radicals. More severe discrepancies have been shown to exist: benzoquinone can undergo transfer with styrene chains, reducing molecular weight but not rate. With less reactive monomers (e.g., vinyl acetate) benzoquinone presumably acts only as an inhibitor. Triphenylmethyl initiates chains as well as inhibiting them (the latter much more rapidly). Diphenylpicrylhydrazyl was long presumed to act only as an inhibitor, but this is not certain.

Configuration of monomer units in vinyl polymer chains

The addition of a vinyl monomer to a free radical can take place in either of two ways:

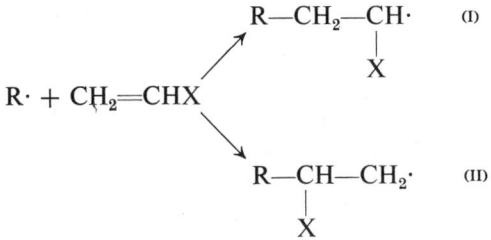

the reaction leading to the more stable product being favored. Since the unpaired electron can participate in resonance with the substituent X in structure I but not in II, reaction I is favored. Steric factors also favor reaction I.

The occurrence of reaction I exclusively would lead to a *head-to-tail* configuration in which the substituents occur on alternate carbon atoms:

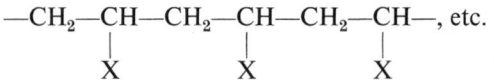

Alternative possibilities are a *head-to-head, tail-to-tail* configuration:

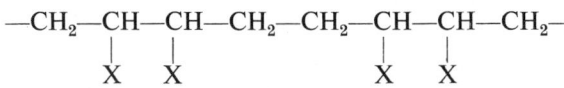

or a random structure containing both arrangements. The possibility of obtaining a regular head-to-head, tail-to-tail configuration exclusively is remote,* and it appears that only an occasional monomer unit enters the chain in a reverse manner to provide a single head-to-head, tail-to-tail linkage.

Evidence for predominantly head-to-tail polymerization was first found by Staudinger (9) and was confirmed for several polymer types in a series of studies by Marvel and coworkers (3, 4, 5) using chemical methods. It is safe to conclude that all polymers so far studied contain a major proportion of head-to-tail linkages.

GENERAL REFERENCES

Baxendale, J. H., "Introduction to Free Radical Polymerization," Chap. I in C. E. Schildknecht, ed., *Polymer Processes*, Interscience Publishers, New York, 1956.

Bevington, J. C., *Radical Polymerization*, Academic Press, New York, 1961.

Burnett, G. M., *Mechanism of Polymer Reactions*, Interscience Publishers, New York, 1954, Chap. III.

Flory, P. J., *Principles of Polymer Chemistry*, Cornell University Press, Ithaca, N.Y., 1953, Chaps. II-4, IV-4, VI-1.

Marvel, C. S., *An Introduction to the Organic Chemistry of High Polymers*, John Wiley & Sons, New York, 1959, pp. 34–45.

* The analog of a completely head-to-head, tail-to-tail polymer has been made (6) by the coordination polymerization of ethylene and 2-butene to a 1 : 1 alternating copolymer with the structure

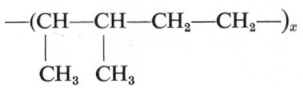

Price, C. C., *Mechanisms of Reactions at Carbon-Carbon Double Bonds*, Interscience Publishers, New York, 1946.

Walling, C., *Free Radicals in Solution*, John Wiley & Sons, New York, 1957, Chaps. 3.1, 3.2, 4.3, 5.1; *Chem. Eng. News* **39**, #40, 98 (Oct. 2, 1961).

BIBLIOGRAPHY

1. Bevington, J. C., H. W. Melville, and R. P. Taylor, *J. Polymer Sci.* **12**, 449 (1954); **14**, 463 (1954).
2. Flory, P. J., *J. Am. Chem. Soc.* **59**, 241 (1937).
3. Marvel, C. S., and C. E. Denoon, Jr., *J. Am. Chem. Soc.* **60**, 1045 (1938).
4. Marvel, C. S., and C. L. Levesque, *J. Am. Chem. Soc.* **60**, 280 (1938).
5. Marvel, C. S., J. H. Sample, and M. F. Roy, *J. Am. Chem. Soc.* **61**, 3241 (1939).
6. Natta, G., paper presented at the International Symposium on Macromolecular Chemistry, Montreal, Can., July, 1961.
7. Schulz, G. V., *Ber.* **80**, 232 (1947).
8. Staudinger, H., *Ber.* **53**, 1073 (1920).
9. Staudinger, H., and A. Steinhofer, *Ann.* **517**, 35 (1935).
10. Whitby, G. S., and M. Katz, *J. Am. Chem. Soc.* **50**, 1160 (1928).

B. Laboratory Methods in Vinyl Polymerization

This section is limited to the discussion of small-scale laboratory polymerizations carried to low conversion with pure monomer and minor amounts of initiator and possibly solvent, retarder, or transfer agent. Such experiments are typical of those used to obtain kinetic data. Further discussion of polymerization methods is found in Chapters 10–12 and Part IV.

As developed in the following sections, the experimental data of interest include the over-all rate of polymerization, the rate constants for initiation, propagation, and termination (and transfer where present), and the molecular weight of the polymer. The rate constants for propagation and termination cannot be separated in the simple experiments described here and are discussed in Section *D*; measurement of molecular weight is discussed in Chapter 3.

Rate of polymerization

Chemical methods or isolation and weighing of polymer can be used to measure the rate of polymerization, but it is customary to work in an evacuated and sealed system to avoid extraneous effects due to the presence

of oxygen. Polymerization rate is then followed through changes in a physical property of the system, such as density, refractive index, or ultraviolet or infrared absorption. Measurement of density is usually the most sensitive and convenient technique. The reaction is carried out in a *dilatometer*, a vessel equipped with a capillary tube in which the liquid level can be measured precisely. The decrease in volume on polymerization is relatively large, being 21% for methyl methacrylate and 27% for vinyl acetate. In practice a few hundredths per cent polymerization can be detected. The experiments are usually carried out in the region below 5% polymer so as to avoid deviations from constant polymerization rate due to depletion of reactants, etc. The dilatometer is placed in a constant-temperature bath to ensure isothermal conditions throughout the experiment.

Types of initiation

In the usual case, initiation is effected by thermal decomposition of an initiator as described in Section *A*. Direct thermal initiation through decomposition of monomer is known for some cases, notably styrene, but the exact mechanism is not clear (Section *C*).

Thermal decomposition of an initiator has a disadvantage in that the rate of generation of free radicals cannot be controlled rapidly because of the heat capacity of the system. *Photoinitiated polymerization*, on the other hand, can be controlled with high precision, since the generation of radicals can be made to vary instantaneously by controlling the intensity of the initiating light.

Light of short enough wavelength (i.e., high enough energy per quantum) can initiate polymerization directly. It is customary, however, to use a *photochemical initiator* such as benzoin or azobisisobutyronitrile, which is decomposed into free radicals by ultraviolet light in the 3600 A region, where direct initiation through decomposition of monomer does not occur. In photopolymerization these initiators are used at temperatures low enough so that they do not undergo appreciable thermal decomposition. Convenient sources of 3600 A light are available in the form of mercury-arc lamps or fluorescent lamps with special phosphors.

For photopolymerization the dilatometer is made in such a form as to serve as an absorption cell for the ultraviolet light. The distribution of free radicals throughout the reaction vessel must be uniform; this means that only a small fraction of the ultraviolet light may be absorbed. Also, the photoinitiator must not undergo any reaction other than to act as a source of free radicals which add rapidly to the monomer.

Rate of initiation

The number of polymer chains initiated per unit time can be calculated from the number of free radicals produced in the system, if the initiator efficiency is known. In photopolymerization, the number of free radicals produced is twice the number of quanta absorbed, since one quantum of light decomposes a single initiator molecule into two free radicals. The number of quanta absorbed in the reaction vessel can be measured in the following way. The fraction of the incident light absorbed in the reaction vessel, with monomer and initiator present exactly as in the polymerization experiment, is determined. The vessel is then filled with a solution, such as uranyl oxalate, which absorbs ultraviolet light by means of a known chemical reaction. The fraction of the light absorbed is again measured, and after exposure for a known time the number of quanta of light which have reacted with uranyl oxalate is determined by chemical analysis of the solution. The quanta absorbed under polymerization conditions can be calculated in a straightforward manner.

The initiation rate obtained in this way can be checked by computing the number of polymer molecules formed per unit time from the over-all rate of polymerization and $\bar{x}_n$, provided that chain transfer is absent and that the mechanism of termination is known. Since this is usually not the case, an independent check on the rate of initiation is desirable. This may be obtained by the use of an inhibitor or retarder for the polymerization.

GENERAL REFERENCES

Burnett, G. M., *op. cit.*, Chap. IIB.
Sorenson, W. R., and T. W. Campbell, *Preparative Methods of Polymer Chemistry*, Interscience Publishers, New York, 1961.

C. Steady-State Kinetics of Vinyl Radical Polymerization

Conversion of monomer to polymer

The chemical equations of Section *A* for initiation, propagation, and termination of vinyl radical polymerization can be generalized in the following mathematical scheme.

Initiation in the presence of an initiator I may be considered in two

steps: first, the rate-determining decomposition of the initiator into free radicals, R·,

$$\mathrm{I} \xrightarrow{k_d} 2\mathrm{R}\cdot \tag{9-1}$$

and, second, the addition of a monomer unit to form a chain radical $M_1\cdot$,

$$\mathrm{R}\cdot + \mathrm{M} \xrightarrow{k_a} \mathrm{M}_1\cdot \tag{9-2}$$

where the k's in these and subsequent equations are *rate constants*, with subscripts designating the reactions to which they refer.

The successive steps in propagation,

$$\mathrm{M}_1\cdot + \mathrm{M} \xrightarrow{k_p} \mathrm{M}_2\cdot$$

$$\mathrm{M}_2\cdot + \mathrm{M} \xrightarrow{k_p} \mathrm{M}_3\cdot \tag{9-3}$$

or in general $$\mathrm{M}_x\cdot + \mathrm{M} \xrightarrow{k_p} \mathrm{M}_{x+1}\cdot$$

are assumed all to have the same rate constant k_p, since radical reactivity is presumed to be independent of chain length.

The termination step involves combination

$$\mathrm{M}_x\cdot + \mathrm{M}_y\cdot \xrightarrow{k_{tc}} \mathrm{M}_{x+y} \tag{9-4}$$

or disproportionation

$$\mathrm{M}_x\cdot + \mathrm{M}_y\cdot \xrightarrow{k_{td}} \mathrm{M}_x + \mathrm{M}_y \tag{9-5}$$

Except where it is necessary to distinguish between the two mechanisms the termination rate constant is denoted k_t.

The rates of the three steps may be written in terms of the concentrations (in brackets) of the species involved and the rate constants. The rate of initiation is

$$v_i = (\mathrm{d}[\mathrm{M}\cdot]/\mathrm{d}t)_i = 2fk_d[\mathrm{I}] \tag{9-6}$$

where the factor f represents the fraction of the radicals formed by Eq. 9-1 which is successful in initiating chains by Eq. 9-2. The rate of termination is

$$v_t = -(\mathrm{d}[\mathrm{M}\cdot]/\mathrm{d}t)_t = 2k_t[\mathrm{M}\cdot]^2 \tag{9-7}$$

For many cases of interest the concentration of free radicals [M·] becomes essentially constant very early in the reaction, as radicals are formed and destroyed at identical rates. In this *steady-state* condition $v_i = v_t$ and Eqs. 9-6 and 9-7 may be equated to solve for [M·]:

$$[\mathrm{M}\cdot] = (fk_d[\mathrm{I}]/k_t)^{1/2} \tag{9-8}$$

The rate of propagation is essentially the same as the over-all rate of

disappearance of monomer, since the number of monomers used in Eq. 9-2 must be small compared to that used in Eq. 9-3 if polymer is obtained. Then

$$v_p = -\mathrm{d}[\mathrm{M}]/\mathrm{d}t = k_p[\mathrm{M}][\mathrm{M}\cdot] \tag{9-9}$$

or, substituting from Eq. 9-8,

$$v_p = k_p(fk_d[\mathrm{I}]/k_t)^{1/2}[\mathrm{M}] \tag{9-10}$$

Thus the over-all rate of polymerization should, in the early stages of the reaction, be proportional to the square root of the initiator concentration and, if f is independent of [M], to the first power of the monomer concentration. This is true if the initiator efficiency is high. With very low efficiency, f may be proportional to [M], making v_p proportional to $[\mathrm{M}]^{3/2}$.

The proportionality of the over-all rate to the square root of initiator concentration has been confirmed experimentally in a large number of cases. Typical experimental data are shown in Fig. 9-2. The straight lines are drawn with the theoretical slope of $\frac{1}{2}$. The deviation for polystyrene at low initiator concentration (line c) is due to the presence of some thermal initiation (see page 277).

Evaluation of rate constants If the rate constant for initiator decomposition and the initiator efficiency are known, the ratio of rate

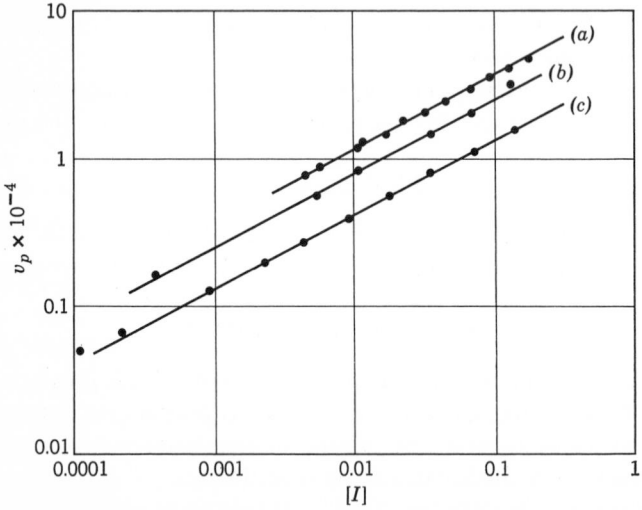

Fig. 9-2. Dependence of rate of polymerization v_p on concentration of initiator [I] for: (a) methyl methacrylate with azobisisobutyronitrile (1); (b) methyl methacrylate with benzoyl peroxide (4); (c) styrene with benzoyl peroxide (8).

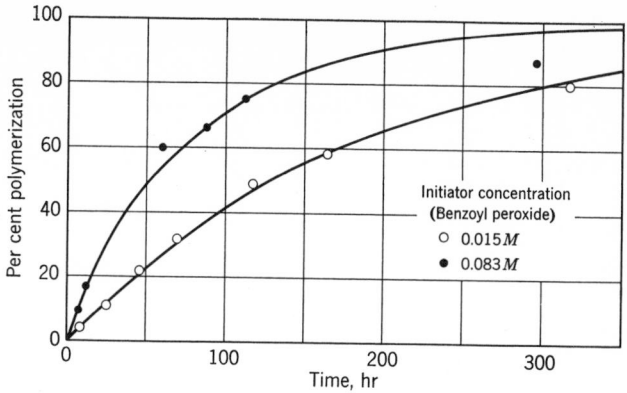

Fig. 9-3. Course of the polymerization of a 40% solution of styrene in toluene with benzoyl peroxide initiator at 50°C (10).

constants k_p^2/k_t can be evaluated from the over-all polymerization rate: from Eqs. 9-6 and 9-10,

$$k_p^2/k_t = 2v_p^2/v_i[\mathrm{M}]^2 \tag{9-11}$$

These rate constants can be separated only by the nonsteadystate methods described in Section *D*. Measurements of k_p^2/k_t as a function of temperature lead to useful information about the thermochemistry of polymerization (Section *E*).

Over-all rate as a function of conversion If the initiator concentration does not vary much during the course of polymerization and the initiator efficiency is independent of monomer concentration, polymerization proceeds by first-order kinetics, i.e., the polymerization rate is proportional to monomer concentration. In some systems, such as the benzoyl peroxide-initiated polymerization of styrene (Figs. 9-3 and 9-4), the reaction is accurately first order up to quite high conversions.

The polymerization of certain monomers undiluted or in concentrated solution is accompanied by a marked deviation from first-order kinetics in the direction of an increase in reaction rate and molecular weight termed *autoacceleration* or the *gel effect*.* The effect is particularly pronounced with methyl methacrylate (7), methyl acrylate, or acrylic acid (Fig. 9-5). It is independent of initiator and is due to a decrease in the rate at which the polymer molecules diffuse through the viscous medium, thus lowering the ability of two long-chain radicals to come together and

* Although often called also the *Trommsdorff effect* (12), this phenomenon was first described by Norrish and Brookman (6).

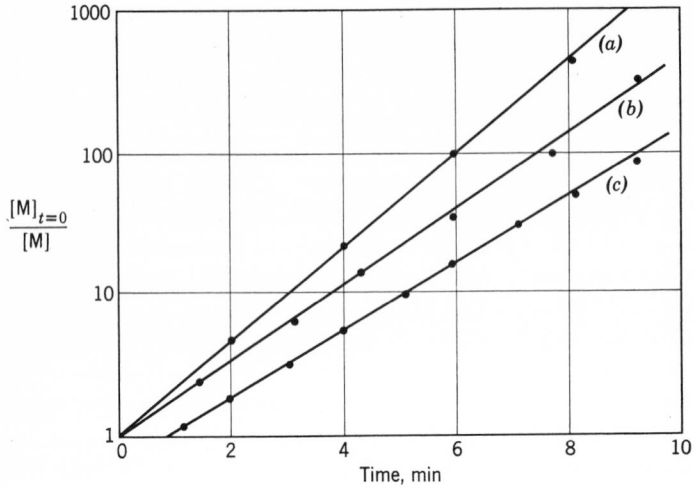

Fig. 9-4. Polymerization of vinyl phenylbutyrate in dioxane with benzoyl peroxide initiator at 60°C plotted as a first-order reaction (3). At $t = 0$, monomer concentration was 2.4 g/100 ml for curve (*a*); 6.0 g/100 ml for (*b*); and 7.3 g/100 ml for (*c*).

terminate. Although termination is in fact diffusion controlled for most liquid-phase polymerizations, even at low conversion (9), the dependence of diffusion rate on the viscosity of the medium leads to the gel effect at high polymer concentrations, with high-molecular-weight polymer, or in the presence of inert solutes increasing the viscosity of the medium.

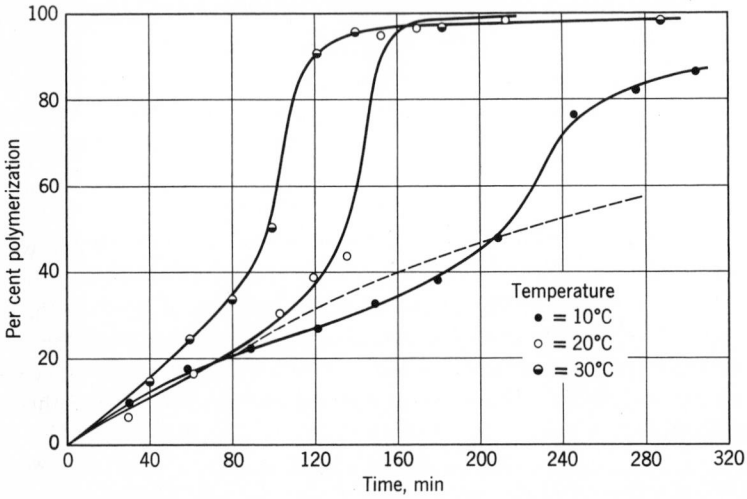

Fig. 9-5. Autoacceleration of polymerization rate in poly(methyl methacrylate) (5).

The decrease in termination rate leads to an increase in over-all polymerization rate and in molecular weight, since the lifetime of the growing chains is increased. At quite high conversions (70–90%), but long after propagation has become diffusion controlled, the rate of polymerization drops to a very low value.

Kinetics of thermal polymerization Certain monomers, including styrene, which have been carefully purified to remove all traces of possible initiators, still undergo polymerization at elevated temperatures. In these cases the initiation is presumably due to thermal decomposition of the monomer to radicals. This decomposition takes place by a mechanism as yet unknown, but the existence of diradicals which grow chains simultaneously from both ends can be ruled out (11). The rate of thermal initiation is usually proportional to monomer concentration, making the over-all polymerization rate proportional to $[M]^2$. Initiation rates are many times smaller than those for polymerization in the presence of an initiator.

Degree of polymerization and chain transfer

Kinetic chain length The *kinetic chain length* ν is defined as the number of monomer units consumed per active center. It is therefore given by $v_p/v_i = v_p/v_t$:

$$\nu = \frac{k_p}{2k_t}\frac{[M]}{[M\cdot]} \tag{9-12}$$

Eliminating the radical concentration by means of Eq. 9-9,

$$\nu = \frac{k_p^2}{2k_t}\frac{[M]^2}{v_p} \tag{9-13}$$

For initiated polymerization, use of Eq. 9-8 leads to

$$\nu = \frac{k_p}{2(fk_dk_t)^{1/2}}\frac{[M]}{[I]^{1/2}} \tag{9-14}$$

If no reactions take place other than those already discussed, the kinetic chain length should be related to $\bar{x}_n$, the degree of polymerization: for termination by combination, $\bar{x}_n = 2\nu$, and for disproportionation, $\bar{x}_n = \nu$. This is found to be precisely true for some systems, but for others wide deviations are noted in the direction of more polymer molecules than active centers. These deviations are the result of chain transfer reactions:

$$M_x\cdot + XP \longrightarrow M_xX + P\cdot$$

$$P\cdot + M \longrightarrow PM\cdot, \text{ etc.}$$

where P may be monomer, initiator, solvent, or other added chain transfer agent. (Transfer to polymer is omitted since no new polymer molecule is produced.)

The degree of polymerization is therefore

$$\bar{x}_n = \frac{\text{rate of growth}}{\sum \text{rates of all reactions leading to dead polymer}}$$

$$= v_p/\{fk_d[\mathrm{I}] + k_{tr,\mathrm{M}}[\mathrm{M}][\mathrm{M}\cdot] + k_{tr,\mathrm{S}}[\mathrm{S}][\mathrm{M}\cdot] + k_{tr,\mathrm{I}}[\mathrm{I}][\mathrm{M}\cdot]\} \quad (9\text{-}15)$$

where the terms in the denominator represent termination by combination and transfer to monomer, solvent, and initiator, respectively. If termination is by disproportionation, the first term becomes $2fk_d[\mathrm{I}]$. If transfer constants are defined as

$$C_\mathrm{M} = k_{tr,\mathrm{M}}/k_p, \qquad C_\mathrm{S} = k_{tr,\mathrm{S}}/k_p, \qquad C_\mathrm{I} = k_{tr,\mathrm{I}}/k_p \quad (9\text{-}16)$$

then (assuming termination by combination)

$$\frac{1}{\bar{x}_n} = \frac{k_t}{k_p^2}\frac{v_p}{[\mathrm{M}]^2} + C_\mathrm{M} + C_\mathrm{S}\frac{[\mathrm{S}]}{[\mathrm{M}]} + C_\mathrm{I}\frac{k_t}{k_p^2 f k_d}\frac{v_p^2}{[\mathrm{M}]^3} \quad (9\text{-}17)$$

The above analysis assumes that the radical formed in the transfer process is approximately as reactive as the original chain radical. Otherwise, retardation or inhibition results.

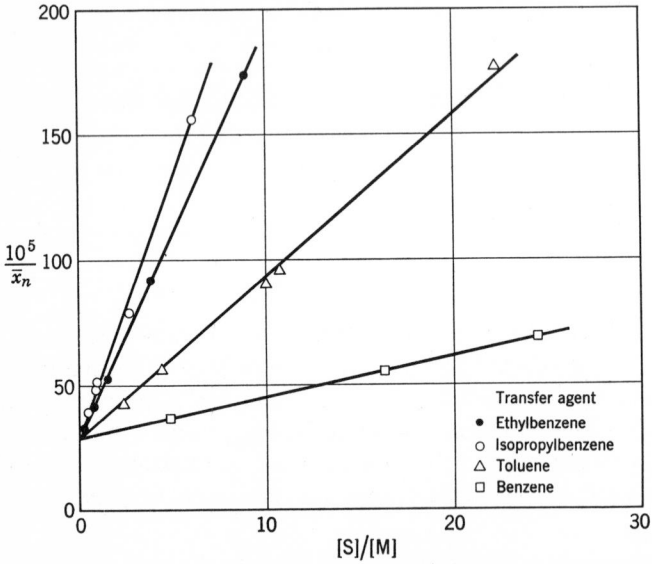

Fig. 9-6. Effect of chain transfer to solvent on the degree of polymerization of polystyrene (2).

Transfer to solvent In the presence of a solvent, and by properly choosing conditions to keep other types of chain transfer to a minimum, Eq. 9-17 reduces to

$$\frac{1}{\bar{x}_n} = \left(\frac{1}{\bar{x}_n}\right)_0 + C_S \frac{[S]}{[M]} \tag{9-18}$$

where $(1/\bar{x}_n)_0$ combines the polymerization and transfer to monomer terms. The linear dependence of $1/\bar{x}_n$ on [S]/[M] is illustrated in Fig. 9-6. In experiments of this type the ratio $v_p/[M]^2$ must be held constant throughout if $(\bar{x}_n)_0$ is identified with the observed degree of polymerization in the absence of solvent.

Transfer to monomer and initiator If only the terms corresponding to polymerization, transfer to monomer, and transfer to initiator are kept in Eq. 9-17, the resulting equation for $\bar{x}_n$ is quadratic in v_p. This behavior has been observed, e.g., in the benzoyl peroxide-initiated polymerization of styrene (4). The contribution of transfer to monomer is constant and independent of polymerization rate; that of transfer to initiator increases rapidly with increasing rate, since high rate requires high initiator concentration.

Control of molecular weight by transfer Chain transfer agents with transfer constants near unity are quite useful in depressing molecular weight in polymerization reactions. This is often of great commercial importance, e.g., in the polymerization of dienes to synthetic rubbers, where chain length must be controlled for ease of processing.

The choice of transfer constant near unity ensures that the transfer agent, or *regulator*, is consumed at the same rate as the monomer so that the ratio [S]/[M] remains constant throughout the reaction. Too large

TABLE 9-1. Chain transfer constants for various solvents and radicals at 80°C

Solvent	$C_S \times 10^4$ for		
	Styrene	Methyl Methacrylate	Vinyl Acetate
Benzene	0.059	0.075	...
Cyclohexane	0.066	0.10	...
Toluene	0.31	0.52	92
Ethylbenzene	1.08	1.35	...
Carbon tetrachloride	130	2.39	10,000
Tetrachloroethane	18	0.20	...

quantities are needed of chain transfer agents with constants much lower than unity; and agents with transfer constants greater than about five are used up too early in the polymerization. Aliphatic mercaptans are suitable transfer agents for several common monomers.

Table 9-1 lists transfer constants for several solvents and radicals.

GENERAL REFERENCES

Bamford, C. H., W. G. Barb, A. D. Jenkins, and P. F. Onyon, *The Kinetics of Vinyl Polymerization by Radical Mechanisms*, Academic Press, New York, 1958, Chaps. 1, 3, 6.

Burnett, G. M., *op. cit.*, Chaps. III, IVA, B, C, E, F, and VI.

Flory, P. J., *op. cit.*, Chap. IV-1, IV-2.

Walling, C., *op. cit.*, Chaps. 3.3, 4.2.

BIBLIOGRAPHY

1. Arnett, L. M., *J. Am. Chem. Soc.* **74,** 2027 (1952).
2. Gregg, R. A., and F. R. Mayo, *Disc. Faraday Soc.* **2,** 328 (1947).
3. Marvel, C. S., J. Dec, and H. G. Cooke, Jr., *J. Am. Chem. Soc.* **62,** 3499 (1940).
4. Mayo, F. R., R. A. Gregg, and M. S. Matheson, *J. Am. Chem. Soc.* **73,** 1691 (1951).
5. Naylor, M. A., and F. W. Billmeyer, Jr., *J. Am. Chem. Soc.* **75,** 2181 (1953).
6. Norrish, R. G. W., and E. F. Brookman, *Proc. Roy. Soc.* (*London*) **A171,** 147 (1939).
7. North, A. M., and G. A. Reed, *Trans. Faraday Soc.* **57,** 859 (1961).
8. Schulz, G. V., and F. Blaschke, *Z. physik. Chem.* **B51,** 75 (1942).
9. Schulz, G. V., and G. Harborth, *Makromol. Chem.* **1,** 106 (1947).
10. Schulz, G. V., and E. Husemann, *Z. physik. Chem.* **B39,** 246 (1938).
11. Tobolsky, A., and D. H. Johnson, *J. Am. Chem. Soc.* **74,** 938 (1952).
12. Trommsdorff, E., H. Kohle, and P. Lagally, *Makromol. Chem.* **1,** 169 (1948).

D. Absolute Reaction Rates

It was shown in Section *C* that the rate constants for propagation and termination occur together as the ratio k_p^2/k_t in both the equations for over-all polymerization rate and those for the kinetic chain length. Thus these rate constants cannot be evaluated separately through steady-state measurements. Recourse must be had to transient phenomena, such as the rate of polymerization before the free radical concentration has reached its steady-state value. The duration of this transient region depends on the length of time a free radical exists from its formation in the initiation step to its demise in termination.

Determination of individual rate constants

The sector method If a photopolymerization is begun by turning on light of intensity I, the concentration of free radicals grows gradually to its steady-state value over a time interval which is a function of the lifetime of the radicals. When the light is turned off, the radical concentration decays in a similar manner. At the steady state the rate of polymerization is proportional to the square root of the light intensity: $v_p \simeq I^{1/2}$.

If the light is flashed on and off quite slowly, so that the steady-state concentration is reached early in each flash, the rate of polymerization is proportional to the fraction of the time the light is on. If, e.g., the light and dark periods are equal, $v_p \simeq \frac{1}{2}I^{1/2}$.

On the other hand, if the flashes are so rapid that the radical concentration changes very little during a single flash, the effect is the same as that produced by reducing the intensity by a constant factor but leaving it on all the time: if the light and dark periods are equal, $v_p \simeq (I/2)^{1/2}$. The rates for long and short flashes differ, in this case, by $\sqrt{2}$.

The rate of flashing at which the transition occurs can be related to the mean lifetime τ_s of the free radicals. This is defined as

$$\tau_s = \frac{\text{number of radicals}}{\text{number disappearing per unit time}} \tag{9-19}$$

$$= \frac{[\mathrm{M}\cdot]}{2k_t[\mathrm{M}\cdot]^2} = \frac{1}{2k_t[\mathrm{M}\cdot]}$$

or, if the radical concentration is substituted from Eq. 9-9,

$$\tau_s = \frac{k_p}{2k_t}\frac{[\mathrm{M}]}{v_p} \tag{9-20}$$

Thus τ_s, [M], and the over-all polymerization rate v_p suffice to calculate k_p/k_t. With the ratio k_p^2/k_t obtained from other data, the individual rate constants may be evaluated.

The relation between τ_s and the flashing rate is complex; in practice the experimental curves of polymerization rate (or radical concentration) against flashing rate are compared to the theoretical relation (Fig. 9-7). In the figure, the ratio t/τ_s, where t is the time light is on, is plotted against $\bar{v}_p/(v_p)_s$, where $\bar{v}_p$ is the rate with flashing illumination and $(v_p)_s$ is the rate with steady illumination, for several values of the ratio r of dark to light intervals.

In practice the light is usually flashed by placing a motor-driven *rotating sector* between the light source and the reaction vessel.

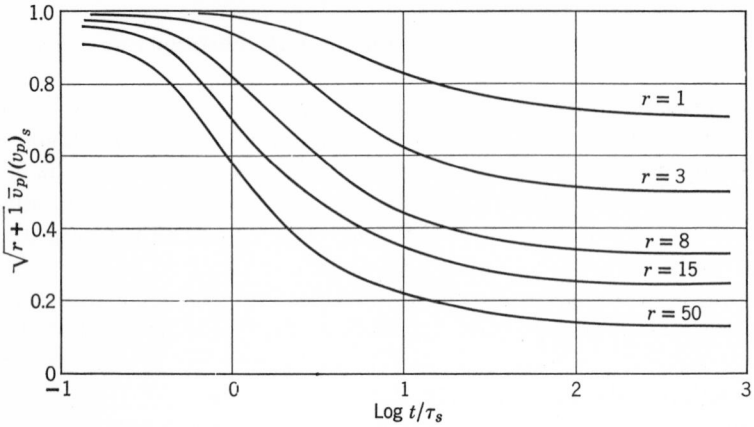

Fig. 9-7. Ratio of polymerization rates for flashing and steady illumination plotted against log (t/τ_s), for several values of the ratio r of dark to light periods (2).

In the case of vinyl acetate, the radical lifetime is of the order of 4 sec, depending, of course, upon polymerization rate, according to Eq. 9-20. The rate constants show that the growing chain adds about 10^4 monomers per second; the kinetic chain length is the product of this figure and τ_s, or about 4×10^4. The value of $\bar{x}_n$ is about ten times less than this because of chain transfer to monomer; thus about ten transfers must occur during the life of a growing chain. This represents a relatively large amount of transfer to monomer.

Other methods In the *viscosity method* (1, 3), the polymerization is carried out in a viscometer, measurement of molecular weight by viscosity being substituted for measurement of rate. The change in viscosity after photoinitiation is stopped provides data obtained under transient conditions. The kinetic scheme for *emulsion polymerization* (4), described in Chapter 12, allows separation of the individual rate constants, since the growth of a single chain effectively takes place in an isolated system.

Values of rate constants

Values of the propagation and termination constants for several monomers are tabulated in Table 9-2. The rate constants vary over a wide range. Since they depend upon the reactivity of both the radical and the monomer, it is difficult to draw conclusions about the effect of structure on reactivity from these data alone. Copolymerization experiments (Chapter 11) are more useful in this respect.

TABLE 9-2. Rate constants for propagation and termination at 60°C

Monomer	k_p, liters/mole sec	k_t, liters/mole sec $\times 10^{-7}$
Vinyl acetate	2,300	2.9
Styrene	145	0.130
Methyl methacrylate	705	1.8
Methyl acrylate	2,090	0.47
Vinyl chloride	12,300	2300
Methacrylonitrile	184	2.3

GENERAL REFERENCES

Bamford, C. H., *et al.*, *op. cit.*, Chap. 2.
Burnett, G. M., *op. cit.*, Chap. VII.
Flory, P. J., *op. cit.*, Chap. IV-3.
Walling, C., *op. cit.*, Chap. 3.4.

BIBLIOGRAPHY

1. Bamford, C. H., and M. J. S. Dewar, *Proc. Roy. Soc.* (*London*) **A192,** 309 (1948).
2. Burnett, G. M., and H. W. Melville, *Proc. Roy. Soc.* (*London*) **A189,** 456 (1947).
3. Onyon, P. F., *Trans. Faraday Soc.* **51,** 400 (1955).
4. Smith, W. V., *J. Am. Chem. Soc.* **70,** 3695 (1948); W. V. Smith and R. H. Ewart, *J. Chem. Phys.* **16,** 592 (1948).

E. Molecular Weight and Its Distribution

The distribution of molecular weights in vinyl polymers can readily be calculated if the reaction is restricted to low conversion, where polymer made throughout the reaction has the same average molecular weight. If termination is by disproportionation (or transfer), let p be defined as the probability that a growing chain radical will propagate rather than terminate: $p = v_p/(v_p + v_t)$.

The probability of formation of an x-mer, as a result of $x - 1$ propagations and a termination, is $p^{x-1}(1 - p)$. The situation is in complete analogy with that for linear stepwise polymerization, and the expressions for $\bar{x}_n$, N_x, w_x, and $\bar{x}_w$ given in Eqs. 8-5 and 8-8 to 8-10 apply. If high polymer is formed, i.e., $p \rightarrow 1$, $\bar{x}_w/\bar{x}_n = 2$.

Termination by combination is somewhat more complex, but it can be shown that, in analogy to the case of multichain polymer (Eq. 8-13) with $f = 2$, $\bar{x}_w/\bar{x}_n = 1.5$ if transfer is absent.

The measurement of $\bar{x}_w/\bar{x}_n$ would distinguish between the possibilities of termination by combination and by disproportionation. However, this measurement is not easily carried out with the desired precision; a preferable alternative procedure involves measuring the number of initiator fragments per molecule of polymer.

For values of p near 1, the most probable distribution of Eq. 8-9 reduces to

$$\begin{aligned} w_x &= xp^x(\ln p)^2 \\ \bar{x}_n &= -1/\ln p \\ \bar{x}_w &= -2/\ln p \end{aligned} \tag{9-21}$$

It can be shown that both this distribution, which describes low-conversion chain polymers terminating by disproportionation, and that describing termination by combination, are special cases of a more general distribution function (8):

$$\begin{aligned} w_x &= (y^{z+1}/z!)x^z e^{-xy} \\ \bar{x}_n &= z/y \\ \bar{x}_w &= (z+1)/y \end{aligned} \tag{9-22}$$

For $z = 1$, Eq. 9-22 reduces to 9-21 with $y = -\ln p$. For $z = 2$, it describes the case of termination by combination. Higher values of z correspond to narrower distributions such as might be found in fractionated samples. In contrast to the most probable distribution, the number distribution corresponding to Eq. 9-22 does not indicate a large number of very small molecules. Hence, it is preferred for describing polymers from which the small molecules have been removed, e.g., by precipitation of the polymer from solution.

As polymerization proceeds, the degree of polymerization of the polymer being formed changes according to Eq. 9-14. If the rate of initiation is constant, as is usual, $\bar{x}_n$ decreases throughout the reaction as [M] decreases, approaching zero at complete conversion. At the same time, $\bar{x}_w/\bar{x}_n$ increases with increasing conversion, approaching infinity as a limit. Typical values of $\bar{x}_w/\bar{x}_n$ for high-conversion vinyl polymers seldom exceed 5, however, except in special cases, such as autoacceleration in the polymerization of methyl methacrylate, where values of $\bar{x}_w/\bar{x}_n$ as high as 10 have been found (2). The above distribution functions are not well suited to the description of these broader distributions. Among empirical functions suggested for these cases are a somewhat artificial function corresponding to equal parts by weight of two simple exponential (most probable) distributions (2):

$$\begin{aligned} w(x) &= (b^2x/2)(e^{-bx} + a^2e^{-abx}) \\ \bar{x}_w/\bar{x}_n &= (1+a)^2/2a \end{aligned} \tag{9-23}$$

and the following form (5), which has a high-molecular-weight "tail" when the exponent ρ is positive and less than 1. Here Γ is the gamma function:

$$w(x) = \left[\rho/\Gamma\left(\frac{(z+1)}{\rho}\right)\right] y^{(z+1)/\rho} x^z e^{-yx^\rho}$$

$$\bar{x}_n = \Gamma\left(\frac{z+1}{\rho}\right) \Big/ y^{1/\rho}\Gamma\left(\frac{z}{\rho}\right) \tag{9-24}$$

$$\bar{x}_w = \Gamma\left(\frac{z+2}{\rho}\right) \Big/ y^{1/\rho}\Gamma\left(\frac{z+1}{\rho}\right)$$

Where chain transfer to polymer is involved, extremely broad molecular weight distributions can result (1). The abnormal breadth arises from the fact that each transfer reaction adds a branch point to the molecule; after the branch has grown, the molecule has increased in molecular weight. But the probability of branching is proportional (approximately) to weight-average molecular weight. Hence, branched molecules tend to become more highly branched and still larger. The result is that the distribution of molecular weights has a long, high-molecular-weight "tail."

Among the empirical distribution functions suitable for describing fractions from which low-molecular-weight species have been removed is the so-called logarithmic normal distribution (4):

$$w(x) = \frac{e^{-\sigma^2/2}}{\sigma\sqrt{\pi}} \frac{1}{x_0} e^{-(\ln x - \ln x_0)^2/2\sigma^2}$$

$$\bar{x}_n = x_0 e^{\sigma^2/2} \tag{9-25}$$

$$\bar{x}_w = x_0 e^{3\sigma^2/2}$$

In this distribution the logarithm of the molecular weight follows a normal distribution function. The ratio $\bar{x}_w/\bar{x}_n$ is identical with $\bar{x}_z/\bar{x}_w$.

A function proposed (3) to describe polymers made by coordination polymerization (Chapter 10*D*) and applicable to unusually broad distributions of molecular weight is more simply expressed in terms of the number distribution:

$$N(x) \simeq 1/[x^{1/2}(x^{1/2} + a)^{b+1}]$$

$$\bar{x}_n = 2a^2/(b-1)(b-2) \tag{9-26}$$

$$\bar{x}_w = 12a^2/(b-3)(b-4)$$

where a and b are related to the kinetic rate constants. This distribution is shown (3) to fit data previously interpreted by empirical functions (6, 7).

GENERAL REFERENCES

Bamford, C. H., *et al.*, *op. cit.*, Chap. 7.
Burnett, G. M., *op. cit.*, Chap. IVD.
Flory, P. J., *op. cit.*, Chap. VIII-2.
Walling, C., *op. cit.*, Chap. 5.2a, b.

BIBLIOGRAPHY

1. Beasley, J. K., *J. Am. Chem. Soc.* **75,** 6123 (1953).
2. Billmeyer, F. W., Jr., and W. H. Stockmayer, *J. Polymer Sci.* **5,** 121 (1950).
3. Gordon, M., and R.-J. Roe, *Polymer* **2,** 41 (1961); R.-J. Roe, *ibid.*, p. 60.
4. Lansing, W. D., and E. O. Kraemer, *J. Am. Chem. Soc.* **57,** 1369 (1935).
5. Muus, L. T., and W. H. Stockmayer, private communication.
6. Tung, L. H., *J. Polymer Sci.* **20,** 495 (1956).
7. Wesslau, H., *Makromol. Chem.* **20,** 111 (1956).
8. Zimm, B. H., *J. Chem. Phys.* **16,** 1099 (1948).

F. Thermochemistry of Chain Polymerization

Thermochemical and thermodynamic data give valuable insight into the mechanisms of vinyl polymerization reactions. By the usual Arrhenius treatment the temperature dependence of the rates of the various steps in radical chain polymerization can be separated into an energy of activation representing the amount of energy which the reactant molecules must have to be able to react on collision, and a frequency factor which allows estimation of the fraction of the collisions which lead to reaction. Heats of polymerization are readily measured and may also be calculated from bond dissociation energies. Differences between the two values give an indication of the amount of steric hindrance in the polymer. Free energies of polymerization reactions can be calculated for some simple cases and give information on the thermodynamic equilibria involved in polymerization.

Activation energies and collision factors

Over-all polymerization The ratio of rate constants k_p^2/k_t or $k_p/\sqrt{k_t}$ is obtained from the over-all polymerization rate and the rate of initiation (Eq. 9-11). Expressing a rate constant by an Arrhenius type equation gives, e.g.,

$$k_p = A_p e^{-E_p/RT} \tag{9-27}$$

where A_p is the *collision frequency factor* for propagation and E_p the *energy of activation* for propagation. Application to the rate constant ratio gives

$$k_p/\sqrt{k_t} = (A_p/\sqrt{A_t})e^{-(E_p-E_t/2)/RT} \tag{9-28}$$

By plotting $\log(k_p/\sqrt{k_t})$ against $1/T$, $E_p - E_t/2$ and $A_p/\sqrt{A_t}$ can be evaluated. For many polymerization reactions, $E_p - E_t/2$ is about 5–6 kcal/mole. This is equivalent to an increase in $k_p/\sqrt{k_t}$ of about 30–35% for every 10°C near room temperature. Equation 9-28 refers only to the ratio of rate constants $k_p/\sqrt{k_t}$ and does not include the effect of temperature on the rate of initiation. In photopolymerization, where the rate of initiation is independent of temperature, Eq. 9-28 represents the complete activation energy.

In an initiated polymerization, v_p is proportional to $k_p\sqrt{k_d/k_t}$, and the observed activation energy is $E_d/2 + (E_p - E_t/2)$. E_d, the activation energy for spontaneous decomposition of the initiator, is about 30 kcal/mole for benzoyl peroxide or azobisisobutyronitrile. The activation energy for initiated polymerization is therefore slightly greater than 20 kcal/mole. This corresponds to a two- to three-fold increase in rate for a 10°C temperature change.

The activation energy for thermal polymerization is of the same order of magnitude, yet this reaction is much slower than normal because of the extremely low collision factor for thermal initiation. The normal value of A for a bimolecular reaction is 10^{11}–10^{13}; for the thermal initiation of polystyrene it is about 10^4–10^6.

Degree of polymerization If transfer reactions are negligible, the temperature coefficient of molecular weight depends on the initiation process. If an initiator is used, it follows* from Eq. 9.14 that

$$\frac{d \ln \bar{x}_n}{dT} = \frac{E_p - \dfrac{E_t}{2} - \dfrac{E_d}{2}}{RT^2} \tag{9-29}$$

Since $E_p - E_t/2$ is 5–6 kcal and E_d is about 30 kcal, this quantity is negative and the molecular weight decreases with increasing temperature. The same is true in thermal polymerization.

In photopolymerization in the absence of transfer, $d \ln \bar{x}_n/dT = (E_p - E_t/2)/RT^2$, which is positive. This is the only case in which molecular weight increases with temperature.

If transfer reactions are controlling, it is found that molecular weight decreases with increasing temperature. By plotting $\log C_S$ or $-\log \bar{x}_n$

* By virtue of the differential form $d \ln k/dT = E/RT^2$ of the Arrhenius equation.

against $1/T$, the quantities $E_{tr,S} - E_p$ and $A_{tr,S}/A_p$ may be evaluated. $E_{tr,S}$ usually exceeds E_p by 5–15 kcal/mole, higher activation energies going with poorer transfer agents. Since the frequency factor for transfer usually exceeds that for propagation, the higher activation energy is responsible for the slower rate of transfer compared to monomer addition.

Absolute rates The activation energies and frequency factors for propagation and termination can be separated if the individual rate constants are known. For most monomers E_p is near 7 kcal/mole, and E_t varies from 3 to 5 kcal/mole. The frequency factors vary rather widely, suggesting that steric effects are more important than activation energies in setting the values of the rate constants. For example, A_p is much lower for methyl methacrylate with two substituents on the same ethylenic carbon atom than it is for less hindered monomers. All the values of A_p are considerably lower than normal, because, it is likely, of low probability of the transition leading to unpairing of the double bond electrons in the monomer.

Polymerization-depolymerization equilibrium Some polymers decompose by stepwise loss of monomer units in a reaction which is essentially the reverse of polymerization (see Chapter 12). The depolymerization reaction can be incorporated into the kinetic scheme for chain polymerization (3). At ordinary temperatures the scheme is not changed because the rate constant for depolymerization is small. The activation energy of depolymerization is quite high (10–26 kcal/mole) compared to that of propagation, however, and at some elevated temperature the rates of polymerization and depolymerization become equal. This temperature is the *ceiling temperature* above which polymer free radicals, in the presence of monomer at 1 atm pressure, depolymerize rather than grow. For polystyrene, the ceiling temperature is about 300°C.

Heats and free energies of polymerization

Heats of polymerization These may be obtained from the heats of combustion of monomer and polymer or directly by a calorimetric experiment. They may also be calculated from bond dissociation energies. When the two results are compared (Table 9-3), it is seen that the observed heat of polymerization is equal to or somewhat lower than the calculated heat. For vinyl compounds with two substituents on the same carbon atom the difference is 3–9 kcal/mole. This is attributed to steric repulsion between the substituent groups in the polymer. For monosubstituted

TABLE 9-3. Heats of polymerization

Monomer	Heat of Polymerization, kcal/mole		
	Observed	Calculated	Difference
Styrene	16.4	18.7	2.3
Vinyl acetate	21.3	21.3	0
Isobutylene	12.6	18.9	6.3
Methyl methacrylate	13.0	19.2	6.2
Isoprene	17.9	17.9	0

vinyl derivatives the difference is much lower; only for styrene is it definitely greater than 1 kcal/mole.

Free energies of polymerization Free energy is a more precise indication than enthalpy of the thermodynamic plausibility of a chemical reaction. The heats of formation of free radicals and of their over-all reactions to polymer are of major importance in establishing the corresponding free energies, but the contribution of the entropy term in

$$\Delta F = \Delta H - T\Delta S$$

is also significant.

The calculation of the heats and entropies of free radicals has been carried out for hydrocarbons and fluorocarbons and applied to the synthesis equations for polyethylene (1) and polytetrafluoroethylene (2). The following examples refer to polyethylene; results for the fluoropolymer are qualitatively similar. The over-all reaction of gaseous ethylene going to solid polyethylene at 25°C has the following calculated heat and free energy in kilocalories per mole:

	ΔH^0	ΔF^0
$xC_2H_4 \longrightarrow (C_2H_4)_x$	−22.3	−12.2

The negative free energy change is, of course, an indication that the reaction is favorable.

Initiation by low-molecular-weight free radicals is also favorable:

	ΔH^0	ΔF^0
$H\cdot + C_2H_4 \longrightarrow C_2H_5\cdot$	−39.4	−33.3
$CH_3\cdot + C_2H_4 \longrightarrow n\text{-}C_3H_7\cdot$	−27.0	−16.8
$C_2H_5\cdot + C_2H_4 \longrightarrow n\text{-}C_4H_9\cdot$	−25.6	−15.2

There is no fundamental difference between the initiation and the propagation steps. After the growing free radical reaches a length of about

eight carbon atoms, all successive steps have the same heat and free energy of reaction, e.g.,

	ΔH^0	ΔF^0
$n\text{-}C_{10}H_{21}\cdot + C_2H_4 \longrightarrow n\text{-}C_{12}H_{25}\cdot$	−22.3	−12.2

The figures are, of course, the same as those for the over-all polymerization step.

Termination, either by combination or by disproportionation, is extremely favorable. This thermodynamic advantage is, of course, offset by the low concentration of free radicals usually present in a polymerizing mixture:

	ΔH^0	ΔF^0
$2\ n\text{-}C_{12}H_{25}\cdot \longrightarrow n\text{-}C_{24}H_{50}$	−71.8	−59.1
$2\ n\text{-}C_{12}H_{25}\cdot \longrightarrow n\text{-}C_{12}H_{24} + n\text{-}C_{12}H_{26}$	−52.2	−51.1

Transfer reactions are also considered in the original papers.

GENERAL REFERENCES

Burnett, G. M., *op. cit.*, Chap. IIID.
Evans, M. G., "Energetics of Polymerization Reactions," Chap. 3 in R. Hill, ed., *Fibres from Synthetic Polymers*, Elsevier Publishing Co., New York, 1953.
Flory, P. J., *op. cit.*, Chap. IV-2, IV-3.
Walling, C., *op. cit.*, Chap. 5.2.

BIBLIOGRAPHY

1. Bryant, W. M. D., *J. Polymer Sci.* **6,** 359 (1951).
2. Bryant, W. M. D., *J. Polymer Sci.* **56,** 277 (1962).
3. Dainton, F. S., and K. J. Ivin, *Trans. Faraday Soc.* **46,** 331 (1950).

10

Nonradical Chain (Addition) Polymerization

A. Chemistry of Nonradical Chain Polymerization

Chain-reaction polymerization is known to occur by several mechanisms other than those involving free radicals discussed in Chapter 9. Prominent among these are reactions in which the chain carriers are *carbonium ions* (*cationic polymerization*, Section *B*) or *carbanions* (*anionic polymerization*, Section *C*). In addition, polymerization may take place by mechanisms, not yet confirmed in detail, involving coordination compounds among the monomer, possibly the growing chain, and a catalyst, usually a solid (*coordination polymerization*, Section *D*). Still other chain polymerization reactions are known and are discussed in Section *E*.

The mechanisms of these polymerizations are less thoroughly understood than that of radical polymerization, for several reasons. Reaction systems are often heterogeneous, involving inorganic catalysts* and organic monomers. Unusually large effects may be produced by a third component (*cocatalyst*) present in very low concentration. Polymerization often leads to very-high-molecular-weight polymer at an extremely high rate, increasing enormously the difficulty in obtaining kinetic data or even reproducible results.

As indicated in Table 10-1 (1), the type of monomer polymerizing best by each mechanism is related to the polarity of the monomer and the acid-base strength of the ion formed. Monomers with electron-donating

* The word *catalyst* is preferred to *initiator* in reference to ionic polymerization, since the agent often exerts its influence throughout the growth of the chain, rather than at the instant of initiation only.

***TABLE 10-1. Type of chain polymerization suitable for common monomers* (1)**

Cationic Only	Free Radical Only	Anionic Only
Isobutylene and derivatives, $CH_2{=}C(CH_3)_2$, $CH_2{=}C(CH_3)R$ Alkyl vinyl ethers, $CH_2{=}CHOR$, and related ether types, $CH_2{=}C(R)OR$, $CH_2{=}C(OR)_2$, $CH_3OCH{=}CHOCH_3$ Coumarone, indene Derivatives of α-methylstyrene, $CH_2{=}C(CH_3)$ 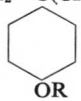 OR	Halogenated vinyls, $CH_2{=}CHX$, $CH_2{=}CX_2$, $CF_2{=}CFX$, $CF_2{=}CX_2$ where X = halogen, hydrogen (but $CH_2{=}CH_2$ not included) Vinyl esters, $CH_2{=}CHOCOR$	Vinylidene cyanide, $CH_2{=}C(CN)_2$, and related cyano derivatives, $CH_2{=}C(CN)Y$, where Y = SO_2R, CF_3, COOR Nitroethylenes, $CH_2{=}C(NO_2)R$

Cationic or Free Radical	Free Radical or Anionic
N-vinylcarbazole, 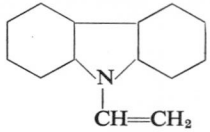	Acrylic and methacrylic esters Vinylidene esters, $CH_2{=}C(COOR)_2$ Derivatives of acrylonitrile, $CH_2{=}CRCN$ $CH_2{=}CRCONH_2$
N-vinylpyrrolidone, 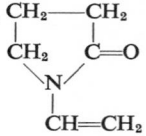	

Cationic, Free Radical, or Anionic
Ethylene, $CH_2{=}CH_2$; butadiene, $CH_2{=}CH{-}CH{=}CH_2$; styrene, $CH_2{=}CH$  ; α-methylstyrene; methyl vinyl ketone

Coordination or Supported Metal Oxide Catalyst
α-Olefins including ethylene, dienes, alkyl vinyl ethers

groups attached to the double-bonded carbons form stable carbonium ions and polymerize best with cationic catalysts. Conversely, monomers with electron-withdrawing substituents form stable anions and require anionic catalysts. In what might be considered an intermediate case, free radical polymerization is favored by moderate electron withdrawal from the double bond plus conjugation in the monomer. Many monomers can polymerize by more than one mechanism, including coordination polymerization.

Initiation of ionic polymerization usually involves the transfer of an ion or an electron to or from the monomer, with the formation of an ion pair. It is thought that the counterion of this pair stays close to the growing chain end throughout its lifetime, particularly in media of low dielectric constant. In contrast to radical polymerization, termination never involves the reaction between two growing chains, but usually involves the unimolecular reaction of a chain with its counterion or a transfer reaction leaving a species too weak to propagate. An agent which retards radical polymerization, such as oxygen, often has little influence on ionic polymerization; but impurities that neutralize the catalysts prevent the reaction, and many aromatic, heterocyclic, olefinic, or acetylenic compounds retard or terminate ionic polymerization.

GENERAL REFERENCES

Burnett, G. M., *Mechanism of Polymer Reactions*, Interscience Publishers, New York, 1954, Chap. XI.

Flory, P. J., *Principles of Polymer Chemistry*, Cornell University Press, Ithaca, N.Y., 1953, Chap. V-4.

Marvel, C. S., *An Introduction to the Organic Chemistry of High Polymers*, John Wiley & Sons, New York, 1959, pp. 46–58.

Schildknecht, C. E., "Ionic Polymerizations," Chap. VI in C. E. Schildknecht, ed., *Polymer Processes*, Interscience Publishers, New York, 1956.

Sorenson, W. R., and T. W. Campbell, *Preparative Methods of Polymer Chemistry*, Interscience Publishers, New York, 1961.

BIBLIOGRAPHY

1. Schildknecht, C. E., *Ind. Eng. Chem.* **50,** 107 (1958).

B. Cationic Polymerization

Typical catalysts for cationic polymerization are Friedel-Crafts catalysts such as $AlCl_3$, $AlBr_3$, BF_3, $SnCl_4$, H_2SO_4, and other strong Lewis acids.

All these are strong electron acceptors. Most of them, possibly excepting the strong protonic acids, require a cocatalyst to initiate polymerization, the cocatalyst supplying H^+, which is presumed to be the effective initiator.

High rate of polymerization at low temperatures is a characteristic of ionic polymerizations. It is often difficult to establish uniform reaction conditions before the reactants are consumed. The polymerization of isobutylene by $AlCl_3$ or BF_3 takes place within a few seconds at $-100°C$, producing polymer of molecular weight up to several million. Both rate and molecular weight are much lower at room temperature.

Mechanism

The most satisfactory theory developed for Friedel-Crafts polymerization involves the carbonium ion (3, 6, 7, 11). For example, in the polymerization of isobutylene with $TiCl_4$ catalyst, the first step is the reaction of catalyst and cocatalyst to form a complex acid,

$$TiCl_4 + RH \longrightarrow TiCl_4R^-H^+$$

which donates a proton to an isobutylene molecule to give a carbonium ion,

$$H^+ + (CH_3)_2C{=}CH_2 \longrightarrow (CH_3)_3C^+$$

This ion then reacts with monomer with the reformation of a carbonium ion at the end of each step. The "head-to-tail" addition of monomer to ion is the only one possible for energetic reasons. Since the reaction is in general carried out in a hydrocarbon medium of low dielectric constant, the anion, $TiCl_4R^-$, cannot be separated from the growing cationic end to any appreciable distance; rather, they form an ion pair.

The termination reaction can take place by the rearrangement of the ion pair to yield a polymer molecule with terminal unsaturation, plus the original complex,

$$CH_3[(CH_3)_2CCH_2]_xC(CH_3)_2^+ + RTiCl_4^- \longrightarrow$$

$$CH_3[(CH_3)_2CCH_2]_xC\begin{matrix} \diagup\!\!\!\!\diagdown CH_2 \\ \diagdown\!\!\!\!\diagup CH_3 \end{matrix} + HRTiCl_4$$

An alternative mechanism, in which the catalyst or cocatalyst combines with the growing chain, is unlikely in this example (3) but can take place

when a covalent bond is formed, as in the polymerization of styrene with trifluoroacetic acid (10):

$$H[CH_2CH]_xCH_2CH^+\ {}^-OCOCF_3 \longrightarrow H[CH_2CH]_xCH_2CHOCOCF_3$$

Chain transfer to monomer can also take place:

$$CH_3[(CH_3)_2CCH_2]_xC(CH_3)_2^+\ RTiCl_4^- + (CH_3)_2C{=}CH_2 \longrightarrow$$

$$CH_3[(CH_3)_2CCH_2]_xC\begin{matrix} CH_2 \\ \diagdown\!\!\!\!\diagup \\ CH_3 \end{matrix} + (CH_3)_3C^+\ RTiCl_4^-$$

Many polymer chains are formed for each catalyst molecule present (2, 5). Chain transfer to polymer is known and leads to branched polymers (1, 8).

The efficiency of the catalyst is dependent on the acid strength of the catalyst-cocatalyst complex. Also, the efficiency of the catalyst as a terminator should be related to the base strength of the complex anion. Thus, the more active a molecule is as a catalyst, the less active it is as a terminator. Although the requirements of the isobutylene-$TiCl_4$ system are fully satisfied by the mechanism just given, generalization of the scheme should be applied with caution. The experimental data on other systems, e.g., do not conclusively prove the necessity of a cocatalyst.

Kinetics

Although many cationic polymerizations proceed so rapidly that it is difficult to establish the steady state, the following kinetic scheme appears to be valid. Let the catalyst be designated by A and the cocatalyst by RH. Initiation, propagation, termination, and transfer may be represented thus:

$$\begin{aligned} A + RH &\rightleftharpoons A{-}RH \\ A{-}RH + M &\xrightarrow{k_i} HM^+\ {}^-AR \\ HM_x^+\ {}^-AR + M &\xrightarrow{k_p} HM_{x+1}^+\ {}^-AR \\ HM_x^+\ {}^-AR &\xrightarrow{k_t} M_x + A{-}RH \\ HM_x^+\ {}^-AR + M &\xrightarrow{k_{tr}} M_x + HM^+\ {}^-AR \end{aligned} \qquad (10\text{-}1)$$

The rate of initiation is

$$v_i = k_i[\mathrm{C}][\mathrm{M}] \tag{10-2}$$

where [C] is the catalyst concentration. (If the formation of A—RH is the rate-determining step, v_i may be independent of [M] and the kinetic scheme should be appropriately modified.)

Since, in strong contrast to radical polymerization, termination is first order,

$$v_t = k_t[\mathrm{M}^+] \tag{10-3}$$

By the steady-state assumption,

$$[\mathrm{M}^+] = (k_i/k_t)[\mathrm{C}][\mathrm{M}] \tag{10-4}$$

The over-all polymerization rate is

$$v_p = k_p[\mathrm{M}^+][\mathrm{M}] = (k_p k_i/k_t)[\mathrm{C}][\mathrm{M}]^2 \tag{10-5}$$

If termination predominates over transfer,

$$\bar{x}_n = \frac{v_p}{v_t} = \frac{k_p[\mathrm{M}^+][\mathrm{M}]}{k_t[\mathrm{M}^+]} = \frac{k_p}{k_t}[\mathrm{M}] \tag{10-6}$$

whereas if transfer predominates,

$$\bar{x}_n = \frac{v_p}{v_{tr}} = \frac{k_p[\mathrm{M}^+][\mathrm{M}]}{k_{tr}[\mathrm{M}^+][\mathrm{M}]} = \frac{k_p}{k_{tr}} \tag{10-7}$$

This scheme appears to account for the experimental facts in the polymerization of styrene with $SnCl_4$ in ethylene dichloride at 25°C (9) and for the polymerization of alkyl vinyl ethers with various catalysts (4).

Since v_p is proportional to $k_i k_p/k_t$ and, if termination predominates, $\bar{x}_n$ is proportional to k_p/k_t, the over-all rate of polymerization increases with decreasing temperature if $E_t > E_i + E_p$, and the molecular weight increases as the temperature of reaction decreases if $E_t > E_p$. Since the propagation step involves the approach of an ion to a neutral molecule in a medium of low dielectric constant, no activation energy is necessary. The termination step, on the other hand, requires the rearrangement of an ion pair and thus involves an appreciable activation energy. These conditions are exactly the reverse of those found in free radical polymerization.

Although some investigators suggest that polymerization also occurs by carbanion chains initiated by the anion of the initiator complex, the concept must be taken with great reservation. The charge separation

necessary for independent growth of oppositely charged chains is much too great to occur in media of low dielectric constant.

GENERAL REFERENCES

Burnett, G. M., *op. cit.*, Chap. XIA.
Eastham, A. M., *Fortschr. Hochpolym.-Forsch.* (*Advances in Polymer Science*) **2,** 18 (1960).
Flory, P. J., *op. cit.*, Chap. V-4a.
Plesch, P. H., ed., *Cationic Polymerization and Related Complexes*, Academic Press, New York, 1954.
Schildknecht, C. E., *op. cit.*, pp. 201–220.

BIBLIOGRAPHY

1. Atkins, J. T., and F. W. Billmeyer, Jr., *J. Phys. Chem.* **63,** 1966 (1959).
2. Colclough, R. O., *J. Polymer Sci.* **8,** 467 (1952).
3. Dainton, F. S., and G. B. B. M. Sutherland, *J. Polymer Sci.* **4,** 37 (1949).
4. Eley, D. D., and D. C. Pepper, *Trans. Faraday Soc.* **43,** 112 (1947); D. D. Eley and A. W. Richards, *ibid.* **45,** 425, 436 (1949).
5. Evans, A. G., and G. W. Meadows, *Trans. Faraday Soc.* **46,** 327 (1950).
6. Evans, A. G., and M. Polanyi, *J. Chem. Soc.* **1947,** 252.
7. Norrish, R. G. W., and K. E. Russell, *Trans. Faraday Soc.* **48,** 91 (1952).
8. Overberger, C. G., and G. F. Endres, *J. Am. Chem. Soc.* **75,** 6349 (1953); **77,** 2201 (1955); *J. Polymer Sci.* **16,** 283 (1955).
9. Pepper, D. C., *Trans. Faraday Soc.* **45,** 397 (1949).
10. Throssell, J. J., S. P. Sood, M. Szwarc, and V. Stannett, *J. Am. Chem. Soc.* **78,** 1122 (1956).
11. Whitmore, F. C., *Ind. Eng. Chem.* **26,** 94 (1934).

C. Anionic Polymerization

Anionic polymerization was carried out on a commercial scale for many years (17) before the nature of the polymerization was recognized, in the production of the buna-type synthetic rubbers in Germany and Russia by the polymerization of butadiene with sodium or potassium as the catalyst. The first anionic chain reaction to be so identified was the polymerization of methacrylonitrile by sodium in liquid ammonia at $-75°C$ (1).

In addition to the monomer types listed in Table 10-1, monomers which polymerize by ring scission (Chapter 8*E*) may do so by anionic mechanisms (16). Typical catalysts for anionic polymerization include the alkali metals, alkali metals alkyls, such compounds as sodium naphthalene and triphenyl methyl sodium, and Grignard reagents.

Mechanism

The conventional method of initiation of ionic chains involves the addition of a negative ion to the monomer, with the opening of a bond or ring, and growth at one end:

$$NH_2^- + CH_2{=}CHX \longrightarrow H_2NCH_2CHX^-$$

$$CH_3O^- + \underset{\diagdown O \diagup}{CH_2CH_2} \longrightarrow CH_3OCH_2CH_2O^-$$

(In the equations of this section the counterions are omitted; however, their role is essentially the same as in cationic polymerization.) Simultaneous growth from more than one center can be obtained from polyvalent ions such as those derived from

or $N(CH_2CH_2O^-Na^+)_3$.

The more basic the ion, the better it serves to initiate chains. Thus, OH^- will not initiate the anionic polymerization of styrene, NH_2^- initiates fairly well, but  CH_2^- is powerful. Similarly, more acid monomers require less basic ions, with the acidity of the monomer depending on the strength of the X-M$^-$ bond formed in initiation and the stability of the resulting ion.

Initiation can also occur by the transfer of an electron to a monomer of high electron affinity. If D or D$^-$ is an electron donor,

$$D + M \longrightarrow D^+ + M^-$$

or

$$D^- + M \longrightarrow D + M^-$$

Presumably, M$^-$ is less reactive than a true carbanion or a free radical, but the addition of a monomer to M$^-$ gives a species which contains one radical end and one anion end:

$$CH_2{=}\bar{C}HX + CH_2{=}CHX \longrightarrow \underset{\text{anion end}}{:\bar{C}H_2}CHXCH_2\underset{\text{radical end}}{\dot{C}HX}$$

This species can add monomer from the two ends by different mechanisms (14). Two radical ends may dimerize, however, leaving a divalent anion to propagate.

Propagation in anionic polymerization may be conventional or may be more complex, as in the elimination of CO_2 from N-carboxyanhydrides. In contrast to radical polymerization, the β (unsubstituted) carbon atom at the end of the growing chain is the site of addition of the next monomer.

As in cationic polymerization, termination is always unimolecular, usually by transfer. The recombination of a chain with its counterion or the transfer of a hydrogen to give terminal unsaturation, frequent in cationic polymerization, is unlikely in anionic mechanisms, as may be recognized by considering the small likelihood of transferring H^- when the counterion is Na^+. Thus, in anionic polymerization, termination usually involves transfer, and the kinetic chain is broken only if the new species is too weak to propagate.

Kinetics

The kinetics of anionic polymerization may be illustrated by the polymerization of styrene with potassium amide in liquid ammonia (10):

$$\begin{aligned} KNH_2 &\longrightarrow K^+ + NH_2^- \\ NH_2^- + M &\xrightarrow{k_i} NH_2M^- \\ NH_2M_x^- + M &\xrightarrow{k_p} NH_2M_{x+1}^- \\ NH_2M_x^- + NH_3 &\xrightarrow{k_{tr}} NH_2M_xH + NH_2^- \end{aligned} \tag{10-8}$$

The usual kinetic analysis leads for high degree of polymerization to

$$v_p = \frac{k_p k_i}{k_{tr}} \frac{[NH_2^-]}{[NH_3]} [M]^2 \tag{10-9}$$

and

$$\bar{x}_n = \frac{k_p}{k_{tr}} \frac{[M]}{[NH_3]} \tag{10-10}$$

Examples of anionic polymerization

"Living" polymers (Szwarc, gen. ref.) Since the termination step usually involves transfer to some species not essential to the reaction, anionic polymerization with carefully purified reagents may lead to systems in which termination is lacking. The resulting species, called "living" polymers, can be prepared by polymerizing styrene with sodium naphthalene (15). Kinetic analysis (2) shows that the polymer can have an extremely narrow distribution of molecular weight and for all practical purposes be essentially monodisperse; this has been confirmed (11, 12) by ultracentrifugation (Chapter 3*D*) in some cases. The polymer can be "killed" by addition of a terminating agent, e.g., water, at the end of the reaction.

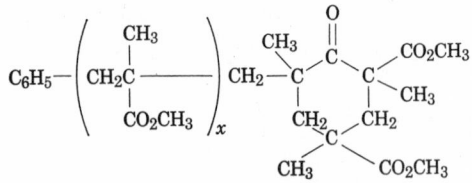

Fig. 10-1. Stabilized β-keto ester ring postulated (8) to occur in the polymerization of methyl methacrylate with phenylmagnesium bromide.

"Alfin" polymerization The "Alfin" catalyst (13) consists of a suspension in an inert solvent like pentane of a mixture of an alkylenylsodium compound (such as allyl sodium), an alkoxide of a secondary alcohol (such as sodium isopropoxide), and an alkali halide (such as sodium chloride). The catalyst is highly specific for the polymerization of dienes into the 1,4 forms.

In a typical "Alfin" polymerization (3), the steady state was achieved. It was shown that termination involved transfer to monomer in some fashion leading to species too weak to propagate. As in Eq. 10-7, this leads to $\bar{x}_n = k_p/k_t$, in accord with the experimental facts. Thus $\bar{x}_n$ should be independent of the course of the reaction, and the distribution of molecular weights narrow. In contrast, broad distributions of molecular weight were observed. It was concluded that they result from differing values of k_p/k_t from site to site on the heterogeneous catalyst (see, however, page 306).

Homogeneous anionic polymerization of methyl methacrylate (4-9) In hydrocarbon media initiators such as 9-fluorenyllithium and Grignard reagents give isotactic polymer, while in solvating media syndiotactic polymer is formed.

The polymerization with a Grignard reagent, such as phenylmagnesium bromide, is terminated by the formation of a six-membered β-keto ester ring (Fig. 10-1) (8), whereas in the 9-fluorenyllithium-initiated reaction a similar open-chain stabilized complex (Fig. 10-2) is formed at the trimer

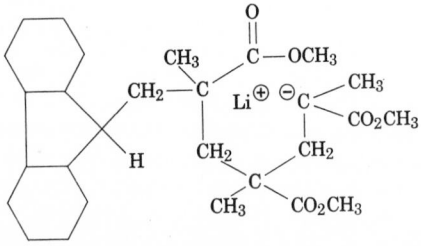

Fig. 10-2. Stabilized complex postulated (8) to occur in the polymerization of methyl methacrylate with 9-fluorenyllithium.

stage, accounting for about 90% of the chains initiated (5). About one in ten of these complexes continues to propagate, yielding isotactic polymer.

On a number basis, the distribution of molecular weights in these polymers has a large peak corresponding to the trimer. If the chains yielding high-molecular-weight polymer propagated freely once past the trimer stage, that part of the distribution corresponding to high polymer should be narrow (5); it is observed, however, to be broad. To account for this it is postulated that, after a sequence of eight to ten isotactic units has added, the chain takes on a relatively stable helical conformation in the solution, possibly as the result of adding a monomer in the syndiotactic configuration and cyclization. Since this retarding step occurs at random, the resulting distribution of molecular weights is broadened.

GENERAL REFERENCES

Burnett, G. M., *op. cit.*, Chap. XIC.
Flory, P. J., *op. cit.*, Chap. V-4b.
Gaylord, N. G., and H. F. Mark, *Linear and Stereoregular Addition Polymers*, Interscience Publishers, New York, 1959, Chap. VIII.
Mulvaney, J. E., C. G. Overberger, and A. M. Schiller, *Fortschr. Hochpolym.-Forsch.* (*Advances in Polymer Science*) **3,** 106 (1961).
Schildknecht, C. E., *op. cit.*, pp. 220–229.
Szwarc, M., *Makromol. Chem.* **35,** 132 (1960).

BIBLIOGRAPHY

1. Beaman, R. G., *J. Am. Chem. Soc.* **70,** 3115 (1948).
2. Brown, W. B., and M. Szwarc, *Trans. Faraday Soc.* **54,** 416 (1958).
3. Cleland, R. L., R. L. Letsinger, E. E. Magat, and W. H. Stockmayer, *J. Polymer Sci.* **39,** 249 (1959).
4. Fox, T. G., and coworkers, *J. Am. Chem. Soc.* **80,** 1768 (1958).
5. Glusker, D. L., I. Lysloff, and E. Stiles, *J. Polymer Sci.* **49,** 315 (1961).
6. Glusker, D. L., E. Stiles, and B. Yoncoskie, *J. Polymer Sci.* **49,** 297 (1961).
7. Goode, W. E., F. H. Owens, R. P. Fellman, W. H. Snyder, and J. E. Moore, *J. Polymer Sci.* **46,** 317 (1960).
8. Goode, W. E., F. H. Owens, and W. L. Myers, *J. Polymer Sci.* **47,** 75 (1960).
9. Graham, R. K., D. L. Dunkelberger, and E. S. Cohn, *J. Polymer Sci.* **42,** 501 (1960).
10. Higginson, W. C. E., and N. S. Wooding, *J. Chem. Soc.* **1952,** 760.
11. McCormick, H. W., *J. Polymer Sci.* **36,** 341 (1959).
12. Meyerhoff, G., and M. Cantow, *J. Polymer Sci.* **34,** 503 (1959).
13. Morton, A. A., E. E. Magat, and R. L. Letsinger, *J. Am. Chem. Soc.* **69,** 950 (1947).
14. O'Driscoll, K. F., and A. V. Tobolsky, *J. Polymer Sci.* **31,** 123 (1958).
15. Szwarc, M., M. Levy, and R. Milkovich, *J. Am. Chem. Soc.* **78,** 2656 (1956).
16. Wichterle, O., J. Šebenda, and J. Králiček, *Fortschr. Hochpolym.-Forsch.* (*Advances in Polymer Science*) **2,** 578 (1961).
17. Ziegler, K., *Angew. Chem.* **49,** 455, 499 (1936).

D. Coordination Polymerization

Since 1954, a major development in polymer chemistry has been a new technique leading to polymers with unusual structures. From the fact that these structures may involve stereospecificity, the technique is sometimes called *stereospecific* or *stereoregular polymerization*, but the term *coordination polymerization* is used here to suggest the essential feature of a coordination complex or, at the very least, a directing force which governs the orientation with which a monomer approaches the growing chain end. Whether stereospecificity is involved depends on the nature of the monomer and the catalyst system.

Coordination polymerization may be defined as polymerization using a reduced transition metal catalyst, generally in the form of a slurry of very small solid particles in an inert medium (a *fluid-bed* process). The earliest disclosures of what has since become known as a coordination catalyst for the polymerization of ethylene stem from the investigations of Roedel, Pease, and coworkers (Du Pont Company) in the period 1949–54. British (6) and United States (1) patents applied for in this period describe, among others, the combination of a metal alkyl with a metal having more than one valence, such as titanium or vanadium. The use of these catalysts (7, 10) was explored in considerable detail by Ziegler and coworkers (16); they are sometimes called Ziegler catalysts. Coordination polymerization of ethylene yields products shown to be less highly branched than those resulting from the usual free radical syntheses (Chapter 13*A*). Natta and coworkers (13) utilized coordination catalysts to polymerize α-olefins and proposed the sytem of nomenclature (Chapter 5*A*) now accepted to describe their stereoregular structure. Their use to polymerize dienes (11) also led to stereoregular products.

Mechanism

Coordination catalysts are complexes formed by the interaction of alkyls of metals of groups I–III in the periodic table with halides and other derivatives of transition metals of groups IV–VIII. A typical coordination catalyst (2, 14) is a complex between an aluminum alkyl and a titanium halide having the structure postulated in Fig. 10-3. The initial step in the formation of the catalyst involves the reduction of the titanium to the +3 or possibly +2 state (12). The polymerization involves first the formation of a complex between the monomer and the titanium in which the configuration of the monomer is maintained (Fig. 10-4). Rearrangement of

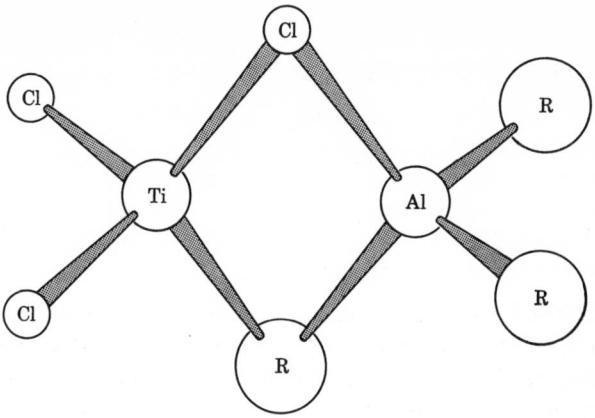

Fig. 10-3. Structure postulated for the metal halide-metal alkyl complex of a typical coordination catalyst.

this complex leads to insertion of the monomer into the chain $-CH_2R$. The steric configuration is maintained provided that the mode of initial approach of the monomer to the reaction site is always the same; presumably this is guaranteed if the catalyst complex is part of a solid surface.

Many mechanisms have been proposed (Gaylord and Mark, gen. ref., Chap. VIIE) to account for the action of the coordination complex in the propagation step of polymerization, but none has been substantiated.

Some coordination catalysts (3, 15) may be soluble rather than existing

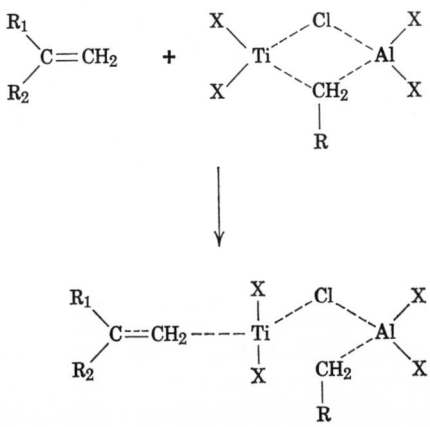

Fig. 10-4. Mechanism proposed for coordination polymerization (2). The symbol X represents either alkyl or halide.

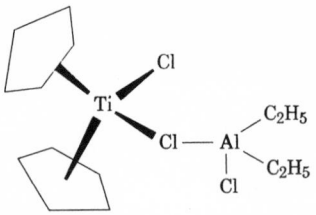

Fig. 10-5. Structure proposed for a soluble coordination catalyst (3), showing the bis(cyclopentadienyl) "sandwich" complex.

as slurries of solid particles. An example is the complex between bis(cyclopentadienyl)titanium dichloride and an aluminum alkyl having the structure proposed in Fig. 10-5. The steric influence of the large planar rings presumably insures the approach of monomer to the reaction site in a uniform manner. It should be noted, however, that soluble coordination catalysts almost certainly do not yield stereoregular polymers.

A further proposal (5) attributes the stereospecificity of coordination catalysts in either homogeneous or heterogeneous reaction to asymmetric inductive forces as indicated in Fig. 10-6. Here the polymer chain is in a configuration which leaves its bond to the catalyst most exposed to electrophilic attack by the more substituted end of the approaching monomer. Steric repulsions between groups on the old and the new substituted carbons (designated by asterisks) are between substituent R and hydrogen, rather than between R and R, H and H, a less likely occurrence. The isotactic configuration results. When the catalyst system is more strongly

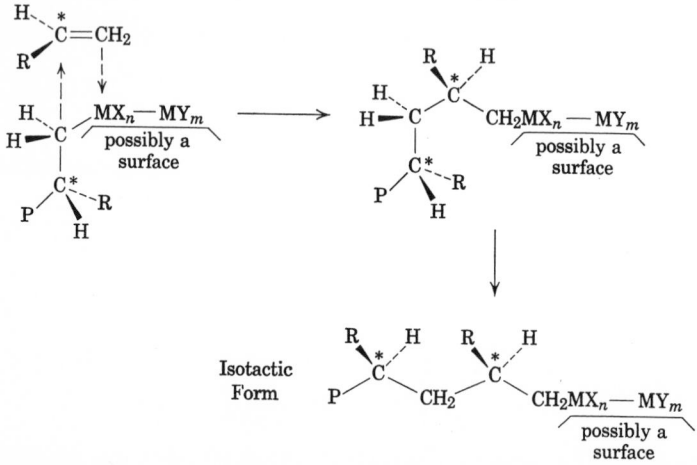

Fig. 10-6. Mechanism for the formation of isotactic polymer based on asymmetric induction (5), applicable to coordination polymerization.

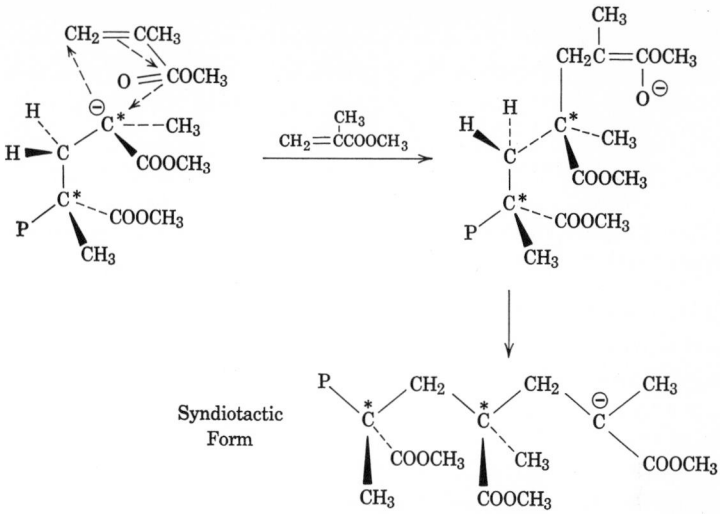

Fig. 10-7. Mechanism for the formation of syndiotactic polymer based on asymmetric induction (5), applicable to more strongly anionic polymerization.

anionic (Section *C*), as in the polymerization of methyl methacrylate by organolithium compounds in dissociating solvents (8), asymmetric induction leads by the scheme of Fig. 10-7 to the syndiotactic form.

Kinetics

Complete kinetic schemes for coordination polymerization are not substantiated. It is known that

$$v_p = k_p[\mathrm{M}][\mathrm{C}] \tag{10-11}$$

where [C] is the effective catalyst concentration (12). In the case of soluble catalyst (4) the following reaction scheme was proposed:

$$\mathrm{Catalyst} \xrightarrow{\text{fast}} \mathrm{AT}$$

$$\mathrm{AT} + \mathrm{CH_2{=}CHX} \xrightarrow{k_i} \mathrm{ATR{=}CH_2}$$

$$\mathrm{ATR}_x\mathrm{{-}CH_2} + \mathrm{CH_2{=}CHX} \xrightarrow{k_p} \mathrm{ATR}_{x+1}\mathrm{{-}CH_2}$$

$$2\mathrm{ATR}_x\mathrm{{-}CH_2} \xrightarrow{k_t} 2\ \text{polymers}$$

where ATR is the metal alkyl complex. The steady state was not achieved, but the kinetic equations could be solved to give expressions for $\bar{x}_n$ as a function of time which were in accord with the experimental facts. The

kinetic scheme is noteworthy for the inclusion of a bimolecular termination step, strongly suggesting that coordination polymerization is indeed different from ionic polymerization. However, it is likely that the termination step is more complex than the simple second-order dependence would indicate.

For heterogeneous systems, a scheme has been proposed (9) which assumes a randomly coiled chain of polymer adsorbed on the surface of the catalyst particle. It considers the effect of reversible sorption of the growing chain end and the individual chain segments, and the rate of diffusion of the chain from the surface. It is found that k_t depends on chain length, and a distribution function (Eq. 9-26) is derived which fits experimental data well. It thus appears that the broad distributions of molecular weight occasionally observed in coordination polymers can be explained without resort to the assumption of variable activity among catalyst sites.

GENERAL REFERENCES

Cooper, W., "Stereospecific Polymerization," pp. 279–339 in J. C. Robb and F. W. Peaker, eds., *Progress in High Polymers*, Vol. 1, Academic Press, N.Y., 1961.
Gaylord, N. G., and H. F. Mark, *op. cit.*, Chap. VII.
Roha, M., *Fortschr. Hochpolym.-Forsch.* (*Advances in Polymer Science*) **1,** 512 (1960).
Stille, J. K., *Chem. Revs.* **58,** 541 (1958).

BIBLIOGRAPHY

1. Anderson, A. W., J. M. Bruce, Jr., N. G. Merckling, and W. L. Truett, U.S. Patent 2,905,645 (1959); R. C. Schreyer, U.S. Patent 2,962,451 (1960); D. C. Pease and M. J. Roedel, U.S. Patent 2,762,791 (1956); M. J. Roedel, U.S. Patent 2,983,704 (1961).
2. Bawn, C. E. H., *Chem. & Ind.* **1960,** 388.
3. Breslow, D. S., and N. R. Newburg, *J. Am. Chem. Soc.* **81,** 81 (1959).
4. Chien, J. C. W., *J. Am. Chem. Soc.* **81,** 86 (1959).
5. Cram, D. J., and K. R. Kopecky, *J. Am. Chem. Soc.* **81,** 2748 (1959); D. J. Cram, *J. Chem. Educ.* **37,** 317 (1960).
6. Du Pont de Nemours and Co., E. I., British Patent 682,420 (1952) (U.S. Application, 1949); British Patent 776,326 (1957) (U.S. Application, 1954); British Patent 777,538 (1957) (U.S. Application, 1954); British Patent 778,639 (1957) (U.S. Application, 1954).
7. Fischer, M. (to Badische Anilin & Soda-Fabrik Akt.-Ges.), German Patent 874,215 (1953).
8. Fox, T. G., and coworkers, *J. Am. Chem. Soc.* **80,** 1768 (1958).
9. Gordon, M., and R.-J. Roe, *Polymer* **2,** 41 (1961); R.-J. Roe, *ibid.*, p. 60.
10. Herman, D. F., and W. K. Nelson, *J. Am. Chem. Soc.* **75,** 3877 (1953).
11. Horne, S. E., Jr., and coworkers, *Ind. Eng. Chem.* **48,** 784 (1956).
12. Ludlum, D. B., A. W. Anderson, and C. E. Ashby, *J. Am. Chem. Soc.* **80,** 1380 (1958).

13. Natta, G., and coworkers, *J. Am. Chem. Soc.* **77,** 1708 (1955); G. Natta, *J. Polymer Sci.* **16,** 143 (1955).
14. Natta, G., *Angew Chem.* **68,** 3939 (1956); G. Natta, I. Pasquon, and E. Giachetti, *ibid.* **69,** 213 (1957).
15. Natta, G., P. Pino, G. Mazzanti, and U. Giannini, *J. Am. Chem. Soc.* **79,** 2975 (1957); G. Natta, P. Corradini, and I. W. Bassi, *ibid.* **80,** 755 (1958).
16. Ziegler, K., E. Holzkamp, H. Breil, and H. Martin, *Angew. Chem.* **67,** 426, 541 (1955); K. Ziegler, Belgian Patent 533,362 (1955).

E. Polymerization with Supported Metal Oxide Catalysts

Catalysts consisting of metals or metal oxides adsorbed on or complexed with the surface of a solid carrier are of commercial interest for the low-pressure polymerization of olefins. Although often used in fixed-bed processes, the catalysts can be slurried and employed in fluid-bed operation like those of different composition described in Section *D*.

Catalyst types

Supported chromium oxide (1, 5) A catalyst consisting of 1–3% of chromium oxide on a supporting base of 90% silica and 10% alumina is widely utilized for the polymerization of ethylene. Numerous other oxides and support materials can be used but are less active. In addition to ethylene, α-olefins and diolefins can be polymerized.

Activated group VA and VIA metal oxide catalysts (3, 8, 10) The most popular catalyst of this type consists of reduced molybdenum oxide combined with alumina, titania, or zirconia. Ethylene is polymerized to a high molecular weight, essentially linear polymer. Propylene and a few other olefins can be polymerized.

Group VIII metal catalysts (2, 7, 8) Nickel, cobalt, and mixtures of these metals, supported on activated carbon, polymerize ethylene and propylene to high-molecular-weight products.

Mechanisms

The mechanisms by which the catalysts which have been described produce polymer are almost entirely unknown. It is postulated (9) that the chromium oxide catalysts involve a cationic mechanism, and the molybdenum oxide and nickel-cobalt catalysts, an anionic mechanism, but this is uncertain.

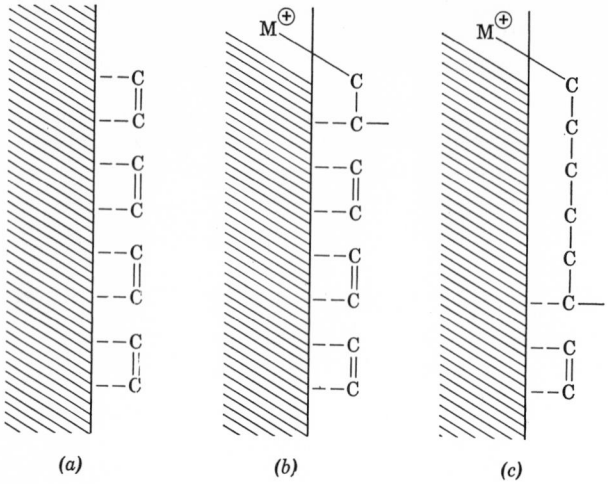

Fig. 10-8. "Bound ion" mechanism for polymerization on a catalyst surface (4), showing (*a*) adsorbed monomer, (*b*) initiation, and (*c*) propagation.

The importance of the surface to the course of the polymerization is made clear by the rather specific requirements for the preparation of the catalysts. The following mechanisms emphasize the importance of a surface, but represent little more than proposals.

In one scheme (4), which may apply also to coordination polymerization, the catalyst is assumed to adsorb monomer on its surface, the first layer being held in a special and uniform fashion (Fig. 10-8*a*). Initiation occurs when an adsorbed monomer is polarized sufficiently by some constituent of the catalyst to convert it to an ion (or a radical or an ion-radical pair) (Fig. 10-8*b*). Propagation follows (Fig. 10-8*c*) along the surface, with the polymer chain eventually desorbing and being replaced by fresh monomer. The chain may be terminated by ion rearrangement or by detachment from the surface.

An alternative scheme (6) envisions propagation taking place at a single site rather than along the surface (Fig. 10-9), with fresh monomer continuously approaching the site in a specified orientation and the growing chain extending away from the surface. This proposal is more satisfying in that the ion pair never becomes widely separated.

In either case, the crucial step in the process is the addition of a monomer molecule held in a fixed orientation at the instant of reaction. This feature is responsible for the stereospecificity of the polymer and can also account for the high rate of reaction, since oriented sorption of the monomer can greatly reduce the activation energy necessary for the propagation step.

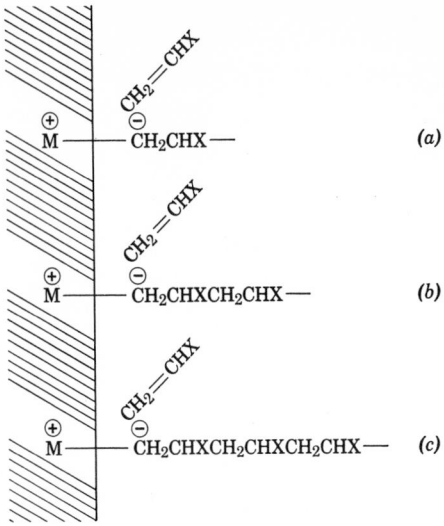

Fig. 10-9. "Bound ion" mechanism for polymerization on a catalyst surface (6) with growth from a single active site and replenishment of monomer from the liquid phase. Consecutive propagation steps are represented in (*a*), (*b*), and (*c*).

GENERAL REFERENCES

Cooper, W., *loc. cit.*
Gaylord, N. G., and H. F. Mark, *op. cit.*, Chap. IX.
Stille, J. K., *Chem. Revs.* **58,** 541 (1958).
Werber, F. X., *Fortschr. Hochpolym.-Forsch.* (*Advances in Polymer Science*) **1,** 180 (1959).

BIBLIOGRAPHY

1. Clark, A., J. P. Hogan, R. L. Banks, and W. C. Lanning, *Ind. Eng. Chem.* **48,** 1152 (1956).
2. Feller, M., and E. Field (to Standard Oil Co. of Indiana), U.S. Patents 2,717,888 and 2,717,889 (1955).
3. Field, E., and M. Feller, *Ind. Eng. Chem.* **49,** 1883 (1957).
4. Friedlander, H. N., and K. Oita, *Ind. Eng. Chem.* **49,** 1885 (1957).
5. Hogan, J. P., and R. L. Banks (to Phillips Petroleum Co.), U.S. Patent 2,825,721 (1958).
6. Natta, G., *Makromol. Chem.* **16,** 213 (1955).
7. Peters, E. F.; E. F. Peters and B. L. Evering (to Standard Oil Co. of Indiana), U.S. Patents 2,658,059 (1953); 2,692,261 (1954); 2,692,295 (1954).
8. Peters, E. F., A. Zletz, and B. L. Evering, *Ind. Eng. Chem.* **49,** 1879 (1957).
9. Schildknecht, C. E., *Ind. Eng. Chem.* **50,** 107 (1958).
10. Zletz, A.; A. K. Roebuck and A. Zletz (to Standard Oil Co. of Indiana), U.S. Patents 2,692,257 (1954); 2,692,258 (1954); 2,780,617 (1957).

11

Copolymerization

A. Kinetics of Copolymerization

Early experiments in copolymerization

Although the polymerization of organic compounds has been known for over one hundred years, the simultaneous polymerization (*copolymerization*) of two or more monomers was not investigated until about 1911, when *copolymers* of olefins and diolefins were found to have rubbery properties. For a number of years the study of the kinetics of copolymerization lagged behind that of the properties of copolymers, which were often found to be more useful than *homopolymers* made from single monomers.

In the 1930's it was found that monomers differed markedly in their tendencies to enter into copolymers. Staudinger (10) fractionated a vinyl chloride-vinyl acetate copolymer made from a mixture of equimolar quantities of the two monomers. He found no polymer containing equal amounts of each monomer, but instead ratios of vinyl chloride : vinyl acetate of 9:3, 7:3, 5:3, and 5:7 among the fractions.

At about the same time, acrylic esters were found to enter copolymers with vinyl chloride faster than did the second monomer. The first polymer formed was rich in the acrylate; later, as the amount of acrylate in the monomer system was depleted, the polymer became richer in vinyl chloride. It was found empirically that this change in composition could be reduced by adding the more active monomer gradually to the reaction mixture throughout the course of the polymerization (5).

Maleic anhydride and other monomers which homopolymerize with great difficulty were found to copolymerize readily with such monomers as styrene and vinyl chloride (11). Pairs of monomers, such as stilbene-maleic anhydride (12) and isobutylene-fumaric ester (6), in which neither

monomer polymerizes alone, readily gave high-molecular-weight copolymers in which the monomers appeared in a 1:1 ratio no matter which was in excess in the monomer feed.

The kinetic scheme for chain-reaction copolymerization is developed for radical reactions; extension to ionic systems is straightforward and is treated by Alfrey, Bohrer, and Mark, and by Burnett (gen. refs.).

The copolymer equation

In 1936 Dostal (3) made the first attack on the mechanism of copolymerization by assuming that the rate of addition of monomer to a growing free radical depends only on the nature of the end group on the radical chain. Thus monomers M_1 and M_2 lead to radicals of types $M_1\cdot$ and $M_2\cdot$. There are four possible ways in which monomer can add:

Reaction	*Rate*	
$M_1\cdot + M_1 \longrightarrow M_1\cdot$	$k_{11}[M_1\cdot][M_1]$	
$M_1\cdot + M_2 \longrightarrow M_2\cdot$	$k_{12}[M_1\cdot][M_2]$	(11-1)
$M_2\cdot + M_1 \longrightarrow M_1\cdot$	$k_{21}[M_2\cdot][M_1]$	
$M_2\cdot + M_2 \longrightarrow M_2\cdot$	$k_{22}[M_2\cdot][M_2]$	

Since four independent rate constants were involved, Dostal made no attempt to test his assumption experimentally.

After several unsuccessful approaches, the kinetics of copolymerization was elucidated in 1944 by Alfrey (1), Mayo (7), Simha (9), and Wall (13). To Dostal's reaction scheme they added the assumption of the steady state applied to each radical type separately, i.e., the concentrations of $M_1\cdot$ and $M_2\cdot$ must each remain constant. It follows that the rate of conversion of $M_1\cdot$ to $M_2\cdot$ must equal that of conversion of $M_2\cdot$ to $M_1\cdot$:

$$k_{21}[M_2\cdot][M_1] = k_{12}[M_1\cdot][M_2] \tag{11-2}$$

The rates of disappearance of the two types of monomer are given by

$$\begin{aligned} -d[M_1]/dt &= k_{11}[M_1\cdot][M_1] + k_{21}[M_2\cdot][M_1] \\ -d[M_2]/dt &= k_{12}[M_1\cdot][M_2] + k_{22}[M_2\cdot][M_2] \end{aligned} \tag{11-3}$$

By definining $r_1 = k_{11}/k_{12}$ and $r_2 = k_{22}/k_{21}$, and combining Eqs. 11-2 and 11-3, it can be shown that the composition of copolymer being formed at any instant is given by

$$\frac{d[M_1]}{d[M_2]} = \frac{[M_1]}{[M_2]}\,\frac{r_1[M_1] + [M_2]}{[M_1] + r_2[M_2]} \tag{11-4}$$

This is known as the *copolymer equation*; it has been verified by many experimental investigations of copolymer composition.

Monomer reactivity ratios

The *monomer reactivity ratios* r_1 and r_2 are the ratios of the rate constant for a given radical adding its own monomer to that for its adding the other monomer. Thus $r_1 > 1$ means that the radical $M_1\cdot$ prefers to add M_1; $r_1 < 1$ means that it prefers to add M_2. In the system styrene (M_1)-methyl methacrylate (M_2) (7), e.g., $r_1 = 0.52$ and $r_2 = 0.46$; each radical adds the other monomer about twice as fast as its own.

Since the rate constants for initiation and termination do not appear in Eq. 11-4, the composition of the copolymer is independent of over-all reaction rate and initiator concentration. The reactivity ratios are unaffected in most cases by the presence of inhibitors, chain transfer agents, or solvents. Even in heterogeneous systems they remain unchanged unless the availability of the monomers is altered by their distribution between phases. A change from a free radical to an ionic mechanism, however, changes r_1 and r_2 markedly.

A few typical values of monomer reactivity ratios are given in Table 11-1. Alfrey, Bohrer, and Mark; Burnett; Mayo and Walling; Walling; and Young (gen. refs.) give extensive compilations of r_1 and r_2.

Types of copolymerization

Ideal A copolymer system is said to be *ideal* when the two radicals show the same preference for adding one of the monomers over the other: $k_{11}/k_{12} = k_{21}/k_{22}$, or $r_1 = 1/r_2$, or $r_1r_2 = 1$. In this case the end group on

TABLE 11-1. Typical monomer reactivity ratios at 60°C

Monomer 1	Monomer 2	r_1	r_2	r_1r_2
Styrene	Butadiene	0.78	1.39	1.08
Styrene	*p*-Methoxystyrene	1.16	0.82	0.95
Styrene	Methyl methacrylate	0.52	0.46	0.24
Styrene	Vinyl acetate	55	~0.01	...
Vinyl acetate	Vinyl chloride	0.23	1.68	0.39
Vinyl acetate	Diethyl maleate	0.17	0.043	0.007
Vinyl acetate	Methyl methacrylate	0.015	20	0.3
Vinyl acetate*	Maleic anhydride	0.055	0.003	0.00017

* 75°C.

the growing chain has no influence on the rate of addition, and the two types of units are arranged at random along the chain in relative amounts determined by the composition of the feed and the relative reactivities of the two monomers. The copolymer equation reduces to $d[M_1]/d[M_2] = r_1[M_1]/[M_2]$.

Alternating Here each radical prefers to react exclusively with the other monomer: $r_1 = r_2 = 0$. The monomers alternate regularly along the chain regardless of the composition of the monomer feed. The copolymer equation simplifies to $d[M_1]/d[M_2] = 1$.

Most actual cases lie between the ideal and the alternating systems; $0 < r_1r_2 < 1$. A third possibility, with both r_1 and r_2 greater than unity, would exist if each radical preferred to add its own monomer: in the extreme, the two monomers would simultaneously homopolymerize. No such case is known; indeed there is no established instance of free radical polymerization in which the product r_1r_2 is significantly greater than unity.

Instantaneous composition of feed and polymer

Let F_1 and F_2 be the mole fractions of monomers 1 and 2 in the polymer being formed at any instant:

$$F_1 = 1 - F_2 = d[M_1]/d([M_1] + [M_2]) \tag{11-5}$$

If f_1 and f_2 similarly represent mole fractions in the monomer feed,

$$f_1 = 1 - f_2 = [M_1]/([M_1] + [M_2]) \tag{11-6}$$

the copolymer equation can be written

$$F_1 = (r_1f_1^2 + f_1f_2)/(r_1f_1^2 + 2f_1f_2 + r_2f_2^2) \tag{11-7}$$

It is apparent that in general F_1 does not equal f_1, and both f_1 and F_1 change as the polymerization proceeds. The polymer obtained over a finite range of conversion consists of many increments, each differing in composition (Section *B*).

Equation 11-7 may be used to calculate curves of feed vs. instantaneous polymer composition for various monomer reactivity ratios. Such curves for a series of ideal copolymerizations ($r_1r_2 = 1$) are shown in Fig. 11-1. Except for pairs of monomers having very similar reactivities, only a small range of feeds gives copolymers containing appreciable amounts of both components.

Several curves for nonideal cases are shown in Fig. 11-2. These curves show the effect of increasing tendency toward alternation. As alternation increases, more and more feeds yield a copolymer containing a good deal

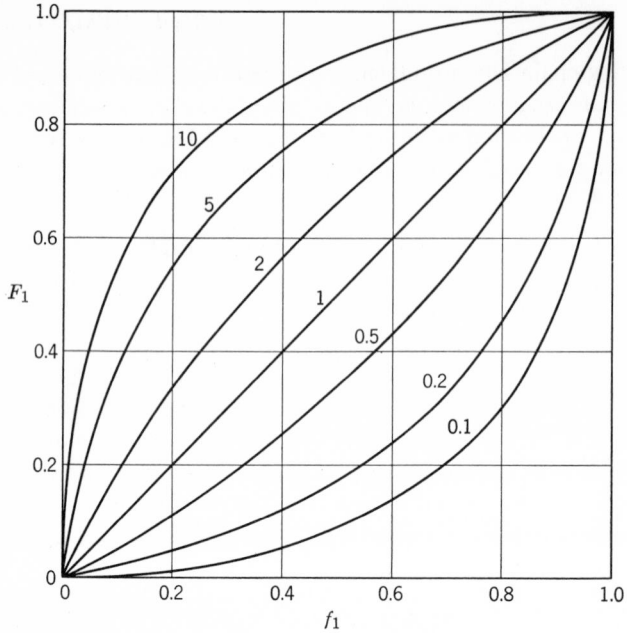

Fig. 11-1. Instantaneous composition of copolymer (mole fraction F_1) as a function of monomer composition (mole fraction f_1) for ideal copolymers with the values of $r_1 = 1/r_2$ indicated.

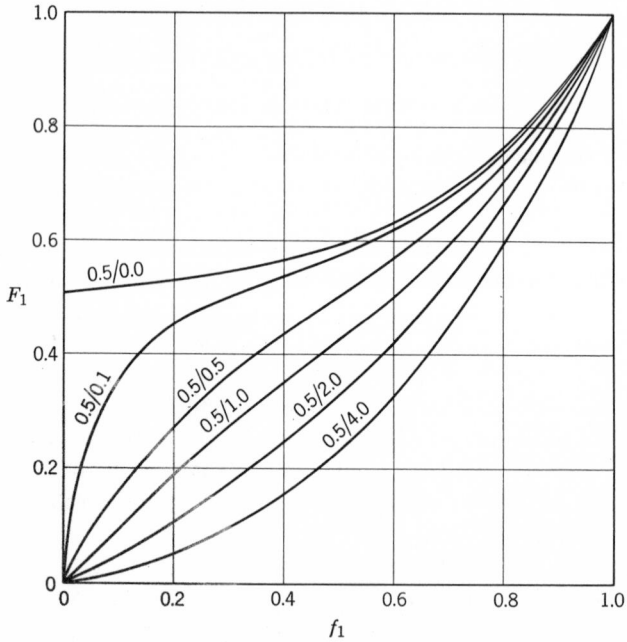

Fig. 11-2. Instantaneous composition of copolymer F_1 as a function of monomer composition f_1 for the values of the reactivity ratios r_1/r_2 indicated.

of each component. This tendency makes practical the preparation of many important copolymers.

If one of the monomers is very much more reactive than the other, the first polymer formed contains mostly the more reactive monomer. Later in the polymerization this monomer is largely used up, and the last polymer formed consists mainly of the less reactive monomer. Styrene and vinyl acetate form such a system.

For cases in which both r_1 and r_2 are less than unity (or, hypothetically, greater than unity), the curves of Fig. 11-2 cross the line representing $F_1 = f_1$. At the point of intersection polymerization proceeds without a change in the composition of feed or polymer. This is known as *azeotropic copolymerization*. Solution of Eq. 11-4 with $d[M_1]/d[M_2] = [M_1]/[M_2]$ gives the critical composition for the azeotrope:

$$[M_1]/[M_2] = (1 - r_2)/(1 - r_1)$$

or, from Eq. 11-7,

$$(f_1)_c = (1 - r_2)/(2 - r_1 - r_2) \tag{11-8}$$

Evaluation of monomer reactivity ratios

The usual experimental determination of r_1 and r_2 involves polymerizing to low conversion for a variety of feed compositions. The polymers are isolated and their compositions measured. Elemental analysis, chemical analysis for reactive groups, or physical analysis (e.g., by refractive index when styrene is one component) may be used.

Three methods of analyzing the data are available:

a. Direct curve fitting on polymer-monomer composition plots. This is a poor method, since the composition curve is rather insensitive to small changes in r_1 and r_2.

b. Plots of r_1 vs. r_2. The copolymer equation may be solved for one of the reactivity ratios:

$$r_2 = \frac{[M_1]}{[M_2]}\left[\frac{d[M_2]}{d[M_1]}\left(1 + \frac{[M_1]}{[M_2]} r_1\right) - 1\right] \tag{11-9}$$

Each experiment with a given feed gives a straight line; the intersection of several of these allows the evaluation of r_1 and r_2 (Fig. 11-3). If the experimental errors are high, the lines may not intersect in a single point; the region within which the intersections occur gives information about the precision of the experimental results.

c. Plots of F_1 vs. f_1. Equation 11-7 can be rearranged to

$$\frac{f_1(1 - 2F_1)}{(1 - f_1)F_1} = r_2 + \frac{f_1^2(F_1 - 1)}{(1 - f_1)^2 F_1} r_1 \tag{11-10}$$

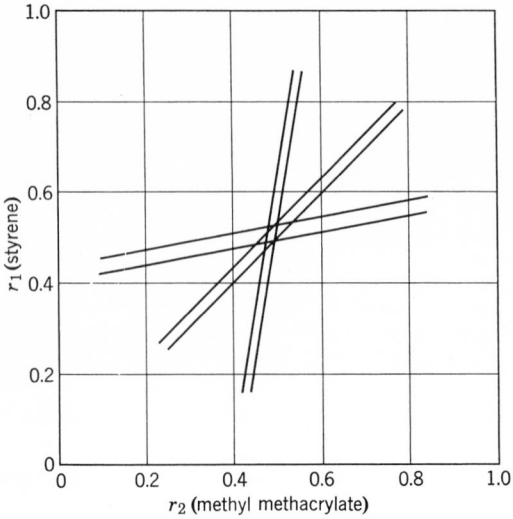

Fig. 11-3. Evaluation of monomer reactivity ratios by graphic solution of the copolymer equation [Burnett, data of (7)].

This is the equation of a straight line with slope r_1 and intercept r_2. Each experiment gives one point on the line. The least-squares treatment of a series of such points gives the best values of r_1 and r_2 in a straightforward way (4).

Rate of copolymerization

The discussion so far has considered only the relative rates of the four possible propagation steps in a binary system. The over-all rate of copolymerization depends also on the rates of initiation and termination.

If the initiator releases radicals which have a high efficiency of combination with each monomer, the two types of initiation step need not be considered separately. Steady-state conditions apply both to the total radical concentration and to the separate concentrations of the two radicals. Three types of termination must be considered, involving all possible pairs of radical types. The steady-state condition for all the radicals requires that the initiation and termination rates be equal:

$$v_i = 2k_{t_{11}}[\mathrm{M_1}\cdot]^2 + 2k_{t_{12}}[\mathrm{M_1}\cdot][\mathrm{M_2}\cdot] + 2k_{t_{22}}[\mathrm{M_2}\cdot]^2 \qquad (11\text{-}11)$$

Here $k_{t_{11}}$ and $k_{t_{22}}$ are the termination rate constants for pairs of like radicals, and $k_{t_{12}}$ is the constant for the cross reaction. If this condition

is applied to the equation for the over-all polymerization rate,

$$-\frac{d([M_1]+[M_2])}{dt} = v_p$$

$$= \frac{(r_1[M_1]^2 + 2[M_1][M_2] + r_2[M_2]^2)(v_i^{1/2}/\delta_1)}{\{r_1^2[M_1]^2 + 2(\phi r_1 r_2 \delta_2/\delta_1)[M_1][M_2] + (r_2\delta_2/\delta_1)^2[M_2]^2\}^{1/2}} \quad (11\text{-}12)$$

where

$$\delta_1 = (2k_{t_{11}}/k_{11}^2)^{1/2}, \qquad \delta_2 = (2k_{t_{22}}/k_{22}^2)^{1/2} \quad (11\text{-}13)$$

and

$$\phi = k_{t_{12}}/2(k_{t_{11}})^{1/2}(k_{t_{22}})^{1/2} \quad (11\text{-}14)$$

The δ values are simply related to quantities of the type $k_p^2/2k_t$ as seen in Chapter 9; ϕ compares the cross termination rate constant to the geometric mean of the termination rate constants for like pairs of radicals. A value of ϕ greater than 1 means that cross termination is favored, and conversely.

If r_1 and r_2 are known, measurement of the rate of copolymerization allows ϕ to be determined. A typical result (14) is shown in Fig. 11-4. The value of ϕ greater than unity is typical of most copolymerizations; indeed, cross termination is markedly favored except for nearly ideal copolymerizations. It is likely that differences in polarity are responsible for the preference for cross termination as well as for deviations from ideal copolymerization. As a result, the copolymerization of pairs of monomers of quite different polarity is likely to be slower than the mean of the rates

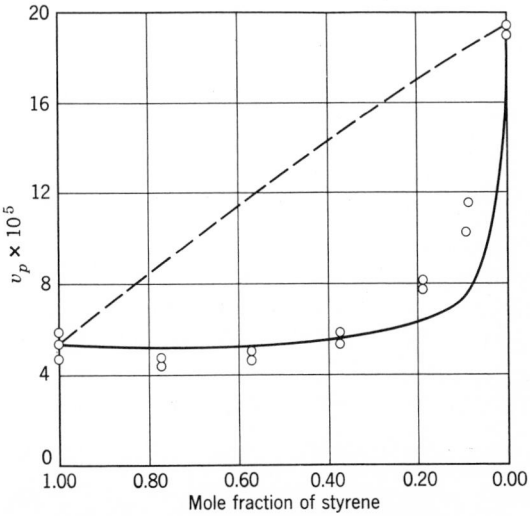

Fig. 11-4. Copolymerization rate data for the system styrene-methyl methacrylate at 60°C (14). The broken line assumes $\phi = 1$; the solid line is calculated for $\phi = 13$.

for the separate homopolymerizations. The fact that termination is almost always diffusion controlled also favors large values of ϕ, since $k_{t_{12}}$ is closer to the arithmetic than to the geometric mean of $k_{t_{11}}$ and $k_{t_{22}}$.

Extensions of copolymer kinetics

Systems of more than two monomers (15) The mathematics of the copolymer equation can be extended formally to systems with more than two monomers, but the interpretation of results quickly becomes complicated. For example, six reactivity ratios are required to describe a three-component system. Several systems of three and four monomers have been studied experimentally (Mayo and Walling, gen. ref.).

Effect of chain units before the last (8) There remains the possibility of failure of the assumption that the end unit alone in a chain radical determines its reactivity. Evidence has been found that this assumption does break down in some cases (2). The effect is particularly pronounced with monomers of type CHX=CHX, where X is a polar substituent. It has been detected in copolymers of styrene with maleic anhydride and fumaronitrile.

When the penultimate monomer unit does have some influence on the subsequent addition of monomer, four reactivity ratios, corresponding to the four possible types of radicals, must be considered (8).

GENERAL REFERENCES

Alfrey, T., Jr., J. J. Bohrer, and H. Mark, *Copolymerization*, Interscience Publishers, New York, 1952, Chaps. I, II, VI, XI, XII.

Bamford, C. H., W. G. Barb, A. D. Jenkins, and P. F. Onyon, *The Kinetics of Vinyl Polymerization by Radical Methods*, Academic Press, New York, 1958, Chap. 5.

Burnett, G. M., *Mechanism of Polymer Reactions*, Interscience Publishers, New York, 1954, Chaps. IVG, VIA-C, VII, XIB.

Flory, P. J., *Principles of Polymer Chemistry*, Cornell University Press, Ithaca, N.Y., 1953, Chaps. V-1a, 1b, 2.

Mayo, F. R., and C. Walling, *Chem. Revs.* **46,** 191 (1950).

Walling, C., *Free Radicals in Solution*, John Wiley & Sons, New York, 1957, Chap. 4.1a, b, d, e, j.

Young, L. J., *J. Polymer Sci.* **54,** 411 (1961).

BIBLIOGRAPHY

1. Alfrey, T., Jr., and G. Goldfinger, *J. Chem. Phys.* **12,** 205 (1944).
2. Barb, W. G., *J. Polymer Sci.* **11,** 117 (1953).
3. Dostal, H., *Monatsh.* **69,** 424 (1936).

4. Fineman, M., and S. D. Ross, *J. Polymer Sci.* **5,** 259 (1950).
5. Fikentscher, H., and J. Hengstenberg, U.S. Patent 2,100,900 (1937).
6. Hopff, H., and G. Steinbrunn, German Patent 671,378 (1939).
7. Mayo, F. R., and F. M. Lewis, *J. Am. Chem. Soc.* **66,** 1594 (1944).
8. Merz, E., T. Alfrey, and G. Goldfinger, *J. Polymer Sci.* **1,** 75 (1946).
9. Simha, R., and H. Branson, *J. Chem. Phys.* **12,** 253 (1944).
10. Staudinger, H., and J. Schneiders, *Ann.* **541,** 151 (1939).
11. Voss, A., and E. Dickhauser (I. G. Farbenindustrie), German Patent 540,101 (1930).
12. Wagner-Jauregg, T., *Ber.* **63,** 3213 (1930).
13. Wall, F. T., *J. Am. Chem. Soc.* **66,** 2050 (1944).
14. Walling, C., *J. Am. Chem. Soc.* **71,** 1930 (1949).
15. Walling, C., and E. R. Briggs, *J. Am. Chem. Soc.* **67,** 1774 (1945).

B. Composition of Copolymers

The copolymer equation (Eq. 11-4) predicts the average composition of the polymer formed at any instant in the polymerization. Not only may statistical fluctuations of composition about this average occur, but also polymer formed during a finite interval may contain a range of compositions resulting from variation of feed composition. In this case, the over-all polymer composition can be calculated by integrating the copolymer equation.

Statistical fluctuation of instantaneous copolymer composition

Stockmayer (3) showed that the spread in copolymer composition due to statistical fluctuations is small in practical cases. Alternating copolymers give the minimum spread in composition; even in ideal copolymers, where the spread is greater and is a function of the degree of polymerization, the effect is not large: for a copolymer containing on the average 50% of each monomer, only 12% of the polymer molecules contain less than 43% of either monomer at $\bar{x}_n = 100$, while only 12% of the molecules contain less than 49.3% of either monomer at $\bar{x}_n = 10{,}000$. The distribution of compositions follows a Gaussian curve. This small spread in composition is neglected in the discussion to follow.

The probability of occurrence of sequences of n similar monomer units can be calculated statistically for an ideal copolymer (Alfrey, Bohrer, and Mark, gen. ref.): the number $N(n)$ of such sequences is

$$N(n) = p^{n-1}(1 - p) \tag{11-15}$$

where p is the probability of addition of that monomer:

$$p = r_1[M_1]/(r_1[M_1] + [M_2]) \tag{11-16}$$

Except for a monomer present in large excess, the sequences are always short. Any tendency toward alternation clearly leads to shorter sequences than those in an ideal copolymer.

Integration of the copolymer equation

The direct integration (1) of the copolymer equation gives a result which is not convenient for calculations. The most convenient method (2) for calculating copolymer composition and distribution involves the use of Eq. 11-7. For a system in which $F_1 > f_1$, when dM moles of monomer have polymerized, the polymer contains F_1 dM moles of monomer 1, and the feed contains $(f_1 - df_1)(M - dM)$ moles of monomer 1. For a material balance,

$$Mf_1 - (M - dM)(f_1 - df_1) = F_1\, dM \tag{11-17}$$

Combining this with Eq. 11-7 gives

$$dM/M = df_1/(F_1 - f_1) \tag{11-18}$$

which can easily be integrated by numerical or graphic means to give the desired composition of polymer as a function of conversion.

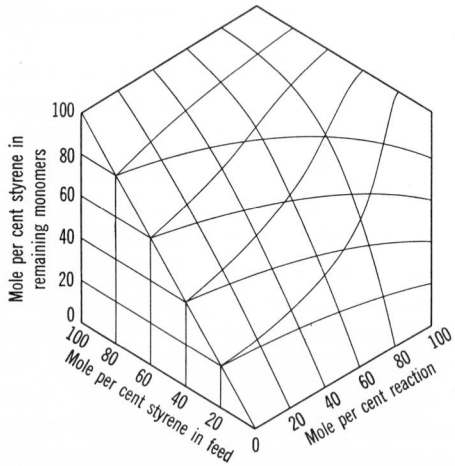

Fig. 11-5. Variation in composition of remaining monomers with initial feed and conversion for the system styrene-2-vinylthiophene (Mayo and Walling, gen. ref.).

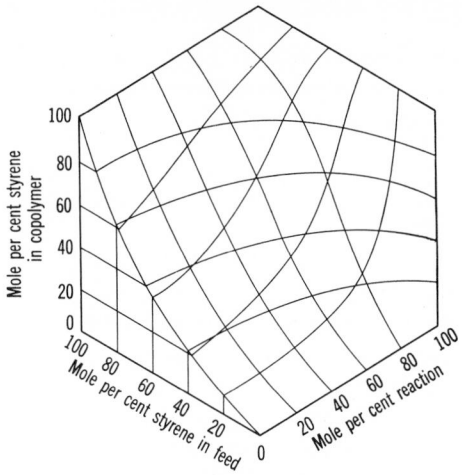

Fig. 11-6. Variation in instantaneous composition of copolymer with initial feed and conversion for the system styrene-2-vinylthiophene (Mayo and Walling, gen. ref.).

Variation of copolymer composition with conversion

Ideal system The system styrene (M_1)-2-vinylthiophene (M_2) is nearly ideal with $r_1 = 0.35$ and $r_2 = 3.10$. The variation of the composition of the remaining monomers with conversion and initial feed composition is shown in Fig. 11-5. The greater reactivity of the vinylthiophene causes it to be depleted until at the last only styrene remains. The corresponding variation in instantaneous polymer composition is shown in Fig. 11-6; pure polystyrene is the final product formed. The over-all polymer composition varies less with conversion, as indicated in Fig. 11-7, and ultimately approaches the initial feed composition as required for a material balance.

The polymer composition distribution for 100% conversion for several initial feeds is shown in the bar graphs in Fig. 11-8. Each block represents a 5% interval in copolymer composition. The U-shaped distribution curves arise whenever the monomer reactivity ratios are less than 0.5 and greater than 2, respectively. They indicate that appreciable quantities of two distinct compositions predominate, with little material of intermediate composition.

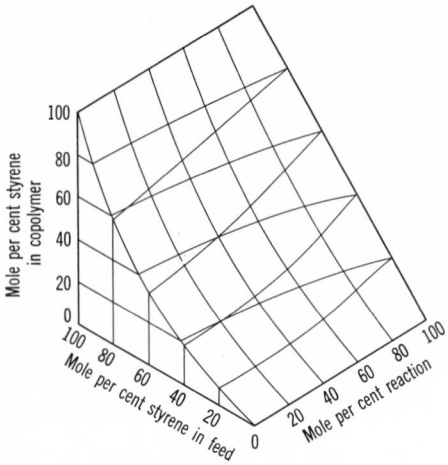

Fig. 11-7. Variation of total polymer composition with initial feed and conversion for the system styrene-2-vinylthiophene (Mayo and Walling, gen. ref.).

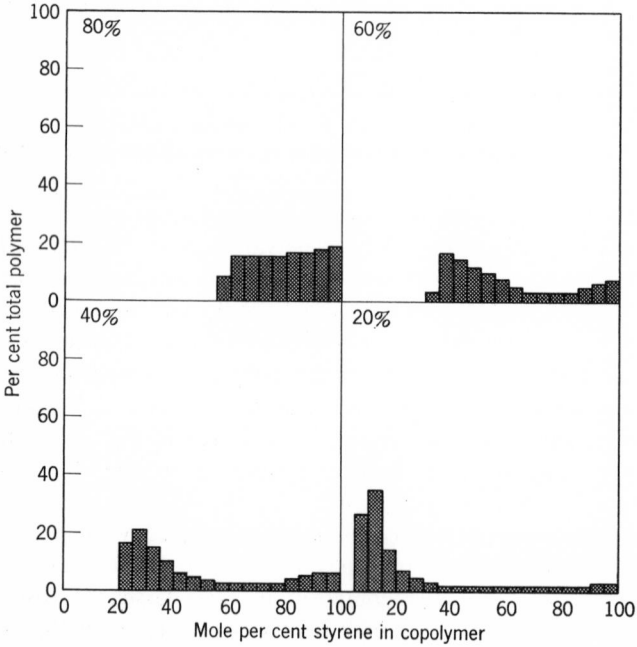

Fig. 11-8. Distribution of copolymer composition at 100% conversion for the indicated values of initial mole per cent styrene in the feed, for the system styrene-2-vinylthiophene (Burnett, after Mayo and Walling, gen. ref.).

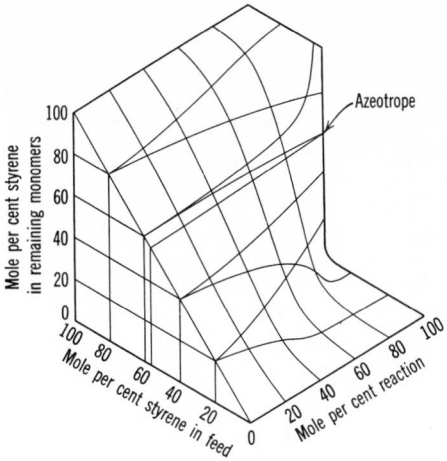

Fig. 11-9. Variation in composition of remaining monomers with initial feed and conversion for the system styrene-diethyl fumarate (Mayo and Walling, gen. ref.).

Alternating system The system styrene (M_1)-diethyl fumarate (M_2) with $r_1 = 0.30$ and $r_2 = 0.07$ has a strong tendency to alternate and an azeotropic composition at 57 mole per cent styrene. The diagrams for this copolymer are more complex than those for the previous system. As

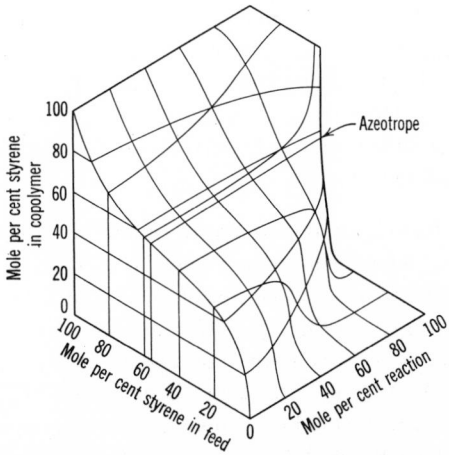

Fig. 11-10. Variation in instantaneous composition of copolymer with initial feed and conversion for the system styrene-diethyl fumarate (Mayo and Walling, gen. ref.).

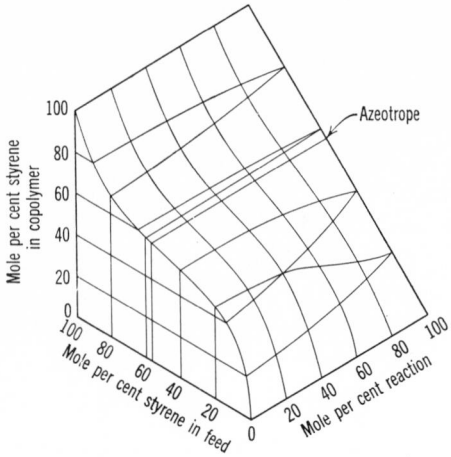

Fig. 11-11. Variation of total polymer composition with initial feed and conversion for the system styrene-diethyl fumarate (Mayo and Walling, gen. ref.).

shown in Fig. 11-9, feeds near the azeotropic composition remain almost unchanged up to rather high conversion, but eventually all feeds containing more than 57 mole per cent styrene drift toward pure styrene and those containing less styrene than the azeotropic amount yield pure diethyl fumarate. Similarly (Fig. 11-10), pure polystyrene is formed at the last from all initial feeds higher in styrene than the azeotropic, and pure poly(diethyl fumarate) from all initial feeds containing less than 57 mole per cent styrene. The over-all polymer composition (Fig. 11-11) shows markedly less change with conversion for all feeds and is of course invariant at the azeotropic composition. The bar graphs of the distribution of copolymer composition (Fig. 11-12) show that feeds containing more styrene than the azeotrope give only polymer containing more styrene than the azeotrope, and vice versa.

Attainment of homogeneity in copolymers It is often desirable to produce a polymer which is homogeneous in composition rather than one having a broad distribution of compositions. The foregoing discussion suggests two ways of doing this. (*a*) The reaction can be stopped short of complete conversion. For example, the styrene-diethyl fumarate copolymer from feed containing 40% styrene, when stopped at 75% conversion, contains only polymer having between 44 and 52% styrene. (*b*) The polymerization can be started with a feed giving the desired polymer and the feed composition maintained by adding an increment of more reactive monomer as the polymerization proceeds. Both methods find commercial use.

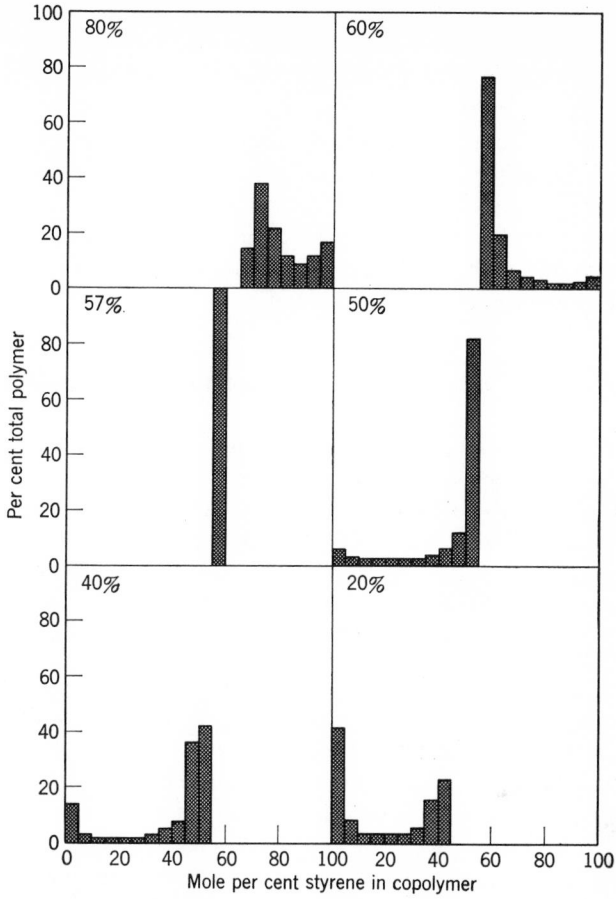

Fig. 11-12. Distribution of copolymer composition at 100% conversion for the indicated values of initial mole per cent styrene in the feed, for the system styrene-diethyl fumarate (Burnett, after Mayo and Walling, gen. ref.).

GENERAL REFERENCES

Alfrey, T., Jr., J. J. Bohrer, and H. Mark, *op. cit.*, Chaps. VII, VIII.
Burnett, G. M., *op. cit.*, Chap. VIIIB2.
Mayo, F. R., and C. Walling, *Chem. Revs.* **46,** 191 (1950).
Walling, C., *op. cit.*, Chap. 4.1c.

BIBLIOGRAPHY

1. Mayo, F. R., and F. M. Lewis, *J. Am. Chem. Soc.* **66,** 1594 (1944).
2. Skeist, I., *J. Am. Chem. Soc.* **68,** 1781 (1946).
3. Stockmayer, W. H., *J. Chem. Phys.* **13,** 199 (1945).

C. Chemistry of Copolymerization

The order of reactivity of monomers toward free radicals not only is a function of the reactivity of the monomers, but also depends on the nature of the attacking radical. This is illustrated by the tendency of many monomers to alternate in a copolymer chain. The two factors, general reactivity and alternating tendency, are predominant in determining the behavior of monomers in copolymerization.

Reactivity and structure of monomers and radicals

The reactivity of monomers and radicals in copolymerization is determined by the nature of the substituents on the double bond of the monomer. These substituents influence reactivity in three ways: they may activate the double bond, making the monomer more reactive; they may stabilize the resulting radical by resonance; or they may provide steric hindrance at the reaction site.

Reactivity of monomers The relative reactivities of monomers to a reference radical can be derived from the monomer reactivity ratio. The inverse of this ratio is the rate of reaction of the reference radical with another monomer, relative to that with its own monomer. If the latter rate is taken as unity, relative monomer reactivities can be examined. A few such rates are listed in Table 11-2; more extensive tabulations are given in the general references. As a different reference point is taken for each radical, values in different columns are not comparable, although ratios of values can be compared from column to column.

TABLE 11-2. Relative reactivities of monomers to reference radicals at 60°C

Monomer	Reference Radical					
	Styrene	Methyl Methacrylate	Acrylonitrile	Vinylidene Chloride	Vinyl Chloride	Vinyl Acetate
Styrene	(1.0)	2.2	20	12	30	>50
Methyl methacrylate	1.9	(1.0)	5.5	4	...	70
Acrylonitrile	2.4	0.74	(1.0)	2.7	>15	18
Vinylidene chloride	0.54	0.4	1.1	(1.0)	5	>30
Vinyl chloride	0.05	0.07	0.30	0.5	(1.0)	3.5
Vinyl acetate	0.02	0.05	0.2	0.25	0.5	(1.0)

Although there are some exceptions, the effectiveness of substituents in enhancing the reactivity of the monomer lies in about the order

$$-C_6H_5 > -CH{=}CH_2 > -COCH_3 > -CN > -COOR > -Cl > CH_2Y > -OCOCH_3 > -OR$$

The effect of a second substituent on the same carbon atom is usually additive.

This order of reactivity corresponds to the resonance stabilization of the radical formed after the addition. In the case of styrene the radical can resonate among forms of the type

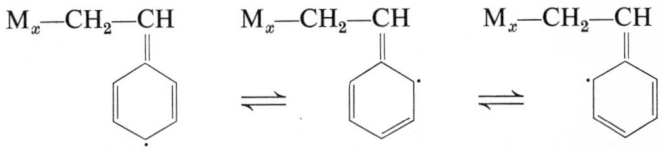

The radical is stabilized with a resonance energy of about 20 kcal/mole. At the other extreme, substituents which have no double bonds conjugated with the ethylenic double bond give radicals having very low resonance energies (1–4 kcal/mole), since only polar or unbonded forms can contribute to the resonance.

Substituents which stabilize the product radical tend also to stabilize the monomer, but the amount of stabilization is much smaller than for the radical; in styrene the monomer is stabilized to the extent of about 3 kcal/mole. Thus the stabilizing effect of a substituent on the product radical is compensated for only to a limited extent by stabilization of the monomer.

Resonance stabilization Resonance stabilization of monomers and radicals can be considered (2) in terms of a potential energy diagram of the type shown in Fig. 11-13. Curve A represents the combined energy of the growing chain radical and the monomer before reaction, while curve B represents the potential energy of the new bond formed by their combination. Resonance stabilization in the original radical or monomer is indicated by a shift from curve A to A^*; in the new radical, by a shift from B to B^*. In either case, the energy of the transition state, represented by one of the intersections 1, 2, 3, is reduced. The activation energy E_p is reduced correspondingly, but the heat evolved in the reaction $-\Delta H_p$ is increased more than E_p is reduced. The conclusion stated above is confirmed: the effect of a substituent in stabilizing a monomer through resonance is outweighed by its effect in stabilizing the resulting radical. By lowering the activation energy of the addition process, the substituent enhances the reactivity of the monomer.

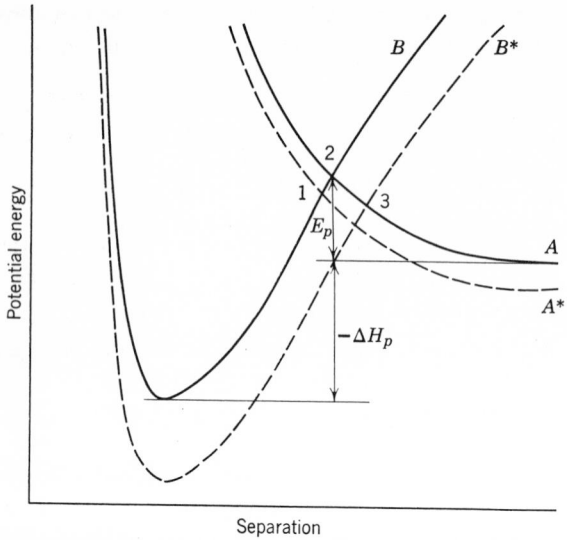

Fig. 11-13. Potential energy of a radical-monomer pair as a function of their distance apart; see text for further explanation (2).

Reactivity of radicals The reactivity of radicals to a reference monomer can be derived by combining absolute propagation rate constants k_{11} with reactivity ratios to obtain the rate constants k_{12} for radical 1 adding monomer 2. Typical data are listed in Table 11-3 (Walling, gen ref.).

As expected, resonance stabilization depresses the reactivity of radicals so that their order of reactivity is the reverse of that of their monomers; styrene is one of the least reactive radicals and vinyl acetate one of the most reactive. The effect of a substituent in depressing the activity of a

TABLE 11-3. Radical-monomer propagation rate constants at 60°C, liters/mole sec

Monomer	Radical					
	Butadiene	Styrene	Methyl Methacrylate	Methyl Acrylate	Vinyl Chloride	Vinyl Acetate
Butadiene	100	190	2,820	42,000	...	...
Styrene	70	145	1,520	11,500	(370,000)	>100,000
Methyl methacrylate	130	278	705	...	...	>100,000
Methyl acrylate	130	194	...	2,090	(150,000)	11,500
Vinyl chloride	...	7	49	230	(12,300)	8,000
Vinyl acetate	...	3	35	230	6,150	2,300

radical is much greater than its effect in enhancing the activity of the monomer. For example, the styrene radical is about 1000 times less reactive than the vinyl acetate radical to a given monomer, if alternation effects are disregarded, but styrene monomer is only about 50 times more reactive than vinyl acetate to a given radical.

Steric effects The effect of steric hindrance in reducing reactivity may be demonstrated by comparing the reactivities of 1,1- and 1,2-disubstituted olefins to reference radicals. As indicated in the preceding discussion, the addition of a second 1-substituent usually increases reactivity three- to ten-fold; however, the same substituent in the 2-position usually *decreases* reactivity two- to twenty-fold. The extent of reduction in reactivity depends on energy differences between *cis* and *trans* forms and on the possibilities for resonance (5).

Polarity and alternation

By examining the tendency for alternation as given by the product r_1r_2, it is possible to tabulate monomers in a series arranged so that two monomers farther apart in the series have a greater tendency to alternate. Such an arrangement is shown in Table 11-4. Values of r_1r_2 generally decrease as monomers farther apart in the table are compared. There are some exceptions suggesting specific interactions, probably steric in nature. For example, vinyl acetate alternates more than styrene with vinylidene chloride, and less than styrene with acrylonitrile. More complete tabulations are given by Mayo and Walling, and Burnett (gen. refs.).

There is little doubt that alternation is primarily due to the polarity of the double bond in the monomer: the order of monomers in Table 11-4 closely parallels the order of the tendency of substituents around the double bond to donate electrons to the bond (hydrocarbon, acetoxy) or to withdraw them (carbonyl, cyano).

As might be expected, the tendency toward alternation (if not too great) parallels the tendency toward decreased reactivity induced by substitution of electron-withdrawing groups at the double bond.

The Alfrey-Price treatment Alfrey and Price (1) have attempted to express the factors of general reactivity and polarity quantitatively. They write the rate constants for copolymerization in the form

$$k_{12} = P_1Q_2 \exp -e_1e_2 \tag{11-19}$$

where exp denotes the exponential, P_1 is the general reactivity of the radical $M_1\cdot$, Q_2 is the reactivity of the monomer M_2, and e_1 and e_2 are proportional to the electrostatic interaction of the permanent charges on

TABLE 11-4. Product of reactivity ratios with monomers arranged in order of alternating tendency

Vinyl acetate									
	Butadiene								
~0.5	1.0	Styrene							
0.6		0.35	Vinyl chloride						
0.3	0.2	0.25	~0	Methyl methacrylate					
0.1	0.1	0.15	~0.35	0.6	Vinylidene chloride				
~1	0.04	0.15	0.75		0.8	Methyl acrylate			
0.25	0.01	0.05		0.4			Methacrylo-nitrile		
0.25	0.02	0.02	0.1	0.25	0.35		1.1	Acrylo-nitrile	
0.004		0.02	0.06		0.55			~0	Diethyl fumarate

the substituents in polarizing the double bond. This form is analogous to Hammett's equation for the effects of nuclear substituents on the reactivity of aromatic compounds (3). From the definitions of the monomer reactivity ratios it follows that

$$r_1 = (Q_1/Q_2) \exp -e_1(e_1 - e_2)$$
$$r_2 = (Q_2/Q_1) \exp -e_2(e_2 - e_1) \tag{11-20}$$

and

$$r_1 r_2 = \exp -(e_1 - e_2)^2 \tag{11-21}$$

The Alfrey-Price equation assumes that the same value of e applies to both monomer and radical. This assumption is justified only by the success of the Q-e scheme in interpreting relative reactivities.

It should be possible to assign Q and e values to a series of monomers from the data on a few copolymerization experiments and compute values of r_1 and r_2 for all the possible arrangements of the monomers in pairs (8). Unfortunately, this practical application of the scheme is hampered by large uncertainties in the values of Q and e which can be assigned with the present experimental error. The scheme is not entirely sound from the theoretical standpoint and is best regarded as semiempirical. In a sense it represents the transcription of the reactivity and polarity series of Tables 11-2 and 11-4 to equation form.

Ionic copolymerization

The order of monomer reactivity in *cationic copolymerization* is quite different from that in free radical polymerization and is distinguished by the high reactivities of vinyl ethers and isobutylene. However, the reactivities correspond to the anticipated effect of substituents upon the reactivity of double bonds toward electrophilic reagents. The differences of reactivity arise from the change in electron availability in the double bond and the resonance stabilization of the resulting carbonium ion.

Compared with radical-induced copolymerization, the carbonium ion reactions show a much wider range of reactivity among the common monomers, and no tendency toward alternation in the copolymer. These characteristics mean that relatively few monomer pairs easily yield copolymers containing large proportions of both monomers.

The data on *anionic copolymerization* suggest that the order of monomer reactivity is much different from both free radical and carbonium ion polymerization. As anticipated, the order of reactivity in the anionic process is determined by the ability of substituents to withdraw electrons from the double bond and to stabilize the carbanion formed. Anionic and coordination polymerizations involving monomers of like electronegativity

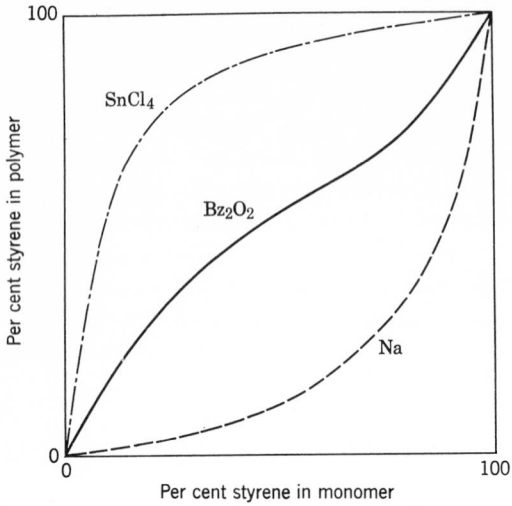

Fig. 11-14. Instantaneous composition of copolymer as a function of monomer composition for the system styrene-methylmethacrylate polymerized by cationic ($SnCl_4$), free radical (benzoyl peroxide, Bz_2O_2) and anionic (Na) mechanisms (4, 7).

are usually ideal, since the unsubstituted carbon atom at the end of the growing chain has little influence on the course of the reaction. When the electronegativities of the monomers are widely different, however, the rates of the cross-propagation reactions may be quite different also. The steady-state assumption leading to the copolymer equation is, therefore, not fulfilled (6). In extreme cases, only one monomer can add at all.

The contrast among the three methods of polymerization is brought out very clearly by the results of copolymerization studies using the three modes of initiation. Figure 11-14 shows the composition of the initial polymer formed from a feed of styrene and methyl methacrylate using different initiators. These results verify the different polymerization mechanisms and illustrate the extreme variations of reactivities that monomers exhibit with different type initiators.

Step-reaction copolymerization

Since the reactivities of all functional groups in simple bifunctional stepwise polymerization are essentially identical, irrespective of the length of the molecule to which they are attached, comonomers in such systems are randomly distributed along the chain in amounts proportional to their concentrations in the feed. Thus, the "reactivity ratios" for stepwise

polymerization are unity. Departures from this generalization are occasionally encountered, however; for example, the esterification rate of succinic acid, $HOOC(CH_2)_2COOH$, differs from that of adipic acid, $HOOC(CH_2)_4COOH$, because of the proximity of the carboxyl groups in the former compound. For similar reasons, ethylene glycol and 1,4-butanediol have different esterification rates. Also, 1,3-butanediol differs in reactivity from 1,4-butanediol because one of the hydroxyl groups of the former is secondary whereas both of the latter are primary. Where such reactivity differences exist, the tendency is for the more reactive species to enter into the polymer first.

GENERAL REFERENCES

Alfrey, T., Jr., J. J. Bohrer, and H. Mark, *op. cit.*, Chaps. III–V.
Burnett, G. M., *op. cit.*, Chap. VIIID-I.
Flory, P. J., *op. cit.*, Chap. V-1c–e.
Marvel, C. S., *An Introduction to the Organic Chemistry of High Polymers*, John Wiley & Sons, New York, 1959, pp. 58–64.
Mayo, F. R., and C. Walling, *Chem. Revs.* **46,** 191 (1950).
Walling, C., *op. cit.*, Chap. 4.1f–i.

BIBLIOGRAPHY

1. Alfrey, T., Jr., and C. C. Price, *J. Polymer Sci.* **2,** 101 (1947).
2. Evans, M. G., *Disc. Faraday Soc.* **2,** 271 (1947).
3. Hammett, L. P., *Physical Organic Chemistry*, McGraw-Hill Book Co., New York, 1940.
4. Landler, Y., *Compt. rend.* **230,** 539 (1950).
5. Lewis, F. M., and F. R. Mayo, *J. Am. Chem. Soc.* **70,** 1533 (1948).
6. O'Driscoll, K. F., *J. Polymer Sci.* **57,** 721 (1962).
7. Pepper, D. C., *Chem. Soc. Quart. Revs.* **8,** 88 (1954).
8. Price, C. C., *J. Polymer Sci.* **3,** 772 (1948).

D. Block and Graft Copolymers

It is shown in Section *B* that the probability of finding long sequences of one monomer in an ordinary random copolymer is very small, except in the trivial case where one monomer is present in large excess. Methods of synthesis of polymers containing such long sequences are of interest, however, since they may lead to polymers with properties widely different from those of either homopolymers or random copolymers (as discussed in Chapter 7). Polymers with long sequences of two monomers can have

two arrangements of chains: in *block copolymers*, the sequences follow one another along the main polymer chain,

—AABB—BBAA—AABB—, etc.,

whereas in *graft copolymers* sequences of one monomer are "grafted" onto a "backbone" of the second monomer type,

```
—AAA—AAA—AAA—, etc.
  B    B    B
  B    B    B
  |    |    |
  |    |    |
```

The ultimate aim in preparing either type is to obtain the desired copolymer free from homopolymer or other unwanted species, on a scale which allows evaluation of physical properties. Few of the techniques described in the literature satisfy these requirements.

Block copolymers

Polymers with labile end groups The usual method of preparing block copolymers utilizes a polymer with end groups which can be made to react under different conditions. In stepwise polymerization, the method is trivially simple: two polyesters, e.g., of different type are separately prepared, mixed, and allowed to polymerize further. Ring-scission polymerization can be used effectively.

By means of free radical techniques, block copolymers can be made in several ways:

a. A polymer is synthesized with end groups which can later be induced to form free radicals. For example, a polystyrene with bromo and tribromomethyl end groups was made by polymerizing styrene in the presence of CBr_4, an efficient chain transfer agent (7). The polymer was then dissolved in methyl methacrylate monomer and irradiated with ultraviolet light. This produced free radicals by the photochemical removal of a bromine atom from the polymer. A block copolymer of styrene and methyl methacrylate resulted.

In a modification of this technique (3), a tertiary amine such as triethylamine is used as the transfer agent. The resulting polymer

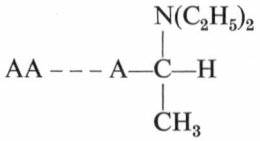

where A was methyl methacrylate, was used as a transfer agent with acrylonitrile to give a block polymer of the structure

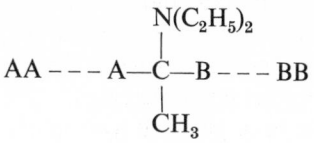

b. During the course of a polymerization, growing polymer radicals are isolated from their monomer and placed in the presence of a second monomer. In one case (6), droplets of styrene were suspended in an aqueous solution of methacrylic acid. The polymerization was initiated in the aqueous phase. Some of the polymer radicals diffused into the styrene droplets, where polymerization continued. In a second case (8, 14), polymer radicals were generated in a capillary tube by intense ultraviolet irradiation. The monomer was flowed continuously through the capillary and into a large vessel filled with a second monomer. By making the flow time from initiation to the point of discharge into the second monomer less than the radical lifetime, a block copolymer again resulted. In this system very little of either homopolymer is formed.

c. An initiator decomposing in two separate steps may be used. Styrene was polymerized (23) with phthaloyl peroxide under conditions mild enough so that only one of the peroxide groups on an initiator molecule was decomposed. Later the polymer was heated more vigorously in the presence of methyl methacrylate. The remaining peroxide groups on the initiator residues in the polystyrene decomposed to free radicals, and a block copolymer was formed.

d. A polymer chain can be caused to break into free radicals, which can add on a different monomer. A poly(vinyl ketone) was degraded to free radicals by photolysis in the presence of acrylonitrile to form a block copolymer (10). Polymer chains can also be broken by mechanical degradation, as by ultrasonic irradiation (13) or milling (2). When the degradation is carried out in the presence of a second monomer, a block copolymer results.

e. The technique of Szwarc (22) for producing unterminated "living polymer" by an anionic mechanism (Chapter 10*C*) can be used to produce a block copolymer by adding a second monomer to the still active polymer.

Graft copolymers

Chain transfer to polymer The chain transfer reaction described in Chapter 9*C* and 9*E* can be used to produce graft copolymers. In one

example, the backbone polymer was a styrene-vinylidene chloride copolymer. When vinyl acetate was polymerized in the presence of this copolymer, some of the vinyl acetate radicals abstracted chlorine from the polymer chain, after which vinyl acetate branches were produced. (Some vinyl acetate homopolymer was formed; this could be separated from the graft copolymer by solubility differences.) The resulting graft copolymer had a hydrophobic backbone and hydrophilic branches. The reverse situation was obtained by polymerizing styrene in the presence of a maleic ester-vinyl trichloroacetate copolymer. Many variations of the process give other possible combinations (21).

Creation of radical sites on polymer chains In addition to the chain transfer method, radical sites can be created on polymer chains by photolysis, as in the scission of side chain ketone groups (10) or carbon-halogen bonds; decomposition of hydroperoxides or other oxidation products (4, 17, 18); or exposure of polymer to high-energy irradiation (1, 5, 10, 20).

Ionic mechanisms Sites for ionic polymerization have been created on preformed polymer chains by reaction with Friedel-Crafts catalysts (15, 19) and by ionic chain transfer (12).

Step-reaction mechanisms Branches may be formed by stepwise polymerization onto functional groups attached to preformed polymer (often made by a radical process) (9, 11, 16).

GENERAL REFERENCES

Burnett, G. M., *Ann. Rev. Phys. Chem.* **9,** 103 (1959).

Burlant, W. J., and A. S. Hoffman, *Block and Graft Copolymers*, Reinhold Publishing Corp., New York, 1960.

Ceresa, R. J., "The Characterization of Block and Graft Copolymers," Chap. 8 in P. W. Allen, ed., *Techniques of Polymer Characterization*, Academic Press, New York, 1959.

Ceresa, R. J., *Block and Graft Copolymers*, Butterworths, Washington, 1962.

Immergut, E. H., and H. Mark, *Makromol. Chem.* **18/19,** 322 (1956).

Smets, G., and R. Hart, *Fortschr. Hochpolym.-Forsch.* (*Advances in Polymer Science*) **2,** 173 (1960).

BIBLIOGRAPHY

1. Angier, D. J., and D. T. Turner, *J. Polymer Sci.* **28,** 265 (1958).
2. Angier, D. J., and W. F. Watson, *J. Polymer Sci.* **18,** 129 (1955); **25,** 1 (1957); D. J. Angier, R. J. Ceresa, and W. F. Watson, *ibid.* **34,** 699 (1959).
3. Bamford, C. H., and E. F. T. White, *Trans. Faraday Soc.* **54,** 268 (1958).
4. Chapiro, A., *J. Polymer Sci.* **29,** 321 (1958).
5. Chapiro, A., *J. Polymer Sci.* **34,** 439 (1959).

6. Dunn, A. S., and H. W. Melville, *Nature* **169,** 699 (1952).
7. Dunn, A. S., B. D. Stead, and H. W. Melville, *Trans. Faraday Soc.* **50,** 279 (1954).
8. Funt, B. L., and E. Collins, *J. Polymer Sci.* **28,** 359 (1958).
9. Graham, R. K., *J. Polymer Sci.* **24,** 367 (1957).
10. Guillet, J. E., and R. G. W. Norrish, *Nature* **173,** 625 (1954).
11. Haas, H. C., S. G. Cohen, A. C. Oglesby, and E. R. Carlin, *J. Polymer Sci.* **15,** 427 (1955).
12. Haas, H. C., P. M. Kamath, and N. W. Schuler, *J. Polymer Sci.* **24,** 85 (1957); H. C. Haas and P. M. Kamath, *ibid.*, p. 143.
13. Henglein, A., *Makromol. Chem.* **14,** 128 (1954).
14. Hicks, J. A., and H. W. Melville, *Nature* **171,** 300 (1953); *J. Polymer Sci.* **12,** 461 (1954); *Proc. Roy. Soc.* (*London*) **A226,** 314 (1954).
15. Kockelbergh, G., and G. Smets, *J. Polymer Sci.* **33,** 227 (1958).
16. Melville, H. W., F. W. Peaker, and R. L. Vale, *J. Polymer Sci.* **30,** 29 (1958).
17. Metz, D. J., and R. B. Mesrobian, *J. Polymer Sci.* **16,** 345 (1955).
18. Natta, G., E. Beati, and F. Severini, *J. Polymer Sci.* **34,** 685 (1959).
19. Plesch, P. H., *Chem. & Ind.* **1958,** 954.
20. Sebban-Danon, J., *J. Polymer Sci.* **29,** 367 (1958).
21. Smets, G., and M. Claesen, *J. Polymer Sci.* **8,** 289 (1952).
22. Szwarc, M., *Makromol. Chem.* **35,** 132 (1960).
23. Woodward, A. E., and G. Smets, *J. Polymer Sci.* **17,** 51 (1955).

12

Polymerization Conditions and Polymer Reactions

The discussion of polymerization kinetics in Chapters 8–11 has dealt with polymerization of pure monomer or of homogeneous solutions of monomer and polymer in a solvent. Certain other types of polymerizing systems are of great interest because they offer practical advantages in industrial applications. Among these are polymerizations in which the monomer is carried in an *emulsion* or in *suspension* in an aqueous phase. Some of the advantages and disadvantages of these systems are itemized in Table 12-1. For purposes of discussion, it is convenient to classify polymerization systems as homogeneous or heterogeneous.

In the chain polymerization of monomers with more than one double bond, crosslinking can occur as in polyfunctional stepwise polymerization. The chain case is important because of the possibility of delaying the crosslinking reaction to a later time and situation, as in the vulcanization of elastomers.

In the same class of postpolymerization reactions are two other phenomena closely related to polymerization: degradation, which frequently involves just the reverse reactions of polymerization, and reactions induced by high-energy radiation, which often involve radicals and lead to crosslinking or degradation.

A. Polymerization in Homogeneous Systems

Bulk polymerization Polymerization in bulk, perhaps the most obvious method of synthesis of polymers, is widely practiced in the manufacture of condensation polymers, where the reactions are only mildly exothermic,

TABLE 12-1. Comparison of polymerization systems

Type	Advantages	Disadvantages
Homogeneous		
Bulk (batch type)	Minimum contamination. Simple equipment for making castings.	Strongly exothermic. Broadened molecular weight distribution at high conversion. Complex if small particles required.
Bulk (continuous)	Lower conversion per pass leads to better heat control and narrower molecular weight distribution.	Requires agitation, material transfer, separation, and recycling.
Solution	Ready control of heat of polymerization. Solution may be directly usable.	Not useful for dry polymer because of difficulty of complete solvent removal.
Heterogeneous		
Suspension	Ready control of heat of polymerization. Suspension or resulting granular polymer may be directly usable.	Continuous agitation required. Contamination by stabilizer possible. Washing, drying, possibly compacting required.
Emulsion	Rapid polymerization to high molecular weight and narrow distribution, with ready heat control. Emulsion may be directly usable.	Contamination with emulsifier, etc., almost inevitable, leading to poor color and color stability. Washing, drying, and compacting may be required.

and most of the reaction occurs when the viscosity of the mixture is still low enough to allow ready mixing, heat transfer, and bubble elimination. Control of such polymerizations is relatively easy.

Bulk polymerization of vinyl monomers is more difficult, since the reactions are highly exothermic and, with the usual thermally decomposed initiators, proceed at a rate which is strongly dependent on temperature. This, coupled with the problem in heat transfer incurred because viscosity increases early in the reaction, leads to difficulty in control and a tendency to the development of localized "hot spots" and "runaways." Except in the preparation of castings, e.g., of poly(methyl methacrylate), bulk polymerization is seldom used commercially for the manufacture of vinyl polymers. Polystyrene is, however, sometimes made in bulk.

Solution polymerization Polymerization of vinyl monomers in solution is advantageous from the standpoint of heat removal (e.g., by allowing the solvent to reflux) and control but has two potential disadvantages. First, the solvent must be selected with care to avoid chain transfer, and, second, the polymer should preferably be utilized in solution, as in the case of poly(vinyl acetate) to be converted to poly(vinyl alcohol) and some acrylic ester finishes, since the complete removal of solvent from a polymer is often difficult to the point of impracticality.

GENERAL REFERENCES

Burnett, G. M., *Mechanism of Polymer Reactions*, Interscience Publishers, New York, 1954, Chap. VI.

Schildknecht, C. E., *Polymer Processes*, Interscience Publishers, New York, 1956, Chaps. II, V.

B. Polymerization in Heterogeneous Systems

Gas-phase polymerization

The polymerization of gaseous vinyl monomers with polar substituents takes place rapidly by photoinitiation with the formation of a cloud of polymer. The kinetics of the process is complicated, however, since the polymer separates as a second phase almost immediately after initiation, and most of the propagation and termination must take place in or on the solid particles.

Gas-phase polymerization is not used commercially. The high-pressure polymerization of ethylene (Chapter 13*A*), though technically a vapor-phase reaction, is better considered as taking place in a highly swollen liquid polymer phase with a vapor phase present.

Bulk polymerization with polymer precipitating

In the preparation of a polymer insoluble in its monomer, or in polymerization in the presence of a nonsolvent for the polymer, marked deviations from the kinetics of homogeneous radical polymerization may occur. An example of particular interest is the preparation of polyacrylonitrile, which, like poly(vinyl chloride), poly(vinylidene chloride), and polychlorotrifluoroethylene, is insoluble in its monomer.

The bulk polymerization of acrylonitrile (7) exhibits marked autoacceleration, the rate increasing continuously up to at least 20% conversion. The initial rate is proportional to a power of the initiator concentration only slightly less than 1. These facts, which suggest that the normal bimolecular termination reaction is not effective, are explained by the trapping or occlusion of radicals in the unswollen, tightly coiled precipitating polymer. The theory has been confirmed by the demonstration of the presence of radicals in the polymer both by chemical methods and by electron paramagnetic resonance (Chapter 4*D*). The lifetime of the trapped radicals is many hours at room temperature, and if polymer containing such radicals is heated in the presence of monomer to about 60°C, where the mobility of the radicals is increased, extremely rapid polymerization takes place.

The polymerization of vinyl chloride in bulk or in the presence of a nonsolvent (2, 5) follows a rate equation of the form

$$v_p = k_p\left(\frac{fk_d[\mathrm{I}]}{k_t}\right)^{1/2}\{[\mathrm{M}] + f(\mathrm{P})\} \tag{12-1}$$

where the first term inside the braces arises from the normal rate equation for polymerization in the homogeneous liquid phase, and the second term represents the increment in rate due to polymerization in the precipitated polymer particles. The function $f(\mathrm{P})$ is proportional to the polymer concentration [P] at low conversion and to $[\mathrm{P}]^{2/3}$ at later times. Radical occlusion occurs, but the depth of radical penetration into the polymer particles is small, limiting radical activity to thin surface layers in larger particles present at higher conversions. Termination occurs primarily in the liquid phase, probably as a result of transfer to monomer within the polymer particles, and subsequent diffusion of the monomer radicals to the liquid phase. In contrast, transfer to monomer in the polymerization of acrylonitrile is so slow that permanent radical occlusion occurs.

Popcorn polymerization A troublesome form of heterogeneous polymerization occurs when tough, insoluble nodules of crosslinked polymer precipitate from systems containing diene monomers. Once formed, popcorn polymer can proliferate rapidly until the supply of monomer is exhausted. It has been a serious cause of the fouling of monomer supply lines in synthetic rubber plants.

Suspension polymerization

The term *suspension polymerization* refers to polymerization in an aqueous system with monomer as a dispersed phase, resulting in polymer

as a dispersed solid phase. The process is distinguished from superficially similar emulsion polymerization by the location of the initiator and the kinetics obeyed: in typical suspension polymerization the initiator is dissolved in the monomer phase, and the kinetics is the same as that of bulk polymerization.

The dispersion of monomer into droplets, typically 0.01–0.5 cm in diameter, is maintained by a combination of agitation and the use of water-soluble stabilizers. These may include finely divided insoluble organic or inorganic materials which interfere with agglomeration mechanically, electrolytes to increase the interfacial tension between the phases, and water-soluble polymers to increase the viscosity of the aqueous phase. The tendency to agglomerate may become critical when the polymerization has advanced to the point where the polymer beads become sticky. At the completion of the reaction, the polymer is freed of stabilizer by washing and is dried. For some applications (Chapter 20), the polymer beads can be used directly, whereas for others compaction is required. A typical recipe for suspension polymerization is given in Table 12-2. The method is used commercially to prepare hard, glassy vinyl polymers such as polystyrene, poly(methyl methacrylate), poly(vinyl chloride), poly(vinylidene chloride), and polyacrylonitrile.

Dispersion polymerization Although not entirely satisfactory, this term has been applied to systems in which the use of less water and larger quantities of stabilizers than in Table 12-2 leads to the production of stable suspensions, as in the so-called poly(vinyl acetate) "emulsions." These suspensions may be creamlike, with particles 1–10 μ in diameter, and are more resistant to coagulation than the latices resulting from typical emulsion polymerization. The kinetics is, however, that of bulk polymerization (Trommsdorff and Schildknecht, gen. ref.).

TABLE 12-2. Recipe for the suspension polymerization of methyl methacrylate at 80-130°C*

Aqueous Phase	Parts by Weight	Monomer Phase	Parts by Weight
Water	4700	Methyl methacrylate	1400
Sodium lauryl sulfate	1.0	Stearic acid (lubricant)	15
Sodium polyacrylate	12.5	Benzoyl peroxide	7.5
Sodium sulfate	38		

* E. Trommsdorff and C. E. Schildknecht, gen. ref.

Emulsion polymerization

Emulsion polymerization differs from suspension polymerization in two important respects: the initiator is located in the aqueous phase, and the polymer particles produced are typically of the order of 0.1 μ in diameter, some ten times smaller than the smallest encountered in suspension or "dispersion" polymerization. As demonstrated below, these differences lead to different kinetics for emulsion polymerization, removing the limitation set by Eq. 9-13—for combination,

$$\bar{x}_n = \frac{k_p^2}{k_t} \frac{[\mathrm{M}]^2}{v_p} \tag{12-2}$$

—on the molecular weight obtainable at a given rate of polymerization. As a result, emulsion systems allow higher-molecular-weight polymer to be produced at higher rates than do bulk or suspension systems.

A typical recipe for an emulsion polymerization is given in Table 12-3. This is the famous "Mutual" recipe used in the polymerization of styrene-butadiene synthetic rubber during World War II. The ingredients, in addition to monomers and water, are a fatty acid soap, a mercaptan-type chain transfer agent, and the water-soluble persulfate initiator.

The soap plays an important role in emulsion polymerization. At the beginning of the reaction, it exists in the form of *micelles*, aggregates of 50–100 soap molecules probably having a layered structure like that shown in Fig. 12-1. Part of the monomer enters the micelles, but most of it exists as droplets a micron or more in diameter.

It can be demonstrated experimentally that no polymer is formed in the monomer droplets. Polymerization can take place (at a very low rate) in

TABLE 12-3. "Mutual" recipe for the emulsion copolymerization of styrene and butadiene at 50°C

Ingredient	Parts by Weight
Butadiene	75
Styrene	25
Water	180
Soap	5.0
Lorol mercaptan*	0.50
Potassium persulfate	0.30

* Crude dodecyl mercaptan.

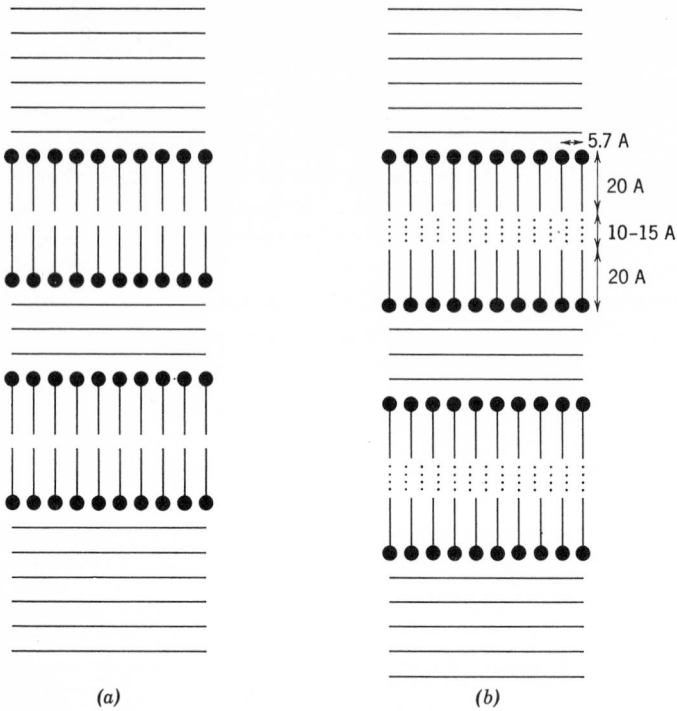

Fig. 12-1. Idealized structure of a soap micelle (*a*) without and (*b*) with solubilized monomer (Burnett, gen. ref.).

the homogeneous phase in the absence of soap, but this cannot account for the bulk of the polymer formed. At the beginning of the reaction, polymer is formed in the soap micelles; these represent a favorable environment for the free radicals generated in the aqueous phase, because of the relative abundance of monomer and the high surface/volume ratio of the micelles compared to the monomer droplets. As polymer is formed, the micelles grow by the addition of monomer from the aqueous phase (and ultimately from the monomer droplets).

Soon (2–3% polymerization) the polymer particles grow much larger than the original micelles and absorb almost all the soap from the aqueous phase. Any micelles not already activated disappear; further polymerization takes place within the polymer particles already formed. The monomer droplets are unstable at this stage; if agitation is stopped, they coalesce into a continuous oil phase containing no polymer. The droplets act as reservoirs of monomer which is fed to the growing polymer particles by diffusion through the aqueous phase. The polymer particles may

contain about 50% monomer up to the point at which the monomer droplets disappear, at 60–80% polymerization. As indicated in Fig. 12-2, rate of polymerization is constant over most of the reaction up to this point, but then falls off as monomer is depleted in the polymer particles. Rate increases with increasing soap (and initial micelle) concentration.

In a typical emulsion system (6), free radicals are generated in the aqueous phase at a rate of about 10^{13} per cubic centimeter per second. There are about 10^{14} polymer particles per cubic centimeter. Simple calculations show that termination of the free radicals in the aqueous phase is negligible and that diffusion currents are adequate for the rapid diffusion of free radicals into the polymer particles. Thus essentially all

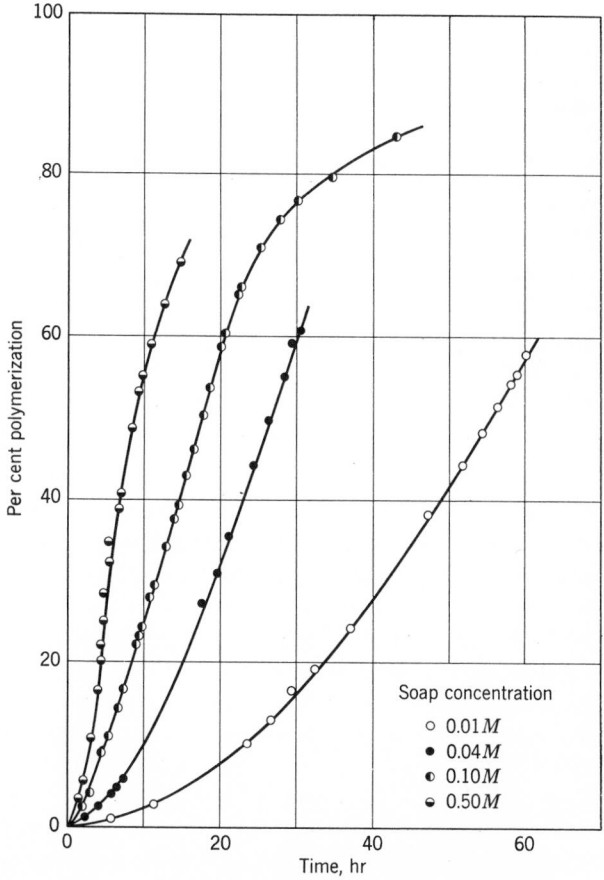

Fig. 12-2. Rate of the emulsion polymerization of isoprene at 50°C as a function of molar concentration M of soap, potassium laurate (Harkins, gen. ref.).

the free radicals get into the polymer particles—on the average about one per particle every 10 sec.

It can also be calculated from the known termination rate constants that two free radicals within the same polymer particle would mutually terminate within a few thousandths of a second. Therefore, each polymer particle must contain over most of the time either one or no free radicals. At any time, half of the particles (on the average) contain one free radical, the other half none. The rate of polymerization per cubic centimeter of emulsion is

$$v_p = k_p[\mathrm{M}](N/2) \tag{12-3}$$

where N is the number of polymer particles per cubic centimeter. Since the monomer concentration is approximately constant, the rate depends principally on the number of particles present and not on the rate of generation of radicals.

The degree of polymerization also depends upon the number of particles:

$$\bar{x}_n = k_p N[\mathrm{M}]/\rho \tag{12-4}$$

where ρ is the rate of generation of radicals. Unlike v_p, $\bar{x}_n$ is a function of the rate of free radical formation. In bulk polymerization, rate can be increased only by increasing the rate of initiation. This, however, causes a decrease in the degree of polymerization. In emulsion polymerization, the rate may be increased by increasing the number of polymer particles. If the rate of initiation is kept constant, the degree of polymerization increases rather than decreases as the rate rises. Since the number of polymer particles is determined by the number of soap micelles initially present, both rate and molecular weight increase with increasing soap concentration.

"Redox" initiation The decomposition of peroxide-type initiators in aqueous systems is greatly accelerated by the presence of a reducing agent. This acceleration allows the attainment of high rates of radical formation at low temperatures in emulsion systems.

A typical redox system is that of ferrous iron and hydrogen peroxide. In the absence of a polymerizable monomer the peroxide decomposes to free radicals as follows (4):

$$\mathrm{H_2O_2 + Fe^{+2} \longrightarrow HO\cdot + OH^- + Fe^{+3}}$$
$$\mathrm{HO\cdot + Fe^{+2} \longrightarrow OH^- + Fe^{+3}}$$

Some of the peroxide is lost by the following chain cycle:

$$\mathrm{HO\cdot + H_2O_2 \longrightarrow H_2O + HO_2\cdot}$$
$$\mathrm{HO_2\cdot + H_2O_2 \longrightarrow HO\cdot + H_2O + O_2}$$

With monomer present the HO· radicals are also used up by the reactions (I):

$$HO\cdot + M \longrightarrow HOM\cdot$$

$$HOM\cdot + M \longrightarrow HOM_2\cdot, \text{ etc.}$$

Another widely used peroxide-type initiator is the persulfate ion. With a reducing agent R the reaction is:

$$S_2O_8^{-2} + R \longrightarrow SO_4^{-2} + SO_4^{-}\cdot + R^+$$

A further possible reaction is:

$$SO_4^{-}\cdot + H_2O \longrightarrow HO\cdot + HSO_4^-$$

The reducing agent is often the thiosulfate ion:

$$S_2O_8^{-2} + S_2O_3^{-2} \longrightarrow SO_4^{-2} + SO_4^{-}\cdot + S_2O_3^{-}\cdot$$

or the bisulfite ion:

$$S_2O_8^{-2} + HSO_3^- \longrightarrow SO_4^{-2} + SO_4^{-}\cdot + HSO_3\cdot$$

Or the bisulfite ion may be activated by *oxidation* with ferric iron (3):

$$HSO_3^- + Fe^{+3} \longrightarrow HSO_3\cdot + Fe^{+2}$$

(There is some evidence that the persulfate-bisulfite reaction takes place only in the presence of traces of iron.) Any or all of these free radicals may compete in the initiation of polymerization.

GENERAL REFERENCES

Gas-phase polymerization

Burnett, G. M., *op. cit.*, Chap. V.

Polymerization with polymer precipitating

Bamford C. H., W. G. Barb, A. D. Jenkins, and P. F. Onyon, *The Kinetics of Vinyl Polymerization by Radical Mechanisms*, Academic Press, New York, 1958, pp. 103–136.

Suspension polymerization

Burnett, G. M., *op. cit.*, Chap. IXA.

Trommsdorff, E., and C. E. Schildknecht, "Polymerization in Suspension," Chap. III in C. E. Schildknecht, ed., *op. cit.*

Emulsion polymerization

Bovey, F. A., I. M. Kolthoff, A. I. Medalia, and E. J. Meehan, *Emulsion Polymerization*, Interscience Publishers, New York, 1955.

Burnett, G. M., *op. cit.*, Chap. IXB-F.
Flory, P. J., *Principles of Polymer Chemistry*, Cornell University Press, Ithaca, N.Y., 1953, Chap. V-3.
Fryling, C. F., "Emulsion Polymerization Systems," Chap. 7 in G. S. Whitby, C. C. Davis, and R. F. Dunbrook, eds., *Synthetic Rubber*, John Wiley & Sons, New York, 1954.
Harkins, W. D., *The Physical Chemistry of Surface Films*, Reinhold Publishing Co., New York, 1952, Chap. 5, "Mechanism of Emulsion Polymerization." See also W. D. Harkins, *J. Am. Chem. Soc.* **69,** 1428 (1947).
Williams, H. L., "Polymerizations in Emulsion," Chap. IV in C. E. Schildknecht, ed., *op. cit.*

BIBLIOGRAPHY

1. Baxendale, J. H., M. G. Evans, and G. S. Park, *Trans. Faraday Soc.* **42,** 155 (1946).
2. Bengough, W. I., and R. G. W. Norrish, *Proc. Roy. Soc.* (*London*) **A200,** 301 (1950).
3. Berry, K. L., and J. H. Peterson, *J. Am. Chem. Soc.* **73,** 5195 (1951).
4. Haber, F., and P. Weiss, *Proc. Roy. Soc.* (*London*) **A147,** 332 (1934).
5. Mickley, H. S., A. S. Michaels, and A. L. Moore, *J. Polymer Sci.* **60,** 121 (1962).
6. Smith, W. V., *J. Am. Chem. Soc.* **70,** 3695 (1948); W. V. Smith and R. H. Ewart, *J. Chem. Phys.* **16,** 592 (1948).
7. Thomas, W. M., *Fortschr. Hochpolym.-Forsch.* (*Advances in Polymer Science*) **2,** 401 (1960).

C. Diene and Divinyl Polymerization

Monomers with two double bonds can arbitrarily be divided into 1,3-*diene* and *divinyl* compounds. With the exception of cases in which the entrance of one double bond in the monomer into a polymer chain leaves the second double bond so unreactive that it will not polymerize, these monomers can produce branched or crosslinked polymers.

By the proper choice of relative reactivity of the two double bonds, it is possible to reduce the reactivity of one just enough so that it will not enter polymerization under the same conditions as the other, but can be made to react under more drastic conditions. This leads to postpolymerization crosslinking reactions of which *vulcanization* reactions are an example.

If the two double bonds are well separated, the reactivity of one is not affected by the polymerization of the other: an example is 2,11-dodecadiene. If the two double bonds are close enough together that the polymerization of one can shield the other sterically, or if they are conjugated, as in the dienes, a difference in reactivity can be expected. Some examples are given in Table 12-4.

TABLE 12-4. Classification of monomers with two double bonds

Dependency of Reactivity	Symmetrical	Unsymmetrical
Independent	Ethylene glycol dimethacrylate	Allyl acrylate
Intermediate	Divinylbenzene	*p*-Isopropenylstyrene
Interdependent	Butadiene	Chloroprene

Vulcanization

This section is concerned only with the statistics of vulcanization (2) and the distribution of molecular weights and crosslinks in the resulting polymer. Chemical aspects of vulcanization are discussed in Chapter 22.

Monodisperse polymer The random crosslinking of double bonds bears formal resemblance to the formation of three-dimensional networks from polyfunctional monomers by stepwise polymerization. In keeping with previous nomenclature, the fraction of the monomer units on a chain which can be crosslinked is defined as q and the degree of polymerization of the chain as x. The "functionality" of the chain is its total number of vulcanizable groups qx. The extent of reaction p is the fraction of the total available crosslinks that has been formed.

At any extent of reaction p the probability that a given chain is still unvulcanized is equal to the weight fraction w_1 of 1-mers, the probability that it is linked to one other chain is equal to w_2, and so on, where

$$\begin{aligned} w_1 &= (1-p)^{qx-1} \simeq e^{-pqx} \\ w_2 &\simeq pqxe^{-2pqx} \\ w_3 &\simeq \tfrac{3}{2}(pqx)^2 e^{-3pqx}, \text{ etc.} \end{aligned} \tag{12-5}$$

Gelation occurs at a critical value of p, p_c, where there is on the average one crosslink for every two chains:

$$p_c qx = 1 \tag{12-6}$$

As in the case of three-dimensional step-reaction polymers, after gelation the finite species in amounts w_1, w_2, etc., constitute only part of the material, the rest being in the form of gel networks. As p increases from p_c to 1, the weight fraction of the finite species drops from unity to zero.

The gel networks consist of chain segments of length n_c, both ends of

which are crosslinked, and "tails" of length n_t, only one end of which is crosslinked. The numbers of these segments are given by

$$N(n_c) \simeq pq\left(\frac{x - n_c}{x}\right)e^{-pqn_c}$$
$$N(n_t) \simeq \frac{1}{x}\,e^{-pqn_t} \tag{12-7}$$

Heterodisperse polymer The case above does not correspond to actual vulcanization because it neglects the molecular weight distribution of the chains initially present. If this is considered, the detailed analysis becomes complicated, but some important features remain about the same as for the monodisperse case. The gel point is given by the equation

$$p_c q \bar{x}_w = 1 \tag{12-8}$$

where $\bar{x}_w$ is the weight-average degree of polymerization of the starting polymer.

Chemical heterogeneity Differences in chemical composition introduced by copolymerization or mixing may have more effect on the type of network structure than does molecular weight distribution. If, e.g., a small amount of high-functionality polymer is mixed with a polymer of low functionality, the resulting vulcanizate will have regions of very tight crosslinking as inclusions in a looser network, leading to pronounced nonuniformity of physical properties.

Dilute solution vulcanization The foregoing treatment assumes that the points of crosslinking are distributed at random. This is not always true; e.g., if vulcanization is carried out in dilute solution, intramolecular crosslinking may occur before two molecules come near enough to form intermolecular links. Thus gelation is delayed considerably beyond the point predicted by random statistics.

Vinyl-divinyl copolymerization

All groups with equal reactivity Consider the copolymerization of monomer A with monomer B—B, where all groups are equally reactive, as in the copolymerization of methyl methacrylate and ethylene glycol dimethacrylate. Let the concentrations of A, B—B, and B groups be [A], [BB], and [B], respectively, where [B] = 2[BB]. Since the A and B groups are equally reactive, their ratio in the copolymer, b/a, is

$$b/a = [\mathrm{B}]/[\mathrm{A}] \tag{12-9}$$

a simplification of the copolymer equation, Eq. 11-4. At extent of reaction p, here the fraction of the double bonds reacted, there are present p[A] reacted A groups, $(1 - p)$[A] unreacted A groups, p^2[BB] doubly reacted BB molecules, $2p(1 - p)$[BB] singly reacted BB's, and $(1 - p)^2$[BB] unreacted BB's. In terms of B groups, the latter quantities become $(1 - p)^2$[B] (unreacted) B groups on unreacted BB's, $p(1 - p)$[B] unreacted or reacted B groups on singly reacted BB's, and p^2[B] (reacted) B groups on doubly reacted BB's.

In this mixture the number of crosslinks is p^2[BB] and the number of chains is derived in terms of the degree of polymerization x: (number of chains) = (total number of reacted units)/x = ([A] + [B])p/x. At the critical extent of reaction p_c for the onset of gelation, the number of crosslinks per chain is $\frac{1}{2}$, whence

$$p_c = ([A] + [B])/[B]\bar{x}_w \tag{12-10}$$

since Flory has shown that the appropriate average of x is $\bar{x}_w$. Equation (12-10) holds (4) except where gelation occurs at very low extent of reaction; here the distribution of crosslinks is not random.

Vinyl and divinyl monomers of different reactivity If the B groups are r times as reactive as the A groups, they enter the polymer r times as rapidly:

$$\frac{b}{a} = \frac{r[B]}{[A]} = \frac{r(1 - rp)[B]_0}{(1 - p)[A]_0} \tag{12-11}$$

where the subscript denotes initial concentrations. By a derivation similar to that above,

$$p_c = \frac{[A]_0 + r[B]_0}{r^2\bar{x}_w[B]_0} \tag{12-12}$$

For the experimentally interesting case of $r \ll 1$ and $[B]_0 \ll [A]_0$,

$$p_c = \frac{[A]_0}{[B]_0 r^2\bar{x}_w} \tag{12-13}$$

One group of divinyl monomer having lower activity Similar considerations apply to unsymmetrical divinyl monomers and dienes. In either case, the important practical result of delaying the onset of gelation until high conversions can be achieved in the same ways:

a. By reducing the mole fraction of the divinyl or diene monomer.
b. By reducing the weight-average chain length, possibly by the use of chain transfer agents.
c. By reducing the relative reactivity of the second double bond.

Structure of diene polymers and copolymers

The polymerization behavior of dienes as a class may be illustrated with butadiene as an example. This monomer is quite reactive, since each double bond activates the other by conjugation. After one bond has polymerized, the second is at least 100 times less active because of the lack of the resonance.

1,2 vs. 1,4 addition The radical formed when a diene has added to a growing chain is of the allylic type, with the odd electron resonating between the 2 and the 4 positions. When the next monomer unit adds onto the radical, these two resonating forms give rise to two possible configurations (Chapter 5*A*) of the resulting chain unit: 1,2 addition, with a pendant vinyl group; and 1,4 addition, with a double bond in the chain:

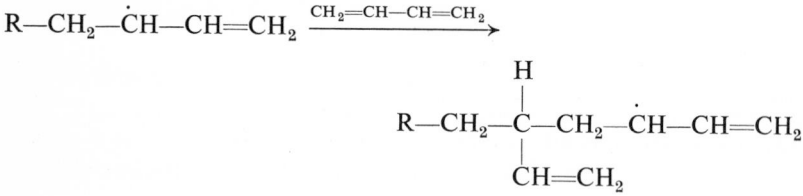

or

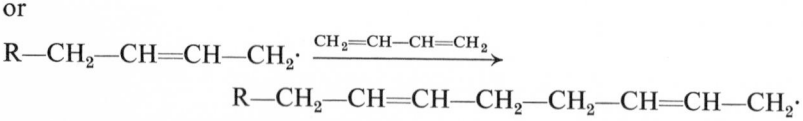

In 1,4 addition, the chain can pass through the residual double bond in either the *cis* or the *trans* configuration:

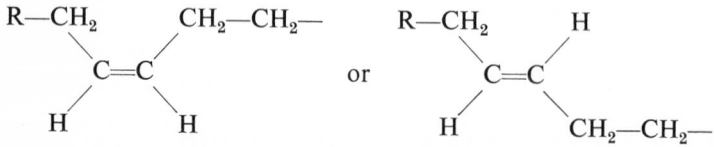

The relative amounts of 1,2 and 1,4 addition can be determined either by chemical means or by infrared spectroscopy (Chapter 4*B*). Since competition among the various propagation steps seems due mainly to differences in energy of activation, an increase in temperature tends to equalize the rates of competing reactions. Thus dienes polymerized at higher temperatures tend to contain more nearly equal numbers of the different types of units, whereas in those polymerized at lower temperatures the preferred form (in this case *trans*-1,4 addition) is favored. This leads to a more regular structure in low-temperature diene polymers.

Coordination polymerization of dienes As discussed in Chapter 10*D*, coordination polymerization can lead to amounts of the various types of addition in diene polymers which differ widely from those obtained in radical polymerization.

Alternating intra-intermolecular polymerization (1, 3) Nonconjugated dienes which have the double bonds separated by three atoms can be polymerized to give soluble, noncrosslinked polymers by an alternating intra-intermolecular mechanism. Successive propagation steps involve, alternately, the addition of a monomer and the formation of a six-membered ring:

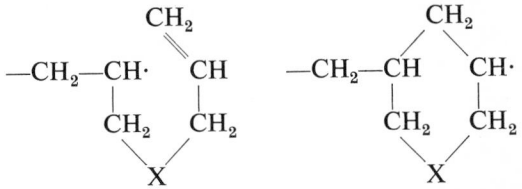

where X is one of a variety of groups. Five- and seven-membered rings form from the appropriate polymers with more difficulty, and some open-chain unsaturated groups occur in the polymer.

GENERAL REFERENCES

Alfrey, T., Jr., J. J. Bohrer, and H. Mark, *Copolymerization*, Interscience Publishers, New York, 1952, Chaps. IX, X.

Marvel, C. S., *An Introduction to the Organic Chemistry of High Polymers*, John Wiley & Sons, New York, 1959, pp. 68–75.

BIBLIOGRAPHY

1. Butler, G. B., and R. J. Angelo, *J. Am. Chem. Soc.* **79,** 3128 (1957); G. B. Butler, A. Crawshaw, and W. L. Miller, *ibid.* **80,** 3615 (1958); G. B. Butler, *J. Polymer Sci.* **48,** 279 (1960).
2. Flory, P. J., *J. Am. Chem. Soc.* **63,** 3083, 3091, 3096 (1941).
3. Marvel, C. S., *J. Polymer Sci.* **48,** 101 (1960).
4. Walling, C., *J. Am. Chem. Soc.* **67,** 441 (1945).

D. Chemical Reactions of Polymers

The concept of the reactivity of functional groups being independent of molecular weight, used in developing the kinetics and statistics of stepwise polymerization (Chapter 8), applies to all functional groups

regardless of their location on the polymer chain. As a result, polymers undergo chemical reactions much as do low-molecular-weight compounds provided that reactants can be supplied to the sites of reaction. To accomplish this, most polymer reactions of importance are carried out in solution. Many of these reactions are discussed elsewhere in this book, as indicated by cross reference.

A well-known sequence of polymer reactions is the conversion of poly(vinyl acetate) through poly(vinyl alcohol) to a poly(vinyl acetal) (Chapter 15*C*). In the latter reaction, as in the reaction of metals to remove chlorine from poly(vinyl chloride) (1), functional groups react in pairs along the chain, with occasional groups isolated and incapable of reaction.

Polyesters are readily hydrolyzed unless low solubility or steric hindrance interferes. Thus, linear condensation polyesters and polyacrylates hydrolyze readily, whereas poly(ethylene terephthalate) and crosslinked alkyd and "polyester" resins are insoluble. Polymethacrylates are inert because of steric hindrance.

The nitration, sulfonation, and reduction of styrene polymers are widely used to produce ion-exchange resins (Chapter 15*A*). The acetylation, nitration, and xanthation of cellulose (Chapter 18) are important reactions, as are vulcanization and other reactions of natural and synthetic rubbers (Chapters 14 and 22).

GENERAL REFERENCE

Marvel, C. S., *op. cit.*, pp. 75–79.

BIBLIOGRAPHY

1. Marvel, C. S., J. H. Sample, and M. F. Roy, *J. Am. Chem. Soc.* **61,** 3241 (1939).

E. Degradation of Polymers

There are two general types of polymer degradation processes, corresponding roughly to the two types of polymerization, step-reaction and chain-reaction.

Random degradation is analogous to stepwise polymerization. Here chain rupture or scission occurs at random points along the chain, leaving fragments which are usually large compared to a monomer unit.

Chain depolymerization involves the successive release of monomer units

from a chain end in a *depropagation* or *unzippering* reaction which is essentially the reverse of chain polymerization.

These two types may occur separately or in combination, may be initiated thermally or by ultraviolet light, oxygen, ozone, or other foreign agent, and may occur entirely at random or preferentially at chain ends or at other weak links in the chain.

It is possible to differentiate between the two processes in some cases by following the molecular weight of the residue as a function of the extent of reaction. Molecular weight drops rapidly as random degradation proceeds but may remain constant in chain depolymerization, as whole molecules are reduced to monomer which escapes from the residual sample as a gas. Examination of the degradation products also differentiates between the two processes: the ultimate product of random degradation is likely to be a disperse mixture of fragments of molecular weight up to several hundred, whereas chain depolymerization yields large quantities of monomer.

Kinetics of random degradation

The kinetics and statistics of random degradation can be treated in exact analogy with the kinetics of linear stepwise polymerization. If p is defined, as before, as the extent of reaction, then $1 - p =$ (number of broken bonds)/(total number of bonds) is the extent of degradation. If the degradation is random, the number of bonds broken per unit time is constant as long as the total number of bonds present is large compared to the number broken. It follows that the number of chain ends increases linearly with time; hence, $1/\bar{x}_n$ increases linearly with time. The acid-catalyzed homogeneous degradation of cellulose is a random degradation of this type.

An example of random degradation initiated by attack at a "weak link" in a chain is the ozonolysis of the isobutylene-isoprene copolymer butyl rubber. Here the initial attack is at the double bonds of the isoprene residues.

Chain depolymerization

Chain depolymerization is a free radical process which is essentially the reverse of chain polymerization. The point of initial attack may be at the chain end or at a "weak link" which may arise from a chain defect, such as an initiator fragment or a peroxide or ether link arising from polymerization in the presence of oxygen. The slightly higher activity of a tertiary hydrogen atom may be enough to provide a site for the initiation of the degradation process.

Poly(methyl methacrylate) degrades thermally by this process. The yield of monomer is 100% of the weight of polymer lost over a large fraction of the reaction. Polystyrene shows an intermediate behavior: the depropagation reaction ceases before the chain is completely destroyed. In some cases, such as olefin-SO_2 copolymers (1), an equilibrium can easily be reached between propagation and depropagation.

Kinetics of chain depolymerization (3) A general kinetic scheme has been formulated which appears to cover all types of depolymerization. It is based on the concept of inverse chain polymerization and includes the steps of initiation, depropagation, termination, and chain transfer. The important feature of this scheme is the inclusion of the chain transfer step, for it can be shown that the kinetics of random degradation result if the kinetic chain length before transfer is just the breaking of one bond.

The transfer reaction probably occurs rapidly by the abstraction of a hydrogen atom from a polymer. The chain which was attacked is likely to split into a radical and one or more inactive fragments at the elevated temperatures where degradation is rapid. Evidence for the transfer reaction includes the observations that tertiary hydrogen atoms at branch points in polyethylene are preferentially attacked and that the degradation of poly(methyl methacrylate) which has been copolymerized with a little acrylonitrile is quite different from that of pure poly(methyl methacrylate) because of the activity of the α-hydrogens on the acrylonitrile units.

The two factors which appear to be important in determining the course of degradation are the reactivity of the depropagating radical and the availability of reactive hydrogen atoms for transfer. With the possible exception of styrene, where the radical is stabilized by resonance, all polymers containing α-hydrogens, such as the polyacrylates and polyacrylonitrile, give poor yields of monomer.

Conversely, the methacrylates give high yields of monomer because of the active radical and the α-methyl group which blocks the possibility of chain transfer. Polytetrafluoroethylene gives high yields of monomer because the strong C-F bond is not easily broken to allow transfer.

The scheme does not apply to polymers such as poly(vinyl acetate) and poly(vinyl chloride), where degradation results from the removal of side groups rather than from chain scission.

Only two parameters beside the rate constants are important in the kinetic analysis. They are the transfer constant, σ, defined as

$$\sigma = \frac{\text{probability of transfer}}{\text{probability of initiation}} = \frac{k_{tr}[\mathrm{R}\cdot]}{k_i} \tag{12-14}$$

and the kinetic chain length. This is not defined in the same way as in

TABLE 12-5. Per cent monomer resulting from thermal degradation

Polymer	Per Cent Monomer	
	Weight	Mole
Poly(methyl methacrylate)	100	100
Poly(α-methylstyrene)	100	100
Polyisobutylene	32	78
Polystyrene	42	65
Polybutadiene	14	57
SBR (butadiene-styrene copolymer)	12	52
Polyisoprene	11	44
Polyethylene	3	21

polymerization, but as the number-average number of monomers produced from one chain. The kinetic chain length is $(1/\epsilon) - 1$, where

$$\frac{1}{\epsilon} = \frac{\text{probability of (propagation + termination + transfer)}}{\text{probability of (termination + transfer)}} \tag{12-15}$$

The kinetic chain length is approximately measured by the ratio of weight of monomer to that of other volatile fragments when depolymerization is completed. In polyethylene, where almost no monomer is produced, $1/\epsilon \simeq 1$. In polystyrene or polyisobutylene, $1/\epsilon \simeq 4$; in poly(methyl methacrylate) $1/\epsilon$ is at least as high as 200.

Degradation products Study of the products of thermal depolymerization in vacuum has shown (2) that the chemical nature and relative amounts of these products are remarkably independent of the temperature and extent of the degradation reaction. Polystyrene, e.g., degraded to 40% styrene, 2.4% toluene, and other products having an average molecular weight of 264, at temperatures between 360° and 420°C, and extents of degradation from 4 to 100%. The amounts of monomer produced by various polymers are shown in Table 12-5.

GENERAL REFERENCES

Burnett, G. M., *op. cit.*, Chap. X.

Grassie, N., *Chemistry of High Polymer Degradation Processes*, Interscience Publishers, New York, 1956.

Jellinek, H. H. G., *Degradation of Vinyl Polymers*, Academic Press, New York, 1955.

Society of the Chemical Industry, *High Temperature Resistance and Thermal Degradation of Polymers*, S.C.I. Monograph No. 13, The Macmillan Co., New York, 1961.

Wall, L. A., *SPE J.* **16**, 810, 1031 (1960).

BIBLIOGRAPHY

1. Dainton, F. S., and K. J. Ivin, *Trans. Faraday Soc.* **46,** 331 (1950).
2. Madorsky, S. L., *J. Polymer Sci.* **9,** 133 (1952).
3. Simha, R., and L. A. Wall, *J. Phys. Chem.* **56,** 707 (1952).

F. Radiation Chemistry of Polymers

The interaction of high-energy radiation with molecular substances involves the following sequence of events, regardless of the source of energy (photons, protons, electrons, neutrons, etc.). The molecules are first excited and ionized. Secondary electrons are emitted with relatively low speeds, and produce many more ions along their tracks. Within 10^{-12} sec or so, molecular rearrangements take place in the ions and excited molecules, accompanied by thermal deactivation or the dissociation of valence bonds. As far as subsequent reactions are concerned, bond dissociation is the more important. It leads to the production of ions or radicals whose lifetimes depend on diffusion rates and may be weeks or months in solids at low temperatures.

The major effects in polymers arise from the dissociation of primary valence bonds into radicals, whose existence can be demonstrated by chemical methods or by EPR spectroscopy (Chapter 4*D*). The dissociations of C-C and C-H bonds lead to different results, degradation and crosslinking, which may occur simultaneously.

Degradation The major result of radiation is degradation by chain scission in 1,1-disubstituted polymers, such as poly(methyl methacrylate) and its derivatives, polyisobutylene, and poly(α-methylstyrene), and in polymers containing halogen, such as poly(vinyl chloride), poly(vinylidene chloride), and polytetrafluoroethylene. The tendency toward degradation is related to the absence of tertiary hydrogen atoms (6), a weaker than average C-C bond [low heat of polymerization (7)], or unusually strong bonds (such as C-F) elsewhere in the molecule.

Degradation is, of course, evidenced by decrease in molecular weight, the weight-average molecular weight being inversely proportional to the amount of radiation received. In polymers with bulky side chains, such as poly(methyl methacrylate) (1), extensive degradation of the side chains to gaseous products also occurs.

Crosslinking Crosslinking is the predominant reaction on the irradiation of polystyrene, polyethylene and other olefin polymers, polyacrylates and

their derivatives, and natural and synthetic rubbers. It is accompanied by the formation of gel (2, 8) and ultimately by the insolubilization of the entire specimen. Radiation crosslinking has a beneficial effect on the mechanical properties of some polymers (3) and is carried out commercially to produce polyethylene with enhanced form stability and resistance to flow at high temperatures (Chapter 13*D*).

Other reactions Radiation crosslinking is often accompanied by the formation of *trans*-vinylene unsaturation (4), both reactions resulting in the formation of hydrogen gas. If the irradiation is carried out in the presence of air, surface oxidation (2, 5) results. The resulting peroxides may be decomposed later in graft copolymerization (Chapter 11*D*).

GENERAL REFERENCES

Bovey, F. A., *The Effects of Ionizing Radiation on Natural and Synthetic High Polymers*, Interscience Publishers, New York, 1958.

Chapiro, A., *Radiation Chemistry of Polymer Systems*, Interscience Publishers, a Division of John Wiley & Sons, New York, 1962.

Charlesby, A., *Atomic Radiation and Polymers*, Pergamon Press, New York, 1960.

Dole, M., *J. Chem. Educ.* **36,** 353 (1959).

BIBLIOGRAPHY

1. Alexander, P., A. Charlesby, and M. Ross, *Proc. Roy. Soc.* (*London*) **A223,** 392 (1954).
2. Charlesby, A., *Proc. Roy. Soc.* (*London*) **A215,** 187 (1952).
3. Charlesby, A., *Nucleonics* **14,** #9, 86 (September, 1956).
4. Dole, M., C. D. Keeling, and D. G. Rose, *J. Am. Chem. Soc.* **76,** 4304 (1954); M. Dole, D. C. Milner, and T. F. Williams, *ibid.* **79,** 4809 (1957).
5. Matsuo, H., and M. Dole, *J. Phys. Chem.* **63,** 837 (1959).
6. Miller, A. A., E. J. Lawton, and J. S. Balwit, *J. Polymer Sci.* **14,** 503 (1954).
7. Wall, L. A., *J. Polymer Sci.* **17,** 141 (1955).
8. Williams, T. F., and M. Dole, *J. Am. Chem. Soc.* **81,** 2919 (1959).

IV

Properties of Commercial Polymers

Introduction

In the chapters of Part IV are described the polymerization, structure, properties, fabrication, and applications of commercially important polymers, including those used as plastics, fibers, and elastomers. Classification is largely by polymer structure, with some exceptions individually noted. Descriptions are of necessity brief; the reader is referred to the general references listed at the end of each section and to the following books which contain more extensive descriptions of commercial polymers.

GENERAL REFERENCES

Fisher, H. L., *Chemistry of Natural and Synthetic Rubbers*, Reinhold Publishing Corp., New York, 1957.

Golding, B., *Polymers and Resins*, D. Van Nostrand Co., Princeton, N.J., 1959.

Modern Plastics Magazine, including annual Encyclopedia Issue, Breskin Publications, New York.

Moncrieff, R. W., *Man-Made Fibres*, 3rd ed., John Wiley & Sons, New York, 1957.

Morton, M., ed., *Introduction to Rubber Technology*, Reinhold Publishing Corp., New York, 1959.

Press, J. J., ed., *Man-Made Textile Encyclopedia*, Interscience Publishers, New York, 1959.

Simonds, H. R., *Concise Guide to Plastics*, Reinhold Publishing Corp., New York, 1957.

Simonds, H. R., ed., *Source Book of the New Plastics*, Reinhold Publishing Corp., New York, Vol. 1, 1959; Vol. 2, 1961; other volumes to follow.

Winding, C. C., and G. D. Hiatt, *Polymeric Materials*, McGraw-Hill Book Co., New York, 1961.

13

Olefin Polymers

A. Low-Density (Branched) Polyethylene

The first commercial ethylene polymer was branched polyethylene, commonly designated as low-density or high-pressure material to distinguish it from the essentially linear material described in Section *B*. After a period of relatively slow growth, culminating in the production in the United States of less than 100 million lb in 1952, the annual volume of low-density polyethylene has expanded rapidly. Polyethylene became in 1959 the first plastic with production exceeding 1 billion lb, with some ten producers operating in this country. Consumption in 1961 was about 1.5 billion lb, with prices in the range of \$0.25–\$0.38/lb.

Polyethylene was first produced in the laboratories of Imperial Chemical Industries, Ltd. (ICI), England, in a fortuitous experiment in which ethylene (and other chemicals which remained inert) was subjected to 1400 atm of pressure at 170°C (4). Traces of oxygen caused polymerization to take place.

Polymerization

Ethylene (boiling point −104°C) can be obtained from the dehydration of ethanol or by the hydrogenation of acetylene. Where supplies of natural or petroleum gas are available, however, the monomer is produced by cracking ethane or propane followed by purification processes.

High polymers of ethylene are made commercially at pressures between 1000 and 3000 atm (15,000–45,000 psi) or possibly higher, and temperatures as high as 250°C.

Traces of oxygen initiate the polymerization of ethylene readily. Rapid exothermic reactions can occur, and violent explosions have taken place. Many other possible impurities in the monomer, such as hydrogen and

acetylene, act as chain transfer agents and must be carefully removed if high-molecular-weight products are to be obtained. Besides oxygen, peroxides (benzoyl, diethyl), hydroperoxides, and azo compounds have been used as initiators.

Ethylene polymerization can be carried out with benzene or chlorobenzene as solvent. At the temperatures and pressures used, both polymer and monomer dissolve in these compounds so that the reactions are true solution polymerizations. Water or other liquids may be added to dissipate the heat of reaction.

Batch polymerizations of ethylene cannot be carried out rapidly with reproducibility and good control. Long reaction times, consistent with good control, are not economical. In addition, chain branching becomes excessive at high conversion and results in poor physical properties of the product. As a result, balanced, continuous polymerization systems are preferred. Emulsion polymerization has had little success.

One continuous process utilizes tubular reactors, which may have diameters of less than 1 in. and lengths up to 100 ft. The stainless steel tube may be filled with water, and ethylene containing initiator and possibly benzene is introduced. Additional initiator and water or benzene can be injected at one or more points along the tube to keep the initiator concentration more nearly constant throughout the reactor. Ten or more per cent of the ethylene is polymerized at the far end of the reactor. Here the gas and liquid phases are taken off continuously, the polymer is separated, and the ethylene is recycled after purification.

Another process utilizes bulk polymerization in a tower-type reactor. Ethylene containing trace amounts of oxygen is charged to the reactor at 1500 atm and 190°C. The reaction is kept essentially isothermal and carried to 10–15% conversion. The effluent from the reactor passes to a separatory vessel in which unconverted ethylene is removed for recycling. The molten polyethylene is chilled below its crystalline melting point and passed through the usual finishing steps.

Low polymers of ethylene (oils, greases, and waxes) can be made at pressures between 100 and 500 atm and at temperatures between room temperature and 250°C.

Structure

Low-density polyethylene is a partially (50–60%) crystalline solid melting at about 115°C, with density in the range 0.91–0.94. It is soluble in many solvents at temperatures above 100°C, but no room-temperature solvents exist.

In 1940 (5) infrared spectroscopy (Chapter 4*B*) revealed that low-density

polyethylene contains branched chains. In a series of papers beginning in 1953 (10), it was shown that these branches are of two distinct types. Branching due to intermolecular chain transfer, arising from reactions of the type

$$\underset{\text{propagating chain}}{R_1CH_2CH_2\cdot} + \underset{\text{dead polymer molecule}}{R_2CH_2CH_2R_3} \xrightarrow[\text{hydrogen transfer}]{\text{intermolecular}} \underset{\text{dead polymer molecule}}{R_1CH_2CH_3} + \underset{\text{propagating chain}}{R_2\dot{C}HCH_2R_3},$$

leads to branches which are, on the average, as long as the main polymer chain. This sort of branching has an observable effect on the solution viscosity of the polymer (Chapter 3*E*) and can be detected by comparing the viscosity of a branched polyethylene with that of a linear polymer of the same molecular weight.

The second branching mechanism in polyethylene is postulated to produce short-chain branching by intramolecular chain transfer:

$$RCH_2CH_2CH_2CH_2CH_2CH_2\cdot \xrightarrow[\text{ring formation}]{\text{transient six-membered}}$$

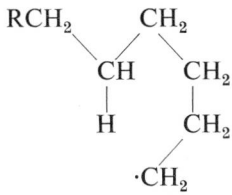

$$\xrightarrow[\text{hydrogen transfer}]{\text{intramolecular}} \underset{\text{propagating secondary free radical}}{RCH_2\dot{C}HCH_2CH_2CH_2CH_3}$$

The transient ring mechanism suggests four carbon atoms as the most probable length of the short branches. Infrared absorption studies (2, 12), as well as studies (6, 8) of degradation under bombardment with high-energy radiation of polyethylene and substituted polyethylenes of known branch structure, suggest that both ethyl and butyl branches are present. A mechanism proposed (13) to account for the ethyl branches assumes a transfer reaction of the type

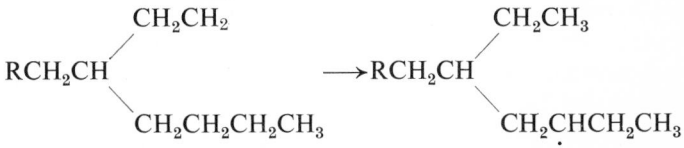

after addition of one monomer unit to the radical resulting from a short-chain branching step by the previous mechanism.

Since the transient ring configurations are relatively probable during propagation, the short-chain branching mechanism accounts for the large majority of the chain ends observed in the infrared. A typical low-density polyethylene molecule may contain fifty short branches and less than one long branch on a number average basis.

The molecular weight distribution of typical polyethylenes has been found experimentally (1, 3, 7) and theoretically (10) to be very much broadened by the long-chain branching mechanism. Weight- to number-average molecular weight ratios $\bar{M}_w/\bar{M}_n$ of 20 to 50 are considered typical (7). The distribution of long branches among the molecules is also very broad: even for highly branched polymer many molecules contain no long branches, while most of the branches are concentrated on a few very large molecules.

Infrared spectroscopy (13) provides much information about the chemical and physical structure of polyethylene (see also gen. refs., Section 4*B*). Structural features associated with the crystallinity of polyethylene are described in Chapter 5.

Properties

Effect of structure on properties (9, 11) The physical properties of polyethylenes are functions of three independent structural variables: molecular weight, molecular weight distribution or long-chain branching, and short-chain branching.

Short-chain branching has a predominant effect on the degree of crystallinity and therefore on the density of polyethylene. (Actually these properties are influenced by total chain branching, but the number of long-chain branch points per molecule in typical polyethylenes is so much less than the number of short-chain branch points that the former can be neglected.) Therefore, as discussed in Chapter 7*D*, properties dependent on crystallinity, such as stiffness, tear strength, hardness, chemical resistance, softening temperature, and yield point, increase with increasing density or decreasing amount of short-chain branching in the polymer, whereas permeability to liquids and gases, toughness, and flex life decrease under the same conditions.

The effect of molecular weight is largely evidenced (Chapter 7*B*) in properties of the melt and properties involving large deformations of the solid. As molecular weight increases, so do tensile strength, tear strength, low-temperature toughness, softening temperature, impact strength, and resistance to environmental stress cracking, while melt fluidity, melt "drawability," and coefficient of friction (film) decrease. These properties are commonly compared on the basis of changes in melt index (Chapter 6*A*), which varies inversely with molecular weight.

The effect of long-chain branching on the properties of polyethylene is often evaluated in terms of the breadth of the molecular weight distribution $\bar{M}_w/\bar{M}_n$. With other structural parameters held constant, a decrease in $\bar{M}_w/\bar{M}_n$ causes a decrease in ease of processing but an increase in tensile

strength, toughness and impact strength, softening temperature, and resistance to environmental stress cracking.

The interrelation of density, melt index, and $\bar{M}_w/\bar{M}_n$ in producing physical properties desirable for specific end uses has been described (9). For film uses polymers producing tough and flexible films are needed. Injection-molding applications require polymers characterized by rigidity and good flow; pipe, by strength; and wire insulation, by good processing characteristics and resistance to stress cracking.

General physical properties The mechanical properties of low-density polyethylene are between those of rigid materials like polystyrene and limp plasticized polymers like the vinyls. Polyethylene has good toughness and pliability over a wide temperature range. Its density falls off fairly rapidly above room temperature, and the resulting large dimensional changes cause difficulty in some fabrication methods. The relatively low crystalline melting point (about 115°C for typical materials) limits the temperature range of good mechanical properties.

The electrical properties of polyethylene are outstandingly good, probably ranking next to those of polytetrafluoroethylene for high-frequency uses. In thick sections polyethylene is translucent because of its crystallinity, but high transparency is obtained in thin films.

Chemical properties Polyethylene is very inert chemically. It does not dissolve in any solvent at room temperature, but is slightly swelled by liquids such as benzene and carbon tetrachloride which are solvents at higher temperatures. It has good resistance to acids and alkalis. At 100°C it is unaffected in 24 hr by sulfuric or hydrochloric acid but charred by concentrated nitric acid. It is often used in containers for acids, including hydrofluoric.

Polyethylene ages on exposure to light and oxygen, with loss of strength, elongation, and tear resistance. The probable point of attack is the tertiary hydrogens on the chain at branch points. Stabilizers retard the deterioration, but few are compatible enough with the polymer to do much good. The weathering of carbon black pigmented material is quite good. The polymer also undergoes some crosslinking when heated or worked at elevated temperatures. Few plasticizers or other additives are compatible with polyethylene in amounts larger than 1% or so.

Applications

Film and sheeting About a third of the polyethylene produced has gone into film and sheeting uses for the last several years. Few competitive film materials have polyethylene's desirable combination of low density,

flexibility without a plasticizer, resilience, high tear strength, moisture and chemical resistance, and little tendency for nicks or cuts to propagate.

Film may be made by extrusion through a slit die at 200–250°C and drawing into a water quench bath. Alternatively, blown film may be produced by extruding a tube of polymer and expanding it by means of internal pressure of inert gas, thus drawing the polymer. The tube may be slit to produce flat film or left as a seamless tube. Film thicknesses are usually 0.001–0.005 in.

Over three-fourths of the polyethylene film produced goes into packaging applications, including bags and pouches and wrappings for produce, textile products, merchandise, frozen and perishable foods, and many other products. Other film uses include drapes and tablecloths, and extensive application in agriculture (greenhouses, ground cover, tank, pond, and canal liners, etc.) and construction (moisture barriers and utility covering material).

Injection molding Production of housewares and toys by injection molding consumes some 15% of the polyethylene produced.

Wire and cable insulation Polyethylene has filled a long-standing need for a material which would effectively insulate electrical cables without introducing electrical losses at high frequencies. The nonpolar nature of the polymer makes it ideal for this purpose. Television, radar, and multicircuit long-distance telephony might well have been impossible without such insulating materials. Weathering introduces polar impurities such as carbonyl groups into the polymer and must be carefully guarded against. In addition to the high-frequency uses, polyethylene is being more generally utilized for mechanical protection of wire and cables, where its chemical inertness and light weight are advantageous. About 10% of the polyethylene produced is used for wire and cable insulation.

Coating About 5% of polyethylene is employed for coatings. These may be applied by dipping from hot solutions, melts, or emulsions, or by flame spraying. The extrusion coating of foils, papers, and other films (Mylar, cellophane) consumes large quantities of polyethylene.

Blow molding The production of "squeeze" bottles and industrial containers by blow molding accounts for about 5% of polyethylene production. Flexibility, low cost, and resistance to corrosion and breakage are important advantages in this use.

Pipe About 5% of polyethylene is consumed in the production of pipe, largely for agricultural and domestic water supply. For this application flexibility is an important advantage over other materials.

GENERAL REFERENCES

Kresser, T. O. J., *Polyethylene*, Reinhold Publishing Corp., New York, 1957.
Raff, R. A. V., and K. W. Doak, eds., *Crystalline Olefin Polymers*, Interscience Publishers, a Division of John Wiley & Sons, New York, 1963.
Renfrew, A., and P. Morgan, eds., *Polythene*, 2nd ed., Interscience Publishers, New York, 1960.

BIBLIOGRAPHY

1. Baskett, A. C., *Ricerca sci.* (*Suppl. A*) **25,** 379 (1955).
2. Boyd, D. R. J., R. C. Voter, and W. M. D. Bryant, paper presented at the 132nd National Meeting of the American Chemical Society, New York, September, 1957.
3. Charlesby, A., and S. H. Pinner, *Proc. Roy. Soc.* (*London*) **A249,** 367 (1959).
4. Fawcett, E. W., R. O. Gibson, and M. W. Perrin, U.S. Patent 2,153,553 (1939); E. W. Fawcett, R. O. Gibson, M. W. Perrin, J. G. Paton, and E. G. Williams, British Patent 471,590 (1936). See note by E. W. Fawcett, p. 119 in H. Staudinger, *Trans. Faraday Soc.* **32,** 97 (1936).
5. Fox, J. J., and A. E. Martin, *Proc. Roy. Soc.* (*London*) **A175,** 208 (1940).
6. Harlen, F., W. Simpson, F. B. Waddington, J. D. Waldron, and A. C. Baskett, *J. Polymer Sci.* **18,** 589 (1955).
7. Kokle, V., F. W. Billmeyer, Jr., L. T. Muus, and E. J. Newitt, *J. Polymer Sci.* **62,** 251 (1962).
8. Lawton, E. J., P. D. Zemany, and J. S. Balwit, *J. Am. Chem. Soc.* **76,** 3437 (1954).
9. McGrew, F. C., *Modern Plastics* **35,** #7, 125 (March, 1958).
10. Roedel, M. J., and coworkers, *J. Am. Chem. Soc.* **75,** 6110–6127 (1953); **79,** 5079–5089 (1957); **81,** 3219 (1959).
11. Sperati, C. A., W. A. Franta, and H. W. Starkweather, Jr., *J. Am. Chem. Soc.* **75,** 6127 (1953).
12. Willbourn, A. H., *J. Polymer Sci.* **34,** 569 (1959).
13. Wood, D. L., and J. P. Luongo, *Modern Plastics* **38,** #7, 132 (March, 1961).

B. High-Density (Essentially Linear) Polyethylene

Polyethylene with an essentially linear structure (2) (hereafter referred to as linear polyethylene) can be produced in several ways, including radical polymerization of ethylene at extremely high pressures (1), coordination polymerization of ethylene (Chapter 10*D*), and polymerization of ethylene with supported metal oxide catalysts (Chapter 10*E*). Commercial production of linear polyethylene, using the second and third routes named, began in 1957 and reached a volume of close to 250 million lb in 1961, at a price of \$0.32–\$0.38/lb. Over twelve companies were involved as actual or potential producers at the end of 1960, and both production capacity and consumption have expanded rapidly.

Polymerization

Coordination polymerization (Chapter 10*D*) A coordination catalyst is prepared as a colloidal dispersion by reacting, typically, an aluminum alkyl and $TiCl_4$ in a solvent such as heptane. Ethylene is added to the reaction vessel under slight pressure, at a temperature of 50–75°C. Heat of polymerization is removed by cooling. Polymer forms as a powder or granules, insoluble in the reaction mixture. At the completion of the reaction, the catalyst is destroyed by the admission of water or alcohol, and the polymer is filtered or centrifuged off, washed, and dried.

Polymerization with supported metal oxide catalysts (Chapter 10*E*) These catalysts can be used in a variety of operating modes, including fixed-bed, moving-bed, fluid-bed, or slurry processes. Ethylene is fed with a paraffin or cycloparaffin diluent, at 60–200°C and around 500 psi pressure. The polymer is recovered by cooling or by solvent evaporation.

Structure

Typical linear polyethylenes (2) are highly (over 90%) crystalline polymers, containing less than one side chain per 200 carbon atoms in the main chain. Melting point is above 127°C (typically about 135°C), and density is in the range of 0.95–0.97. Infrared spectroscopy (3) gives detailed information on the chemical and physical structure of the polymer. Structural features associated with the crystallinity of linear polyethylene are discussed in Chapter 5.

Properties

Most of the differences in properties between branched and linear polyethylenes can be attributed to the higher crystallinity of the latter polymers. Linear polyethylenes are decidedly stiffer than the branched material (modulus of 100,000 psi vs. 20,000 psi), and have a higher crystalline melting point and greater tensile strength and hardness. The good chemical resistance of branched polyethylene is retained or enhanced, and such properties as low-temperature brittleness and low permeability to gases and vapors are improved in the linear material.

Applications

Injection molding Almost half of the linear polyethylene produced is used in the injection molding of housewares and toys. The higher stiffness

and heat resistance of the linear material have led to its replacement of branched polyethylene in applications where these properties are important.

Blow molding The production of bottles and other containers by blow molding accounts for about 25% of the linear polyethylene made. The adjustment of structure variables to obtain high resistance to environmental stress cracking, allowing the material to be used in detergent bottles, produced a large expansion in this field.

Pipe Manufacture of pipe utilizes almost 10% of the linear polyethylene produced and is growing rapidly. The higher strength of the linear material allows lighter and less expensive thin-wall construction in comparison to branched polyethylene.

Filaments The consumption of linear polyethylene in textile and other filament applications is significant. Uses fall into two classes: conventional fabrics (drapes, upholstery, industrial cloths and filters) which, however, cannot be ironed; and products utilizing the low melting point of polyethylene compared to other textile fibers. The latter category includes seersucker and other fabrics in which puckering arises from shrinkage of the polyethylene yarn on controlled heat treatment, and fusible interlinings (as for collars), in which the polyethylene serves to bond other fabrics together.

GENERAL REFERENCES

Gloor, W. E., *Modern Plastics* **39,** #3, 131 (November, 1961).
Raff, R. A. V., and K. W. Doak, eds., *loc. cit.*
Renfrew, A., and P. Morgan, eds., *loc. cit.*

BIBLIOGRAPHY

1. Hines, R. A., W. M. D. Bryant, A. W. Larchar, and D. C. Pease, *Ind. Eng. Chem.* **49,** 1071 (1957).
2. Larchar, A. W., and D. C. Pease (to E. I. du Pont de Nemours & Co.), U.S. Patent 2,816,883 (1957).
3. Wood, D. L., and J. P. Luongo, *Modern Plastics* **38,** #7, 132 (March, 1961).

C. Polypropylene

With the commercial utilization of coordination polymerization (Chapter 10*D*) in 1957, the production of polypropylene became possible.

Consumption in this country was estimated at about 33 million lb in 1960, at a price of about $0.42/lb.

Polymerization

The polymerization of propylene, a by-product gas of alkylate gasoline refining, is carried out with coordination catalysts essentially as described in Section *B* for linear polyethylene. Ethylene, propylene, and other α-olefins can be polymerized in the same equipment with very little modification, leading to highly flexible operation. Catalysts and operating conditions must be selected with care to ensure that isotactic polypropylene (see the next paragraph) is produced.

Structure

Polypropylene can be made in isotactic, syndiotactic or atactic form (2, Chapter 5*A*). The crystallizability of isotactic polypropylene makes it the sole form with properties of commercial interest. Isotactic polypropylene is an essentially linear, highly crystalline polymer, with a melting point of 165°C. Its crystal structure is described in Chapter 5*B*.

Properties

Polypropylene is the lightest plastic known, with a density of 0.905. Its high crystallinity imparts to it high tensile strength, stiffness, and hardness (3). The resulting high strength-to-weight ratio is an advantage in many applications. Finished articles usually have good gloss and high resistance to marring. The high melting point of polypropylene allows well-molded parts to be sterilized, and the polymer retains high tensile strength at elevated temperatures.

The low-temperature impact strength of polypropylene is somewhat sensitive to fabrication and test conditions. This sensitivity results from the presence of a dominating α-transition (Chapter 6*D*) in polypropylene at about 0°C (1), resulting in a marked loss in stiffness near this temperature. In high-density polyethylene, the dominant transition is the lower-temperature β-transition. Thus, the restriction of molecular motion leading to brittle behavior takes place not far below room temperature in polypropylene, but at a much lower temperature in polyethylene.

Polypropylene has excellent electrical properties and the chemical inertness and moisture resistance typical of hydrocarbon polymers. It is

completely free from environmental stress cracking. However, it is inherently less stable than polyethylene to heat, light, and oxidative attack (presumably because of the presence of tertiary hydrogens) and must be stabilized with antioxidants and ultraviolet light absorbers for satisfactory processing and weathering.

Applications

Filament Polypropylene rope was perhaps the largest single end use for this polymer in 1960, utilizing, together with other filament applications such as seat covers, about one-third of the polymer produced. Light weight and high strength are important properties for this application.

The future of polypropylene as a textile fiber is still in doubt because of problems involving dyeability and stability to heat and to ultraviolet light. Because of its low cost and low density (high covering power per unit weight), polypropylene could ultimately compete heavily with cotton and the rayons.

Injection moldings Applications involving molded-in one-piece hinges, including large pieces such as luggage, are successfully met by polypropylene, and its use for large injection-molded pieces is expected to increase rapidly.

Film Biaxially oriented polypropylene film, having high strength and low water-vapor transmission, has potential high utility for packaging applications.

GENERAL REFERENCES

Grassie, V. R., R. D. Ullrich, and R. D. Hanna, *Modern Plastics* **38,** #1A (Encyclopedia Issue), 117 (1960).

Kresser, T. O. J., *Polypropylene*, Reinhold Publishing Corp., New York, 1960.

Raff, R. A. V., and K. W. Doak, eds., *loc. cit.*

Simonds, H. R., ed., *Source Book of the New Plastics*, Reinhold Publishing Corp., New York, Vol. 1, 1959, pp. 177–190, 232–240.

BIBLIOGRAPHY

1. Muus, L. T., N. G. McCrum, and F. C. McGrew, *SPE J.* **15,** 368 (1959).
2. Natta, G., and coworkers, *J. Am. Chem. Soc.* **77,** 1708 (1955); G. Natta, I. Pasquon, and A. Zambelli, *ibid.* **84,** 1488 (1962).
3. Shearer, N. H., Jr., J. E. Guillett, and H. W. Coover, Jr., *SPE J.* **17,** 83 (1961).

D. Other Olefin Polymers

Poly(α-olefins) and olefin copolymers

Polymers of α-olefins, including propylene, 1-butene, and higher homologs, and copolymers of these monomers with ethylene, can be prepared by coordination polymerization (Chapter 10*D*). Although homopolymers of olefins higher in chain length than propylene have not been offered commercially, a number of ethylene copolymers are sold, including those with propylene, 1-butene, and isobutylene. Copolymers of ethylene with vinyl acetate, maleic anhydride, and ethyl acrylate are made commercially by radical polymerization.

The major interest in the copolymers of ethylene with about 5% of 1-butene lies in their improved resistance to stress cracking compared to linear polyethylene. Ethylene-maleic anhydride copolymers have the advantage of potential reactivity through the maleic anhydride groups. Copolymers of 60% ethylene and 40% propylene have elastomeric properties much like those of natural rubber (3, 4, 7). They are cured by chemical crosslinking (see next page) with organic peroxides. Ethylene-vinyl acetate copolymers have softness and pliability like those of the plasticized vinyl resins. In other composition ranges, they are used as wax additives and in wax coatings and adhesives.

Chlorinated olefin products

Chlorinated polyethylene Polyethylene can be chlorinated in solution in CCl_4 at about 60°C or in suspension. Products containing up to 30% chlorine by weight are softer, more rubbery, and more soluble than polyethylene because of loss of crystallinity. In the range of 40–50% chlorine the polymer becomes harder and higher softening. As high as 73% chlorine by weight can be incorporated into the chain.

Chlorinated polyethylene can be made to have properties like those of plasticized poly(vinyl chloride), but cannot compete with it in cost. Chlorinated polyethylene exhibits chemical instability similar to that of poly(vinyl chloride).

Chlorinated polypropylene A polymer of propylene chlorinated to about 65% chlorine by weight is offered as a film-forming material for protective coatings, inks, adhesives, paper coatings, etc.

Chlorosulfonated polyethylene (6) When polyethylene is treated with a mixture of chlorine and sulfur dioxide, some chlorine atoms are

substituted on the chains and some sulfonyl chloride groups ($—SO_2Cl$) are formed. The chlorosulfonation can be carried out either on the solid material or in solution. There are two results of these modifications: (*a*) the chlorine atoms break up the regularity of the polyethylene chain structure so that crystallization is no longer possible, thus imparting an elastomeric character to the polymer; and (*b*) the sulfonyl chloride groups provide sites for crosslinking. A typical polymer contains 25–30% chlorine (one chlorine for every seven carbon atoms) and about 1.5% sulfur (one $—SO_2Cl$ for every ninety carbon atoms).

The elastomer can be crosslinked by a large variety of compounds, including many rubber accelerators. Metallic oxides are recommended for commercial cures. Fillers are not needed to obtain optimum strength properties.

Chlorosulfonated polyethylene is resistant to ozone, being better than neoprene and butyl rubber in this respect. Oxidative resistance and heat resistance are good. Chemical resistance is better than that of the common elastomers. The material is poor in "snap" and rebound and has low elongation and some permanent set. Its abrasion resistance, flex life, low-temperature brittleness, and resistance to crack growth are good.

Proposed uses include tire sidewalls, blends with other elastomers to improve aging properties, ozone resistance, and abrasion resistance, and as a component of light-colored vulcanizates and paints.

Crosslinked polyethylene

There has been considerable interest in converting polyethylene to a thermosetting material, in order to combine its low cost, easy processing, and good mechanical properties with the enhanced form stability at elevated temperatures, resistance to stress crack, and tensile strength expected in a crosslinked polymer.

Chemical crosslinking (2) Incorporation of relatively stable peroxides, such as dicumyl peroxide and di-*t*-butyl peroxide, provides a chemical means of crosslinking polyethylene. The peroxides are stable at normal processing temperatures but decompose to provide free radicals for crosslinking at higher temperatures in a postprocessing vulcanization or curing reaction. Chemically crosslinked polyethylene is used in the wire and cable industry and is of interest for pipe, hose, and molded articles. Ethylene-propylene rubbers are also vulcanized by peroxide curing systems.

Radiation crosslinking (1, 5; Chapter 12*F*) The crosslinking of polyethylene by irradiation with high-energy electrons has been used in the

commercial production of films combining the properties typical of polyethylene with form stability up to 200°C and a significant increase in tensile strength. The film can be made heat shrinkable by biaxial stretching. It is used for insulating (by wrapping) electrical power cables, coils, transformers, and motors and generators.

BIBLIOGRAPHY

1. Bockhoff, F. J., and J. A. Neumann, *Chem. Eng.* **62,** 228 (September, 1955).
2. Carlson, B. C., *SPE J.* **17,** 265 (1961).
3. D'Ianni, J. D., *Rubber Chem. Tech.* **34,** 361 (1961).
4. Gladding, E. K., B. S. Fisher, and J. W. Collette, *Ind. Eng. Chem. Prod. Res. & Dev.* **1,** 65 (1962); J. J. Verbanc, M. S. Fawcett, and E. J. Goldberg, *ibid.*, p. 70.
5. Goodwin, P. A., and J. W. Wetzel, *Gen. Elec. Rev.* **58,** 48 (September, 1955).
6. Keeley, F. W., "Hypalon Synthetic Rubber," Chap. 14 in M. Morton, ed., *Introduction to Rubber Technology*, Reinhold Publishing Corp., New York, 1959.
7. Natta, G., and coworkers, *Rubber Age* (*N.Y.*) **89,** 636 (1961).

14

Polymers Derived from Dienes

In describing polymers derived from 1,3-dienes, this chapter encompasses almost all the major commercial elastomers (other elastomers are discussed in Chapters 13, 15–17, and 19). About 3.5 billion lb of rubber was used in the United States in 1961, representing a gradual over-all increase from an annual level of about 2.5 billion lb in the early 1950's, to an estimated level of perhaps 4.5 billion lb by 1970. World consumption (excluding the Soviet nations) is about 8.5 billion lb per year.

In the United States, synthetic rubbers account for about 70% of the total consumed, distributed roughly according to the figures in Table 14-1.

TABLE 14-1. Approximate United States consumption of rubber, 1961*

Type	Consumption, million lb/year	
SBR†	1950	
Neoprene	175	
Butyl	135	
Nitrile	75	
Polybutadiene‡	50	
Polyisoprene‡	10	
Total synthetic		2400
Natural	950	
Total		3350

* W. S. Fedor, *Chem. Eng. News* **40,** #1, 88 (March 12, 1962).
† Styrene-butadiene rubber.
‡ Stereoregular.

TABLE 14-2. United States new rubber consumption by end use, 1960*

End Use	% of Total
Tires	63
Mechanical goods	12
Natural latex foam	5.5
Footwear	5
Wire and cable	1.5
All others	13

* *Rubber Age* (*N.Y.*) **88,** 627 (1961).

The consumption of natural rubber has remained relatively steady over the last 10 years at 1–1.5 billion lb, approximately 15% of the world's supply.

As indicated in Table 14-2, automobile tires remain the major single end use of both natural and synthetic rubbers. Over 40 million original-equipment tires of all types and almost 80 million replacement tires were produced in 1960. About a third of the rubber so used was natural rubber, since (with the possible exception of the "stereoregular" rubbers) the synthetics cannot meet the performance of natural rubber in this end use.

The price of natural rubber fluctuates considerably, ranging between \$0.27 and \$0.44/lb in 1959–61. Styrene–butadiene rubber (SBR) was priced at \$0.23/lb in the same period, and the "stereoregular" polymers sold at about \$0.27/lb.

The properties and processing of elastomers are discussed in Chapter 22.

A. Natural Rubber and Other Polyisoprenes

Natural rubber

Natural rubber is a high-molecular-weight polymer of isoprene, in which essentially all the isoprenes have the *cis*-1,4 configuration (Chapter 5*A*). The natural polymer has a number-average degree of polymerization of about 5000 and a broad distribution of molecular weights.

Source Natural rubber can be obtained from nearly five hundred different species of plants. The outstanding source is the tree *Hevea brasiliensis*, from which comes the name *Hevea rubber*. The tree grows

best in a damp, warm climate. It was first cultivated in Brazil and then imported to the Malay peninsula, whence the bulk of the world's rubber came before World War II.

Latex Rubber is obtained from a latex which exudes from the bark of the *Hevea* tree when it is cut. A cut is made halfway through the bark, extending about two-thirds around the tree. The flow of latex from the cut diminishes in time, necessitating the removal of another thin layer of bark. This process, called tapping, is continued at intervals throughout the life of the tree, which may be as long as 40 years.

Latex is an aqueous dispersion of rubber, containing 25–40% rubber hydrocarbon, stabilized by a small amount of protein material and fatty acids. As the latex is gathered, a small amount of ammonia is added as a preservative. The latex is later coagulated by the addition of a 5% solution of acetic acid. The coagulum is washed and dried. Two different processes are used.

Crepe rubber results if a small amount of sodium bisulfite is added to bleach the rubber. The coagulum is rolled out into sheets about 1 mm thick and dried in air at about 50°C. If *smoked sheets* are to be made, the bleach is omitted and somewhat thicker sheets are rolled. These are dried in smokehouses at about 50°C in the smoke from burning wood or coconut shells.

Mastication It was discovered by Hancock in 1824 that rubber becomes a soft, gummy mass when subjected to severe mechanical working. This process is known as *mastication*. The addition of compounding ingredients is greatly facilitated by this treatment, which is usually carried out on roll mills or in internal mixers or plasticators (similar to extruders). Mastication is accompanied by a marked decrease in the molecular weight of the rubber. Oxidative degradation is an important factor in mastication, since the decrease in viscosity and the other property changes do not take place if the rubber is masticated in the absence of oxygen.

After mastication is complete, compounding ingredients are added, and the rubber mix is prepared for vulcanization. These processes are discussed in Chapter 22.

Properties The properties of vulcanized natural rubber form the model for the ideal elastomeric properties discussed in Chapter 6*B*, including rapid extensibility to great elongations, high stiffness and strength when stretched, and rapid and complete retraction on release of the external stress. The properties of natural rubber are discussed further in comparison with those of the major synthetic elastomers in Chapter 22.

Hard rubber, ebonite As mentioned in Chapter 22, the final product of the reaction of rubber with an excess of sulfur is a hard, inextensible solid called ebonite, containing about 32% combined sulfur. Ebonite was first made in the 1840's and has been produced on a commercial scale since about 1860.

To minimize difficulties in handling the rubber-sulfur mix before vulcanization, fillers are usually employed in the production of ebonite. Since reinforcement is not necessary, inert fillers are satisfactory as well as carbon black. Ebonite dust, made by grinding scrap ebonite, is often used. Accelerators reduce the curing time from several hours to several minutes. Activators must be used with caution, as they may retard the hard rubber cure.

Ebonite stocks may be prepared for vulcanization by calendering or extrusion. Vulcanization, as with soft rubber, is effected by heat and pressure. In some cases the rubber may be given a soft cure, formed into a desired shape, and then cured to the hard-rubber state.

Ebonite can be machined well and is often produced in bar, tube, or sheet stock for this purpose. Its chief uses depend upon its chemical inertness and corrosion resistance and its electrical and thermal insulating properties. The material softens at about 50°C, and hence is not suitable for high-temperature applications.

Chlorinated rubber The reaction between chlorine and rubber has been known since 1801. It has traditionally been carried out by adding chlorine gas to a solution of rubber in a chlorinated solvent. To keep the viscosity of the solution low, the rubber is usually masticated beforehand. The chlorination is continued to completion, when the product contains about 65% chlorine, or about 3.5 chlorine atoms per isoprene residue. Newer methods of producing chlorinated rubber involve the direct chlorination of latex or the passage of chlorine over thin sheets of rubber swollen with a solvent such as carbon tetrachloride.

The mechanism of the chlorination reaction and the structure of chlorinated rubber are not known in detail. Several mechanisms have been proposed to account for the final composition involving seven chlorines per two isoprene residues. It is likely that substitution, addition, and cyclization (see next page) are all involved.

Chlorinated rubber is used chiefly in the production of heat- and chemical-resistant paints, varnishes, and lacquers. Films, impregnating solutions, adhesives, and (with the addition of plasticizers) molding powders can also be made.

Rubber hydrochloride Hydrogen chloride adds to the double bonds of rubber to give a material of the structure

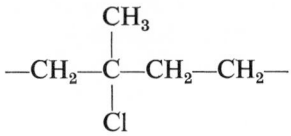

The presence of the chlorine atom on the tertiary carbon is in accord with Markownikoff's rule and has been confirmed by x-ray studies. The crystal structure of rubber hydrochloride is similar to that of natural rubber.

Rubber hydrochloride is produced by adding hydrogen chloride gas to a solution of previously milled rubber in benzene. The chlorine content is controlled to 28–30%, slightly less than the theoretical 34% for complete substitution. As in the formation of chlorinated rubber, methods involving latex or swollen sheets of rubber are also used.

Stretched and plasticized films of rubber hydrochloride have good mechanical properties, including high tear resistance. They are quite impermeable to air and water. The polymer is susceptible to thermal and photochemical decomposition with the loss of hydrogen chloride by olefin formation or by hydrolysis. Stabilizers similar to those used in vinyl chloride plastics are helpful in retarding the decomposition. Rubber hydrochloride is highly resistant to chemical attack.

The chief use of rubber hydrochloride is in film for protective wrapping purposes. It finds extensive application for wrapping foods and precision machines and machine parts.

Oxidized rubber Rubber can be oxidized in a controlled way by mastication in air in the presence of a catalyst. Various grades contain 0.25–1.0 oxygen atom per isoprene residue, distributed among hydroperoxide, acid, ester, ketone, alcohol, and epoxide groups. The material is useful for impregnating paper and cardboard and for protective coatings. It has advantages over chlorinated rubber in that its residual unsaturation can be used for vulcanization or thermal hardening reactions. Varnishes prepared with oxidized rubber have excellent electrical insulating properties.

Cyclized rubber When rubber is heated slowly or treated with acidic reagents, it becomes hard and brittle. Its unsaturation is reduced, but the empirical formula $(C_5H_8)_x$ remains unchanged. The changes are thought

to result from the condensation of isoprene residues in pairs to give cyclic structures of the type

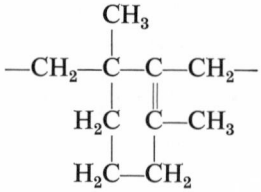

Cyclized rubbers are made commercially by treating rubber with either sulfuric acid or various sulfonic acids or sulfonyl chlorides, or chlorostannic acid. The products are nonelastic. They are used primarily as compounding ingredients in shoe soles and heels and for rubber-to-metal bonding adhesives.

Gutta-percha and balata These two natural resins or gums come from trees indigenous to Malaya and Central America, respectively. Both are impure forms of *trans*-polyisoprene. The purified hydrocarbon polymer is useful for cable covering, tissue for adhesive and surgical purposes, and golf-ball covers.

Stereoregular synthetic polyisoprene

In 1955, the production was announced (1, 2) of essentially *cis*-1,4-polyisoprenes with structures closely duplicating that of natural rubber. These polymers can be made by two processes which are almost identical except for the catalyst used: one is based on coordination polymerization (Chapter 10*D*), using a catalyst of titanium tetrachloride and an aluminum alkyl such as triisobutyl aluminum; the other is an anionic polymerization (Chapter 10*C*) with butyl lithium as the catalyst.

In a typical operation (3), isoprene (derived from petroleum) is mixed with a hydrocarbon solvent such as *n*-pentane. The catalyst is added, and the reaction allowed to take place at about 50°C and moderate pressures until a solids content of about 25% is reached. At this point the reaction mixture is a highly viscous "cement." A catalyst deactivator and an antioxidant are added, and solvent is removed in an extruder or a drum dryer. The polymer made by coordination polymerization has a molecular weight distribution similar to that of masticated natural rubber and requires no further processing before compounding, whereas that made with butyl lithium is higher in molecular weight and requires even longer mastication than natural rubber.

The properties of the *cis*-1,4-polyisoprenes are, commensurate with their structure, very nearly identical with those of natural rubber, and the

synthetic polymers are considered to be complete replacements for the natural product.

GENERAL REFERENCES

Cole, O. D., "Natural Rubber," Chap. 3 in M. Morton, ed., *Introduction to Rubber Technology*, Reinhold Publishing Corp., New York, 1959.
Davies, B. L., and J. Glazer, *Plastics Derived from Natural Rubber*, Plastics Monograph No. C.8, Plastics Institute, London, 1955.
D'Ianni, J. D., *Rubber Chem. Tech.* **34,** 361 (1961).
Wren, W. G., *Rubber Chem. Tech.* **34,** 378 (1961).

BIBLIOGRAPHY

1. Horne, S. E., Jr., and coworkers, *Ind. Eng. Chem.* **48,** 784 (1956).
2. Stavely, F. W., and coworkers, *Ind. Eng. Chem.* **48,** 778 (1956).
3. Winchester, C. T., *Ind. Eng. Chem.* **51,** 19 (1959).

B. Rubbers Derived from Butadiene

Although also made from alcohol during World War II, butadiene is now derived exclusively from petroleum. Fractionation of the products of cracking petroleum, either for producing olefins or for obtaining high-octane gasoline, yields a cut containing largely hydrocarbons of the butane and butene family. 1-Butene is separated and catalytically dehydrogenated in the vapor phase to butadiene. Current production of butadiene is about 2 billion lb/year, roughly 35% of which is used in styrene-butadiene rubber. Minor applications include other styrene copolymers, nitrile rubber, adiponitrile, and polybutadiene rubbers. The current price of butadiene is about $0.13/lb.

Styrene-butadiene rubber (SBR)

History of synthetic rubber The first attempts to produce synthetic rubbers centered around the homopolymerization of dienes, particularly isoprene because it was known to be the monomer for natural rubber. It was found in the late nineteenth century that rubberlike products could be made from isoprene by treating it with hydrogen chloride or allowing it to polymerize spontaneously on storage. These materials could be vulcanized with sulfur, becoming more elastic, tougher, and more heat resistant.

Around 1900 it was discovered that other dienes such as butadiene and 2,3-dimethylbutadiene could be polymerized to rubberlike materials spontaneously by alkali metals or by free radicals. Application was made of these facts during World War I in Germany, where 2,3-dimethylbutadiene was polymerized spontaneously. If the monomer was held at 30°C for 3 months, a hard rubber was produced; if it was heated under pressure to 70°C for 6 months, a soft rubber resulted. Production of hard rubber, useful for such items as battery cases, reached 30 tons a month in 1918.

After World War I this research on rubberlike products continued, with the emphasis shifting to butadiene because of its more ready availability, at that time from acetaldehyde via the aldol synthesis. Alkali metals were used as initiators, and the products were called *buna rubbers* from the first letters of butadiene and the symbol Na for sodium. The polymerizations were slow and difficult to control, and the polymers were poor in solubility, milling behavior, and tack. Vulcanizates were poor in tensile strength, aging characteristics, and low-temperature brittleness.

Improvements in processing and properties were made in two ways: copolymers of butadiene with vinyl monomers, notably styrene, were introduced, and emulsion polymerization was adopted. The latter system had been studied for many years in attempts to produce latices similar to natural rubber latex. Generally, water, butadiene, and a dispersing agent were shaken together for days. Fortuitously, the autoxidation of the monomer gave some peroxides which initiated the polymerizations.

Through many years of experimentation, the practice of making emulsion polymers evolved, although the principles were not understood. The significance of initiators, reduction activators, and oxygen became known. The importance of maintaining low conversion or utilizing modifiers such as CCl_4 and long-chain mercaptans was discovered. By the beginning of World War II, acceptable polymers (buna-S) were being produced in Germany containing 68–70% butadiene and 30–32% styrene. They were made in a multistage batch emulsion process with nine to twelve reaction vessels arranged in series. A persulfate initiator was used. The total polymerization time was about 30 hr. After 10% conversion was reached, bisisopropylxanthogen was added as a chain transfer agent, and conversion was carried to 60%. Phenyl-β-naphthylamine was then added as a "shortstop" and antioxidant, and the latex was filtered and coagulated.

Polymerization of SBR In the United States, production during World War II of styrene-butadiene rubber, then known as GR-S, was supervised and coordinated by the Rubber Reserve Company, a government agency. Plants, processes, and products were standardized as closely as possible to

obtain a uniform-quality product. It was decided to make a styrene-butadiene polymer of the same composition as the German buna-S, but of lower viscosity so as to be used directly in existing rubber-processing machinery without a preliminary heat-softening step such as the German material needed. The emulsion polymerization was to be carried out using the so-called Mutual recipe (Table 12-3). The original GR-S plants were operated batchwise, using 3750-gallon glasslined steel vessels which were agitated and designed to work at a pressure of 125 psi. The water phase was charged in at such a temperature that the reactor was at 50°C when the ingredients were all in place. The reaction was followed by the withdrawal of samples and the measurement of solids content (conversion) and melt viscosity of the polymer. When these properties reached the specified values, hydroquinone and sodium sulfite were added in a "shortstop" solution. The latex was then stripped of volatiles, treated with antioxidant, and coagulated by adding first brine and then dilute acid. It was then washed and dried.

As the need for GR-S became greater, some plants were converted to multistage batch operation. Each of the twelve reactors in a plant was baffled into two compartments, and all reactors were connected in series. Production increases up to 20% were realized because of better utilization of equipment, and polymer quality was more uniform. By the end of World War II all the plants had been so converted.

After the war it was found that the German production of buna-S had been carried out at increasingly lower temperatures with improvements in polymer quality. German recipes were adopted and improved, and, beginning in 1948, most SBR production has been converted from 50° to 5°C (41°F) with some being made at temperatures as low as −10° or −18°C. These changes were brought about by the use of more active initiators, such as cumene hydroperoxide and *p*-menthane hydroperoxide, and the addition of antifreeze components to the mixture. The product is known as *cold rubber*.

Structure of SBR By virtue of its free radical polymerization, SBR is a random copolymer. The butadiene units are found to be about 20% in the 1,2 configuration, 20% in the *cis*-1,4, and 60% in the *trans*-1,4 for polymer made at 50°C, with the percentage of trans-1,4 becoming higher for polymer made at lower temperatures (1). In consequence of its irregular structure, SBR does not crystallize.

Branching reactions due to chain transfer to polymer and to polymerization of both double bonds of a diene unit become extensive if conversion is allowed to become too high or a chain transfer agent is not used in SBR polymerization. However, SBR has been shown to have exactly one

double bond per butadiene unit. Thus no extensive side reactions occur during its formation, at least up to about 75% conversion.

Processing of SBR In general, the differences in mastication and vulcanization between SBR and natural rubber are minor. A reinforcing filler is essential to the achievement of good physical properties in SBR. However, some fillers other than carbon black reinforce it moderately well. SBR is compatible with the other major elastomers and can be used in blends. The techniques of oil extension and masterbatching are widely employed. These techniques, as well as the processing of SBR in general, are discussed further in Chapter 22.

Properties of SBR Tire tread stocks made from regular SBR are inferior in tensile strength to those from natural rubber (3000 vs. 4500 psi), whereas those from "cold rubber" are almost equivalent to *Hevea* (3800 psi). At elevated temperatures, however, regular and "cold" SBR lose almost two-thirds of their tensile strength whereas natural rubber loses only 25%. The ozone resistance of SBR is superior to that of natural rubber, but when cracks or cuts start in SBR they grow much more rapidly. Perhaps the most serious defect of both types of SBR for tire uses, however, is its poorer resilience and greater heat buildup. Tread wear of the synthetic material is at least as good as that of natural rubber. The weatherability of SBR is better than that of natural rubber.

"Cold" SBR is superior to the standard product because it contains less low-molecular-weight (nonreinforceable) rubber, less chain branching and crosslinking, and a higher proportion (70%) of the trans-1,4 configuration around the double bond.

Applications of SBR Originally, SBR was used in tires only of necessity, but "cold" SBR appears equal in most respects to natural rubber, especially for lighter-duty tire use. It is definitely inferior to the natural product for truck tires. For many mechanical goods SBR is superior to natural rubber and is preferentially used because of its easier processing and good-quality end product. Such items include belting, hose, molded goods, unvulcanized sheet, gum, and flooring. Rubber shoe soles are now almost universally made of SBR. Extruded goods and coated fabrics are other fields in which SBR offers advantages in processability. It is widely used for electrical insulation, although its properties are not as good as those of butyl rubber.

Nitrile rubbers

The nitrile rubbers are polymers of butadiene and acrylonitrile, having ratios of the two monomers similar to the ratio of butadiene to styrene in

SBR. They are noted for their oil resistance but are not suitable for tires. The first commercial nitrile rubbers were made in Germany.

The oil resistance of the nitrile rubbers varies greatly with their composition. Several grades are available commercially, ranging in acrylonitrile content between 18% (only fair oil resistance) and 40% (extremely oil resistant).

The nitrile rubbers are prepared in emulsion systems similar to those used for SBR. Because the monomer reactivity ratios are quite different, the compositions of the feed and the polymer differ markedly. This fact is usually taken into account by adjusting the monomer composition during the polymerization to achieve the desired polymer composition.

Compounding and vulcanization are similar to that of natural rubber. Mastication is essential. Sulfur vulcanization is ordinarily utilized, and the usual accelerators and activators are satisfactory. Carbon black reinforces the rubber markedly. Softeners are required to control the hardness and low-temperature flexibility of the stocks. Choice of softeners is limited by the oil resistance of the rubber; aromatic oils, esters, and other polar derivatives are preferred to paraffin oils. Nitrile rubber is compatible with natural rubber, but two-phase blends have also been successful.

Nitrile rubber is used primarily for its oil resistance, by which is implied its low solubility, low swelling, and good tensile strength and abrasion resistance after immersion in gasoline or oils. Swelling of nitrile rubbers is greater in polar solvents than in nonpolar solvents, but they can be used in contact with water and antifreeze solutions. Resistance to ethylene glycol is good. The rubbers are inherently less resilient than natural rubber. Their heat resistance is good; and, if properly protected by antioxidants, they show satisfactory resistance to oxidative degradation as well.

Nitrile rubbers are extensively used for gasoline hoses, fuel tanks, creamery equipment, and the like. In addition they find wide application in adhesives and, in the form of latex, for impregnating paper, textiles, and leather.

Butadiene-acrylic acid copolymers

Emulsion-polymerized copolymers of butadiene and acrylic acid serve as binders for solid-fuel rocket propellants. Good elongation, tear resistance, creep resistance, and storage characteristics are required in this application.

Stereoregular polybutadienes

cis*-1,4-*Polybutadiene This polymer is made by coordination or anionic polymerization in the same processes used for *cis*-1,4-polyisoprene

(Section *A*) and has similar properties. It appears likely to be utilized primarily in blends with natural rubber, rather than as a complete replacement for the latter. A small amount of natural rubber appears to be required to prevent the polybutadiene from crumbling during processing and to improve tack. Polybutadiene has high elasticity, low heat buildup, and good resistance to oxidation. For tire uses the blends are at least equivalent in properties to natural rubber.

trans*-1,4-*Polybutadiene By proper selection of catalyst, the processes leading to stereoregular diene polymers can produce a polybutadiene with about 90% *trans*-1,4 units. This polymer has the toughness, resilience, and abrasion resistance of the natural *trans*-polyisoprenes, balata and gutta-percha, which are currently in short supply. The synthetic product is likely to serve as a replacement for the natural rubbers in specialty products such as golf-ball covers.

GENERAL REFERENCES

Daum, G. A., "Nitrile and Polyacrylate Rubbers," Chap. 11 in M. Morton, ed., *op. cit.*, especially pp. 285–296.

D'Ianni, J. D., "Butadiene-Styrene Rubbers," Chap. 10 in M. Morton, ed., *op. cit.*

D'Ianni, J. D., *Rubber Chem. Tech.* **34,** 361 (1961).

Semon, W. L., "Nitrile Rubber," Chap. 23 in G. S. Whitby, C. C. Davis, and R. F. Dunbrook, eds., *Synthetic Rubber*, John Wiley & Sons, New York, 1954.

Whitby, G. S., *et al.*, eds., *op. cit.*, Chaps. 1–23 dealing with various aspects of SBR.

BIBLIOGRAPHY

1. Hampton, R. R., *Anal. Chem.* **21,** 923 (1949).

C. Polyisobutylene and Butyl Rubber

Butyl rubbers are copolymers of isobutylene with a small amount of isoprene added in order to make them vulcanizable. Since the amount of comonomer is small, the methods of polymerization and the properties of the unvulcanized copolymer are similar to those of polyisobutylene itself.

Monomers

Both isobutylene and isoprene are derived from cracked refinery gases. The products from cracking processes to produce high-octane gasoline or

olefins are fractionated. The C_4 cut contains about 10–35% isobutylene, along with normal butenes and saturated hydrocarbons. The isobutylene is extracted by absorption in sulfuric acid, neutralized, washed, and distilled. Isoprene is recovered from the C_5 cut by an extractive distillation process.

Polymerization

The polymerization of isobutylene and its mixtures with diolefins typifies the industrial application of low-temperature cationic polymerization (Chapter 10*B*). Isobutylene polymerizes rapidly at −80°C with Friedel-Crafts catalysts. In a bulk system at −80°C, rapid polymerization is induced by bubbling BF_3 gas through isobutylene. The heat of reaction can be absorbed by adding solid carbon dioxide to the monomer or by adding a low-boiling diluent such as pentane or ethylene which is refluxed. In a typical process, butyl rubber is manufactured by mixing isobutylene with 1.5–4.5% isoprene and methyl chloride as diluent. This mixture is fed to stirred reactors cooled to −95°C by liquid ethylene. Catalyst solution, made by dissolving anhydrous aluminum chloride in methyl chloride, is added. Polymer forms at once as a finely divided product suspended in the reaction mixture. This slurry is pumped out of the reactor continuously as monomer and catalyst are added. The product mixture is passed into a large volume of hot agitated water in a tank where the volatile components are flashed off and recovered. An antioxidant and some zinc stearate to prevent agglomeration of the polymer particles are added at this point. The polymer is then filtered off, dried, and extruded.

Structure and properties

Polyisobutylene and butyl rubber are amorphous under normal conditions, but crystallize on stretching. Most, if not all, of the isoprene units are present in the 1,4 structure. It is usually assumed that the polymers are linear, although the possibility of a small amount of branching has not been investigated in detail. The molecular weights of the polymers made by low-temperature ionic polymerization can be quite high; chain transfer agents such as diisobutylene are often added to control molecular weight at the 200,000–300,000 level.

Unstabilized polyisobutylenes are degraded by heat or light to sticky low-molecular-weight products. The usual rubber antioxidants or retarders of free radical reactions stabilize the polymers well.

Low-molecular-weight polyisobutylenes are liquids. As molecular

weight increases, they change to balsamlike solids, and then to rubberlike polymers. Unless low-molecular-weight material is removed, even polyisobutylene of 100,000 molecular weight is sticky. Unlike natural rubber, polyisobutylene and butyl rubber do not crystallize on cooling and hence remain flexible to as low as −50°C.

The response of isobutylene polymers and copolymers to stress is quite different from that of natural rubber. Polyisobutylene is sluggish, showing large viscoelastic and viscous components in its response (Chapter 6*C*). The polymer has been widely used in the study of viscoelasticity.

Applications of polyisobutylene

The strong tendency toward cold flow in polyisobutylene prevents its direct application as an elastomer. It is used in adhesives, caulking compounds, pressure-sensitive tapes, and coatings for paper.

The following paragraphs of this section refer only to butyl rubber.

Vulcanization

The limited unsaturation of butyl rubber (about 2 mole per cent diolefin units) offers fewer sites for crosslinking than are available in natural rubber or some of the other synthetics. Correspondingly, a large fraction of the available double bonds must be utilized in crosslinks to produce useful vulcanizates. Ultrarapid accelerators are therefore required, and vulcanization temperatures are higher (up to 175°C) and vulcanization cycles longer than those for natural rubber. Good accelerators for butyl rubber include the dithiocarbamates and thiuram sulfides. A typical mix may contain 100 parts rubber, 1–1.5 parts of one of the above accelerators, 1–2.5 parts of sulfur, possibly 0.5–1 part of thiazole, guanidine, or aldehydeamine accelerator, and 5–20 parts of zinc oxide.

Reinforcement and compounding

Carbon blacks, fillers, or pigments do not increase the tensile strength of butyl rubber. The highest tensile strengths are found in pure gum vulcanizates. Channel and furnace blacks, and to some extent other inorganic fillers, do improve modulus and tear resistance, however.

Plasticizers are often used to control the viscosity of butyl rubber during

compounding, although mastication can be made to take place if temperatures are kept low and viscosities correspondingly high. Any of a number of saturated low-polarity materials, notably paraffin oils and esters, is a satisfactory plasticizer.

Blends of butyl rubber with other elastomers do not cure well since the more unsaturated materials tend to deprive the butyl of the use of the curing agents. Neoprene, which reacts less readily than natural rubber or SBR, is an exception.

Properties of vulcanizates

The properties of butyl vulcanizates are compatible with their structure. The very low residual unsaturation of the rubbers leads to outstanding chemical inertness. The closepacked linear paraffinic chains result in unusualiy low permeability to gases. Steric hindrance of the methyl groups on the chains causes high internal viscosity and viscoelastic response to stresses.

Chemical properties A property of great importance in elastomers is aging in the presence of oxygen. It has been found (1) that the presence of a double bond in the skeletal structure of a polymer is very important in enhancing the rate of absorption of oxygen, and that the presence of methyl side groups is also significant but less so. Butyl rubber is therefore, as expected, less sensitive to oxidative aging than are most other elastomers except the silicones. Since methyl side groups appear to favor chain scission whereas double bonds favor crosslinking, butyl rubber becomes soft rather than brittle on oxidative degradation.

Butyl has much better ozone resistance than natural rubber. Its solvent resistance is typical of that of hydrocarbon elastomers. Its acid resistance is quite good.

Physical properties The stress-strain properties of butyl rubber are similar to those of natural rubber. Both show the importance of crystallization in obtaining high tensile strength. Crystallization does not take place in butyl, however, until higher elongations are reached. The tear resistance of butyl is quite good and is retained well at high temperatures and for long times, in contrast to natural rubber. The electrical properties are quite good, as predicted from its nonpolar, saturated nature. The dynamic and elastic properties of butyl are marked by sluggishness over the temperature range $-30°$ to $+40°C$, characteristic of a polymer with high internal friction and high damping power. Rebound is slow and heat buildup is high.

Applications

Almost all inner tubes are made from butyl rubber because of its superior impermeability to gases. The largest use of butyl outside of tubes has been for wire and cable insulation. Butyl rubber tires, first marketed in 1959, are made from a latex of butyl rubber, permitting good bonding of the rubber to the tire cord. The tires have good ozone resistance, give a smooth ride, do not squeal when turning corners, and stop quickly on wet pavement.

GENERAL REFERENCES

Buckley, D. J., *Rubber Chem. Tech.* **32,** 1475 (1959).

Dunkel, W. L., R. F. Neu, and R. L. Zapp, "Butyl Rubber," Chap. 12 in M. Morton, ed., *op. cit.*

Thomas, R. M., and W. J. Sparks, "Butyl Rubber," Chap. 24 in G. S. Whitby *et al.*, eds., *op. cit.*

BIBLIOGRAPHY

1. Mesrobian, R. B., and A. V. Tobolsky, *J. Polymer Sci.* **2,** 463 (1947).

D. Polychloroprene (Neoprene)

The generic term neoprene denotes rubberlike polymers and copolymers of chloroprene, 2-chloro-1,3-butadiene. Neoprenes were the first synthetic rubbers developed in the United States. Although they are primarily known for their oil resistance, they are good general-purpose rubbers which can replace natural rubber in most of its uses and are satisfactory in a wide variety of applications.

Chloroprene is prepared by the catalytic addition of hydrogen chloride to vinylacetylene, which in turn is made by the catalytic dimerization of acetylene. The neoprenes are produced by emulsion polymerization. Some types are polymerized in the presence of sulfur, which introduces some crosslinking in the polymer. In these cases, the latex is allowed to age in the presence of an emulsion of tetraethylthiuram disulfide, which restores the plasticity of the polymer. The latex is then coagulated by acidification followed by freezing.

Polymerization appears to take place almost entirely in the *trans*-1,4 form. As a result the neoprenes are crystallizable elastomers.

The vulcanization of neoprene is different from that of other elastomers in that it can be vulcanized by heat alone. Zinc oxide and magnesium oxide are the preferred vulcanizing agents. The mechanism by which they cause crosslinking is not known. Sulfur vulcanizes neoprene very slowly, and the usual rubber accelerators are in general not effective—some, in fact, are potent retarders of the cure. A few chemicals are known to accelerate the vulcanization, however, among them antimony sulfide.

Unlike may other elastomers, neoprene vulcanizates have high tensile strength (3500–4000 psi) in the absence of carbon black. No reinforcing effect is found with any filler. Suitably protected neoprene vulcanizates are extremely resistant to oxidative degradation. Weathering resistance and ozone resistance are quite good. Neoprene is slightly inferior to nitrile rubber in oil resistance, but markedly better than natural rubber, butyl, or SBR. The dynamic properties of neoprene are superior to those of most other synthetics and only slightly inferior to those of natural rubber. They are less affected by elevated temperature than those of natural rubber. Neoprene has been shown to make excellent tires but cannot compete with other elastomers in price. Its major uses include wire and cable coatings, industrial hoses and belts, shoe heels, and solid tires. Gloves and coated fabrics are made from neoprene latex.

GENERAL REFERENCES

Munn, J. P., "Neoprene," Chap. 13 in M. Morton, ed., *op. cit.*
Neal, A. M., and L. R. Mayo, "Neoprene," Chap. 22 in G. S. Whitby *et al.*, eds., *op. cit.*

15

Vinyl and Vinylidene Polymers

This chapter includes discussion of polymers made from vinyl monomers, $CH_2{=}CHX$, and vinylidene monomers, $CH_2{=}CY_2$, with the following exceptions: poly(α-olefins), where X is an aliphatic hydrocarbon group (Chapter 13); some fluorocarbon polymers such as poly(vinyl fluoride) derived from $CH_2{=}CHF$ (Chapter 16*C*); and thermosetting resins made in part from vinyl monomers, such as "polyester" resins (Chapter 19*C*). It should be noted that throughout this chapter the term "vinyl" is used in the sense of monosubstituted ethylenic, while in the plastics industry (and in Section *D*) the phrase "vinyl resins" commonly refers to poly(vinyl chloride) and related plastics.

A. Polystyrene and Related Polymers

The family of styrene polymers includes polystyrene, copolymers of styrene with other vinyl monomers, polymers of derivatives of styrene, and physical mixtures of polystyrene and styrene-containing copolymers with elastomers. Total production of these polymers reached 1 billion lb in 1959, with some twenty producers (compared to four in 1955). The production figures for 1961 include the following categories: extrusion and molding uses, polystyrene, 330 million lb; related polymers, 370; export, 130; coating uses, 130; other applications (foam, ion-exchange, rubber reinforcement, etc.), 110.

The price of polystyrene was \$0.18/lb in 1961, making it one of the lowest priced molding and extrusion resins made. Prices of other resins in this class ranged between \$0.20 and \$1.00/lb.

The applications of these materials are considered together at the end of this section.

Polystyrene

Polystyrene is a thermoplastic with many desirable properties. It is clear, transparent, easily colored, and easily fabricated. It has reasonably good mechanical and thermal properties, but is slightly brittle and softens below 100°C.

Monomer Styrene (vinylbenzene),  $CH{=}CH_2$, is made from benzene and ethylene. In one process, ethylene is passed into liquid benzene under pressure, in the presence of aluminum chloride catalyst:

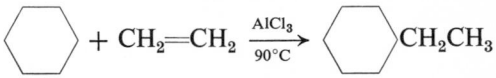

The resulting ethylbenzene is dehydrogenated to styrene by passing it over an iron oxide or magnesium oxide catalyst at about 600°C. The styrene is then refined by distillation. The process is inherently rather expensive, but the large-volume production (nearly 2 billion lb/year) allowed a sales price of about \$0.12/lb in 1961. (The production of styrene is the largest single use for benzene.) Other polymer applications of styrene include elastomers (Chapter 14*B*) and "polyester" resins (Chapter 19*C*).

Polymerization Although solution or emulsion polymerization may occasionally be used, most polystyrene is made either by suspension polymerization or by polymerization in bulk. All but the latter process are typically carried out as described in Chapter 12.

The bulk polymerization of styrene (1) is begun in a "prepolymerizer" (Fig. 15-1), a stirred vessel in which inhibitor-free styrene is polymerized (usually with a peroxide initiator) until the reaction mixture is as concentrated in polymer as is consistent with efficient mixing and heat transfer. Normally, a solution containing about 30% polymer is as viscous as can be handled.

The syrupy mixture from the prepolymerizer then enters a cylindrical tower (about 40 ft long by 15 ft diameter) maintained essentially full of fluid. By cooling the upper part of the tower and heating the lower part, the polymerization is controlled to prevent runaways but to proceed to essentially pure molten polymer at the bottom. This melt is discharged through spinnerets or into an extruder producing small-diameter rod which is chopped, after cooling, into short lengths to provide the finished molding powder.

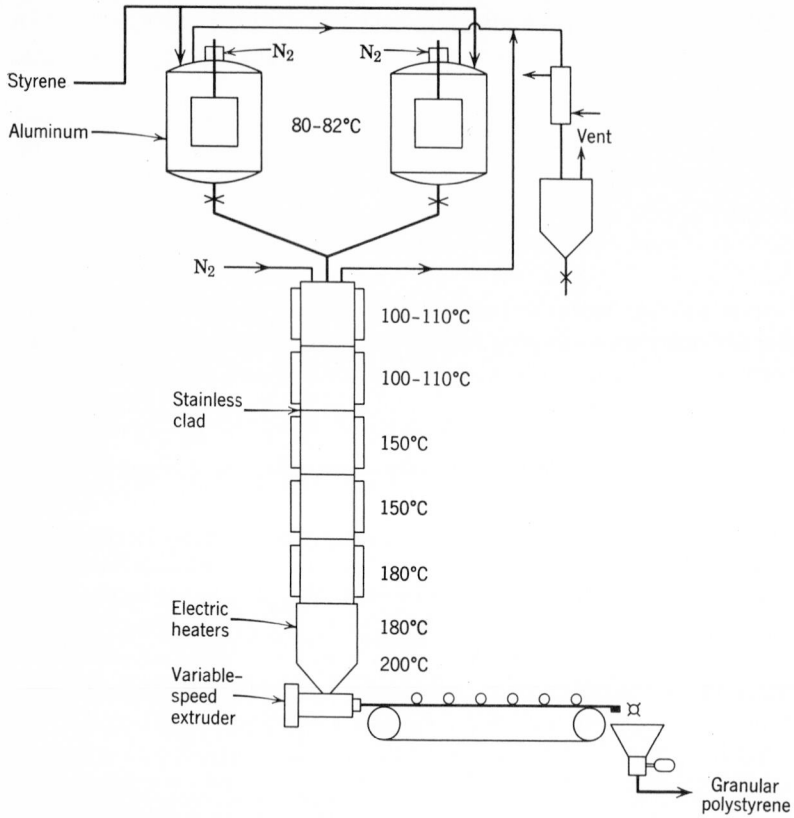

Fig. 15-1. Flow sheet for the "tower" process for polymerizing styrene (Winding and Hiatt, gen. ref.).

Despite being fitted with cooling pipes so that no polymer is more than a few inches from a wall, the tower reactor operates with extremely non-uniform conditions throughout its volume, but once dynamic stability has been established, the properties of the polymer remain remarkably constant. The tower is usually run continuously under almost completely automatic control. Despite residence times of 1–3 days from feed to discharge, the hourly output is high, typically 5000 lb/hr.

Structure and properties Polystyrene is a linear polymer, the commercial product being atactic and therefore amorphous. [Isotactic polystyrene can be produced, but offers little advantage in properties except between the glass transition (about 80°C) and its crystalline melting point (about 230°C), where it is much like other crystalline plastics. Isotactic polystyrene is not available commercially.]

Like most polymers, polystyrene is relatively inert chemically. It is quite resistant to alkalis, halide acids, and oxidizing and reducing agents. It can be nitrated by fuming nitric acid, and sulfonated by concentrated sulfuric acid at 100°C to a water-soluble resin (see ion-exchange resins, page 398). Chlorine and bromine are substituted on both the ring and the chain at elevated temperatures. Polystyrene degrades at elevated temperatures to a mixture of low-molecular-weight compounds about half of which is styrene. The characteristic odor of the monomer serves as an identification for the polymer.

As made, polystyrene is outstandingly easy to process. Its stability and flow under injection-molding conditions make it an ideal polymer for this technique. Its optical properties—color, clarity, and the like—are excellent, and its high refractive index (1.60) makes it useful for plastic optical components. Polystyrene is a good electrical insulator and has a low dielectric loss factor at moderate frequencies. Its tensile strength reaches about 8000 psi.

On the other hand, polystyrene is readily attacked by a large variety of solvents, including dry-cleaning agents. Its stability to outdoor weathering is poor; it turns yellow and crazes on exposure. Two of its major defects in mechanical properties are its brittleness and its relatively low heat-distortion temperature of 82–88°C, which means that polystyrene articles cannot be sterilized. There has been considerable work (see below) aimed at producing high-impact and heat-resistant styrene polymers and copolymers.

Copolymers of styrene

The most important copolymers of styrene in terms of volume are the styrene-butadiene synthetic rubbers, discussed in Chapter 14*B*.

Heat- and impact-resistant copolymers A number of copolymers of styrene with a minor amount of comonomer have enhanced heat and impact resistance without loss of other desirable properties of polystyrene. About 300 million lb of such copolymers was produced in 1961. Typical comonomers are those which increase intermolecular forces of attraction by introducing polar groups, or those which stiffen the chain and reduce rotational freedom through the steric hindrance of bulky side groups. In the first class are acrylonitrile ($CH_2{=}CHCN$), fumaronitrile (*trans*-$NCCH{=}CHCN$), and 2,5-dichlorostyrene. Comonomers with bulky side groups include N-vinylcarbazole and N,N-diphenylacrylamide.

The products of major commercial interest in this group contain, typically, 70% styrene and 30% acrylonitrile. Such copolymers have

heat-distortion temperatures of 90–92°C and more resilience and impact resistance than polystyrene. Their color, however, is slightly yellow.

Styrene also forms 1:1 alternating copolymers with fumaronitrile, fumarate esters (ROCOCH=CHCOOR), maleic anhydride:

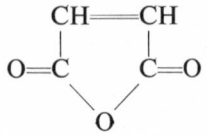

and other monomers which cannot form homopolymers because of steric hindrance. The copolymers soften at too high a temperature to be injection molded.

Crosslinked copolymers By the addition of small amounts of a divinyl compound such as divinylbenzene, $H_2C{=}CH\langle\bigcirc\rangle CH{=}CH_2$, to styrene, controlled branching or crosslinking can be introduced into the resulting copolymer. As little as 0.0025% divinylbenzene gives soluble branched polymers. About 0.1% divinylbenzene results in an insoluble, non-thermoplastic polymer which, however, swells greatly in solvents. If 5–10% divinylbenzene is used, the product is a typical hard, brittle cross-linked material.

Ion-exchange resins Cationic type ion-exchange resins are produced by making a suspension polymerization of styrene with several per cent divinylbenzene. The product, in the form of uniform spheres, is sulfonated to the extent of about one $-SO_3H$ group per benzene ring. Anionic ion-exchange resins are made by copolymerizing styrene with divinylbenzene and vinylethylbenzene. The polymer is treated with chloromethyl ether to put chloromethyl groups on the benzene rings. Quaternary ammonium salts are formed by reaction with tertiary amines such as trimethylamine, methyldiethanolamine, and dimethylpropanolamine.

Copolymers of styrene with methyl methacrylate are discussed in Section *B*.

Polymers of styrene derivatives

Many derivatives of styrene have been polymerized, but few have been of commercial interest. Polymethylstyrene, prepared from mixed isomers (two-thirds *para*, one-third *ortho*) of methylstyrene, is made commercially for use, primarily, in emulsion paints. It has better heat and impact resistance than polystyrene.

Poly-2,5-dichlorostyrene was made commercially at one time. It has a heat-distortion temperature (110–120°C) considerably above that of styrene and better hardness and impact strength, and is fire resistant. Its electrical properties are good, since the C—Cl dipoles are *para* to each other on the benzene ring and thus cancel. In contrast poly-3,4-dichlorostyrene has very poor electrical properties.

ABS resins and polyblends

Composition These materials are blends or copolymers of a plastic, usually polystyrene or styrene-acrylonitrile copolymer, with a rubber, usually a butadiene-acrylonitrile copolymer. The material may be an intimate physical mixture of the two components (polyblend) or a terpolymer (ABS resin), possibly made by block or graft techniques (Chapter 11*D*). In the latter case, a typical example is a graft copolymer of styrene on nitrile rubber. Typical compositions are in the range 20–30% acrylonitrile, 20–30% butadiene, and 40–60% styrene. The term ABS is derived from the initial letters of the three monomers. About 50–60 million lb was produced in 1961.

Properties The ABS resins are rigid and tough, with high tensile strength and thermal stability and good resistance to abrasion. They are easily processed by extrusion or molding. They are classified as engineering plastics, along with many of the heterochain thermoplastics discussed in Chapter 17. In general, the polyblends have somewhat poorer properties.

Applications

As a group, the styrene-based molding and extrusion plastics just discussed are used in the following applications.

Packaging Almost one-fifth of these materials go into packaging in a variety of forms. Polystyrene and, in a much smaller degree, the impact-resistant resins are used widely for small molded containers, lids, and closures and, to an increasing extent, for bottles and jars. Biaxially oriented polystyrene and styrene-acrylonitrile copolymer film is finding increasing application for thermoformed lids and packages. Foamed polystyrene, widely used as a cushioning material, is available in several forms: logs or planks from which shapes are machined, granules which are expanded in place, and thin sheets. Consumption of foamed polystyrene was estimated at 60 million lb in 1961.

Appliance housings and parts Over one-quarter of the styrene-based plastics is used for appliance housings and parts, and a major portion in refrigerator liners and trays and housings for air conditioners, radio and television sets, etc. Polystyrene has been displaced by the impact copolymers in these applications, and the trend is now strongly toward the ABS resins.

Engineering uses Other uses in which the mechanical properties of the impact-resistant copolymers and, particularly, the ABS resins play a part are the following: automotive, including instrument clusters and dash panels, seat, door, and roof panels; boats; telephones; luggage; shoe heels; and pipe.

Miscellaneous Toys, premiums and novelties, housewares, wall tile, and other small molded articles account for almost half of the consumption of these materials.

GENERAL REFERENCES

Boundy, R. H., and R. F. Boyer, eds., *Styrene*, Reinhold Publishing Corp., New York, 1952.

Teach, W. C., and G. C. Kiessling, *Polystyrene*, Reinhold Publishing Corp., New York, 1960.

Thompson, M. S., *Gum Plastics*, Reinhold Publishing Corp., New York, 1958.

Winding, C. C., and G. D. Hiatt, *Polymeric Materials*, McGraw-Hill Book Co., New York, 1961.

BIBLIOGRAPHY

1. Stober, K. E., and J. L. Amos, U.S. Patent 2,530,409 (1950).

B. Acrylic Polymers

In this section are discussed the acrylate and methacrylate plastics, of which the most important is poly(methyl methacrylate), and the acrylic fibers, based largely on polyacrylonitrile.

Poly(methyl methacrylate)

Poly(methyl methacrylate) is a clear, colorless transparent plastic with a higher softening point, better impact strength, and better weatherability

than polystyrene. It is available in molding and extrusion compositions, syrups, and cast sheets, rods, and tubes. The consumption of this polymer for molding and extrusion and in cast form in 1960 was estimated at 85 million lb, at a price of $0.55/lb for molding compositions.

Monomer Methyl methacrylate is made by heating acetone cyanhydrin (from the addition of hydrocyanic acid to acetone) with sulfuric acid to form methacrylamide sulfate. The latter is reacted (without separation) with water and methanol to give methyl methacrylate:

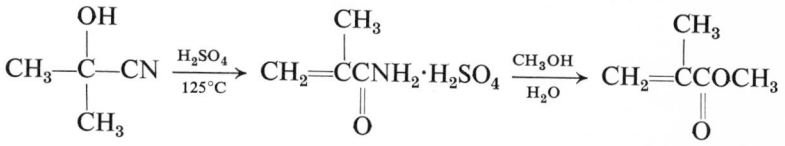

The price of the monomer is about $0.29/lb. Production in 1960 of all acrylic esters was about 125 million lb.

Polymerization Poly(methyl methacrylate) for molding or extrusion is made by bulk or suspension polymerization, as described in Chapter 12. The production of cast sheets, rods, and tubes is carried out by bulk polymerization, starting in most cases with a syrup of partially polymerized methyl methacrylate with a convenient viscosity for handling. Shrinkage and heat evolution during polymerization are reduced by the use of a syrup.

Sheets of poly(methyl methacrylate) are commonly made by extrusion. Alternatively, they may be cast in cells made up of two sheets of heat-resistant glass separated by a coated rubber gasket. The cell is filled with syrup and sealed, and polymerization is carried out at 60–70°C in an air oven or water bath, with a finishing treatment at 100°C. Peroxide or azo initiators may be used. The sheets are easily removed by separating the glass plates after the cell is chilled in cold water. The plastic sheet may be given a heat treatment to relax residual strains and is protected from scratching by covering with adhesive paper. A few per cent of a plasticizer such as dibutyl phthalate may be used in the polymerization to improve the postforming properties of the sheet.

Poly(methyl methacrylate) rods may be formed by polymerizing syrup under pressure in cylindrical metal tubes. The tubes are heated in successive zones by lowering them vertically into hot water in order to control shrinkage and prevent the formation of bubbles. Tubes of the plastic have been prepared by polymerizing syrup inside heated metal cylinders rotated horizontally on their axis, using just enough syrup to cover the walls to the desired thickness and blanketing the interior with an inert gas.

Properties Poly(methyl methacrylate) is a linear atactic thermoplastic. Because of its lack of stereoregularity and its bulky side groups, it is amorphous. Both isotactic and syndiotactic poly(methyl methacrylate) have been prepared (Chapter 5*A*, Chapter 10*C*) but have not been offered commercially. Poly(methyl methacrylate) is resistant to many aqueous inorganic reagents, including dilute alkalis and acids. It is quite resistant to alkaline saponification, in contrast to the polyacrylates. It undergoes pyrolysis almost completely to monomer by a chain reaction.

Perhaps the outstanding property of poly(methyl methacrylate) is its optical clarity and lack of color. Coupled with its unusually good outdoor weathering behavior, its optical properties make it highly useful in all applications where light transmission is important. These considerations apply to colored compositions as well. An unusually wide range of brilliant, light-fast colors is available in this plastic.

The mechanical and thermal properties of the polymer are good also. Tensile strength ranges as high as 10,000 psi. Impact strength is about equal to that of the impact-resistant styrene copolymers. Heat-distortion temperatures are above 90°C for the heat-resistant grades of poly(methyl methacrylate) molding powder. Electrical properties are good but not outstanding. Fabricability is quite good; only slightly higher temperatures are needed for molding poly(methyl methacrylate) than for polystyrene. Poly(methyl methacrylate) is less susceptible to crazing than is polystyrene.

A limitation to the optical uses of the material is its poor abrasion resistance compared to glass. Despite considerable effort, attempts to improve the scratch resistance or surface hardness of poly(methyl methacrylate) have so far been accompanied by deterioration in other properties, such as impact strength.

Applications Automotive uses (largely in tail- and signal-light lenses, dials, and medallions) consume more than half the poly(methyl methacrylate) molding powder produced. Other applications of molding compositions include brush backs, jewelry, lenses, and small signs. Major uses for sheets, both cast and extruded, include signs, glazing, skylights, and decorative purposes in the building industry. Mixtures of about 50% methyl methacrylate monomer and 50% polymer make a slurry or dough which can be reinforced with glass fiber mat and polymerized with a minimum of shrinkage and heat evolution. Glass fiber-reinforced poly(methyl methacrylate) combines outstanding mechanical and weathering properties. Special grades of poly(methyl methacrylate) are available for making dentures by a similar syrup or dough molding process. The use of the polymer in paints and lacquers is growing rapidly, accounting for about 27 million lb in 1960.

Other acrylic plastics

Higher methacrylates The monomers of higher alkyl methacrylates are most conveniently prepared from methyl methacrylate by alcoholysis. The reaction is carried out at 150°C with an excess of the alcohol whose methacrylate ester is desired and a small amount of sulfuric acid catalyst. As described in Chapter 7*C*, the higher alkyl methacrylate polymers have glass transition temperatures lower than that of poly(methyl methacrylate).

Poly(lauryl methacrylate) is widely used as a pour-point depressant and improver of viscosity-temperature characteristics for lubricating oils. In 1960 about 35 million lb of methacrylate polymers was sold for this purpose. Poly(cyclohexyl methacrylate) was developed to be combined with polystyrene in achromatic plastic lenses, the two polymers being similar in refractive index to crown and flint glass, respectively. Poly(cyclohexyl methacrylate) is suitable for this application because of its relatively small shrinkage on polymerization: 12.5 vs. 22% for poly(methyl methacrylate).

Polyacrylates Methyl acrylate is produced from ethylene cyanhydrin, which is made from ethylene oxide:

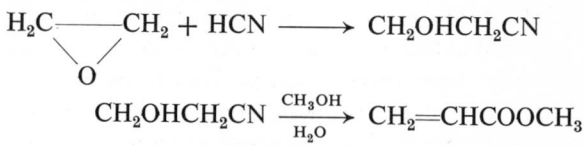

$$H_2C\text{—}CH_2 + HCN \longrightarrow CH_2OHCH_2CN$$

$$CH_2OHCH_2CN \xrightarrow[H_2O]{CH_3OH} CH_2{=}CHCOOCH_3$$

Higher acrylate esters may be prepared by similar processes or by alcoholysis of the lower esters.

The rubbery, adhesive nature of the lower acrylates makes both suspension polymerization and casting less feasible than with the methacrylates. Consequently, solution and emulsion polymerizations are used commercially. Since the polyacrylates contain an easily removed tertiary hydrogen atom, they undergo some chain transfer to polymer when polymerized to high conversion. This leads to highly branched, less soluble materials.

Since the lower acrylate polymers have glass transition points below room temperature, they are typically soft and rubbery. As the size of the ester group increases, the polymers become harder, tougher, and more rigid. The polyacrylates have been used in finishes, in textile sizes, and in the manufacture of pressure-sensitive adhesives.

Copolymers of ethyl acrylate with a few per cent of a chlorine-containing monomer such as 2-chloroethyl vinyl ether have elastomeric properties (2). Vulcanization reactions apparently involve the chlorine atom, ester group,

and α-hydrogens on the chain. Numerous rubber vulcanization agents and accelerators vulcanize these polymers. They are of interest because of their heat resistance and their excellent resistance to oxidation, allowing their use to over 180°C.

Poly(acrylic acid) and poly(methacrylic acid) Acrylic acid, $CH_2{=}CHCOOH$, can be prepared directly from ethylene cyanhydrin. Methacrylic acid can be prepared from acetone cyanhydrin. Salts of the acids can be polymerized directly, then acidified to give poly(acrylic acid) and poly(methacrylic acid). Alternatively, poly(methyl acrylate) or poly(methyl methacrylate) may be saponified to form the acids. Or, polymethacrylonitrile may be treated with hydrogen chloride to give poly(methacrylic acid). The polymers are insoluble in their monomers and in most organic liquids, but are soluble in water and very soluble in dilute bases. Poly(acrylic acid) and poly(methacrylic acid) are too water sensitive to serve as plastics. They are brittle when dry, and on heating do not become thermoplastic but crosslink, char, and decompose. In solution they show typical polyelectrolyte behavior, including abnormally high viscosities. Because of this property they are useful as thickening agents for lattices and for adhesives.

Cyanoacrylate adhesives (1) Methyl cyanoacrylate monomer is an extremely powerful adhesive. Adhesion occurs when the liquid monomer is spread in a thin layer between the surfaces to be bonded. Traces of bases (even as weak as alcohols or water) on the surfaces catalyze polymerization by an anionic mechanism. Adhesion arises in part from mechanical interlocking between polymer and surface, and in part from strong secondary bond forces.

Polyacrylamide This water-soluble polymer is used as a thickening agent and as a flocculant.

Copolymers and blends Copolymers of styrene and methyl methacrylate are less expensive than acrylic resins but have poorer weathering properties. A copolymer of α-methylstyrene, $CH_2{=}C(CH_3)$–C_6H_5, and methyl methacrylate has a heat-distortion temperature and other physical properties similar to those of poly(methyl methacrylate). A modified acrylic polymer [possibly a blend of poly(methyl methacrylate) with a rubber or an olefin polymer] is said to have good impact properties and to be similar to the ABS resins (Section *A*).

Acrylic fibers

The *acrylic* fibers are polymers containing at least 85% acrylonitrile. Other monomers are often used in small amount to make the polymer amenable to dyeing with conventional textile dyes. Common comonomers are vinyl acetate, acrylic esters, and vinyl pyrrolidone. The last may be used as the homopolymer in a blend or grafted onto a copolymer backbone. Generic classes of fibers closely related to the acrylics are the *modacrylic* fibers, containing 35–85% acrylonitrile and, usually, 20% or more vinyl chloride or vinylidene chloride; and *nytril* fibers, one example of which is a 1:1 alternating copolymer of vinylidene cyanide and vinyl acetate.

Polyacrylonitrile Acrylonitrile can be made either by the direct catalytic addition of HCN to acetylene, or by the addition of HCN to ethylene oxide to give ethylene cyanhydrin, followed by dehydration. The monomer is soluble in water to the extent of about 7.5% at room temperature, and polymerization is usually carried out in aqueous solution by means of redox initiation. The polymer precipitates from the system as a fine powder.

During the time that other vinyl polymers were being developed commercially as plastics as well as fibers, polyacrylonitrile was considered a useless material because it could not be dissolved or plasticized. It softens only slightly below its decomposition temperature; and, because the polymer is insoluble in the monomer, it could not be polymerized into useful shapes by casting. This intractability was for a time attributed to crosslinking, but better knowledge of the requirements for solubility of polymers led to an extensive search (3) for molecules which might interact with the highly polar —C≡N groups and cause solution of the polymer. Several such solvents were discovered, among them dimethyl formamide and tetramethylene sulfone. Some concentrated aqueous solutions of salts, such as calcium thiocyanate, also dissolve polyacrylonitrile. With these solvents available, both wet- and dry-spinning techniques (Chapter 21) have been used for this polymer.

Properties of acrylic fibers The acrylic fibers exhibit the properties of high strength, stiffness, toughness, abrasion resistance, resilience, and flex life associated with the synthetic fibers as a class. They are relatively insensitive to moisture and have good resistance to stains, chemicals, insects, and fungi. Their weatherability is outstandingly good. In continuous filament form, they are considered to have superior "feel," while as crimped staple they are noted for bulkiness and a woollike "hand."

In many respects the acrylic fibers are similar to the chlorine-containing

fibers described in Section *D*, and to the condensation fibers discussed in Chapter 17. Further information on fibers and fiber technology is found in Chapters 18 and 21.

GENERAL REFERENCES

Horn, M. B., *Acrylic Resins*, Reinhold Publishing Corp., New York, 1960.
Moncrieff, R. W., *Man-Made Fibres*, 3rd ed., John Wiley & Sons, New York, 1957, Chaps. XXVIII–XXXIII.
Symposium on acrylic fibers: *Textile Research J.* **27**, 701–745 (1957).

BIBLIOGRAPHY

1. Coover, H. W., F. B. Joyner, N. H. Shearer, Jr., and T. H. Wicker, Jr., *SPE J.* **15**, 413 (1959).
2. Daum, G. A., "Nitrile and Polyacrylate Rubbers," Chap. 11 in M. Morton, ed., *Introduction to Rubber Technology*, Reinhold Publishing Corp., New York, 1959, especially pp. 296–308.
3. Houtz, R. C., *Textile Research J.* **20**, 786 (1950).

C. Poly(vinyl Esters) and Derived Polymers

The most widely used polymer of a vinyl ester is poly(vinyl acetate). It is utilized not only as a plastic, primarily in the form of emulsions, but also as the precursor for two polymers which cannot be prepared by direct polymerization, poly(vinyl alcohol) and the poly(vinyl acetals). None of these polymers is useful for molding or extrusion, but each is important in certain special applications.

Poly(vinyl acetate)

Monomer Vinyl acetate, $CH_2{=}CH{-}O{-}\overset{\overset{\displaystyle O}{\|}}{C}{-}CH_3$, is prepared by the addition of acetic acid to acetylene:

$$CH{\equiv}CH + CH_3COOH \longrightarrow CH_3COOCH{=}CH_2$$

The reaction can be carried out either in the liquid or in the vapor phase. In the liquid-phase reaction acetylene is bubbled through hot glacial acetic acid in the presence of catalysts such as mercuric salts plus sulfuric acid. The vapor-phase synthesis is carried out by passing the mixed gases

at about 200°C over a catalyst consisting of zinc or cadmium salts on charcoal.

Polymerization Bulk polymerization of vinyl acetate is difficult to control at high conversions, and in addition the properties of the polymer may deteriorate because of chain branching. Bulk or solution polymerization is usually stopped at low to medium conversion (20–50%), after which the monomer is distilled off. If a solvent (such as methanol, ethanol, ethyl acetate, or benzene) is present, the polymer may be precipitated out or further reacted in solution to one of the derived polymers. The polymerization may be done batchwise or continuously.

Poly(vinyl acetate) may also be made in suspension or emulsion systems. In suspension polymerizations, various additives must be used to coat the beads of polymer to prevent their sticking together during the drying operation. Most of the poly(vinyl acetate) emulsions commercially available are made by emulsion polymerization mechanisms (Chapter 12*B*).

Properties and uses This polymer is atactic and hence amorphous; stereoregular polymers have not been offered commercially. The glass transition temperature of poly(vinyl acetate) is only slightly above room temperature (28°C). As a result the polymer, while tough and form stable at room temperature, becomes sticky and undergoes severe cold flow at only slightly elevated temperatures. Lower-molecular-weight polymers are brittle but become gumlike when masticated and in fact are used in chewing gums. Poly(vinyl acetate) is water sensitive in certain physical properties, such as strength and adhesion, but does not hydrolyze in neutral systems.

Aside from the production of poly(vinyl alcohol), one major use of poly(vinyl acetate) is in the production of water-based emulsion paints. The low cost [poly(vinyl acetate) emulsions sell for about $0.18/lb], stability, quick drying, and quick recoatability of emulsion paints have led to their wide acceptance. Poly(vinyl acetate) is often copolymerized with dibutyl fumarate, vinyl stearate, 2-ethylhexyl acrylate, or ethyl acrylate, or is plasticized, to obtain softer compositions for emulsion use.

Poly(vinyl acetate) is also widely used in adhesives, both of the emulsion type and of the hot-melt type.

Poly(vinyl alcohol)

Since vinyl alcohol is unstable with respect to isomerization to acetaldehyde, its polymer must be prepared by indirect methods.

Preparation Poly(vinyl alcohol) is prepared by the alcoholysis of

poly(vinyl acetate) (the less accurate terms "hydrolysis" and "saponification" are also used):

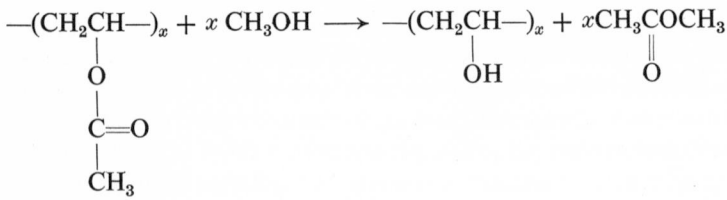

Ethanol or methanol can be used to effect the alcoholysis, with either acid or base as catalyst. Alkaline hydrolysis is much more rapid. The acid hydrolysis is more likely to give some ether linkages in the chain by a mechanism involving the loss of a molecule of water from two adjacent hydroxyl groups. This is an undesirable side reaction. The alcoholysis is usually carried out by dissolving the poly(vinyl acetate) in the alcohol, adding the catalyst, and heating. The poly(vinyl alcohol) precipitates from the solution.

Structure and properties Although poly(vinyl alcohol) is amorphous when unstretched, it can be drawn into a crystalline fiber, the hydroxyl groups being small enough to fit into a crystal lattice despite the atactic chain structure. Poly(vinyl alcohol) does not melt to a thermoplastic, but decomposes by loss of water from two adjacent hydroxyl groups at temperatures above 150°C. Double bonds are left in the chain; and, as more of these are formed in conjugate positions, severe discoloration takes place.

Poly(vinyl alcohol) is water soluble. It dissolves slowly in cold water, but more rapidly at elevated temperatures, and can usually be dissolved completely above 90°C. The aqueous solutions are not particularly stable, especially if traces of acid or base are present. The solutions can undergo a complex series of reversible and irreversible gelation reactions. For example, crosslinking can occur at ether linkages, resulting in increased viscosity through the formation of insoluble products.

Poly(vinyl alcohol) can be reacetylated by heating it with an excess of acetic anhydride in the presence of pyridine. The resulting poly(vinyl acetate) may or may not have the same structure as the parent polymer from which the alcohol was made, owing to the nature of the branched-chain structure in the polymer (1, 2, 4).

The tendency of vinyl acetate to undergo chain transfer can lead to the abstraction of either a hydrogen atom from the acetate group or the tertiary hydrogen on one of the carbon atoms of the main polymer chain. In consequence, poly(vinyl acetate), unless made at quite low conversion

or in a solvent system, contains two types of branches: those attached to the main chain and those attached to the acetate groups.

On alcoholysis of the poly(vinyl acetate) to poly(vinyl alcohol), the branches which resulted from chain transfer to the acetate group are removed, whereas those resulting from transfer to chain hydrogens are not. If both types of branches were present, the poly(vinyl alcohol) obtained from a branched poly(vinyl acetate) is itself branched, but has fewer branches than the parent polymer.

The relative numbers of the two types of branches are not well known. Studies of chain transfer to model compounds suggest that the two types of hydrogen atom are removed with about the same ease; hence both types of branches are probably present in poly(vinyl acetate).

On reacetylation the hydrolyzed branches are, of course, not replaced. Hence a branched poly(vinyl acetate) which is alcoholized and reacetylated decreases in molecular weight, whereas a linear polymer regains its original molecular weight.

Applications The major uses of poly(vinyl alcohol) fall into two categories. In one type of application, use is made of the water solubility of the polymer. It serves as a thickening agent for various emulsion and suspension systems, and as a packaging film where water solubility is desired. A major outlet is wet-strength adhesives.

In the second type of end use, the final form of the polymer is insoluble in water as a result of chemical treatment. The use of poly(vinyl alcohol) as a textile fiber (*vinal* fiber) is the major example. The polymer is wet spun from warm water into a concentrated aqueous solution of sodium sulfate containing sulfuric acid and formaldehyde. The polymer is insolubilized by the formation of formal groups:

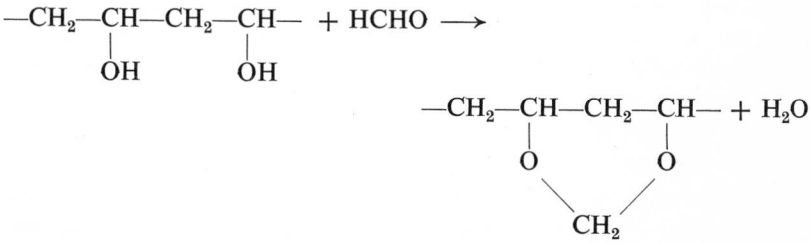

About one-third of the hydroxyl groups is reacted to insolubilize the fiber. Some interchain acetalization is desirable to reduce shrinkage of the fiber, but the amount must be carefully controlled.

Poly(vinyl alcohol) fibers have higher water absorption (about 30%) than other fibers. They can thus replace cotton in uses where the fiber is

in contact with the body. The hand of the fabric can be varied from woollike to linenlike. The fiber washes easily, dries quickly, and has good dimensional stability. Tenacity and abrasion resistance are good.

Poly(vinyl acetals)

An important use of poly(vinyl alcohol) is as the starting material for poly(vinyl acetals). By far the most important of these polymers is poly(vinyl butyral), used as the plastic interlayer for automotive and aircraft safety glazings. Lower poly(vinyl acetals), especially poly(vinyl formal), are utilized in enamels for coating electrical wire and in self-sealing gasoline tanks.

Poly(vinyl butyral) This polymer is made by condensing poly(vinyl alcohol) with butyraldehyde in the presence of an acid catalyst, usually sulfuric acid:

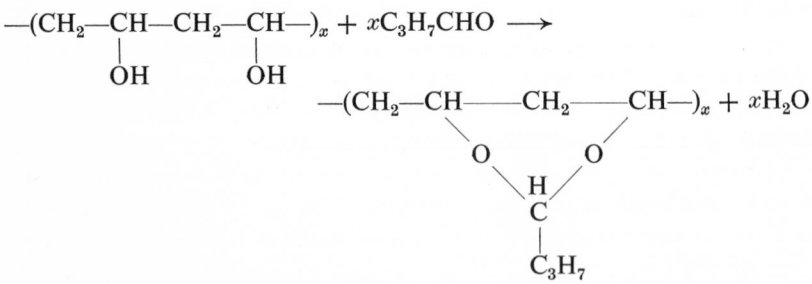

The reaction can be carried out by starting with an aqueous solution of poly(vinyl alcohol) and adding the butyraldehyde and catalyst. The poly(vinyl butyral) precipitates as it is formed. Alternatively, one may start with a suspension of poly(vinyl alcohol) in a water-alcohol mixture, the poly(vinyl butyral) dissolving as it is formed.

The condensation cannot be carried to completion, since single hydroxyl groups will sometimes be isolated between two adjacent acetal groups and so cannot undergo intramolecular condensation. It is possible to calculate the frequency of occurrence of these groups statistically for the equilibrium case (3); about 13.5 mole per cent of the hydroxyls are so isolated and unavailable if the condensation reaction is not reversible. The possibility remains that some of these are used up by intermolecular acetal formation, though the reactions are carried out in dilute solution to minimize the crosslinking. If the condensation is reversible, all the hydroxyls can in theory be consumed after a sufficiently long time.

Poly(vinyl butyral) produced for safety-glass use requires more than this

minimum number of hydroxyl groups in order to have adequate strength and adhesion to glass. Consequently the condensation reaction is stopped when about one-quarter of the hydroxyl groups is left.

Safety glass The earliest safety-glass interlayer was cellulose nitrate. Its poor weathering qualities led to its general replacement in 1935 by cellulose acetate, and in 1940 by poly(vinyl butyral). Some of the advantages of the latter resin as an interlayer are its superior adhesion to glass, toughness, stability on exposure to sunlight, clarity, and insensitivity to moisture.

Poly(vinyl butyral) must be plasticized for safety-glass use. The usual plasticizers are high-molecular-weight esters such as dibutyl sebacate and triethyleneglycol di-2-ethylbutyrate. About 40–45 parts plasticizer per 100 parts resin by weight is used. The plasticizer may be incorporated in the resin in several ways. One common method is to mix the resin, plasticizer, and a solvent such as ethanol to form a plastic mass. This is extruded in sheet form into a salt-water bath where the solvent is leached out, leaving the plasticized sheet. Since the plasticized resin sticks to itself, it must be powdered with talc or sodium bicarbonate before being rolled up for storage.

Safety-glass laminates are made by washing and drying the poly(vinyl butyral) sheet, and placing it between two pieces of glass. These are subjected to mild heat and pressure to seal the assembly, and are then autoclaved at higher temperature and pressure to complete the lamination process. The usual automotive safety glass contains one plastic layer about 0.015 in. thick.

GENERAL REFERENCES

Moncrieff, R. W., *op. cit.*, Chap. XXVII.

Schildknecht, C. E., *Vinyl and Related Polymers*, John Wiley & Sons, New York, 1952, Chap. VI.

Smith, W. M., *Vinyl Resins*, Reinhold Publishing Corp., New York, 1958.

BIBLIOGRAPHY

1. Bevington, J. C., G. M. Guzmán, and H. W. Melville, *Proc. Roy. Soc.* (*London*) **A221,** 437 (1954); H. W. Melville and P. R. Sewell, *Makromol. Chem.* **32,** 139 (1959).
2. Clarke, J. T., R. O. Howard, and W. H. Stockmayer, *Makromol. Chem.* **44/46,** 427 (1961).
3. Flory, P. J., *J. Am. Chem. Soc.* **72,** 5052 (1950).
4. Wheeler, O. L., S. L. Ernst, and R. N. Crozier, *J. Polymer Sci.* **8,** 409 (1952); O. L. Wheeler, E. Lavin, and R. N. Crozier, *ibid.* **9,** 157 (1952).

D. Chlorine-Containing Polymers

Polymers and copolymers of vinyl chloride and vinylidene chloride are widely used as plastics and as fibers. In the trade, these polymers are known as "the vinyls" or "vinyl resins." Once the largest group of thermoplastic materials, the vinyl resins have been surpassed in volume by the olefin polymers in recent years, and about equaled by the styrene-type resins. Consumption of vinyl resins in 1961 was about 940 million lb, at a price of about $0.16/lb.

Poly(vinyl chloride)

Monomer Vinyl chloride is a gas boiling at −14°C. It is produced commercially by the catalytic addition of dry hydrogen chloride to acetylene: the two gases are passed together at 100–250°C over charcoal catalysts containing mercuric or other heavy metal salts. An alternate source is the vapor-phase cracking of ethylene dichloride.

Polymerization Because of the low boiling point of vinyl chloride and the limited stability of the polymer, polymerizations of vinyl chloride must be carried out under conditions giving rapid reactions at low temperatures. Oxygen must be carefully removed, since it retards the polymerization severely. Poly(vinyl chloride) is insoluble in its monomer. Bulk polymerizations cannot be carried to high conversions because of local overheating leading to deterioration of the polymer. Solvents are often used to prevent this effect. Suspension polymerization is becoming more important as a means of obtaining better control of the reaction.

Emulsion polymerization with a redox initiator is a highly suitable method for polymerizing vinyl chloride, especially since the monomer is only very slightly soluble in water. Practical polymerization rates can be obtained at temperatures as low as 20°C. The reaction is carried out under pressure in an autoclave. If the initiator is active enough, the reaction begins without the addition of heat, and the autoclave must be cooled to maintain temperature and pressure at safe levels.

Structure The structure of poly(vinyl chloride) has been studied extensively, but the results of the various investigations have not always been clear cut. The essential head-to-tail nature of the polymer chain has been confirmed. Studies of solution properties have been hampered by the limited number of solvents in which the polymer can be dissolved free from association effects. It seems probable that the molecular structure of the

polymer is complicated by branching arising from chain transfer to polymer during polymerization.

Solid poly(vinyl chloride) is only slightly crystalline, and the crystallites are not so large or well defined as in many other polymers. It is thought that the polymer is partially but not completely syndiotactic, the irregularity of structure accounting for its low crystallinity.

Stability Poly(vinyl chloride) is relatively unstable to heat and light. Degradation involves loss of HCl with the formation of conjugated double bonds along the chain, leading to progressive discoloration. Chain scission also occurs. It is likely (2) that hydrogen and chlorine atoms are successively removed by a free radical mechanism, initiated either by oxidative attack on the chain or by photochemical reactions.

Stabilizers are almost invariably incorporated in the polymer to improve its stability to light and heat. Successful stabilizer types include mild alkalis, such as alkali or alkali earth oxides, hydroxides, or fatty acid salts, and HCl acceptors, such as ethylene oxide compounds. Metallo-organic compounds of lead, barium, cadmium, and tin are useful stabilizers, but zinc and ferric salts (especially the chlorides) catalyze the decomposition strongly.

"Vinyl resins"

"Rigid vinyls" This term usually refers to unplasticized poly(vinyl chloride), which is a tough, rigid, horny material. Certain copolymers of high stiffness and other modified resins, including a "polyvinyl dichloride" of unspecified composition, also fall in this category. The use of the rigid vinyls began in this country around 1950 and consumed about 25 million lb in 1960.

Copolymers The advantages in polymer properties resulting from the copolymerization of small amounts of vinyl acetate with vinyl chloride were discovered around 1928. The lower softening point and higher solubility of the copolymers make fabrication very much easier. Stability is improved, if anything, over that of the homopolymer. Color and clarity are also better. Polymerization methods are similar to those for vinyl chloride homopolymer except that emulsion polymerization has had less success. Commercial compositions containing 5–40% vinyl acetate are available.

Polymers containing around 10% vinylidene chloride have better tensile properties than pure poly(vinyl chloride). Copolymers containing 10–20% diethyl fumarate or diethyl maleate have improved workability and toughness and retain the high softening point of poly(vinyl chloride).

Acrylic esters have also been used to impart improvements in solubility and workability.

Plasticization Many properties of poly(vinyl chloride) and vinyl chloride-vinyl acetate copolymers are improved by plasticization (Chapter 20). The large majority of the commercial production of vinyl resins is in the form of plasticized compositions. The first important plasticizer for the vinyls was tricresyl phosphate, which has since been replaced by other esters because of its tendency to cause low-temperature brittleness in plasticized compounds. Dibutyl phthalate, dibutyl sebacate, and tributyl phosphate have also been used and in turn replaced. Dioctyl phthalate, trioctyl phosphate, dioctyl sebacate and adipate, and various low-molecular-weight polymers such as poly(propylene glycol) esters are now widely utilized as plasticizers for the vinyls. The plasticizers are usually added to the polymers on hot rolls or in a hot mixer such as a Banbury. The plasticizer content varies widely with the end use of the material, but typically may be around 30% by weight.

Plastisols and organisols These liquid compositions are produced by spray-drying latices obtained from the emulsion polymerization of vinyl chloride. The latex particles are dispersed into plasticizers to make plastisols, or into mixtures of plasticizers and volatile organic liquids to make organisols. Other ingredients, such as stabilizers, fillers, colorants, surfactants, and possibly blowing agents or gelling agents, are also present. The polymer particles do not dissolve in the liquids, but remain dispersed until the mixture is heated (1). Fusion (plus loss of solvent from the organisols) then yield the final plastic object.

Applications About one-third of the vinyl resins produced find molding and extrusion applications. Much of this material is used for insulation for electrical wire and cable. A typical composition might be poly(vinyl chloride) with about 30% plasticizer, 5% stabilizer, and some filler or pigment. Such resins are quite fire resistant, especially if phosphate plasticizers are used. They are extruded directly around the wire at about 180°C. The rigid vinyls, used primarily for the production of pipe and of extruded shapes and sheet, fall in this category also. Other major end uses of molded and extruded vinyl resins are phonograph records, garden hose, and articles made by the "slush" molding of plastisols and organisols, including dolls and rainboots.

Compositions for making vinyl film and sheet account for a quarter of the production of the vinyls. These may be poly(vinyl chloride) or copolymers containing about 5% vinyl acetate, plasticized with 20–35% of a plasticizer like dioctyl phthalate. The film and sheeting are usually made

by calendering and are familiar in rain wear, handbags, shower curtains, food covers, etc.

Another large field for the vinyls is coated fabrics. The coating is done by hot rolling the fabric with calendered sheets or by coating from solutions. Upholstery and rain wear are common applications.

Another fifth of the vinyl resin production is used for floor covering. In the usual practice, a plastisol coating is applied to a paper base, covered with a protective vinyl film, and then applied to a felt base.

Chlorine-containing fibers

Compositions The use of poly(vinyl chloride) as a fiber was patented as early as 1913, and the material was commercialized in 1931. It was soon withdrawn in favor of modified polymers with better solubility, including chlorinated poly(vinyl chloride) and copolymers of vinyl chloride with vinylidene chloride and with vinyl acetate. Currently, the materials of most interest in the class of chlorine-containing fibers are copolymers of about 60% vinyl chloride and 40% acrylonitrile, and of about 85% vinyl chloride and 15% vinyl acetate. Other comonomers are occasionally used. The generic classes of fibers in this group include *vinyon*, containing more than 85% vinyl chloride; *saran*, containing more than 80% vinylidene chloride; and *modacrylic*, defined in Section *B*.

Properties and uses The chlorine-containing fibers are usually moderately crystalline and are drawn (typically while hot) five- to fifteen-fold to increase strength and reduce extensibility. Like the acrylic (Section *B*) and condensation (Chapter 17) fibers, they are insensitive to moisture and have high wrinkle resistance and good resistance to chemicals, insects, fungi, and the like. They have poorer dimensional stability at high temperatures, and somewhat less strength, elasticity, and abrasion resistance than the condensation fibers.

Like the condensation fibers, they can be given a wide variety of mechanical properties depending upon degree of orientation, annealing, and crimp. All are affected in about the same way by amount of drawing. The stress-strain curves of the various fibers are similar for products with similar tenacity and ultimate elongation.

Despite subsequent annealing, these fibers are strongly anisotropic because of the stretch introduced in their manufacture. When they are first exposed to an elevated temperature, they shrink rapidly by an amount depending on the temperature. They are thereafter dimensionally stable, however, at all lower temperatures.

The chlorine-containing fibers have a variety of cross-sectional shapes resulting from removal of solvent during manufacture, in contrast to the featureless round cross sections of the melt-spun condensation fibers. It is not clear whether these shapes contribute to the good bulking capacity, stiffness, or resistance to pilling of the fibers.

These fibers are heat sensitive near their annealing or melting temperatures, but share with the condensation fibers the advantages of heat setting, crease retention, and easy ironing. Their stability, especially at elevated temperatures, however, is characteristically poor.

These fibers are used for protective clothing and industrial filter cloths because of their chemical resistance and nonflammability. They find only moderate application in apparel and in blankets, bedspreads, etc. The acrylonitrile copolymer is used for the covers of paint rollers.

GENERAL REFERENCES

Moncrieff, R. W., *op. cit.* Chaps. XXIII–XXVI.
Smith, W. M., *loc. cit.*

BIBLIOGRAPHY

1. Anagnostopoulos, C. E., A. Y. Coran, and H. R. Gamrath, *J. Appl. Polymer Sci.* **4,** 181 (1960).
2. Winkler, D. E., *J. Polymer Sci.* **35,** 3 (1959).

E. Miscellaneous Vinyl and Related Polymers

Poly(vinyl amines)

Poly(vinyl carbazole) The polymer of N-vinyl carbazole

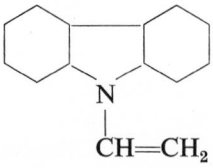

can be oriented into fibrous structures resembling asbestos or mica. It can be pressed into flat sheets but has low plasticity.

Poly(vinyl pyrrolidone) This water-soluble polyelectrolyte

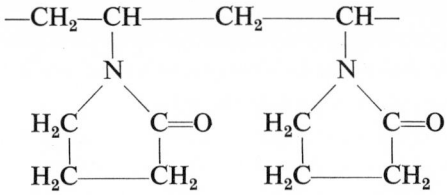

has been extensively studied as a substitute for blood plasma. It is a good protective colloid and suspending agent and forms strongly adhesive films. The monomer, and vinyl pyridine as well, are useful in copolymerization, e.g., to impart dyeability to fibers.

Poly(vinyl oxazolidinone) Polymers and copolymers (with vinyl acetate) of N-vinyl-5-methyl-2-oxazolidinone:

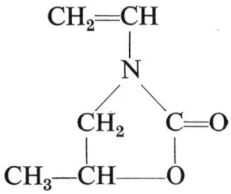

are water-soluble complexing agents for many organic and some inorganic substances.

Poly(vinyl ethers)

The vinyl alkyl ethers are made from the addition of alcohols to acetylene:

$$ROH + CH{\equiv}CH \longrightarrow CH_2{=}CHOR$$

Of the many vinyl ethers polymerized, only the methyl, ethyl, and butyl have had much commercial interest. They polymerize only by ionic mechanisms, giving high polymers by low-temperature (-80 to $-100°C$) bulk or solution polymerization. The higher-molecular-weight polymers are used in adhesives, and lower-molecular-weight (liquid) polymers serve as plasticizers.

Poly(vinyl ketones)

Methyl vinyl ketone, $CH_2{=}CHCOCH_3$, can be made by the hydration of vinylacetylene, $CH_2{=}CH{-}C{\equiv}CH$, or by the condensation of formaldehyde with acetone. The monomer is very reactive and will polymerize in a few hours at room temperature in the presence of light with no added

initiator. The homopolymers have low softening points and are unstable, tending to condense by reactions of the following type:

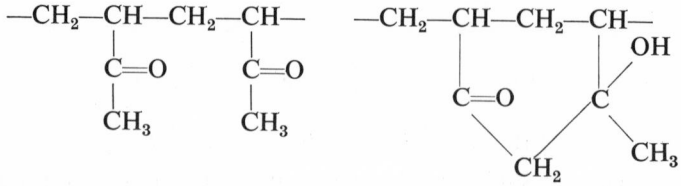

Copolymers are water sensitive and have limited stability.

Coumarone-indene resins

A synthetic resin can be produced from coal-tar oils recovered from coking operations. The oils contain coumarone (I), indene (II), and related

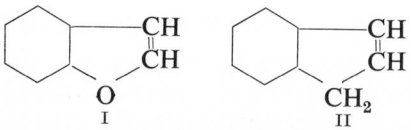

unsaturated compounds. They are polymerized by the addition of sulfuric acid, presumably to a linear polymer with the structure

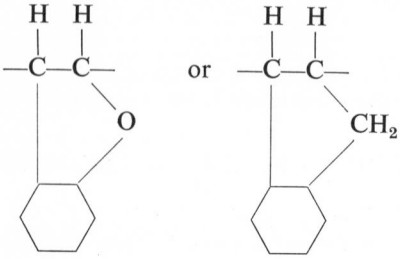

The polymers are low in molecular weight (about 1000 or below) and are completely fluid above their melting points. They are widely used in surface coatings, printing inks, adhesives, waxed papers, and floor tiles, and as softeners and tackifiers in the rubber industry.

Similar resins are made by hydrogenating coumarone-indene resins, or adding cresols, xylenols, or cyclopentadiene to the oil before polymerization. Terpenes and by-products of petroleum cracking and refining operations can be polymerized to similar products.

These resins are among the cheapest commercially available, selling for about $0.15/lb. About 200 million lb was used in 1960.

16

Fluorocarbon Polymers

Fluorocarbon polymers represent in many respects the extremes in polymer properties. Within this family are found materials of high thermal stability and concurrent usefulness at high temperatures (in some cases combined with high crystalline melting points and high melt viscosity), and extreme toughness and flexibility at very low temperatures. Many of the polymers are almost totally insoluble and chemically inert, some have extremely low dielectric loss and high dielectric strength, and most have unique nonadhesive and low friction properties.

The prices of fluorocarbon polymers in 1961 ranged from \$12–\$15/lb for newer materials to \$3–\$7/lb for established products.

A. Polytetrafluoroethylene and Other Fully Fluorinated Polymers

Polytetrafluoroethylene

Monomer Tetrafluoroethylene is a nontoxic gas boiling at −76°C. It can be made by the pyrolysis of chlorodifluoromethane, the dechlorination of *sym*-dichlorotetrafluoroethane, the decarboxylation of sodium perfluoropropionate, or the decomposition of tetrafluoromethane in an electric arc.

Polymerization The first polymer of tetrafluoroethylene was discovered by Plunkett (11) when an apparently empty cylinder of the gas was cut open to see why more material had not been obtained from it. The cylinder was partly filled with a waxy white powder shown to be the polymer.

Tetrafluoroethylene is usually polymerized with free radical initiators at elevated pressure in the presence of water. Redox initiation may be used (2); persulfates and hydrogen peroxide have been employed as initiators.

Alternatively, oxygen or organic peroxides can serve as initiators. Granular powders or aqueous dispersions (7) may be produced. Polymerization can lead to reactions of explosive violence. The heat of polymerization is high, and precautions must be taken to prevent local overheating leading to explosive disproportionation of the monomer into carbon and carbon tetrafluoride.

Structure (3) Polytetrafluoroethylene is a highly crystalline, orientable polymer. These facts indicate a regular structure, which implies the absence of any considerable amount of crosslinking. Branching is presumed to be absent, since branching mechanisms would involve breaking C-F bonds, which are estimated to be too strong to be ruptured easily. The chances are, therefore, that polytetrafluoroethylene consists of linear $—CF_2—CF_2—$ chains.

The crystal structure and crystalline-phase transitions in polytetrafluoroethylene are discussed in Chapter 5*B*. These transitions, occurring near room temperature, involve a 1.3% volume change having an important effect on the mechanical properties of the polymer for some applications. The degree of crystallinity of the polymer as formed from the monomer is generally quite high, 93–98%. The crystalline melting point is 327°C.

The crystalline density of polytetrafluoroethylene is 2.30 g/cm^3. Upon melting the polymer and subsequently cooling it, lower densities are obtained. Subsequent thermal treatment (annealing) generally increases the crystallinity of the solid. A standardized molding and annealing technique under conditions carefully controlled to prevent the formation of voids in the sample (15) gives "standard" specific gravities (1) for various samples ranging between 2.15 and 2.28. This range of densities has been correlated with the molecular weight and thus the melt viscosity of the samples. Higher-molecular-weight polymers have more viscous melts and hence crystallize more slowly and reach lower crystallinities under standardized thermal treatment.

Despite the reputation of polytetrafluoroethylene as a totally insoluble polymer, solvents for it have been found at temperatures not far below its crystalline melting point. Solution viscosities have been measured in perfluorinated kerosene fractions at 300°C, the measurements giving clues as to the molecular weight of the polymer. Other information as to molecular weight is derived from the kinetics of polymerization of tetrafluoroethylene using radioactive initiators (2, 3), and from the flow properties of the polymers (see the next paragraph). All the evidence points to unusually high number-average molecular weights of many millions for this polymer.

Polytetrafluoroethylene does not flow easily above its crystalline melting point. The viscosity of the polymer is very high because of restricted rotation about the chain bonds and high molecular weight. Upon the continued application of stress to the amorphous polymer above the crystalline melting point, the strength of the polymer is exceeded, causing fracture before the stresses necessary to induce rapid flow are reached. This unusual behavior necessitates unconventional fabrication techniques, as discussed on page 422.

Polytetrafluoroethylene decomposes at elevated temperatures. In vacuum, monomer is the chief product (9). The vapor pressure of monomer in equilibrium with the polymer at 500°C is 1 mm of mercury. At low temperatures (250–350°C) degradation appears to start at chain ends; at higher temperatures random cleavage becomes more important (5). In the presence of air the degradation is more complicated.

Properties (4, 12) Polytetrafluoroethylene is extremely resistant to attack by corrosive reagents or solvents. Of many hundreds of reagents tested up to their boiling points, only alkali metals, either molten or dissolved in liquid ammonia, attack the polymer, presumably by removing fluorine atoms from the chain. Fluorine itself degrades the polymer on prolonged contact under pressure. For all practical purposes the polymer is completely unaffected by water. Its thermal stability is such that its electrical and mechanical properties do not change for long intervals (months) at temperatures as high as 250°C.

Molded polytetrafluoroethylene articles have high impact strength but are easily strained beyond the point of elastic recovery. In a tensile test molded articles begin to cold draw at 1500–2000 psi, elongating (with orientation) to 300–450% before breaking at a load of 2500–4500 psi. Highly oriented fibers have tensile strengths as high as 50,000 psi. Polytetrafluoroethylene is subject to cold flow under stress but exhibits some delayed elastic recovery. The polymer is not hard, but is slippery and waxy to the touch, and has a very low coefficient of friction on most substances. Its density is unusually high (2.1–2.3) and its refractive index unusually low (1.375).

Polytetrafluoroethylene has extremely good electrical properties. Its dielectric constant is low (2.0), and its loss factor for all frequencies tested, including those used for television and radar, is one of the lowest known for solids.

Effect of structure on properties (14) The properties of polytetrafluoroethylene can be varied widely as a function of molecular structure and fabrication, as described in general in Chapter 7. The physical

properties of polytetrafluoroethylene depend mainly upon molecular weight, crystallinity, and void content. Not all these variables affect all properties. For example, stiffness is a function of crystallinity independent of molecular weight, whereas both crystallinity and molecular weight affect the flex life of the polymer. Dielectric strength is a function of void content only. Properties of the melt, such as its tensile strength, are affected by molecular weight. Here, of course, crystallinity can have no influence. Tensile strength, ultimate strength, and elongation of the polymer at temperatures below the crystalline melting point are affected by all three structural variables. The major role of molecular weight in determining the physical properties of the polymer lies in its influence on crystallization rate and, therefore, on final levels of crystallinity.

Fabrication Since polytetrafluoroethylene is almost completely insoluble and has impracticably low melt flow rates, most of the fabrication techniques ordinarily used with polymers (Chapter 20) are not suitable for this material. Several unusual techniques have been developed for putting polytetrafluoroethylene into usable shapes. Most of these techniques are variants of two important processing steps: (*a*) pressing the polymer cold into the desired shape, followed by (*b*) sintering at a temperature above the crystalline melting point (say, 380°C) to yield a dense, strong, homogeneous piece.

Techniques for *molding* polytetrafluoroethylene resemble those of powder metallurgy or ceramics processing more than those of polymer fabrication. Granular polytetrafluoroethylene powder is compressed at room temperature and 2000–10,000 psi to give a preform. Sintering of the preform can be carried out in several different ways.

For small pieces for which dimensional tolerances are not critical, the preform is removed from the mold and baked in an oven. For accurate dimensional control, the piece is removed from the sintering oven while hot and pressed in a cold die (*coining*, if pressures are low and the preform has cooled below the crystalline melting point, or *hobbing*, at higher pressures and with the preform still above 327°C when pressure is applied). For thin shapes such as diaphragms, a hot die may be used and the piece cooled in the die. For heavy sections (rods and tubes), the preform may be sintered in the preform mold under pressure.

Polytetrafluoroethylene may be *extruded* at slow rates in a ram or screw extruder. In these devices the ram or screw serves merely to compact the molding powder and feed it to the die, within which the material is heated above the crystalline melting point, shaped, and allowed to solidify. The die has a uniform cross section, with dimensions 14% larger than the finished article (required for thermal contraction on cooling). Pressure

necessary to shape the material is created by friction of the polymer against the walls as a result of its own thermal expansion.

A modification of the extrusion technique involves the use of a processing aid (8) such as naphtha, cetane, or naphtha thickened with low-molecular-weight polyisobutylene. The liquid is mixed with the polymer to form a thick paste, which can be extruded in a ram extruder at room temperature or slightly above. After the polymer is shaped, it is heated in air above 100°C to volatilize the extrusion aid and then sintered.

The high degree of cohesion between cold pressed particles of polytetrafluoroethylene is utilized in a *calendering* process for making tape and coating wire. Molding powder is fed to calender rolls, which compress the particles into a solid structure. This passes into a sintering bath or oven.

Polytetrafluoroethylene is available in the form of aqueous dispersions of ultimate particles about 0.2 μ in diameter (7). They can be used for *casting films* or for *dip coating* or *impregnating* porous structures.

Mechanical properties of polytetrafluoroethylene, including resistance to wear and to deformation under load, stiffness, and compressive strength, can be enhanced by the use of *additives or fillers*. Most of the desirable properties of the polymer (heat resistance, low friction, weatherability, etc.) are retained. Some of the additives used are asbestos, bronze powder, glass fiber, graphite, and zirconium oxide.

Polytetrafluoroethylene can be produced as a fiber by *spinning* an aqueous dispersion of elongated particles into an acid medium. After sintering and cold drawing, the fiber has a tenacity (Chapter 21) of 1.5 g/denier (tensile strength of 50,000 psi) and 13% elongation.

Applications The uses of polytetrafluoroethylene are largely those requiring excellent toughness, electrical properties, heat resistance, low frictional coefficient, or a combination of these. Among the electrical applications of the polymer are wire and cable insulation, insulation for motors, generators, transformers, coils, and capacitors, and high-frequency electronic uses. Chemical equipment such as gaskets, pump and valve packings, and pump and valve parts is made from the polymer. Low-friction and antistick applications include nonlubricated bearings, linings for trays in bakeries and other food-processing equipment, mold-release devices, and covers for heat-sealer plates of packaging machines. Low-molecular-weight polymers of tetrafluoroethylene dispersed as aerosols are effective dry lubricants. Uses for the fiber include gasketing, belting, pump and valve packing, filter cloths, and other industrial functions where essentially complete chemical and solvent resistance, combined with heat resistance to above 250°C, is required (13).

Fluorocarbon copolymers

A copolymer of hexafluoropropylene and tetrafluoroethylene (10) has a crystalline melting point near 290°C. It retains most of the properties of polytetrafluoroethylene, but has melt viscosity in the range allowing fabrication by conventional techniques.

Properties The copolymer is tough at liquid-air temperatures, yet retains adequate mechanical strength for continuous service at temperatures up to 200°C. Like polytetrafluoroethylene, it is chemically inert and has zero water absorption. It maintains a low dielectric constant and loss factor over the unusually wide range of 60 cycles to 60 megacycles. It has excellent weatherability, nonstick and low friction properties, and very low permeability to gases.

The material can be processed by conventional extrusion, injection molding (6), compression, transfer, or blow molding, fluidized bed coating, or vacuum forming of sheets or film. It finds application as wire jacketing, as extruded film, rods, tubes, or complex shapes, and as an encapsulating resin. End uses are similar to those for polytetrafluoroethylene.

GENERAL REFERENCES

Rudner, M. A., *Fluorocarbons*, Reinhold Publishing Corp., New York, 1958.

Sperati, C. A., and H. W. Starkweather, Jr., *Fortschr. Hochpolym.-Forsch.* (*Advances in Polymer Science*) **2,** 465 (1961).

BIBLIOGRAPHY

1. American Society for Testing Materials, *ASTM Standards on Plastics*, Tentative Specifications, ASTM Designation: D1457-56T.
2. Berry, K. L., and J. H. Peterson, *J. Am. Chem. Soc.* **73,** 5195 (1951).
3. Doban, R. C., A. C. Knight, J. H. Peterson, and C. A. Sperati, paper presented at the 130th National Meeting of the American Chemical Society, Atlantic City, N.J., September, 1956. See also Sperati and Starkweather, gen. ref.
4. Doban, R. C., C. A. Sperati, and B. W. Sandt, *SPE J.* **11,** #9, 17 (1955).
5. Florin, R. E., L. A. Wall, D. W. Brown, L. A. Hymo, and J. D. Milhaelsen, *J. Research Natl. Bur. Standards* **53,** 121 (1954).
6. Larsen, H. A., G. R. DeHoff, and N. W. Todd, *Modern Plastics* **36,** #12, 89 (August, 1959).
7. Lontz, J. F., and W. B. Happoldt, Jr., *Ind. Eng. Chem.* **44,** 1800 (1952).
8. Lontz, J. F., J. A. Jaffe, L. E. Robb, and W. B. Happoldt, Jr., *Ind. Eng. Chem.* **44,** 1805 (1952).
9. Madorsky, S. L., V. E. Hart, S. Straus, and V. A. Sedlak, *J. Research Natl. Bur. Standards* **51,** 327 (1953).

10. Mallouk, R. S., and W. B. Thompson, Jr., *SPE J.* **14,** #10, 42 (1958).
11. Plunkett, R. J., U.S. Patent 2,230,654 (1941); A. B. Garrett, *J. Chem. Educ.* **39,** 288 (1962).
12. Renfrew, M. M., and E. E. Lewis, *Ind. Eng. Chem.* **38,** 870 (1946).
13. Rivers, J. T., and R. L. Franklin, *Textile Research J.* **26,** 805 (1956).
14. Sperati, C. A., and J. L. McPherson, paper presented at the 130th National Meeting of the American Chemical Society, Atlantic City, N.J., September, 1956. See also Sperati and Starkweather, gen. ref.
15. Thomas, P. E., J. F. Lontz, C. A. Sperati, and J. L. McPherson, *SPE J.* **12,** #5, 89 (1956).

B. Poly(vinyl Fluoride), Poly(vinylidene Fluoride), and Related Polymers

The polymers described in this section could logically have been discussed in Chapter 15. However, their properties are profoundly altered by the substitution of fluorine for, say, chlorine, making them logical members of the fluorocarbon family.

Poly(vinyl fluoride)

Poly(vinyl fluoride) (3, 5) is a highly crystalline plastic which is commercially available in the form of a tough but flexible film. The polymer has the outstanding chemical resistance of the fluorocarbons, and excellent outdoor weatherability. It is extremely resistant to thermal degradation and maintains usable strength above 150°C while remaining tough at −180°C. It has low permeability to most gases and vapors and resists abrasion and staining.

The film is expected to have wide use as a glazing material and as a protective coating in the building industry. In 0.001–0.002 in. thickness, bonded to wood, metal, or asphalt-based materials, it should last many times longer than paints, enamels, or other surface coatings.

Poly(vinylidene fluoride)

Poly(vinylidene fluoride) (1) is a crystalline plastic with a melting point of about 170°C. It has good strength properties and resists distortion and creep at both high and low temperatures. It has very good weatherability and chemical and solvent resistance.

This polymer can be processed by extrusion and by compression and injection molding. It is expected to find use as a coating, gasketing, and wire- and cable-jacketing material.

Vinylidene fluoride-hexafluoropropylene copolymers

These copolymers (2) (in 70%:30% proportions in one composition) are elastomers combining high resistance to heat and to fluids and chemicals with good mechanical properties. They are cured with amine-type systems (4). These materials can be compounded to remain serviceable for short periods as high as 300°C, and retain useful properties indefinitely at 200°C. Their resistance to the lubricants, fuels, and hydraulic fluids used in jet aircraft is unequalled by other elastomers. Their mechanical properties are respectable for any elastomer, and excellent among the oil-resistant types.

GENERAL REFERENCE

Montermoso, J. C., *Rubber Chem. Tech.* **34,** 1521 (1961).

BIBLIOGRAPHY

1. (*a*) Barnhart, W. S., R. A. Ferren, and H. Iserson; (*b*) N. Capron, A. A. Dukert, M. E. Milville, and L. E. Robb, papers presented at the 17th Annual Technical Conference, Society of Plastics Engineers, Washington, D.C., January, 1961.
2. Dixon, S., D. R. Rexford, and J. S. Rugg, *Ind. Eng. Chem.* **49,** 1687 (1957); J. S. Rugg and A. C. Stevenson, *Rubber Age* (*N. Y.*) **82,** #1, 102 (October, 1957).
3. Kalb, G. H., D. D. Coffman, T. A. Ford, and F. L. Johnston, *J. Appl. Polymer Sci.* **4,** 55 (1960).
4. Moran, A. L., R. P. Kane, and J. F. Smith, *Ind. Eng. Chem.* **51,** 831 (1959); J. F. Smith, *Rubber World* **140,** 263 (1959).
5. Simril, V. L., and B. A. Curry, *J. Appl. Polymer Sci.* **4,** 62 (1960); *Modern Plastics* **36,** #11, 121 (July, 1959).

C. Polymers Based on Chlorotrifluoroethylene

Polychlorotrifluoroethylene

Monomer Chlorotrifluoroethylene is made by dechlorination of trichlorotrifluoroethane. The monomer is less subject to spontaneous explosive polymerization than is tetrafluoroethylene. Unlike tetrafluoroethylene, however, chlorotrifluoroethylene is itself toxic.

Polymerization As in the case of tetrafluoroethylene, the polymerization of chlorotrifluoroethylene is best carried out in an aqueous system using a redox initiator. The details of commercial polymerizations have not been disclosed.

Structure and properties Polychlorotrifluoroethylene is surpassed only by polytetrafluoroethylene and tetrafluoroethylene-hexafluoropropylene copolymers in chemical inertness and resistance to elevated temperatures. Differences in the properties of the polymers are a consequence of the lower symmetry of the chlorine-containing polymer. For example, the crystalline melting point of polychlorotrifluoroethylene is 218°C, and the polymer can be quenched to quite clear sheets in which crystallinity is absent. Slower cooling provides opportunity for growth of crystallites and spherulites and results in quite different optical and mechanical properties.

Polychlorotrifluoroethylene is soluble in a number of solvents above 100°C, and is swelled by several solvents at room temperature. The polymer is tough at −100°C and retains useful properties as high as 150°C. Although high compared to that of many other polymers, its melt viscosity is low enough that the usual fabrication techniques, such as molding and extrusion, are practicable. The electrical properties of polychlorotrifluoroethylene are inferior to those of polytetrafluoroethylene, especially for high-frequency applications, because of the more polar nature of the polymer. Typical applications of polychlorotrifluoroethylene include gaskets, valve seats, tubing, and wire and cable insulation.

Copolymers of chlorotrifluoroethylene

Copolymers of chlorotrifluoroethylene and vinylidene fluoride range from tough, flexible thermoplastics to elastomers, depending on composition. One of their outstanding properties is their resistance to the attack and penetration of powerful oxidizing agents such as propellant-grade red-fuming nitric acid and 90% hydrogen peroxide, typical of the fluorocarbons, coupled with the toughness and flexibility required for use as valves, ring seals, caulking compounds, etc. They also find application in automobile transmissions and brake systems as gaskets and seals.

GENERAL REFERENCES

Montermoso, J. C., *loc. cit.*
Rudner, M. A., *loc. cit.*

D. Other Fluorocarbon Polymers

Perfluoroacrylates

Polymers of acrylate monomers of the type $CH_2{=}CHCO_2CH_2(CF_2)_nCF_3$, where $n = 0$–4, are elastomers. A typical commercial product is the

polymer of perfluorobutyl acrylate, $n = 2$. These polymers exhibit good heat resistance and thermal stability and good resistance to swelling in aircraft fuels and hydraulic fluids.

Fluorosiloxane elastomers

The polysiloxane rubbers (Chapter 19*F*) are useful over a wide temperature range, but are not resistant to organic solvents and aircraft fluids. These properties can be greatly improved by incorporating fluoroalkyl or fluoroaryl groups on the silicon. A typical product is poly(trifluoropropyl methylsiloxane), the polymer of $Cl_2CH_3SiCH_2CH_2CF_3$.

Nitrosofluorocarbon elastomer

A copolymer of trifluoronitrosomethane and tetrafluoroethylene combines the nonflammability and solvent resistance of other fluorocarbon elastomers with improved low-temperature flexibility down to at least −50°C (1). The repeat unit of this polymer is

$$-[N(CF_3)-O-CF_2CF_2-]_x.$$

GENERAL REFERENCES

Montermoso, J. C., *loc. cit.*

Postelnek, W., L. E. Coleman, and A. M. Lovelace, *Fortschr. Hochpolym.-Forsch.* (*Advances in Polymer Science*) **1**, 75 (1958).

BIBLIOGRAPHY

1. Montermoso, J. C., C. B. Griffis, A. Wilson, and G. H. Crawford, *Rubber & Plastics Age* **42**, 514 (1961); J. C. Montermoso, *Chem. Eng. Progr.* **57**, #4, 98 (1961).

17

Heterochain Thermoplastics

Of the various chemical classes of polymers, some, such as polyamides and polyesters, differ from, say, polystyrene and polyethylene in having atoms other than carbon in the chain. Omitting the thermosetting resins discussed in Chapter 19, the majority of the thermoplastic *heterochain* polymers have sufficiently similar properties to warrant their consideration as a group. In this category fall such prominent artificial fibers as the nylons and poly(ethylene terephthalate) (but see also Chapters 13, 15, and 18) and strong, tough "engineering" plastics based on similar and related polymers, as well as several important elastomers.

A. Polyamides

The coined word *nylon* has been accepted as a generic term for synthetic polyamides. The nylons are described by a numbering system which indicates the number of carbon atoms in the monomer chains. Amino acid polymers are designated by a single number, as 6 nylon for poly(ω-aminocaproic acid) (polycaprolactam). Nylons from diamines and dibasic acids are designated by two numbers, the first representing the diamine, as 66 nylon for the polymer of hexamethylenediamine and adipic acid, and 610 nylon for that of hexamethylenediamine and sebacic acid. Of greatest commercial importance are 6, 66, 610, and 11 nylon, poly(ω-aminoundecanoic acid), and their copolymers.

The production of nylon fibers in the United States in 1960 was estimated at about 330 million lb of 66 nylon and 100 million lb of 6 nylon. Production of all types of nylon for plastics uses was thought to be about 30 million lb in 1960. The price of both the plastics and the tire cord and industrial fibers was in the range \$0.90–\$1.30/lb, with textile fiber prices ranging somewhat higher, depending on form and size.

Development of the nylons

The research leading to the production of synthetic fibers from polyamides began (in the Du Pont Company) about 1928, when W. H. Carothers (3) undertook a series of researches dealing with the fundamentals of polymerization processes. These studies were not aimed at the production of any specific product or process. In his fundamental researches, Carothers studied the condensation reactions of glycols and dibasic acids. He made a large number of polyesters ranging in molecular weight between 2500 and 5000. He also studied polymers of ω-amino acids, obtaining polyamides with a degree of polymerization of about 30. These were hard, insoluble, waxy materials.

Recognizing the equilibria involved in stepwise polymerization, Carothers and J. W. Hill (3) introduced the use of a molecular still to remove the water liberated in the condensation process and to shift the equilibrium toward higher molecular weights. In this way they synthesized materials with molecular weights up to 25,000. These substances had properties so different from those previously studied that Carothers coined the term "superpolymers" to describe those having molecular weights above 10,000.

The superpolyesters obtained from glycols and dibasic acids were tough, opaque solids which melted at moderately elevated temperatures to clear, viscous liquids. Filaments could be pulled from the liquids by touching the surface with a rod and withdrawing it. When cool, these filaments could be drawn to several times their original length. They became tough, transparent, lustrous materials of high strength and elasticity, possessing as high tenacity when wet as dry. Examination by x-ray diffraction showed them to be crystalline and highly oriented.

At this point the properties of the new materials were worthy of examination from a commercial standpoint, and the emphasis in this project was shifted away from that of fundamental research.

The properties of the aliphatic superpolyesters made them unsuitable for textile fibers because of their relatively low melting points and high solubility. Mixed aliphatic polyester-polyamides had the same bad features as the polyesters. It was clear that success, if achieved at all, would involve the use of a polyamide fiber.

The next phase of the research was a concentrated effort to find polyamides which could be made into fibers. A large variety of amino acids, diamines, and dibasic acids was studied. Several of the products gave good fibers melting near 200°C and equal to silk in strength and pliability. In 1935, the polyamide of hexamethylenediamine and adipic acid was made.

Like the others, it gave fibers by melt spinning or by dry spinning from phenol solutions. The fibers could be cold drawn to obtain high tenacity and elasticity. They melted at about 265°C and were insoluble in all common solvents.

This polymer was selected for commercial development because of its good balance of properties and the possibilities foreseen for the manufacture of the raw materials. At that time, adipic acid was made only in Germany, and hexamethylenediamine was little more than a laboratory curiosity.

The responsibility for the development of the fiber was then turned over to other groups in Du Pont who carried the processes for the manufacture of the intermediates, the polymerization, and the spinning and utilization of the fibers through the laboratory and semiworks and into full-scale manufacture.

Poly(hexamethylene adipamide), 66 nylon

Production Adipic acid can be obtained from a large number of chemicals, including cyclohexane, acetylene, and tetrahydrofuran. One commercial process involves oxidation of cyclohexane by air to a mixture of cyclohexanone and cyclohexanol. The mixture is oxidized to adipic acid by catalytic treatment with nitric acid. Hexamethylenediamine can be made from adipic acid by catalytic dehydration in the presence of ammonia to give adiponitrile, followed by hydrogenation. It can also be made from acetylene, tetrahydrofuran, or butadiene.

Achievement of the stoichiometric balance needed to obtain high-molecular-weight step-reaction polymers (Chapter 8) is simplified by the tendency of hexamethylenediamine and adipic acid to form a 1:1 salt which can be isolated because of its low solubility in methanol. This salt is dissolved in water and added to an autoclave with 0.5–1 mole per cent acetic acid as viscosity stabilizer. As the temperature is raised, the steam generated purges the air from the vessel. Pressure is kept at 250 psi as the temperature is raised to 270–280°C. Pressure is then reduced, and a vacuum may be applied. After a total time of 3–4 hr, nitrogen pressure is used to extrude the nylon as a ribbon through a valve in the bottom of the autoclave. The ribbon is subsequently cut into cubes, the finished product for the plastics-grade material. For fiber use, the polymer (somewhat lower in molecular weight) is melt spun as described in Chapter 21*B*.

Properties Both as a plastic and as a fiber, 66 nylon is characterized by a combination of high strength, elasticity, toughness, and abrasion resistance.

Good mechanical properties are maintained up to 150°C, although a more conservative limit for plastics use is 125°. Both toughness and flexibility are retained well at low temperatures.

The solvent resistance of nylon is good; only phenols, cresols, and formic acid dissolve the polymer at room temperature. Strong acids degrade it somewhat. The polymer discolors in air at temperatures of about 130°C and is degraded by hydrolysis at elevated temperatures. Its outdoor weathering properties are only fair unless it is especially stabilized or pigmented with carbon black. Nylon has a moderately low specific gravity, 1.14. Its moisture resistance is fair; moisture acts as a plasticizer to increase flexibility and toughness. The electrical uses of nylon are restricted to low frequencies because of the polar groups in the polymer.

Fiber applications About half of all the nylon fiber produced goes into tire cord; this market, however, is shared between 6 nylon and 66 nylon. Tires with nylon cord are used as original equipment for most heavy-duty applications, including airplane and truck tires. Most original tires on passenger automobiles have rayon cord, but nylon cord has most of the replacement market and about half the total volume. Other industrial uses of nylon include rope, thread, and cord, where high tenacity and good elasticity are important, and belting and filter cloths, where resistance to abrasion and to chemical attack is required.

Apparel uses of nylon, consuming about 150 million lb/year, include ladies' hose, undergarments, and dresses. Low modulus and high elasticity combined with strength and toughness are needed for this application. The ability to give nylon a permanent set by an annealing process at about 100°C has been utilized in the production of permanently pleated garments as well as in preshaped hosiery.

About 50 million lb/year of 66 nylon is used in carpets.

Plastics applications The most important use of 66 nylon plastic is as an engineering material—a substitute for metal in bearings, gears, cams, etc. Nylon is well suited for such applications because of its high tensile and impact strength, form stability at high temperatures, good abrasion resistance, and self-lubricating bearing properties. The ability to be injection-molded to close dimensional tolerances, augmented by its light weight, frequently gives the plastic a cost advantage in appliances. Rollers, slides, and door latches are often made of nylon. Bearings and thread guides in textile machinery may be made of nylon, which requires no lubricant that could damage yarn or cloth. A large amount of electrical wire is jacketed with nylon, which provides a tough, abrasion-resistant outer cover to protect the primary electrical insulation.

Other polyamides

Polycaprolactam, 6 nylon Caprolactam can be made in a large number of ways. One which has been used commercially involves oxidation of cyclohexane to cyclohexanone, from which the oxime is made. This oxime reacts by the Beckmann rearrangement to give caprolactam. The polymerization of caprolactam is carried out by adding water to open the rings and then removing the water again at elevated temperature, where linear polymer forms (2). An autoclave or a continuous reactor can be used. Polycaprolactam is in equilibrium with about 10% of the monomer, which must be removed by washing with water before the polymer can be spun. At the spinning temperature, more monomer is formed to restore the equilibrium, and this must again be removed to achieve good properties in the yarn.

Caprolactam can also be polymerized by ionic chain mechanisms (1), as described in Chapters 8*E* and 10*C*. The reaction can be carried out below the melting point of the nylon and at atmospheric pressure, making the technique very attractive for the production of large cast articles.

The properties of polycaprolactam are in general similar to those of 66 nylon, but the former has a lower crystalline melting point (225° vs. 265°C for 66 nylon) and is somewhat softer and less stiff. The major use for this polymer is in tire cord.

Poly(hexamethylene sebacamide), 610 nylon This polymer has not been used as a textile fiber, but has several properties making it suitable for monofilaments (brushes, sports equipment, and bristles). The additional hydrocarbon segments in the sebacic acid give the polymer a lower melting point (215°C) and lower moisture absorption. It thus retains its stiffness and mechanical properties better when wet than does 66 nylon.

Poly(ω-aminoundecanoic acid), 11 nylon The ω-aminoundecanoic acid leading to 11 nylon is obtained from castor oil. The polymer is less water sensitive than the other nylons because of its greater hydrocarbon character. It is used as a textile fiber in Europe.

Alkoxy substituted nylons An alkoxy substituted 66 nylon which has especially good resistance to impact and flexural fatigue is used for wrist-watch straps, valve seats, and the like. It is alcohol soluble but can be crosslinked by heat or by treatment with acids to an alcohol-insoluble form.

Soluble nylons In addition to the above polyamides, some nylons soluble in methanol and ethanol are available. They are copolymers of

undisclosed composition. More flexible than the other nylons, they are used for wire jacketing, sheeting, and textile coating. Although soluble in alcohol, they are very resistant to hydrocarbons.

Vegetable oil polyamides Condensation polymers made from dimerized vegetable oil acids and diamines, such as ethylene diamine, are available as low-molecular-weight (6000–9000), film-forming thermoplastics. These materials form high-gloss, flexible films with good adhesion, weatherability, and impact strength. They are also used in thixotropic paints and for paper coatings. Lower-molecular-weight products, liquid at room temperature, serve as curing agents for epoxy resins (Chapter 19*D*).

GENERAL REFERENCES

Bannermann, D. G., and E. E. Magat, "Polyamides and Polyesters," Chap. VII in C. E. Schildknecht, ed., *Polymer Processes*, Interscience Publishers, New York, 1956.

Floyd, D. E., *Polyamide Resins*, Reinhold Publishing Corp., New York, 1958.

Inderfurth, K. H., *Nylon Technology*, McGraw-Hill Book Co., New York, 1953.

Moncrieff, R. W., *Man-Made Fibres*, 3rd ed., John Wiley & Sons, New York, 1957, Chaps. XX, XXI.

Munden, A. R., "Polyamide Fibres," Chap. 17 in R. Hill, ed., *Fibres from Synthetic Polymers*, Elsevier Publishing Co., New York, 1953.

BIBLIOGRAPHY

1. Hanford, W. E., and R. M. Joyce, *J. Polymer Sci.* **3**, 167 (1948).
2. Hermans, P. H., *J. Appl. Chem.* **5**, 493 (1955).
3. Mark, H., and G. S. Whitby, eds., *Collected Papers of Wallace Hume Carothers on High Polymeric Substances*, Interscience Publishers, New York, 1940.

B. Polyesters

The properties of low melting point and high solubility caused Carothers (see Section *A*) to reject the linear aliphatic polyesters as fiber-forming candidates, and to this day no commercial product is based solely on these polymers. The stiffening action of the *p*-phenylene group in a polymer chain, however, leads to high melting points and good fiber-forming properties, as discussed in Chapter 7*C*. Commercially important polyesters are based on such polymers, of which poly(ethylene terephthalate) is the major product. These polymers are used for fibers, with a production of

about 100 million lb in 1960, and as film (about 20 million lb). The price of the staple fiber was of the order of $1.00/lb.

Polymerization and properties The production of high-molecular-weight polyesters differs somewhat from that of similar polyamides. In the case of the nylons, the chemical equilibrium favors the polyamide under readily achieved polymerization conditions. Stoichiometric equivalence is easily achieved by the use of salts, and amide interchange reactions are slow. In polyester formation, however, the equilibrium is much less favorable, and equivalence is more difficult to achieve, since the salts do not form. Furthermore, the dibasic aromatic acids are very difficult to purify because of low solubility and high melting point.

This situation has been met by taking advantage of the rapidity of ester-interchange reactions. The acid, such as terephthalic acid, is converted to the dimethyl ester, which can easily be purified by distillation or crystallization. This is then allowed to react with the glycol by ester interchange. In practice a low-molecular-weight glycol is used and the reaction takes place in two steps. First, a low-molecular-weight polyester is made with an excess of glycol to ensure hydroxyl end groups. Then the temperature is raised and pressure lowered to effect condensation of these molecules by ester interchange with the loss of the glycol.

In the production of poly(ethylene terephthalate), terephthalic acid is made by the oxidation of *p*-xylene. The polymerization step is similar to that for the polyamides in so far as the equipment and conditions are concerned. The polymer coming from the autoclave is quenched from the molten state to below its glass transition point of about 80°C and is amorphous. Crystallinity develops on heating; the crystalline melting point is 265°C. The polymer is melt spun.

Because of its high crystalline melting point and glass transition temperature, poly(ethylene terephthalate) retains good mechanical properties at temperatures up to 150–175°C. Its chemical and solvent resistance is good, being similar to that of nylon.

Fiber applications The properties of poly(ethylene terephthalate) fiber which influence its uses are its outstanding crease resistance and work recovery and its low moisture absorption. These properties arise from the stiff polymer chain and the resulting high modulus, and the fact that the interchain bonds are not susceptible to moisture. As a result, garments made from the polyester fibers are very resistant to wrinkling and can be washed repeatedly without subsequent ironing. The higher modulus of the polyester fibers and their wrinkle resistance are reminiscent of those of wool, but fully oriented polyester fibers are too stiff to have the crisp but soft hand of wool. It is possible, however, to impart this property to the

polyesters by control of crystallinity and orientation so that their stress-strain curve is similar to that of wool. This is done in the production of staple polyester fiber.

The polyester fibers are seldom used alone, but are blended with cotton or wool to make summer- and medium-weight suiting and other goods. They have set new standards for the performance of wearing apparel through the "wash and wear" concept.

Polyester fibers are unique among the synthetics in their ability to form felts. These products are better than wool felts in resistance to temperature, to abrasion, and to further felting.

Film applications The tensile strength of poly(ethylene terephthalate) film is about 25,000 psi, two to three times that of cellophane or cellulose acetate film (Chapter 18). If the area of the specimen at the break point is considered, the tensile strength of this plastic is about twice that of aluminum and almost equal to that of mild steel.

The stiffness of poly(ethylene terephthalate) film is comparable to that of cellophane and other cellulosic films, but its resistance to failure on repeated flexing is unusually high. In one flex test it lasts for over 20,000 cycles compared to a few hundred for cellulosic films. Its tear strength is also better than that of the cellulosics. The impact strength of this material is three to four times that of any other plastic film. This toughness is a major advantage in typical applications, such as magnetic recording tape.

Other polyester fibers and films The structures of other commercial polyester fibers and films are closely related to that of poly(ethylene terephthalate). They include polymers of terephthalic or isophthalic acid or their mixtures with ethylene glycol, 1,4-cyclohexylene dimethylene glycol (1), and others.

GENERAL REFERENCES

Bannermann, D. G., and E. E. Magat, *loc. cit.*

Bjorksten, J., H. Tovey, B. Harker, and J. Henning, *Polyesters and Their Applications*, Reinhold Publishing Corp., New York, 1956.

Marshall, I., and J. R. Whinfield, "Polyester Fibres," Chap. 18 in R. Hill, ed., *op. cit.*

Moncrieff, R. W., *op. cit.*, Chap. XXII.

Wilfong, R. E., *J. Polymer Sci.* **54,** 385 (1961).

BIBLIOGRAPHY

1. Watson, M. T., *SPE J.* **17,** 1083 (1961).

C. Polyurethans

Polyurethans are polymers containing the group

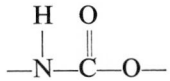

formed typically through the reaction of a diisocyanate and a glycol:

$$x\text{OCNRCNO} + x\text{HOR'OH} \longrightarrow \text{—[OCONHRNHCOOR'—]}_x$$

The polymers formed in this way are useful in four major types of product: foams, fibers, elastomers, and coatings. Polyurethan foams are invariably crosslinked; hence they are discussed with thermosetting resins in Chapter 19*E*.

Fibers

A polyurethan fiber made from hexamethylene diisocyanate and 1,4-butandiol has been used in Europe (1) as a textile fiber and a bristle material. Its strength and abrasion resistance are similar to those of the nylons, but it has lower moisture sensitivity and better resistance to outdoor exposure.

Polyurethan fibers with unusually high elasticity are used for lightweight foundation garments and swimsuits. It is expected that they will replace a large part of the rubber latex thread market in the near future.

Elastomers

Polyurethan elastomers are made in several steps. A *basic intermediate* is first prepared in the form of a low-molecular-weight (1000–2000) polymer with hydroxyl end groups. This may be a polyester (2, 5, 7), such as that made from ethylene glycol and adipic acid; a polyether (4); or a mixed polyester-polyamide (3).

The basic intermediate, which is here designated **B**, is then reacted with an aromatic diisocyanate to give a *prepolymer*. Typical isocyanates are 2,4-tolylenediisocyanate, 4,4-benzidenediisocyanate, and 1,5-naphthylenediisocyanate: using the former as an example, a typical prepolymer can be

represented as

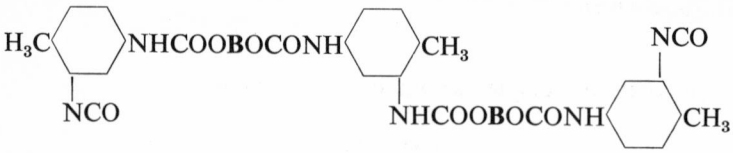

The elastomer is vulcanized through the isocyanate groups by reaction with glycols, diamines, diacids, or amino alcohols. If water is used, carbon dioxide is eliminated during crosslinking, as in the production of urethan foams (Chapter 19*E*).

Polyurethan elastomers are noted for extremely good abrasion resistance and hardness, combined with good elasticity and resistance to greases, oils, and solvents (6). They make tire treads with unusually long life and are widely used in applications requiring outstanding abrasion resistance, such as heel lifts and small industrial wheels.

Coatings

Coatings based on polyurethans have very good resistance to abrasion and solvent attack plus good flexibility and impact resistance. They can be applied by dip, spray, or brush and adhere well to a wide variety of materials. They are suitable in applications for which unusual impact and abrasion resistance is required, such as gymnasium and dance floors and bowling pins, in magnet wire coatings, and in a variety of outdoor and marine uses because of their good weatherability. About 1.5–2 million gallons was sold in 1961.

GENERAL REFERENCES

Arnold, R. G., J. A. Nelson, and J. J. Verbanc, *J. Chem. Educ.* **34,** 158 (1957).
Dombrow, B. A., *Polyurethanes*, Reinhold Publishing Corp., New York, 1957.
Moncrieff, R. W., *op. cit.*, Chap. XXI.

BIBLIOGRAPHY

1. Bayer, O., *Angew. Chem.* **A59,** 257 (1947).
2. Bayer, O., E. Müller, S. Petersen, H.-F. Piepenbrink, and E. Windemuth, *Angew. Chem.* **62,** 57 (1950).
3. Harper, D. A., W. F. Smith, and H. G. White, *Rubber Chem. Tech.* **23,** 608 (1950).
4. Hill, F. B., C. A. Young, J. A. Nelson, and R. G. Arnold, *Ind. Eng. Chem.* **48,** 927 (1956).

5. Müller, E., and coworkers, *Angew. Chem.* **64,** 523 (1952).
6. Saunders, J. H., *Rubber Chem. Tech.* **33,** 1259 (1960).
7. Seeger, N. U., and coworkers, *Ind. Eng. Chem.* **45,** 2538 (1953).

D. Polypeptides

Polypeptides, the basis of natural proteins, are polyamides of α-amino acids. They are characterized by this repeat unit:

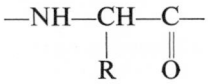

where R varies widely among the thirty or so amino acids in proteins.

Although synthetic products can be made, the polypeptides of greatest interest are based on naturally occurring materials. They include wool and silk and a variety of fibers and plastics made from proteins extracted from nonfibrous natural products.

The importance of wool and silk is discussed in relation to other fibers in Chapter 18. The protein-based fibers and plastics are at present of little more than historical interest.

Wool

Wool is a complex polypeptide, a polyamide made up of about twenty α-amino acids. The acids glycine, leucine and isoleucine, proline, cystine, and arginine, glutamic acid, and aspartic acid make up about two-thirds of its weight. About half of wool's mass is in the main polymer chain and half in the side chains. Since some 30% of the side chains have acid or amine groups along their length, wool is a huge polymeric "zwitterion" whose properties and solubility depend markedly upon *p*H.

Wool can exist in both the α- and β-keratin structures characteristic of the polypeptides: helical and extended sheet structures, respectively. The high reversible extensibility of wool (about 30% compared to 5% for cotton) derives from the fact that it can exist in either of these structures.

The microscopic appearance of wool is easily recognized by the presence of scales in the epidermal or surface layer of the fiber. These scales give wool the ability to felt or pack into dense stable mats.

Wool is noted for its high moisture absorption (up to 10–15%) and good insulating properties, its wrinkle resistance and elastic recovery, and its crimpiness. Its soil resistance, dyeability, and resistance to organic (dry-cleaning) solvents are good. On the other hand, it is attacked by

moths and mildew, mild alkalis, and bleaches. It causes allergies. Its strength and stiffness are low compared to those of cotton. It has poor temperature resistance and weatherability in sunlight, and is adversely affected by hot water, with shrinkage and loss of luster and strength.

Despite these drawbacks, the virtues of wool are so great that it probably will continue to find extensive use, as it has to date, in suitings and other wearing apparel, blankets, yarns, carpets, felts, and upholstery. So far, no artificial fiber has equaled it in resilience, hand, insulating properties, dyeability, and flame resistance. The price of wool is notoriously unsteady, however, and its future may well depend upon its price in a competitive market.

Wool is crosslinked through cystine links:

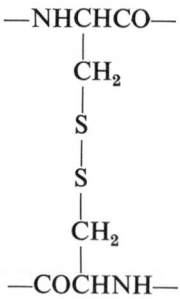

This structure gives swelling but not solubility in the natural product. On the addition of thioglycolic acid, however, the crosslinks are broken with the formation of sulfhydryl groups. The tensile strength of the fiber decreases, its stress-strain curve changes to resemble that of a typical elastomer, and it shrinks readily and becomes soluble in polar solvents.

Wool can be recrosslinked by a number of reagents, including alkyl dihalides. Since the thioethers resulting from the treatment are much more stable than the original disulfides, the new fiber has much better resistance to laundering and chemicals. Since moth damage usually begins with rupture of the disulfide links, the treated material has improved moth resistance. The scission and crosslinking can be made so mild that the appearance and hand of the material are not changed.

Silk

Silk is a polypeptide made up of only four amino acids: glycine, H_2NCH_2COOH; alanine, $H_2NCH(CH_3)COOH$; serine; and tyrosine. Glycine and alanine, present in a 2:1 ratio, account for over 75 mole per cent of the material. Silk crystallizes in an anti-parallel-chain pleated

sheet β-keratin structure, in which only glycine and alanine residues occur. The other amino acids are present only in the amorphous regions of the fiber. There is no structure in silk equivalent to the α-keratin spiral in wool and other polypeptides. Thus the silk fiber has an elongation (20–25%) lower than that of wool but higher than that of cotton.

The silkworm spins a double fiber from which the two single fibers are easily separated as continuous filaments 400–700 yards long. Silk is thus more nearly analogous to continuous filament than to staple synthetic fibers.

Silk has been noted for thousands of years for its strength, toughness, and smooth, soft feel. It is resilient and shapes well. It is a poor conductor of heat and electricity. Its moisture regain is 11%; it withstands heat and hot water better than wool.

Some disadvantages of silk are those associated with its cultivation and gathering, unevenness of the fibers, loss of strength on dyeing, and poor launderability of the dyed fabrics. It lacks the strength of nylon for hosiery use but would otherwise be preferred for its warmer feel. There is relatively little effort in progress to improve the properties of silk.

Protein fibers and plastics

Casein Casein is a protein by-product of the dairy industry. It can be dissolved in dilute alkali, aged, and regenerated by extruding the solution into an acid bath while stretching. It is customary to insolubilize the fibers so formed by treatment with formaldehyde and acid [as with poly(vinyl alcohol) fibers]. Similar fibers can be made from the protein extracted from peanuts.

These protein fibers crystallize in the α-keratin wool structure and have a natural crimp and a soft, woollike hand. Their tenacity is quite low, however, and they are generally used as a cut staple in blends with wool, cotton, or rayon. The blends with wool have good abrasion resistance; and those with viscose rayon, good washability.

Casein plastic dates back to 1900, when its properties of nonflammability and light color gave it advantages over competitive materials. Today these advantages are outweighed in many respects, and the limited form in which the material is available, together with its high water absorption, restricts its uses to a few articles such as buttons and buckles.

Casein is obtained from the curds of skimmed milk. Water and colorants are added to the dried product, and it is extruded into sheets, rods, or tubes. Button blanks are cut or turned from these forms and hardened in 4–5% formaldehyde solution.

Zein (corn protein) fiber The protein zein can be extracted from corn with isopropyl alcohol. It is readily purified and subsequently spun from alkali like the other protein fibers. It is finally crosslinked with formaldehyde. This fiber is more useful, however, for its resistance to acids and alkalis, moths, mildew, and insects. It launders well. The fabrics are similar in hand to wool but are nonallergenic, nonitchy, and odorless. The zein fiber is often used in blends.

GENERAL REFERENCES

Mauersberger, H. R., ed., *Matthews' Textile Fibers*, 6th ed., John Wiley & Sons, New York, 1954.
Moncrieff, R. W., *op. cit.*, Chaps. XVI–XIX.

E. Ether and Acetal Polymers

Polyethers, by definition, contain the ether linkage —R—O—R—O— as part of their chain structure. By convention, the polymers in which R is a methylene group, $—CH_2—$ (i.e., polyoxymethylenes), are described by the generic term *acetal resins*, the name polyethers being applied to polymers in which R is more complex. Epoxy resins, also polyethers, are discussed in Chapter 19*D*.

Polyoxymethylene

Polyoxymethylene (1, 4) is a polymer of formaldehyde or of trioxane. Although polymeric products of formaldehyde have been known for over 100 years, and were studied in detail by Staudinger (5), thermally stable polymers of formaldehyde were only recently prepared (3). The improved stability of these acetal resins allows them to be fabricated into useful articles.

Polymerization (3) Exceptionally pure formaldehyde (better than 99.9% CH_2O) is polymerized by an anionic mechanism in the presence of an inert solvent (e.g., hexane) at atmospheric pressure and a temperature, preferably between −50 and +70°C, where the solvent is liquid. A wide variety of anionic catalysts is suitable, including amines, cyclic nitrogen containing compounds, arsines, stibines, and phosphines. The polymer is insoluble in the reaction mixture and is removed continuously as a slurry.

Poststabilization Thermal stability is improved by acetylation of the hydroxyl end groups of the polymer using acetic anhydride:

$$\begin{aligned}&-CH_2OCH_2OH + CH_3COO^- \longrightarrow \\ &\qquad -CH_2OCH_2O^- + CH_3COOH \xrightarrow{(CH_3CO)_2O} \\ &\qquad\qquad -CH_2OCH_2OCOCH_3 + CH_3COO^-\end{aligned}$$

Properties Acetal resins have properties characteristic of partially crystalline high-molecular-weight polymers. Typical specimens are about 75% crystalline, with a melting point of 180°C. The impact strength of the polymers is high, and both their stiffness (450,000 psi) and yield stress (10,000 psi) are greater than those of other crystalline polymers. Moisture absorption is negligibly small, and the polymers are insoluble in all common solvents at room temperature. They can be processed by conventional molding and extrusion techniques.

Applications Major uses for the acetal resins are expected to be as direct replacements for metals. At their 1961 price of $0.65/lb, they are cheaper than brass per unit volume. Their stiffness, light weight, dimensional stability, and resistance to corrosion, to wear, and to abrasion have led to their replacing brass, cast iron, and zinc in many instances. Typical applications include automobile parts, such as instrument panels, door hardware, and pump housings and mechanisms; pipe, especially for oil field systems; and a wide variety of machine and instrument parts.

Higher aldehyde polymers Higher aldehydes can be polymerized by both anionic and cationic mechanisms. In some cases stereoregular polymers result (6). This is the only known case where polymerization involving a double bond other than C=C yields stereoregular polymers.

Polyglycols

Polymers made from ethylene oxide and its higher homologs $\underset{\diagdown\,\diagup\atop O}{CH_2CHR}$ can be considered to be the condensation products of glycols, although they are in fact produced by ring-scission polymerization of the corresponding oxides (Chapter 8*E*). The product of greatest commercial interest is the polymer of ethylene oxide. This polymer combines the thermoplastic behavior and mechanical properties of a highly crystalline, high-molecular-weight polymer with complete water solubility.

Ethylene oxide polymers have moderate strength and stiffness, similar

to the corresponding properties of polyethylene, at room temperature. Their crystalline melting point is 65°C, however, so that their mechanical properties fall off sharply near this temperature. The polymer can be highly oriented by cold drawing. It retains its properties well up to about 75% relative humidity, despite its water solubility. It can be processed by any of the common thermoplastic processing techniques.

The polymers of ethylene oxide are useful where water solubility must be combined with film-forming, fiber-forming, or thickening properties. Typical applications include sizing for cotton and synthetic fibers, thickeners and stabilizers for pastes and adhesives, and binders and film-formers in pharmaceuticals.

Other polyethers

A chlorinated polyether with the structure $—CH_2C(CH_2Cl)_2CH_2O—$ is obtained from pentaerythritol by preparing a chlorinated oxetane and polymerizing it by ring scission. The polymer is linear and crystalline, and is much more resistant to thermal degradation than are other halogenated polymers, because of the absence of a hydrogen atom beta to the chlorines.

The polymer (2) has good impact strength, low moisture absorption, high dimensional stability, and good chemical resistance, and retains its properties with increasing temperature well enough to allow use at 100–125°C. It can be processed by injection molding or extrusion, and can be applied as a coating from suspensions or in fluidized bed processes.

The polymer is used in pipes, pumps, and processing equipment where corrosion resistance is important, as a liner for pipe and tubing, and as a material to replace metal die castings.

GENERAL REFERENCES

Akin, R. B., *Acetal Resins*, Reinhold Publishing Corp., New York, 1962.

Furukawa, J., and T. Saegusa, *Polymerization of Aldehydes and Oxides*, Interscience Publishers, a Division of John Wiley & Sons, New York, 1962.

Gaylord, N. G., ed., *Polyethers*, Part 1, Interscience Publishers, a Division of John Wiley & Sons, New York, 1963.

BIBLIOGRAPHY

1. Cogdell, J. F., and R. H. Hardesty, *SPE J.* **14,** #4, 25 (1958).
2. Cronin, E. W., *Modern Plastics* **34,** #6, 150 (February, 1957).
3. MacDonald, R. N., U.S. Patent 2,768,994 (1956).

4. Schweitzer, C. E., and coworkers, *J. Appl. Polymer Sci.* **1,** 158–185 (1959).
5. Staudinger, H., *Die Hochmolekularen Organischen Verbindungen*, Springer-Verlag, Berlin, 1932.
6. Vogl, O., *Chem. & Ind.* **1961,** 748; *J. Polymer Sci.* **46,** 261 (1960).

F. Polysulfides

Polysulfide elastomers are the reaction products of ethylene dihalides and alkali sulfides. It was discovered in 1920 that a rubberlike material could be made from these reagents. Commercial production began in 1930, and the products were quite successful despite some disadvantages (notably odor) because of their oil resistance.

The polysulfide rubbers are believed to be linear condensation polymers:

$$x\mathrm{RCl_2} + x\mathrm{Na_2S_y} \longrightarrow -(\mathrm{R{-}S_y{-}})_x + 2x\mathrm{NaCl}$$

In the manufacture of the elastomers, the dihalide is added to an aqueous solution of sulfur in sodium hydroxide, whose specific gravity is adjusted to be equal to that of the dihalide. Mixing is thus simplified, and the product can be isolated as a latex if desired. An excess of the sodium polysulfide is used to obtain high molecular weights.

The physical properties of the materials depend on the length of the aliphatic group and the number of sulfur atoms present. With four sulfurs per monomer all products are rubbery; whereas with only two sulfurs at least four methylene groups are needed in the dihalide to obtain elastomeric properties. The halides ordinarily used are ethylene dichloride, β,β'-dichloroethyl ether, and dichloroethyl formal. Cocondensation can be carried out.

The polysulfide elastomers can be vulcanized by zinc oxide or by a variety of oxidizing agents. Mechanisms of vulcanization are not known. Sulfur does not vulcanize the polymers but does accelerate their cure. Carbon black has a marked reinforcing effect and is always used.

The polysulfide elastomers are outstanding in oil and solvent resistance and in gas impermeability. They have good resistance to aging and ozone. On the other hand, they have disagreeable odors, poor heat resistance, poor abrasion resistance in tire treads, and low tensile strength (1500 psi). Most of their applications as rubbers depend on solvent resistance and low gas permeability. They include gasoline hoses and tanks, diaphragms and gaskets, and balloon fabrics.

Liquid polysulfide elastomers can be cured at room temperature and

without shrinkage to tough, solvent-resistant rubbers. These liquid compounds are widely used as gasoline tank sealants and liners for aircraft and in a variety of other sealing and impregnating applications.

These liquid polymers, when combined with oxidizers, burn with great intensity and generate large volumes of gas. They form the basis of solid-fuel rocket propellants. Inorganic oxidizers, in finely ground form, are mixed with the liquid elastomeric binder and cast directly into the rocket motor, binding to the walls of the vessel for support and insulation of the motor case.

GENERAL REFERENCES

Fettes, E. M., and J. S. Jorczak, "Polysulfide Polymers," Chap. XI in C. E. Schildknecht, ed., *op. cit.*

Gaylord, N. G., *op. cit.*, Part 3, 1962.

Jorczak, J. S., *Polysulfide Polymers*, Chap. 15 in M. Morton, ed., *Introduction to Rubber Technology*, Reinhold Publishing Corp., New York, 1959.

G. Polycarbonates

A polycarbonate plastic, characterized by the —OCOO— heterochain unit, can be made either from phosgene and bisphenol A (4,4′-dihydroxydiphenyl-2,2′-propane) or by ester exchange between bisphenol A and diphenyl carbonate. It has this structure:

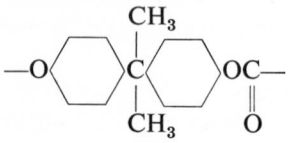

Like the nylon, acetal, and polyether resins, this polymer is a crystalline thermoplastic of good mechanical properties (1, 2). It has unusually high impact strength, even at low temperatures, ascribed in part to a combination of relatively high ordering in the amorphous regions and considerable disordering in the crystalline regions. It has low moisture absorption, good heat resistance (useful to 140°C), and good thermal and oxidative stability in the melt. It is transparent and self extinguishing. It can be processed by conventional injection molding and extrusion. Typical applications include telephone parts, business machine housings, machinery housings, safety equipment, and graphic arts film bases.

GENERAL REFERENCE

Christopher, W. F., and D. W. Fox, *Polycarbonates*, Reinhold Publishing Corp., New York, 1962.

BIBLIOGRAPHY

1. Christopher, W. F., *SPE J.* **14**, #6, 51 (1958).
2. Thompson, R. J., and K. B. Goldblum, *Modern Plastics* **35**, #8, 131 (April, 1958).

18

Cellulosic Polymers

The class of polymers based on cellulose includes the following: native cellulose, including wood and most plant matter, but of interest here primarily because of the widespread popularity of cotton as a fiber; regenerated cellulose, used as a fiber (viscose rayon) and a film (cellophane); chemical derivatives of cellulose, of which the organic esters, particularly the acetate, are most important; and minor polymers with structures similar to that of cellulose.

The cellulose fibers described in this chapter (cotton, viscose, cellulose acetate) account for about 80% of the total fiber consumption in the United States, as indicated in Table 18-1. This total is indeed enormous, fluctuating between 4 and 6 billion lb/year.

Although cotton is still the most important fiber, its share of the total has declined steadily, from 80% in 1920 to its present position of slightly less than two-thirds. This corresponds nevertheless to a per capita

TABLE 18-1. Approximate 1960 consumption of fibers in the United States

Fiber	Consumption, billions of lb	
Cotton		3.8
Viscose (rayon) and acetate		1.2
"Synthetics"		0.9
Nylon	0.45	
Acrylic	0.15	
Polyester	0.10	
Glass	0.20	
Wool		0.3
Total		6.2

consumption of over 20 lb/year. The consumption of viscose and acetate has remained steady at about 1.2 billion lb for several years.

Consumption of the synthetic fibers has continued to expand rapidly, however, and is expected to surpass that of the rayons before 1965.

The consumption of wool in this country has declined steadily over the last decade. Other natural fibers, including silk and the natural cellulosics flax, jute, and ramie, are of very minor importance in this country but are still widely used elsewhere.

The price of cotton and viscose is in the range \$0.30–0.35/lb, while that of acetate (and cellulose acetate plastic) is \$0.40–0.45/lb and that of the synthetic fibers ranges from \$0.90 to \$1.10.

The price of wool fluctuates markedly, ranging upward from \$0.85/lb.

The properties of fibers are discussed further in Chapter 21.

A. Native Cellulose

Cellulose, the major constituent of the vegetable kingdom, is a high polymer of extreme importance in almost all facets of human life. It is beyond the scope of this book to discuss important cellulosic materials such as wood and paper.

Cellulose is a polysaccharide made up of β-D(+)-glucose residues joined in linear chains having the structure:

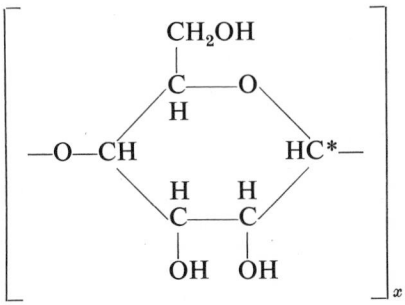

Starch, similarly made up of chains of α-D(+)-glucose, has a quite different chain configuration, resulting from its different steric configuration about the carbon atom starred in the diagram.

With its three hydroxyl groups, cellulose has the opportunity of forming many hydrogen bonds. The resulting high intermolecular forces plus the regular structure of the polymer result in its having an unusually high

degree of crystallinity.* The crystalline melting point of cellulose is far above its decomposition temperature. The solubility of the polymer is very low; it is doubtful that solution ever takes place unless a chemical derivative is formed. Cellulose does swell, however, in hydrogen-bonding solvents, including water. The swelling is, of course, restricted to the amorphous regions of the structure. When native cellulose is dissolved via chemical reaction and then reprecipitated as pure cellulose, the product is known as *regenerated cellulose* (Section *B*).

The highest-molecular-weight cellulose is found in cotton, which is the best source of very pure cellulose. Molecular weights as high as 570,000 are found, corresponding to a degree of polymerization of 3500. The molecular weight of cellulose from wood and other sources is considerably lower (around 160,000). Almost all chemical reactions of cellulose, including solution, are accompanied by some degradation in chain length. The degree of polymerization of cellulose derivatives is usually below 500.

Cotton

Cotton is about 95% cellulose, with small amounts of protein, pectin, and wax. The fiber can easily be recognized under the microscope because of its flattened, twisted shape.

Cotton fabrics launder well and show excellent resistance to alkalis. Since the strength of cotton is 25% greater wet than dry, it withstands repeated washings well, in contrast to most animal and synthetic fibers which lose strength when wet or moist. Cotton is less stiff than flax, hemp, or jute. Fabrics of cotton can be folded repeatedly without loss of strength. With proper construction and treatment, shrinkage and stretching can be controlled. Cotton is a good conductor of heat, better than silk or wool but not as good as linen. This property and its high moisture absorption give it comfort in wearing apparel. Cotton withstands heat better than the other natural fibers. It can be dyed readily, and the colors are good in lightfastness and washfastness. It is not affected by solvents and usually not attacked by moths.

Cotton is thus a versatile fiber with many desirable properties. These properties and its low price have given it an advantage over the years which has only recently been threatened. Rayon (i.e., viscose) and acetate are now available at prices competitive to cotton with advantages of luster, drape, wrinkle and soil resistance, low shrinkage, rapid drying, and good weathering resistance.

* See footnote, page 129.

A number of research groups are actively working toward improvements in the properties of cotton. Agricultural research includes that on new breeds of cotton plants giving longer, more extensible fibers, improvements in soil fertilization, weed, insect, and disease control, and better methods of picking and preparing the fiber.

Mercerization The chemical treatment of cotton known as *mercerization* to improve its luster has been practiced for many years. The treatment consists in swelling the cotton by concentrated alkali solution and then washing out the alkali. The chemical changes during the process are similar to those in the formation of alkali cellulose (Section *B*). At the end of the treatment there is no molecular difference in the product, but it is somewhat more amorphous than native cotton and, unless the fibers were held under tension during the process, somewhat less oriented.

As a result of these changes, mercerized cotton is slightly lower in density than the untreated material and has increased water absorption, better dyeability, lower tensile strength, and higher extensibility (1). The higher luster is presumed to come from the change in cross-sectional shape of the fibers, which swell from their flattened natural form and dry with a more nearly circular cross section.

Other modifications of cotton Treatment of cotton with glacial acetic acid and acetic anhydride partially acetylates the fiber. The treated fibers are much improved in heat, acid, and mildew resistance and have lower moisture regain. The cost of the fiber is increased about three-fold, however, and other disadvantages result, such as impaired launderability and a tendency to develop static electricity.

Treatment of cotton with acrylonitrile leads to the production of the cyanoethyl ether of cellulose. Cyanoethylated cotton containing 3–4% nitrogen shows increased heat, acid, and mildew resistance over the native product with no loss of strength, elongation, moisture regain, or launderability.

The most important after-treatment of cotton is the application of so-called resin finishes which impart wash-and-wear properties to the fabric. About 20% of the cotton products made was so treated in 1960. This after-treatment is discussed in Chapter 21*C*.

GENERAL REFERENCES

Honeyman, J., ed., *Recent Advances in the Chemistry of Cellulose and Starch*, Interscience Publishers, New York, 1959.

Moncrieff, R. W., *Man-Made Fibres*, 3rd ed., John Wiley & Sons, New York, 1957, Chap. XIV.

Ott, E., H. M. Spurlin, and M. W. Grafflin, eds., *Cellulose and Cellulose Derivatives*, Interscience Publishers, New York, Parts I and II, 1954; Part III, 1955.

Rånby, B. G., and S. A. Rydholm, "Cellulose and Cellulose Derivatives," Chap. IX in C. E. Schildknecht, ed., *Polymer Processes*, Interscience Publishers, New York, 1956.

BIBLIOGRAPHY

1. Krassig, H., and W. Kitchen, *J. Polymer Sci.* **51,** 123 (1961).

B. Regenerated Cellulose

The term *regenerated cellulose* describes cellulose which has been dissolved by virtue of the production of a soluble chemical derivative, cellulose xanthate, and subsequently reprecipitated. When prepared as a fiber, regenerated cellulose is known as *viscose* or (viscose) *rayon*. Traditionally both viscose and cellulose acetate (Section *C*) have been known as rayons, but the term is now restricted to viscose in order to avoid confusion. As a film, regenerated cellulose is known by the generic term *cellophane*.

Manufacture The first step in the production of regenerated cellulose, starting with purified cellulose usually obtained from wood pulp, is the formation of *alkali* or *soda cellulose*. This is produced by the reaction of cellulose with strong alkalis at low temperatures, and is a further step in the reaction of mercerization (Section *A*).

The exact structure of alkali cellulose is not completely understood. It is certain that some of the alkali (say NaOH) is held in the swollen amorphous regions of the cellulose by hydrogen bonding of the hydroxyl radicals to the cellulose molecules. It is also possible that some alcoholates of the type R—ONa are formed, where R is the cellulose residue exclusive of one hydroxyl group. Several crystal structure modifications are known in which sodium ions are located interstitially between the cellulose chains.

Alkali cellulose is formed by reacting the cellulose pulp with about 18% aqueous NaOH at room temperature for 15 min to 2 hr. The excess alkali is then pressed out and the mass is shredded into "crumbs" and aged for several days at 25–30°C to promote oxidative degradation of the chains to the desired degree of polymerization.

The alkali cellulose is then converted to cellulose xanthate by the addition of carbon disulfide:

$$R{-}OH + CS_2 + NaOH \longrightarrow R{-}O{-}\overset{\overset{\displaystyle S}{\|}}{C}{-}S{-}Na + H_2O$$

where R is a cellulose residue. The average degree of xanthation is about one CS_2 for every two cellulose residues. Since the reaction takes place heterogeneously by the addition of liquid or gaseous CS_2 to the alkali cellulose crumbs, the xanthation in some regions may correspond to the dixanthate or higher.

The mass retains most of its physical form but becomes more sticky or gelatinous and takes on a deep orange color and a characteristic and highly unpleasant odor. It is held at 25–35°C for 1–3 hr with about 10% CS_2 by weight and then evacuated to remove excess CS_2. It is now dissolved in dilute NaOH, becoming completely soluble in this reagent for the first time (3). The solution is known as viscose. Its color and odor are due to products of side reactions such as Na_2CS_3, NaSH, H_2S, sulfides, and polysulfides.

A fresh viscose solution cannot easily be coagulated and must be allowed to "ripen" for a few days. The changes taking place during ripening are complex, and the reaction must be carefully controlled by time and temperature to avoid premature coagulation.

Cellulose xanthate is essentially unstable and decomposes gradually during the ripening process by hydrolysis and saponification:

$$R{-}O{-}\overset{\overset{\displaystyle S}{\|}}{C}{-}S{-}Na + H_2O \longrightarrow R{-}O{-}\overset{\overset{\displaystyle S}{\|}}{C}{-}S{-}H + NaOH$$

$$R{-}O{-}\overset{\overset{\displaystyle S}{\|}}{C}{-}S{-}H + H_2O \longrightarrow R{-}OH + CS_2 + H_2O$$

Alternatively, the hydrolysis may occur without the formation of cellulose xanthic acid as an intermediate.

At the end of the ripening period, enough cellulose residues have been regenerated so that coagulation is imminent. The viscose is then spun into a bath containing sulfuric acid to effect the regeneration of the remaining cellulose residues and coagulate the polymer:

$$R{-}O{-}\overset{\overset{\displaystyle S}{\|}}{C}{-}S{-}Na + H^+ \longrightarrow R{-}OH + CS_2 + Na^+$$

If the spinning process places no mechanical stress on the polymer, it can be obtained in a completely amorphous form. This form is unstable, however, and the application of rather small stresses by such processes as stretching, working, spinning, and extruding causes the cellulose to crystallize (usually to a different structure from that of the native material). In this case, the ordering of the chains is not as great as in the native material, and the density and stiffness of the regenerated product are

lower. These differences in density, stiffness, strength (2), and other properties between native and regenerated cellulose can be accounted for by differences in chain length, chain length distribution, crystal structure, degree of crystallinity, orientation, crystallite size, nature and amount of impurities and other flaws, degree of swelling, and chemical reactivity. All these changes on regeneration are in line with the structural differences outlined above.

Viscose rayon

Normal tenacity Viscose rayon normally has a tenacity of 2–2.5 g/denier, somewhat lower than that of cotton (3–5 g/denier). Its wet tenacity is only about half the dry value but is still adequate for laundering. The moisture absorption of viscose is about twice that of cotton. The elasticity is not high (typically 15% elongation at break), and the fiber shows considerable viscoelastic character.

Viscose of normal tenacity is primarily an apparel fiber, but its popularity is declining. Several one-time major manufacturers have ceased producing the material, and the others are operating well below capacity.

High tenacity In contrast to apparel rayon, a high-tenacity product (3–6 g/denier) is produced by stretching the fibers just short of their breaking point in the spinning bath. Considerable orientation (and crystallinity) is so introduced. High-tenacity rayon has lower elongation and moisture absorption than the apparel material and is used almost exclusively for tire cords. Here it shares a sizable market (some 400 million lb/year) with nylon; rayon cord is currently used for original-equipment tires for passenger automobiles.

Cellophane

In the manufacture of cellophane (1) the viscose solution is extruded as film and then immersed in a bath of ammonium and sodium sulfate and dilute sulfuric acid, which removes the xanthate groups and precipitates the cellulose. The film is passed through various washing, bleaching, and desulfurizing baths, and finally through a bath of glycerol, glucose, or a polyhydric alcohol, which is imbibed and acts as a plasticizer. The film is then dried. For applications in which low moisture permeability is desired, it is coated with a mixture of nitrocellulose and various plasticizers and waxes.

Cellophane is a thin (0.001–0.002 in.) film of fair physical properties.

Its tensile strength is good, but its tear strength, impact strength, and resistance to flexing are poor compared to those of newer film materials. The permeability of the uncoated film to water vapor and to water-soluble gases is extremely high. The coated material passes about 0.3 gram of water vapor/sq in./hr at 40°C. Cellophane is widely used as a wrapping and packaging material.

Cellulose sponges are made by stirring lumps of a salt such as sodium sulfate into viscose, coagulating, and washing out the salt to leave a porous product. Sponges may also be made by incorporating a blowing agent in the material.

The production of cellophane exceeded 400 million lb in 1960.

Other regenerated cellulose products

Cuprammonium rayon A process which involves dissolving cellulose in an ammoniacal solution of copper oxide and then regenerating yields this unusually fine fiber, in demand at premium prices for high-grade fabrics.

Saponified cellulose acetate An extremely strong regenerated cellulose fiber is made by stretching cellulose acetate in steam under pressure, and then saponifying off the acetate groups. Final tenacity can reach 7 g/denier.

Vulcanized fiber A rigid, lightweight sheet product is made by gelatinizing cellulose in aqueous zinc chloride, then washing it free of chemicals, forming it while still moist, and drying. This yields a rigid, hard material of high impact strength and good machining properties, used for dielectric applications, gaskets, suitcases, etc.

GENERAL REFERENCES

Honeyman, J., *loc. cit.*
Moncrieff, R. W., *op. cit.*, Chaps. VII, VIII, IX, XII, XIII.
Ott, E., *et al.*, *loc. cit.*
Rånby, B. G., and S. A. Rydholm, *loc. cit.*

BIBLIOGRAPHY

1. Inskeep, G. C., and P. Van Horn, *Ind. Eng. Chem.* **44,** 2511 (1952).
2. Krassig, H., and W. Kitchen, *J. Polymer Sci.* **51,** 123 (1961).
3. Treiber, E., *J. Polymer Sci.* **51,** 297 (1961).

C. Cellulose Esters

The cellulose esters of commercial interest include the acetates (diacetate for fiber and plastics use and triacetate for fiber use), higher esters and mixed esters used as plastics (butyrate, propionate), and the nitrate, once popular as a plastic.

Cellulose acetate

Manufacture As a polyhydric alcohol, cellulose can undergo esterification reactions. In the presence of a strong organic acid such as formic acid, equilibrium in the reaction

$$\text{acid} + \text{alcohol} \rightleftharpoons \text{ester} + \text{water}$$

lies partly to the right, and some ester is formed. With other organic acids, including acetic acid, equilibrium is well to the left and ester formation does not take place under normal circumstances. The easiest way to shift the equilibrium to the right is by the removal of water as it is formed in the reaction. To accomplish this, part of the acid is replaced by an acid chloride or anhydride. Sulfuric acid is also utilized in the reaction mixture. It acts as catalyst and also assists in removing water.

Chemical reactions of cellulose must of necessity begin in the (swollen) amorphous regions of the polymer. Unless these heterogeneous reactions are carried to the point where the cellulose is completely dissolved, the chemical homogeneity of the material is usually lower than desired. This is true in the formation of cellulose acetate, and consequently the esterification is carried to the point of complete solution.

Purified cotton linters or wood pulp form the raw material for cellulose acetate. This cellulose is partially dried from its natural moisture content of 5–10% to reduce the water content of the subsequent reaction mixture (complete drying would lead to a product difficult to reswell and low in chemical reactivity). The acetylating mixture of acetic acid, acetic anhydride, and sulfuric acid is added in portions and the reacting mixture is kept at or below 50°C. Since the reaction is exothermic, cooling is required. Higher temperatures are avoided to prevent excessive degradation in molecular weight.

The end of the acetylation is indicated by complete solution of the cellulose. At this point complete acetylation to the triacetate has taken place. The mixture is now held at about 50°C until the desired chain degradation has been achieved, as indicated by the viscosity of the mixture.

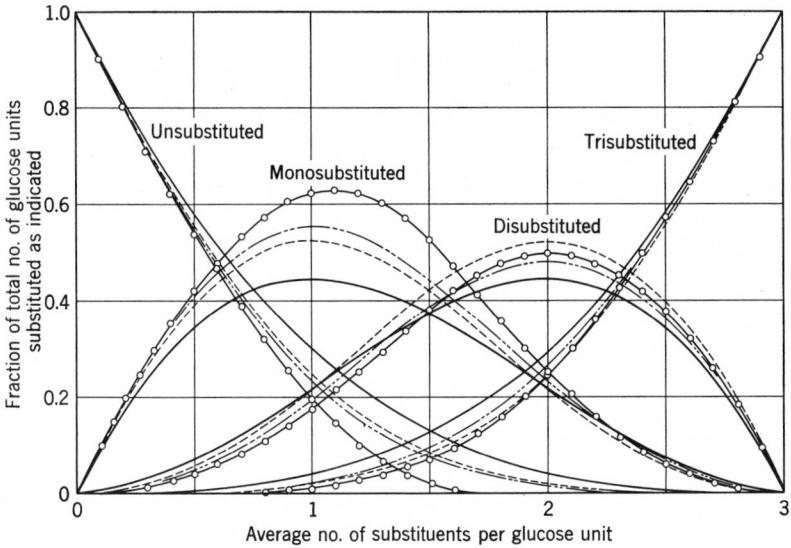

Fig. 18-1. Distribution of substituent groups on cellulose. The various lines refer to different ratios of equilibrium and rate constants for the three hydroxyl groups of cellulose. [Ott *et al.*, gen. ref., after (1).]

To obtain a more easily spinnable composition, it is necessary to reverse the acetylation reaction to the point where the proper solubility relations are obtained. This corresponds to the acetone-soluble diacetate, or an acetyl content of about 30%. This stage of the reaction can be carried out in homogenous solution, thus preserving the maximum amount of chemical homogeneity in the final product. The relative amounts of unsubstituted and mono-, di-, and trisubstituted hydroxyls in the product correspond closely to those predicted in Fig. 18-1.

The acetylation reaction is reversed by the addition of water to the reaction mixture. Since gross addition of water would precipitate the polymer, acetic acid is added to the mixture, which at this point still contains large amounts of acetic anhydride. The final composition of the mixture needed to produce the diacetate corresponds to 90–95% acetic acid and 5–10% water. (Further reversal of the reaction, with a final composition of 60–70% acetic acid, would yield the monoacetate, which is insoluble in acetone and alcohols but is soluble in water.)

At the end of the reaction to produce the diacetate the mixture is cooled and added to a large amount of 25–35% acetic acid. This precipitates the polymer in finely divided form. After the polymer is washed and dried, it is dissolved in acetone for the spinning operation.

Fiber properties and applications The traditional acetate fiber

(diacetate) has a tenacity of 1.2–1.5 g/denier and about the same moisture absorption as cotton. It tends to soil less and wash more easily than viscose. Since acetate is thermoplastic, pleats and creases can be permanently set. On the other hand, acetate is less susceptible to creasing and wrinkling in use than viscose because its resilience is greater.

With the advent of spinning processes using methylene chloride, cellulose triacetate has become popular as an apparel fiber. Its properties are improved over those of the diacetate in moisture absorption, wrinkle resistance and creep retention, dyeability, and speed of drying.

Aided by the success of triacetate and by new household, apparel, and industrial applications, including nonwoven fabrics and cigarette filters, the use of cellulose acetate increased to about 400 million lb/year in 1960, after several years of production at about half this level.

Plastics properties and uses Cellulose acetate was introduced in sheet, rod, and tube form (as well as in lacquers and dopes) around the time of World War I. Its popularity as a substitute for cellulose nitrate (see next page) was not high, and it was not widely used until injection molding became common in the 1930's. It has been for many years a very popular injection-molding material. The usual molding material has 50–55% acetyl content by weight. It is somewhere between the diacetate and the triacetate in composition. Higher acetyl content gives poorer flow but better moisture resistance; lower acetyl content gives better impact strength. A plasticizer is ordinarily used. Higher-softening grades are preferred for compression molding, and softer grades for extrusion. Cellulose acetate molded and extruded articles are extensively used where extreme moisture resistance is not required.

Cellulose acetate is also widely utilized in sheet form. Thick sheets are made by slicing from blocks and polishing (as with cellulose nitrate). Thinner films can be cast from solution onto polished metal surfaces. The films are also widely used in photography, for wrapping, and for making small envelopes, bags, and boxes for packaging.

Cellulose acetate has good impact strength and electrical properties, and the advantage of low flammability not possessed by cellulose nitrate. It has, however, a rather low softening point and high water absorption.

The plastics uses of cellulose acetate plus the mixed organic esters described below amounted to about 150 million lb in 1961.

Mixed organic esters of cellulose

Cellulose acetate-butyrate The mixed ester cellulose acetate-butyrate has several advantages in properties over cellulose acetate. These include

lower moisture absorption, greater solubility and compatibility with plasticizers, higher impact strength, and excellent dimensional stability. The material used in plastics has about 13% acetyl and 37% butyryl content. It is an excellent injection-molding material. Moldings of cellulose acetate-butyrate are widely used, e.g., in automobile hardware. Motion picture safety film is cellulose acetate-butyrate or cellulose propionate-butyrate. The outdoor weathering properties of cellulose acetate-butyrate are superior to those of the other cellulose esters, but not nearly as good, e.g., as those of the acrylic resins.

Cellulose propionate and acetate-propionate These materials are similar to the acetate-butyrate in properties and uses. The acetate-propionate has some application in surface coatings.

Cellulose nitrate

Although cellulose nitrate accounts today for only a minute fraction of the volume of cellulose plastics, it has considerable historical interest (Chapter 1*B*). The major use of plastics-grade cellulose nitrate is in the coating field. The properties and applications of the material are dependent on its degree of nitration; the plastics and lacquer grades contain 10.5–12% nitrogen, corresponding roughly to the dinitrate. Explosives grades contain 12.5–13.5% nitrogen.

Manufacture The esterification reaction of the hydroxyl group on cellulose with nitric acid is an equilibrium

$$ROH + HONO_2 \rightleftharpoons RONO_2 + H_2O$$

which can be shifted to the ester side by the removal of water. The usual reagent for this purpose is sulfuric acid; the nitration mixture ordinarily used consists of sulfuric and nitric acids and a limited amount of water. The composition of the mixture determines the degree of nitration of the product and the solubility of the material. In the nitration process it is desirable to have the cellulose undissolved and unswollen. This requirement plus the desired final nitrogen content and cost considerations fixes the composition of the nitrating mixture at about 25% HNO_3, 55% H_2SO_4, and 20% H_2O.

The nitration reaction is carried out with 1 part of purified cellulose (cotton linters) and about 50 parts of nitrating mixture. The rate of the reaction increases with temperature, but so does the rate of degradation of the cellulose (cleavage of the chains with corresponding drop in degree of polymerization). To control the latter reaction, the temperature is kept at 15–25°C during nitration, which is complete in about half an hour. The

cellulose nitrate is then transferred to a centrifuge where the nitrating mixture is removed and the material is washed, stabilized by alkaline solution, bleached, and finally washed with alcohol to remove the water. Even the plastic-grade material is never allowed to dry completely because of its instability and high flammability.

Fabrication The temperature sensitivity of cellulose nitrate excludes it from all fabrication methods involving heat, such as molding and extrusion. The plastic is handled by so-called "block" methods. Cellulose nitrate, alcohol, and camphor, the plasticizer always used, are mixed in a dough mixer until homogeneous. Here for the first time the fibrous shape of the original cellulose is lost. The colloid is filtered under pressure and rolled on hot rolls (65–80°C) where much of the alcohol is evaporated. It is then sheeted into slabs which are pressed while hot into a homogeneous block about 6″ × 5′ × 2′ in size.

Sheets of any desired thickness are subsequently sliced from the block. They must be aged by storage at about 30°C until the last traces of solvent have diffused out. This may take from 2 hr to 6 months, depending on the thickness of the sheet. The final operation consists of finishing the surface of the sheet by pressing in contact with polished metal sheets or by buffing.

Rods, tubes, and other shapes can be formed by solvent extrusion of the plasticized cellulose nitrate, followed by removal of solvent as above.

Properties and uses Despite its disadvantages of flammability, instability, and poor weathering properties, cellulose nitrate is one of the cheapest and most highly impact-resistant plastics. It still has many uses because of these properties, including fountain pens, tool and brush handles, spectacle frames, and some motion-picture film.

Only about 1 million lb of cellulose nitrate was used as a plastic in 1961, the amount decreasing steadily over the years.

GENERAL REFERENCES

Honeyman, J., *loc. cit.*
Miles, F. D., *Cellulose Nitrate*, Interscience Publishers, New York, 1955.
Moncrieff, R. W., *op. cit.*, Chaps. X, XI.
Ott, E., *et al.*, *loc. cit.*
Paist, W. D., *Cellulosics*, Reinhold Publishing Corp., New York, 1958.
Rånby, B. G., and S. A. Rydholm, *loc. cit.*

BIBLIOGRAPHY

1. Spurlin, H. M., *J. Am. Chem. Soc.* **61,** 2222 (1939).

D. Other Polymers Related to Cellulose

Cellulose ethers

The cellulose ethers are products of the reaction of an organic halide with cellulose swollen by contact with an aqueous base:

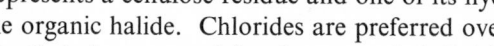

$$ROH + R'Cl + NaOH \longrightarrow ROR' + NaCl + H_2O$$

where ROH represents a cellulose residue and one of its hydroxyl groups, and R′Cl is the organic halide. Chlorides are preferred over bromides or iodides, despite their lower reactivity, because of their higher diffusion rate in the heterogeneous reaction system used.

The substitution of the ether groups onto the individual glucose units in the cellulose chain is, of course, a random process. At any over-all degree of substitution a mixture of nonsubstituted and mono-, di-, and trisubstituted glucose units is present. The relative amounts of these can be calculated if the equilibrium and rate constants for all the reactions are known. Some typical results are shown in Fig. 18-1. These considerations apply to the esterification as well as the etherification of cellulose.

Manufacture Cellulose ethers are made by reacting cellulose, alkali, and the organic chloride at about 100°C. The ether content and the viscosity of the product are controlled by temperature, pressure, reaction time, and composition of the reacting mixture.

Ethyl cellulose The most important of the cellulose ethers is ethyl cellulose. The commercial material has 2.4–2.5 ethoxy groups per glucose residue. It is a molding material which is heat stable and has low flammability and high impact strength. It is flexible and tough at temperatures as low as −40°C. Its electrical, mechanical, and weathering properties are good in comparison with those of other cellulosics, but not generally outstanding. Its softening point is low, and it has a high water absorption (but not as high as cellulose acetate) and some tendency to cold flow.

Other cellulose ethers Methyl cellulose is an edible, water-soluble material widely used as a thickening agent for foods. It serves also as an emulsifying agent and an adhesive. Carboxymethyl cellulose (available as the sodium salt) has important textile uses.

Alginate fiber

Certain seaweeds, mostly found off the coast of Scotland, contain up to 40% of alginic acid, polyanhydro-*d*-mannuronic acid, related to cellulose

by the replacement of the latter's —CH_2OH side chain with —COOH. The acid is extracted from the seaweed by treatment with aqueous sodium carbonate and is marketed as the sodium salt.

The calcium salt of alginic acid is insoluble, and fibers are formed by spinning solutions of the sodium salt into solutions of calcium chloride, in the presence of lubricants and wetting agents.

The properties of the alginate fiber are unusual. It is completely fireproof but is readily soluble in soap solutions and other alkalis. (Fibers insoluble in alkalis can be made from chromium or beryllium salts.) The fiber is used mainly for fireproof fabrics and for applications where solubility is needed, such as surgical dressings and draw threads in the knitting industry.

Lignin

About 30% of woody material consists of lignin, the binder in the growing plant. It is a polymer of various groups having the carbon skeleton of *n*-propylbenzene, linked through oxygen and possibly also carbon bonds. Large quantities of lignin are available as by-products in the manufacture of paper pulp. Little success has been achieved in utilizing the material as a plastic. It lacks the flow properties of a molding material and is used only as a binder or extender.

GENERAL REFERENCES

Honeyman, J., *loc. cit.*
Moncrieff, R. W., *op. cit.*, Chap. XV.
Ott, E., *et al.*, *loc. cit.*
Rånby, B. G., and S. A. Rydholm, *loc. cit.*

19

Thermosetting Resins

Thermosetting resins are those which change irreversibly under the influence of heat from a fusible and soluble material into one which is infusible and insoluble through the formation of a covalently crosslinked, thermally stable network. In contrast, *thermoplastic* polymers, discussed in Chapters 13–18, soften and flow when heat and pressure are applied, the changes being reversible.

In some of the systems considered in this chapter, network formation occurs with little or no heat required, as in the production of urethan foams and the use of unsaturated polyester resins. Furthermore, vulcanized rubbers consist of covalently crosslinked network polymers, usually formed by the application of heat. However, this network is generated in a separate postpolymerization step. With the exception of the silicone rubbers, thermosetting resins discussed in this chapter are those in which crosslinking occurs simultaneously with the final stages of polymerization, regardless of the amount of heat required in this step.

The most important thermosetting resins, both from a historical standpoint and in current commercial application, are formaldehyde condensation products with phenol (*phenolic resins*) or with urea or melamine (*amino resins*). Other thermosetting types include the epoxy resins, the unsaturated polyester resins, urethan foams, the alkyds widely used for surface coating, and minor types.

A. Phenolic Resins

Phenolic resins have been in commercial use longer than any other synthetic polymer except cellulose nitrate. In contrast to the latter, however, the production of phenolics continues to rise, with some fluctuations, being about 660 million lb in 1961, at a price of $0.21/lb.

Reactions of phenol and formaldehyde

Phenols react with aldehydes to give condensation products if there are free positions on the benzene ring *ortho* and *para* to the hydroxyl group. Formaldehyde is by far the most reactive aldehyde and is used almost exclusively in commercial production. The reaction is always catalyzed, either by acids or by bases. The nature of the product is greatly dependent on the type of catalyst and the mole ratio of the reactants.

Methylolation The first step in the reaction is the formation of addition compounds known as methylol derivatives, the reaction taking place at the *ortho* or *para* position:

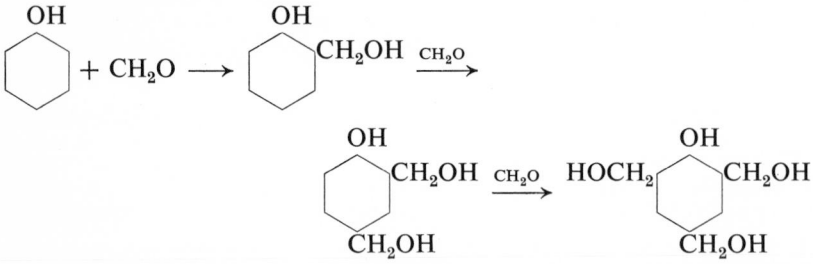

These products, which may be considered the monomers for subsequent polymerization, are formed most satisfactorily under neutral or alkaline conditions.

Novolac formation In the presence of acid catalysts, and with the mole ratio of formaldehyde to phenol less than 1, the methylol derivatives condense with phenol to form, first, dihydroxydiphenyl methane:

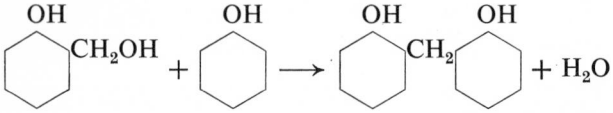

and, on further condensation and methylene bridge formation, fusible and soluble linear low polymers called *novolacs* with the structure

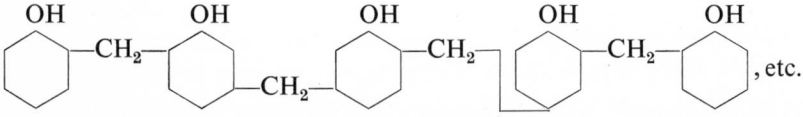

where *ortho* and *para* links occur at random. Molecular weights may range as high as 1000, corresponding to about ten phenyl residues. These materials do not themselves react further to give crosslinked resins, but

must be reacted with more formaldehyde to raise its mole ratio to phenol above unity.

Resole formation In the presence of alkaline catalysts and with more formaldehyde, the methylol phenols can condense either through methylene linkages or through ether linkages. In the latter case, subsequent loss of formaldehyde may occur with methylene bridge formation:

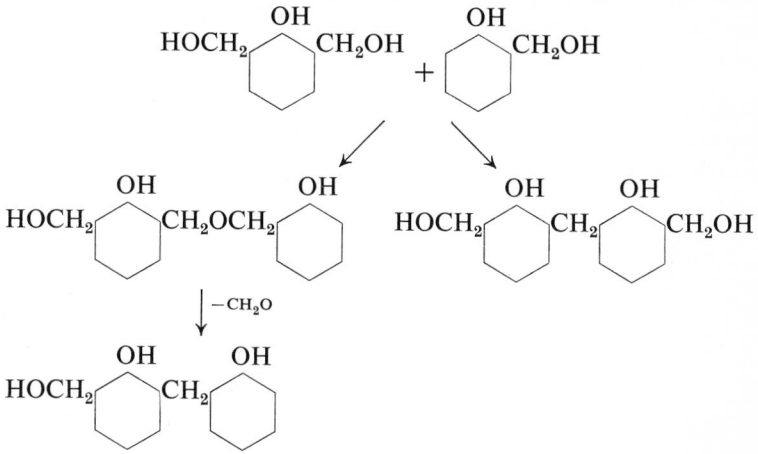

Products of this type, soluble and fusible but containing alcohol groups, are called *resoles*. If the reactions leading to their formation are carried further, large numbers of phenolic nuclei can condense to give network formation, the structure of the final product being like that of Fig. 19-1.

Summary The four major reactions in phenolic resin chemistry are as follows: (*a*) addition to give methylol phenols; (*b*) condensation of a methylol phenol and a phenol to give a methylene bridge; (*c*) condensation of two methylol groups to give an ether bridge; and (*d*) decomposition of ether bridges to methylene bridges and formaldehyde, the latter reacting again by the first mode.

Production of phenolic resins

The formation of resoles and novolacs, respectively, leads to the production of phenolic resins by *one-stage* and *two-stage* processes.

One-stage resin In the production of a one-stage phenolic resin, all the necessary reactants for the final polymer (phenol, formaldehyde, and catalyst) are charged into a resin kettle and reacted together. The ratio of formaldehyde to phenol is about 1.25:1, and an alkaline catalyst is used.

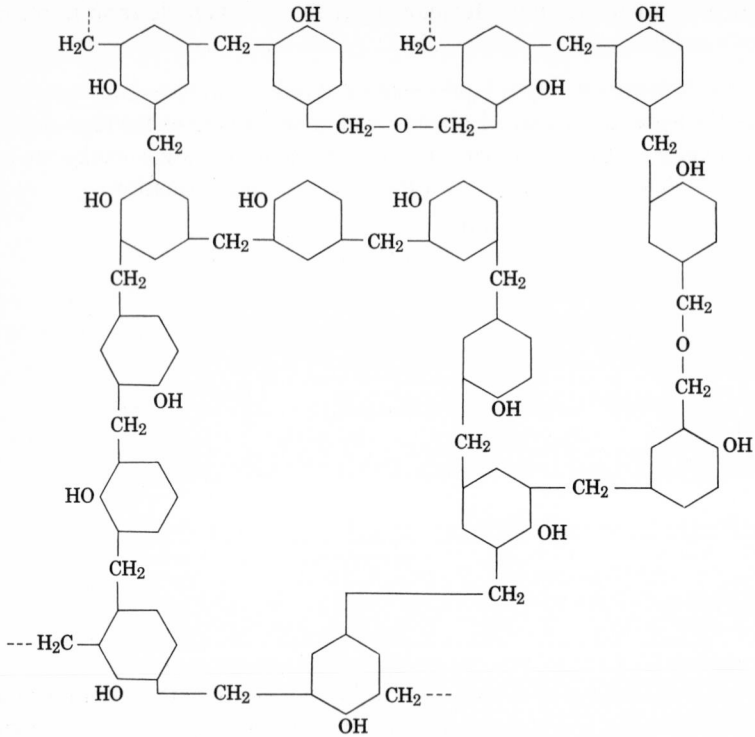

Fig. 19-1. Type of structure present in phenol-formaldehyde resin (Carswell, gen. ref.).

Two-stage resin These resins are made with an acid catalyst, and only part of the necessary formaldehyde is added to the kettle, producing a mole ratio of about 0.8:1. The rest is added later as hexamethylenetetramine (I):

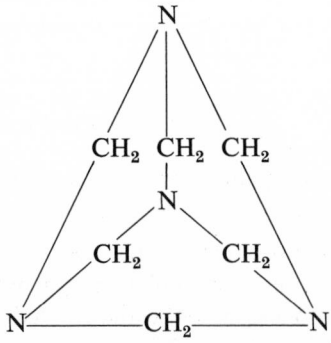

I

which decomposes in the final curing step, with heat and moisture present, to yield formaldehyde and ammonia which acts as the catalyst for curing.

Resin formation The procedures for one- and two-stage resins are similar, and the same equipment is used for both. A typical flow sheet for a two-stage process is shown in Fig. 19-2. The reaction is exothermic and cooling is required. The formation of a resole or a novolac is evidenced

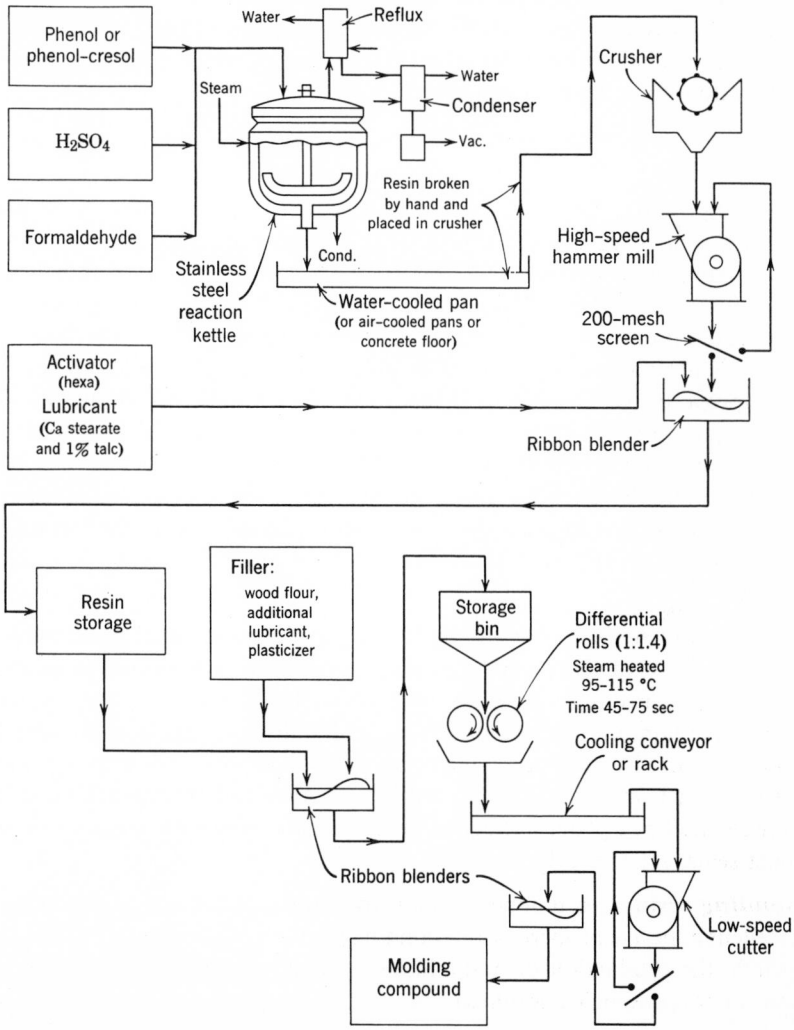

Fig. 19-2. Flow sheet for the production of phenol-formaldehyde molding powders (Winding and Hiatt, gen. ref.).

by an increase in viscosity. Water is then driven off under vacuum, and a thermoplastic *A-stage* resin, soluble in organic solvents, remains. This material is dumped from the kettle, cooled, and ground to a fine powder.

At this point fillers, colorants, lubricants, and (if a two-stage resin) enough hexamethylenetetramine to give a final formaldehyde:phenol mole ratio of 1.5:1 are added. The mixture is rolled on heated mixing rolls, where the reactions are carried further, to the point where the resin is in the *B-stage*, nearly insoluble in organic solvents but still fusible under heat and pressure. The resin is then cooled and cut into final form. The *C-stage*, the final, infusible, crosslinked polymer, is reached on subsequent fabrication—e.g., by molding.

Properties and applications

Molding resins Some 20–25% of the phenolic resins produced in this country is used in compression or transfer molding. The products are outstanding in heat resistance, dimensional stability, and resistance to cold flow. They are widely used for their good dielectric properties in electrical, automotive, radio and television, and appliance parts.

Fillers are almost always used in phenolic molding applications, both to improve properties and to reduce cost. Commonly chosen are wood flour, cotton flock, chopped rags, asbestos, fibrous glass and other fibers, and nitrile rubbers (see also Chapter 20*C*).

The ability of phenolic resins to withstand very high temperatures briefly is important for their use in missile nose cones. Whereas ceramics melt and metals vaporize, phenolics carbonize, maintaining a protective thermal barrier.

Laminating resins For impregnating paper, wood, and other fillers, about 15–20% of the phenolic resins is produced as alcoholic solutions of one-stage resins (essentially varnishes). These are used to produce decorative laminates for counter tops and wall coverings and industrial laminates for electrical parts, including printed circuits. The impregnated filler is dried in an air oven to remove volatiles and then is hot pressed between polished platens. Shaped products, including cutlery handles and toilet seats, are similarly made.

Bonding resins A number of industrial applications are based on the excellent adhesive properties and bonding strength of the phenolics. These include the production of brake linings, abrasive wheels and sandpaper, and foundry molds (sand-filled).

Casting resins Phenolic casting resins are unfilled and are made with specially purified reagents to keep the color light. A high mole ratio of formaldehyde to phenol (2.3:1) is used in an alkaline system. The highly

methylolated resoles thus formed remain dispersed in water, and 5–10% water is left in the honeylike syrup at the end of the reaction. Castings may be cured slowly by heat (2–8 days at 70–80°C) for the best physical properties, or rapidly by the addition of a little acid (1–6 hr at room temperature). Some glycerin is often added to reduce opalescence arising from the residual water content of the castings.

In addition to the temperature and flammability resistance of their class, the cast phenolics have the advantages of good machinability (because of their self-lubricating properties arising from the presence of glycerin and water) and availability in a full range of transparent or opaque colors. They are somewhat more expensive than phenolic molding compositions.

Coatings and adhesives Phenolic resins are widely used in varnishes, electrical insulation, and other protective coatings. Heat-setting adhesives based on phenolics are used in producing most plywood.

Other applications The *extrusion* of phenolic resins gives promise of items equivalent to those produced by laminating at 50–75% of the cost of the latter. Phenolics are widely used in the production of *ion-exchange resins*, with amine, sulfonic acid, hydroxyl, or phosphoric acid functional groups.

Related resins

Resorcinol-formaldehyde The higher reactivity of resorcinol compared to phenol leads to resins capable of rapid curing, but correspondingly more difficult to control. Their outstanding use is in wood adhesives.

Phenol-furfural Furfural, which in itself is a thermosetting resin by virtue of its aldehyde group and reactive furan ring, may replace part of the formaldehyde in phenolic resins. The resulting compounds have a long flow time before setting up at lower temperatures, coupled with rapid curing at higher temperatures. They are used in molding large pieces, such as television cabinets.

GENERAL REFERENCES

Carswell, T. S., *Phenoplasts*, Interscience Publishers, New York, 1947.
Gould, D. F., *Phenolic Resins*, Reinhold Publishing Corp., New York, 1959.
Martin, R. W., *The Chemistry of Phenolic Resins*, John Wiley & Sons, New York, 1956.
Megson, N. J. L., *Phenolic Resin Chemistry*, Academic Press, New York, 1958.
Suen, T. J., "Condensations with Formaldehyde," Chap. VIII in C. E. Schildknecht, ed., *Polymer Processes*, Interscience Publishers, New York, 1956.
Winding, C. C., and G. D. Hiatt, *Polymeric Materials*, McGraw-Hill Book Co., New York, 1961, Chap. 7.

B. Amino Resins

The two important classes of amino resins are the condensation products of urea and of melamine with formaldehyde. They are considered together here because of the similarity in their production and applications. In general, the melamine resins have somewhat better properties but are higher in price. Production of amino resins in the United States in 1961 was about 375 million lb, at \$0.32/lb for the urea resins and \$0.47 for the melamines.

Chemistry and production

Both melamine (I), a trimer of cyanamide, and urea react with formal-

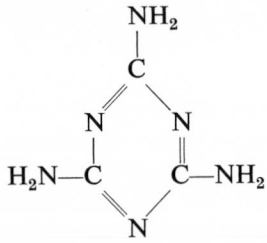

I

dehyde, first by addition to form methylol compounds and then by condensation in reactions much like those of phenol and formaldehyde. The methylol reaction takes the form

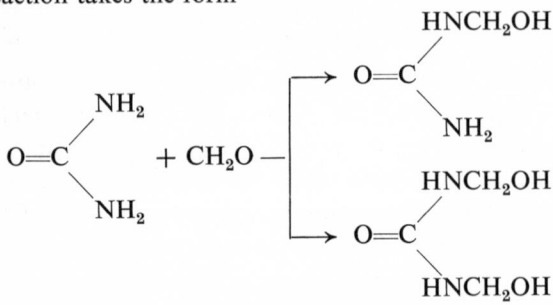

These may polymerize to crosslinked resins by loss of water. Other reactions which may enter into the condensation include the formation of methyleneurea

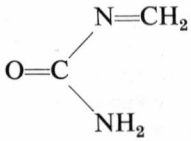

and the following crosslinking reactions (1):

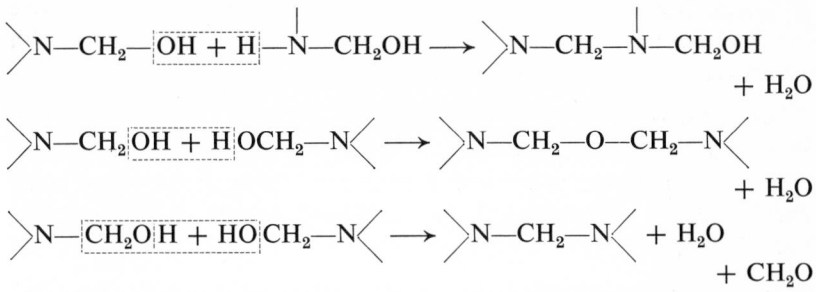

The production of the amino resins is similar to that of phenolic resins (Section *A*). Since the A-stage resin is water-soluble, it is only partially dehydrated, the water solution being used to impregnate the filler. The molding resins are almost always filled with cellulose obtained from good-quality sulfite-bleached paper. Impregnation is carried out in a vacuum mixer, and the subsequent drying step carries the resin to the B-stage. It is then ground to the desired particle size in ball mills.

Properties and applications

A distinct advantage of the amino resins over the phenolics is the fact that they are clear and colorless, so that objects of light or pastel color can be produced. The tensile strength and hardness of the amino resins are better than those of the phenolics, but their impact strength and heat and moisture resistance are lower, although still characteristic of thermosetting resins. The melamine resins have better hardness, heat resistance, and moisture resistance than the ureas.

Molding resins Practically all urea molding compounds are cellulose filled, whereas the melamines, although predominantly cellulose filled, are also used with asbestos, glass or silica, and cotton fabric. Because of their poorer flow characteristics, the urea resins are usually compression molded, but transfer molding is common with the melamines. Both resins can be preheated by high-frequency current because of their high polarity.

Because of their colorability, solvent and grease resistance, surface hardness, and mar resistance, the urea resins are widely used for cosmetic container closures, appliance housings, and stove hardware. The production of high-quality dinnerware from cellulose-filled compounds is the largest single use for the melamine resins. About one-third of the amino resins goes into molding applications.

Adhesives Roughly another third of the amino resins is used for adhesives, largely for plywood and furniture. The melamine resins give excellent, boil-resistant bonds, but for economy are usually blended with the ureas.

Laminating resins Melamine resins are widely used for the production of decorative laminates. These are usually assembled with a core of phenolic-impregnated paper and a melamine-impregnated overlay sheet. They are cured by hot pressing and are widely used for counter, cabinet, and table tops.

Other applications The amino resins modify *textiles* such as cotton and rayon by imparting crease resistance, stiffness, shrinkage control, fire retardance, and water repellency. They also improve the wet strength, rub resistance, and bursting strength of *paper*. Alkylated resins, in which butyl- or amyl-substituted monomethylol ureas or melamines are used, are combined with alkyd resins (Section *G*) to give *baking enamels*. The urea-based enamels are used for refrigerator and kitchen appliances, and the melamine formulations in automotive finishes.

Related resins Other amino resins include those based on ethyleneurea, used in textile applications: guanamine, in surface coatings; and aniline, in molded and laminated electrical insulation.

GENERAL REFERENCES

Blais, J. F., *Amino Resins*, Reinhold Publishing Corp., New York, 1959.
Suen, T. J., *loc. cit.*

BIBLIOGRAPHY

1. De Jong, J. I., and de Jonge, J., *Rec. trav. chim.* **72,** 139, 1027 (1953).

C. Unsaturated Polyester Resins

Polyesters of several diverse types are useful as polymers. This section is concerned only with those in which the dibasic acid or the glycol, or both, contain double bonded carbon atoms. It is further restricted to cases in which radical chain polymerization involving these double bonds and a vinyl monomer, usually styrene, is made to take place in the presence of a fibrous filler, generally glass. Other polyester systems are the fiber-forming saturated polyesters (Chapter 17*B*), the polyester intermediates

in the production of urethan elastomers (Chapter 17*C*) or foams (Section *E*), and the alkyd paint and molding resins (Section *G*). The polyester systems discussed here have been known from time to time as *low-pressure laminating resins*, *contact resins*, *polymerizable polyesters*, and *styrenated polyesters*. The term *reinforced plastics*, however, is more general and includes the use of thermosetting resins other than the unsaturated polyesters, among them the phenolic, epoxy, melamine, dialkyl phthalate, and alkyd resins, all discussed elsewhere in this chapter.

The production of unsaturated polyester resins for reinforced plastics use was approximately 190 million lb in 1960, at about $0.33/lb. About 60 million lb of other resins, mostly epoxy and phenolic, was used with fibrous glass to form reinforced plastic articles.

Chemistry of reinforced polyester systems

Much of the versatility of the reinforced polyester systems lies in the wide variation in resin composition and fabrication methods possible, allowing the properties of the product to be tailored to the requirements of the application. This variability is manifested in the large number of resin components used.

Dibasic acids The unsaturation in the polyester is usually supplied by the inclusion of maleic anhydride or fumaric acid as one component. In addition, a saturated acid or anhydride is often used, such as phthalic anhydride or adipic, azelaic, or isophthalic acid. A higher proportion of unsaturated acid gives a more reactive resin, with improved stiffness at high temperatures, while more of the saturated components give less exothermic cures and less stiff resins, particularly if the aliphatic acids are used.

Dihydric alcohol Ethylene and propylene glycols are perhaps most popular, but 1, 3- and 2,3-butylene, diethylene, and dipropylene glycols are also common.

Monomer Styrene is by far the most widely used monomer in these systems. Others often encountered are vinyl toluene, methyl methacrylate (leading to improved weatherability), diallyl phthalate (often preferred in molding compounds), and triallyl cyanurate (imparting good heat resistance).

Formulation The ingredients of the polyester resin (a typical formulation is given in Table 19-1) are mixed in a resin kettle and polymerized by step reaction to a molecular weight of 1000–5000, which is in the highly

viscous liquid range. After cooling, the mixture is thinned down to a pourable liquid by the addition of the monomer. An inhibitor such as hydroquinone is then added to prevent premature polymerization. When kept cool, the mixture is stable for months to years.

TABLE 19-1. Typical formulation of an unsaturated polyester resin*

Mole	Ingredient	Pounds/100 lb Resin
0.2	Phthalic anhydride	28.86
0.2	Maleic anhydride	19.11
0.2	Propylene glycol	14.83
0.2	Ethylene glycol	12.10
0.3	Styrene	30.00
Trace	Hydroquinone	0.02
		104.92†

* Source: *Chem. Eng. News* **37**, #51, 56 (Dec. 21, 1959).
† About 5 lb of water is eliminated during esterification.

Cure is begun by adding an initiator, usually an organic peroxide, such as benzoyl peroxide, or a hydroperoxide. Typically, promoters or accelerators are used to promote the decomposition of the initiator at room temperature and, thus, rapid low-temperature curing. Common accelerators are cobalt naphthenate or alkyl mercaptans. Cure takes place in two stages: the initial formation of a soft gel is followed by rapid polymerization. The heat evolved can lead to quite high temperatures in large masses of resin.

Fabrication

The fabrication of reinforced plastic articles is usually divided into laminating and molding processes, but the techniques are in some cases quite similar.

Laminating Laminating processes are those in which separate plies of reinforcing material are coated or impregnated with resin and pressed together until cured to a single reinforced structure. The operation can be done batchwise or continuously.

Molding The simplest molding process for reinforced plastics is *contact molding* or open or hand lay-up molding, a process closely resembling laminating. A single mold is used; the reinforcing material is placed on the mold, impregnated with resin, and allowed to cure in air. Variations

include *bag molding*, where a bag or blanket is used to apply low pressure to the open surface of the material, and *spray-up* techniques, in which the resin and sometimes chopped glass are sprayed onto the mold surface.

Matched-die molding includes processes involving two dies which, together, match or conform to all the dimensions of the finished piece. The material can be placed in the die in the form of mats or preforms of reinforcing material with the resin separately added; "pre-pregs," where the preform has been impregnated; or premixes, where the resin, reinforcement, and other extenders have been premixed.

Other fabrication techniques include continuous extrusion, filament winding, casting, and foaming.

Properties and applications

The most important properties of the unsaturated polyester systems include ease of handling, rapid curing with no volatiles evolved, light color, dimensional stability, and generally good physical and electrical properties.

The major applications of glass-reinforced polyester resins fall in the following categories: boat hulls, whose popularity has risen spectacularly; transportation, including passenger car parts and bodies and truck cabs; consumer products, including such diverse items as luggage, chairs, and fishing rods; trays, pipe, and ducts; electrical applications; appliances; construction applications, largely sheet and paneling; and missile and radome uses.

GENERAL REFERENCES

Bjorksten, J., H. Tovey, B. Harker, and J. Henning, *Polyesters and Their Applications*, Reinhold Publishing Corp., New York, 1956.

DeDáni, A., ed., *Glass Fiber Reinforced Plastics*, Interscience Publishers, New York, 1961.

Lawrence, J. R., *Polyester Resins*, Reinhold Publishing Corp., New York, 1960.

D. Epoxy Resins

The epoxy resins are fundamentally polyethers, but retain their name on the basis of their starting material and the presence of epoxide groups in the polymer before crosslinking.

Chemistry of preparation and curing

The epoxy resin used most widely is made by condensing epichlorohydrin with bisphenol A, diphenylol propane. An excess of epichlorohydrin

is used, to leave epoxy groups on each end of the low-molecular-weight (900–3000) polymer:

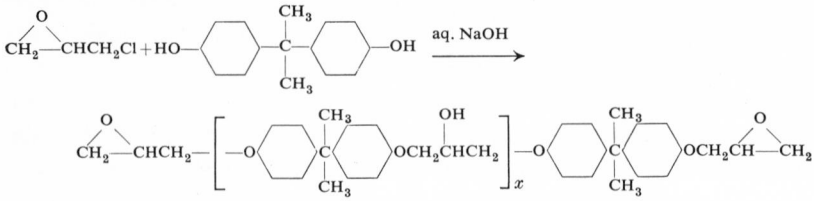

Depending on molecular weight, the polymer is a viscous liquid or a brittle high-melting solid.

Other hydroxyl-containing compounds, including resorcinol, hydroquinone, glycols, and glycerol, can replace bisphenol A. No epoxides other than epichlorohydrin are available at attractive prices, however.

The epoxy resins are cured by many types of materials, including polyamines, polyamides, polysulfides, urea- and phenol-formaldehyde, and acids or acid anhydrides, through coupling or condensation reactions. The reaction with amines involves opening the epoxide ring to give a β-hydroxyamino linkage:

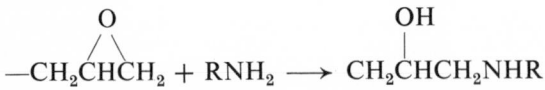

Acids and acid anhydrides react through esterification of the secondary hydroxyl groups on the epoxy resin as well as with the epoxide groups. The phenolic and amino resins may react in several ways, including condensation of methylol groups with the secondary hydroxyls of the epoxy, and reaction of the epoxide groups with phenolic hydroxyls and amino groups.

Epoxy resins can also be cured by cationic polymerization, using Lewis acid catalysts such as BF_3 and its complexes, which form polyethers from the epoxide groups.

Properties and applications

The major use of the epoxy resins is as surface-coating materials, which combine toughness, flexibility, adhesion, and chemical resistance to a nearly unparalleled degree. In addition to curing systems of the types described above, the epoxies can be esterified with drying or nondrying oil fatty acids and then cured by either air drying or baking.

Epoxy resins can be used in both molding and laminating techniques to make glass fiber-reinforced articles with better mechanical strength, chemical resistance, and electrical insulating properties than those obtained with

the unsaturated polyesters. Only the higher price (about $0.65/lb) of the epoxies prevents their wider use in this field.

Casting, potting, encapsulation, and embedment are widely practiced with the epoxy resins in the electrical and tooling industries. Liquid resins are often utilized, while hot-melt solids have some application.

Other important uses include industrial floorings, adhesives and solders, foams, highway surfacing and patching materials, and stabilizers for vinyl resins.

GENERAL REFERENCES

Gaylord, N. G., ed., *Polyethers*, Part 2, Interscience Publishers, a Division of John Wiley & Sons, New York, 1963.
Lee, H., and K. Neville, *Epoxy Resins*, McGraw-Hill Book Co., New York, 1957.
Skeist, I., *Epoxy Resins*, Reinhold Publishing Corp., New York, 1958.
Stivala, S. S., "Epoxy Resins," Chap. X in C. E. Schildknecht, ed., *op. cit.*

E. Urethan Foams

As indicated in Chapter 17*C*, urethan polymers contain the group —NHCOO—, and are formed through the reaction of a diisocyanate and a glycol. In the production of urethan foams, excess isocyanate groups in the polymer react with water or carboxylic acids to produce carbon dioxide, blowing the foam, at the same time that crosslinking is effected.

Urethan foams can be made in either flexible or rigid form, depending on the nature of the polymer and the type of crosslinking produced. Their popularity has increased very rapidly; production rose from 35 million lb in 1958 to 100 million lb of flexible and 15 million lb of rigid foams in 1961. At the same time, the raw materials cost per pound of foam dropped from $0.46 to $0.37.

Chemistry of urethan foams

Like polyurethan elastomers, urethan foams are made in several steps. A *basic intermediate* of molecular weight around 1000 is, traditionally, a hydroxyl-terminated polyester made from ethylene glycol and adipic acid. More recently, the trend has been strongly toward polyether (polyglycol) intermediates made from poly(1,4-butylene glycol), sorbitol polyethers, or others. The basic intermediate is bifunctional if flexible foams are desired and polyfunctional if rigid foams are to be made.

As in elastomer production, the intermediate is reacted with an aromatic diisocyanate, usually tolylenediisocyanate, to give a *prepolymer* of structure similar to that illustrated in Chapter 17*C*. In foam formation, however, the crosslinking step is traditionally carried out by adding water:

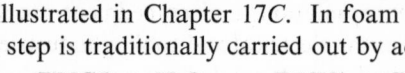

$$RNCO + H_2O \longrightarrow RNH_2 + CO_2$$
$$R'NCO + RNH_2 \longrightarrow RNHCONHR',$$

crosslinking taking place through the formation of urea linkages.

With polyether intermediates, whose activity is lower than that of the polyesters, it is necessary to add catalysts based on tertiary amines or on stannous soaps to achieve rapid production of foam. The use of low-boiling inert liquids, in particular fluorocarbons, to augment or replace the chemical blowing action described above has led to certain property advantages in the final foams, such as low thermal conductivity characteristic of the entrapped fluorocarbon gas. The ingredients of the foam may be partly expanded (frothed) with an inert gas in a preliminary step, and subsequently expanded to form the final object in a separate second step.

Properties and applications

Flexible foams The use of flexible urethan foams for cushions for furniture and automobiles increased greatly in 1958–60; urethan foams have displaced rubber foam in these applications because of improved strength, lower density, and easier fabrication.

The "one-shot" process, wherein polyether intermediate, tolylenediisocyanate, and catalysts are mixed just before foaming, offers several advantages for molding urethan foams. The mix flows enough to aid in covering the mold area before foaming starts. Since the foam does not collapse under pressure even when newly formed, it will flow laterally to fill corners, etc., before setting. The greater amount of heat evolved, compared to prepolymer systems, shortens curing time. The tendency for the foam to stick to the mold is controlled by adding waxes as lubricants.

Rigid foams Rigid urethan foams are resistant to compression and may be used to reinforce hollow structural units with a minimum of weight. In addition, they consist of closed cells and so have low rates of heat transmission. They develop excellent adhesion when they are formed in voids or between sheets of material. Finally, they are resistant to oils and gasoline and do not absorb appreciable amounts of water. These properties make the rigid foams valuable for sandwich structures used in prefabrication in the building industry, for thermal insulation in refrigerators, portable insulated chests, etc., and for imparting buoyancy to boats.

GENERAL REFERENCES

Arnold, R. G., J. A. Nelson, and J. J. Verbanc, *J. Chem. Educ.* **34,** 158 (1957).
Dombrow, B. A., *Polyurethanes*, Reinhold Publishing Corp., New York, 1957.
Saunders, J. H., *Rubber Chem. Tech.* **33,** 1259, 1293 (1960).

F. Silicone Polymers

Like carbon, silicon has the capability of forming covalent compounds. Silicon hydrides (silanes) up to Si_6H_{14} are known. The Si-Si chain becomes thermally unstable at about this length, however, so that polymeric silanes are unknown. The siloxane link:

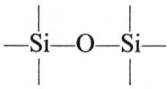

is more stable, and is the one found in commercial silicone polymers. Unlike carbon, silicon does not form double or triple bonds. Thus silicone polymers can be formed only by condensation-type reactions.

Silicone polymers became available commercially during World War II. They are particularly noted for their stability at temperatures as high as 150°C. The variety of products available ranges from liquids (lubricants, water repellents, release agents, defoamers) through greases and waxes to resins and rubbers.

Chemistry of the silicones

The study of the chemistry of silicon and its compounds began with the discovery of the element in 1824. Soon afterwards $SiCl_4$ was prepared by reacting silicon and chlorine, and ethyl silicate was first made in 1844 by the reaction of the tetrachloride with ethanol. The compounds of silicon were studied intensively from that time on.

Silicone polymers are made from organosilicon intermediates prepared in various ways from elemental silicon, which is produced by reducing quartz in an electric furnace.

The intermediates ("monomers") are compounds of the type SiR_nX_{4-n}, where R is an alkyl or aryl group and X is a group which can be hydrolyzed to —SiOH, such as chlorine or alkoxy. The intermediates can be formed in two ways.

a. Direct synthesis. In the direct synthesis of the intermediates, the R and X groups are attached simultaneously to the silicon by a high-temperature reaction of a halide with silicon in the presence of a metal catalyst. The chief reaction is, e.g.,

$$2CH_3Cl + Si \longrightarrow Si(CH_3)_2Cl_2$$

but a number of side reactions occur.

b. Coupling. Coupling reactions involve the alkylation of $SiCl_4$ or $Si(OC_2H_5)_4$ by Wurtz or Grignard reactions:

$$\begin{array}{c}|\\ -\text{Si}\text{OR} \\ |\end{array} + \text{R}'\text{Na} \longrightarrow \begin{array}{c}|\\ -\text{Si}\text{R}' \\ |\end{array} + \text{RONa}$$

$$\begin{array}{c}|\\ -\text{Si}\text{OR} \\ |\end{array} + \text{R}'\text{MgCl} \longrightarrow \begin{array}{c}|\\ -\text{Si}\text{R}' \\ |\end{array} + \text{MgORCl}$$

Polymerization

Silicone polymers are produced by intermolecular condensation of silanols, which are formed from the halide or alkoxy intermediates by hydrolysis:

$$\begin{array}{c}|\\ -\text{Si}\text{Cl} \\ |\end{array} + H_2O \longrightarrow \begin{array}{c}|\\ -\text{Si}\text{OH} \\ |\end{array} + \text{HCl}$$

$$\begin{array}{c}|\\ -\text{Si}\text{OH} \\ |\end{array} + \begin{array}{c}|\\ \text{HOSi}- \\ |\end{array} \longrightarrow \begin{array}{c}|\\ -\text{Si}- \\ |\end{array}\text{O}\begin{array}{c}|\\ -\text{Si}- \\ |\end{array} + H_2O$$

The desired siloxane structure is obtained by using silanols of different functionality, the alkyl (R) groups in the intermediate being unreactive.

Linear low polymers of methyl silicone are widely used as lubricating oils and greases. High-molecular-weight linear polymers show the properties of elastomers. There are no useful thermoplastic polymeric silicones because of the very small change in viscosity of the compounds with temperature. Silicone resins are crosslinked, thermoset structures, containing less than two alkyl groups per atom of silicon.

Silicone resins are polymerized in two ways:

a. Intermediates with lower functionality than that desired in the final resin may be treated to increase their functionality. The resulting soluble polymer can be oxidized by bubbling air through it at 200–300°C to give a soluble high-viscosity product. Instead, aryl groups can be removed by heating the silicone at about 180°C with water or hydrogen chloride present. In either case the product can be crosslinked by further heating.

b. Intermediates having the same functionality as the finished resin can be polymerized. If such intermediates are hydrolyzed directly, they form gels which cannot be processed. To avoid this, the hydrolysis is carried out in dilute solution. The intermediate polymer remains dissolved in the solvent (usually toluene) and is brought to the desired viscosity by mild heat. Final curing is carried out at 150–250°C.

Silicone resins

Coating resins Silicones are used in surface coatings to provide heat resistance, chemical resistance, and water repellency. The silicone resin may be mixed with the other resins in the coating, or the silicone intermediate may replace a part of the fatty acid used to make the coating resin. For the best heat resistance (up to 250°C), the silicone resin may be used alone or with only a drying agent to accelerate its cure.

Laminating resins Silicone laminates are of interest because of their high temperature stability and good electrical properties. Since high-temperature applications are most important, glass fiber is the usual filler for the laminates. The physical properties of the silicone laminates do not equal those of organic laminates at room temperature, but remain relatively constant as temperature is increased.

Release resins The properties which make liquid silicones useful in preparing nonstick surfaces (as in mold-release agents) carry over to the resins. The most common application of silicone resins as release agents is in baking bread. Baking pans are spray- or dip-coated with the resin and baked to cure it. They can subsequently be used hundreds of times without greasing.

Water-repellent coatings Silicone resins make masonry and concrete water repellent. They may be applied in several forms: e.g., ethyl silicate may be used, hydrolyzing to a silicone; or a fully condensed resin may be applied and allowed to set up by loss of solvent.

Molding resins Silicone resins do not have sufficient strength to be used unfilled. As in the production of laminates, inorganic fillers are usually preferred in molding applications. Compression or transfer molding techniques are utilized. The products find most use where high heat and moisture resistance is required.

Foaming resins When mixed with foaming agents, silicone resins can be made to expand on heating to rigid, low-density foams, which may be prepared in place in the final structure. They form a light-weight, heat-stable thermal insulation with good dielectric properties, useful for aircraft and missile applications.

Potting and encapsulating Fluid silicone resins are used where a heat-stable, high dielectric encapsulating or potting liquid is required to protect electrical or electronic parts. A transparent jellylike coating can be made which is self-healing, permitting electrical probes to be inserted and removed without damage. Room-temperature-curing silicone rubbers may also be used for potting and encapsulating.

Silicone rubbers

By careful elimination of impurities which would lead to crosslinking or limit molecular weight, silicone polymers with linear chains of the type

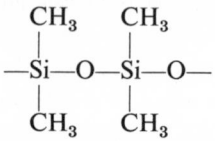

have been produced with number-average molecular weights in the millions. These polymers possess elastomeric properties and have good temperature resistance up to 150°C. Vulcanization is carried out by means of peroxide initiators. Finely divided silica must be utilized as a reinforcing filler if useful properties are to be obtained. The silicone rubbers are outstanding in low-temperature flexibility (to −90°C), weather resistance, and resistance to lubricating oils, in addition to stability at high temperatures.

Other types of silicone rubbers include: (*a*) those with vinyl groups substituted for a few of the methyl groups, which have improved vulcanization properties and are gradually replacing the all-methylated rubbers; (*b*) those with phenyl groups replacing some methyls, which have lower brittle points; and (*c*) those with cyanoalkyl groups replacing 5–50% of the methyls, having good low-temperature flexibility and good oil and solvent resistance (1).

GENERAL REFERENCES

Eaborn, C., *Organosilicon Compounds*, Academic Press, New York, 1960.

McGregor, R. R., *Silicones and Their Uses*, McGraw-Hill Book Co., New York, 1954.

Meals, R. N., and F. M. Lewis, *Silicones*, Reinhold Publishing Corp., New York, 1959.

Montermoso, J. C., "Silicone Rubbers," Chap. 16 in M. Morton, ed., *Introduction to Rubber Technology*, Reinhold Publishing Corp., New York, 1959.

Rochow, E. G., *An Introduction to the Chemistry of the Silicones*, 2nd ed., John Wiley & Sons, New York, 1951.

BIBLIOGRAPHY

1. Williams, T. C., R. A. Pike, and F. Fekete, *Ind. Eng. Chem.* **51**, 939 (1959).

G. Miscellaneous Thermosetting Resins

Alkyd resins

Alkyd resins (the name deriving from *al*cohol + a*cid*) are polyesters used primarily in organic paints, with some molding applications. About 600 million lb was produced for paints, varnishes, and lacquers in 1960.

Among many possible compositions for alkyd resins, perhaps the most common is based on phthalic anhydride and glycerol ("glyptal"). Other polyhydric alcohols commonly used are glycols, pentaerythritol, and sorbitol; other acids include maleic anhydride, isophthalic, and terephthalic.

Many alkyd resins are modified by the addition of fatty acids derived from mineral and vegetable oils. If the acids are unsaturated, the resulting resins are the air-drying type which cure by oxidation of the (saponified) acids (*drying oils*). Baking type alkyds cure by heat alone, or by cocondensation with alkylated amino resins.

Surface coatings may be classified as *lacquer* types, in which drying involves only the evaporation of solvent, and *varnish* types, in which chemical reactions also take place on drying. These reactions may involve free radical crosslinking of drying oils, the triglycerides of unsaturated acids such as oleic, linoleic, and ricinoleic.

Alkyd resins are applied in surface coating uses as follows.

Plasticizing resins These alkyds are of the glycerol-phthalic acid or glycerol-sebacic acid type and are used in lacquers along with natural resins such as shellac to impart flexibility—a plasticizing action.

Drying resins Drying alkyds contain, in addition to the glycerol phthalate components, some drying oils or acids of drying oils. They are, of course, used in varnish-type coatings. Here phenol-aldehyde or, better, urea- or melamine-aldehyde resins may be added to improve hardness.

Hard resins The hard resins are usually made from maleic anhydride combined with glycerol and rosin. Their function is to improve film hardness and gloss in both lacquer and varnish coatings.

Allyl resins

Although allylic monomers are not used directly in chain-reaction or addition polymerization because of the stability and consequent low reactivity of the allyl radical, diallyl esters can be crosslinked by polymerization through their double bonds to give thermosetting resins. Two major types are of commercial interest.

Diallyl phthalate Prepolymers (i.e., partially polymerized but still thermoplastic resins) of diallyl phthalate and diallyl isophthalate are used as molding compounds and in the production of glass fiber-reinforced laminates. They are cured by peroxide catalysts to heat- and chemical-resistant products with good dimensional stability and electrical insulating properties.

Allyl diglycol carbonate The ester diethylene glycol bisallyl carbonate, $(CH_2{=}CHCH_2OCOOCH_2CH_2)_2O$, is used directly for casting clear glass-like products similar to cast poly(methyl methacrylate) but much harder. The castings are cured with peroxide-type catalysts. They are used for special glazing applications and as spectacle lenses and for other optical purposes.

Cold-molded plastics

Cold-molded plastics are highly filled mixtures using either bituminous varnishlike resin or a phenolic resin as a binder. They are compression molded by pressure alone, and subsequently cured by heat without the application of pressure to crosslink the binder. The usual filler is asbestos fiber, and the resins are used in high-temperature-resisting electrical parts for which surface texture and exact dimensions are not critical.

Furan resins

Furan resins are based on furfuraldehyde,

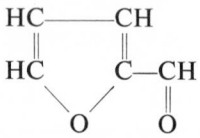

which is derived from waste vegetable matter. It is used in a thermosetting resin in combination with phenol, or converted to furfuryl alcohol, which itself can be thermoset by acids, or reacted with aldehydes or ketones to give polymerizable products. The resins are all dark in color but are strong, have good chemical resistance, and penetrate porous surfaces well.

GENERAL REFERENCES

Martens, C. R., *Alkyd Resins*, Reinhold Publishing Corp., New York, 1961.

Patton, T. C., *Alkyd Resin Technology*, Interscience Publishers, a Division of John Wiley & Sons, New York, 1962.

V

Polymer Processing

20

Plastics Technology

As the earlier sections of this book have suggested, the unique physical and mechanical properties of polymers are responsible in large part for their important place in our modern life. But these properties can be developed and utilized only by fabricating the polymer into useful articles or shapes. The methods by which this fabrication is carried out are described in this chapter.

Fabrication methods are largely determined by the rheological properties of the polymer in question. A primary consideration is whether the material is *thermoplastic*, i.e., retains the ability to flow at elevated temperatures for relatively long times, or *thermosetting*, i.e., subject to (controlled) crosslinking reactions at the temperatures necessary to induce flow, so that the ability to flow is rather quickly lost in favor of form stability. Other considerations of importance in selecting fabrication methods are softening temperature, stability, and, of course, the size and shape of the end product.

In this chapter only fabrication methods pertinent to plastics as distinguished from fibers or rubbers are considered. Techniques specific to the latter classes of material are discussed in Chapters 21 and 22.

A. Molding

Compression molding

Molding processes are those in which a finely divided plastic is forced by the application of heat and pressure to flow into, fill, and conform to the shape of a cavity (mold). One of the oldest methods of polymer processing is *compression molding*. Here the polymer is put between stationary and moveable members of a mold (Fig. 20-1). The mold is closed, and heat

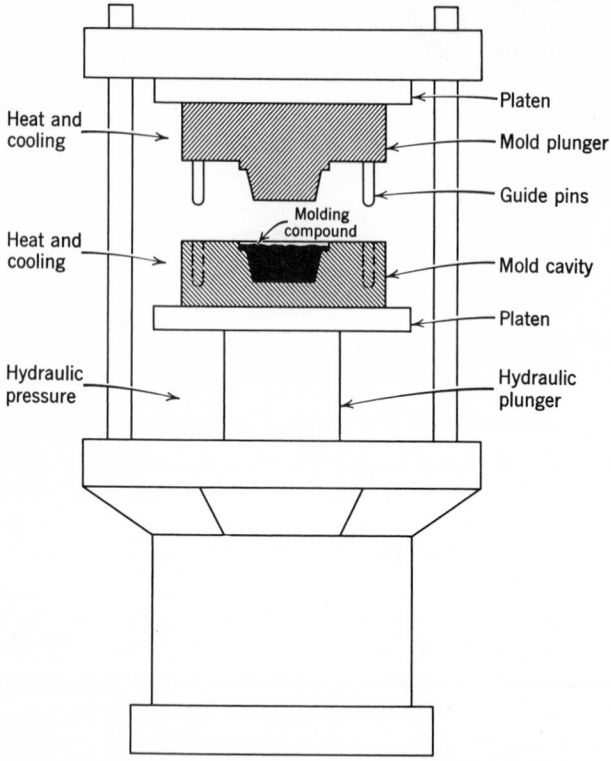

Fig. 20-1. Diagram of a compression-molding press and mold.

and pressure are applied so that the material becomes plastic, flows to fill the mold, and becomes a homogeneous mass. The necessary pressure and temperature vary considerably depending upon the thermal and rheological properties of the polymer. For a typical compression-molding material they may be near 150°C and 1000–3000 psi. A slight excess of material is usually placed in the mold to insure its being completely filled. The rest of the polymer is squeezed out between the mating surfaces of the mold in a thin, easily removed film known as *flash*.

If a thermoplastic material is being molded, the mold is cooled, the pressure released, and the molded article removed. If a thermosetting material is used, the mold need not be cooled at the end of the molding operation or *cycle*, as the polymer will have set up and can no longer flow or distort.

A variation of compression molding suitable for some thermosetting materials is *cold molding* or *preforming*, in which the polymer mix is formed into shape by the application of pressure alone. It is then baked in an

oven to give the crosslinked product. Another variation is *transfer molding*, in which the thermosetting polymer is heated hot enough to flow but not to crosslink in a premolding chamber and then is transferred to a hot mold, where crosslinking takes place at a higher temperature and under pressure.

Injection molding

Most thermoplastic materials are molded by the process of *injection molding*. Here the polymer is preheated in a cylindrical chamber to a temperature at which it will flow and then is forced into a relatively cold, closed mold cavity by means of quite high pressures applied hydraulically through a plunger (Fig. 20-2). Injection-molding temperatures are higher than those for compression molding, rising well above 250°C for many materials. Pressures applied to the plunger may range from 10,000 to 30,000 psi. An outstanding feature of injection molding is the speed with which finished articles can be produced. Cycle times of 10–30 sec are common. The method has been used in the past for small articles, with a few ounces of polymer for a single molding. Today, however, large objects such as television and radio cabinets, and door liners for refrigerators, weighing many pounds, are being successfully injection molded.

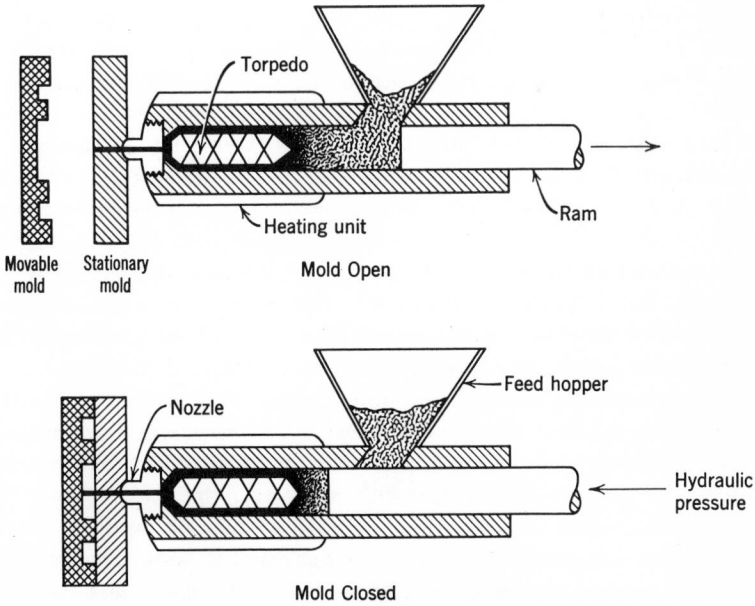

Fig. 20-2. Diagram of an injection-molding machine.

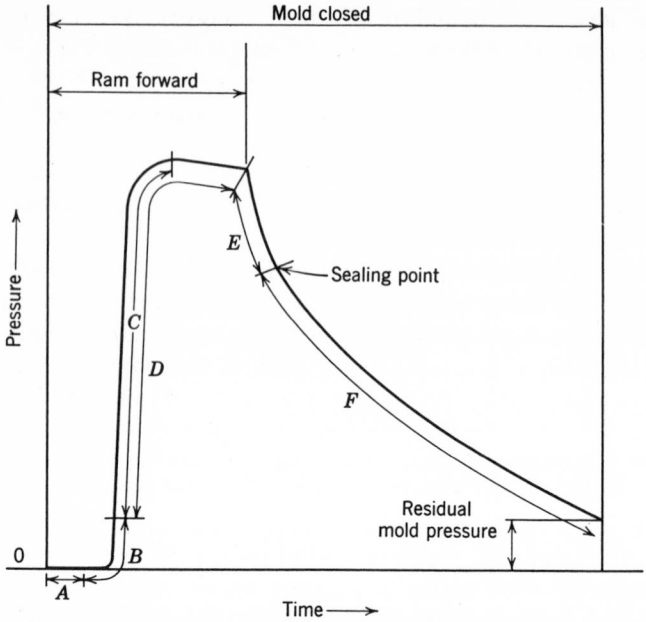

Fig. 20-3. Injection-molding cycle in terms of pressure applied to the polymer in the mold (1). For explanation, see text.

Although the development of injection-molding techniques over the past 25 years or so has been mostly empirical, some studies have been made of the fundamental principles of the process. One of the initial results was an analysis of the molding cycle in terms of the pressure applied to the polymer in the mold (1). The cycle (Fig. 20-3) begins at zero time with the closing of the mold. Before this time (*A*, Fig. 20-3) the ram or plunger has been retracted to allow a charge of molding powder to drop into the rear part of the cylinder. During part *B* of the cycle the ram is moving forward, pushing hot molten polymer into the mold. As the mold fills (*C*, *D*) the pressure rises. Full ram pressure (less frictional losses) is applied while the polymer cools to "pack" more material into the mold and reduce shrinkage of the molded article. If the ram is withdrawn before the mold is sealed by solidification of polymer at its *gate* or entrance, the pressure may drop somewhat (*E*) because of discharge of polymer from the mold. Usually, however, the ram is held forward until solidification occurs. Finally the pressure drops as the polymer cools (*F*), reaching some small residual pressure at the time the mold is opened and the piece removed. The quality of the molded piece is largely controlled by the time, temperature, and pressure variables of this cycle and by such properties of the polymer as thermal expansion and compressibility.

- *Jet molding* is a modification of injection molding which is adaptable to both thermoplastic and thermosetting resins. In this process, after the resin is warmed in a cylinder to incipient fusion, it is forced under very high pressure by a plunger through a very hot nozzle into the mold. The nozzle is heated rapidly by induction at the instant of injection, and then chilled equally rapidly at the completion of the injection stroke. By this means, the resin is almost instantaneously heated to very high temperature immediately before injection, forced into the mold, and allowed to cool. Thermosetting materials can, by proper choice of conditions, be given time to crosslink before cooling.

Plastication

Unlike the typical extruder (Section *B*), the usual injection-molding machine does not provide much mixing of the resin between melting and injection. Such mixing would be advantageous in improving the homogeneity of the melt and of the molded piece, and in eliminating the need for prior compounding steps.

In 1960–61, several devices (2, 3, 4) for mixing or plasticating the melt in an injection-molding cylinder gained considerable popularity. One is a *rotating spreader* (2), in which the spreader or *torpedo* (Fig. 20-2) is fitted with blades and rotated by means of a shaft passing through the ram to an external drive. A second technique is to replace the ram with a movable screw, much like an extruder screw, which runs during part of the cycle to provide plastication and then moves forward to deliver the molten polymer to the die. In a further variation, an essentially conventional extruder feeds plasticated molten polymer into the chamber of an essentially conventional injection-molding machine ahead of the ram. The feed can be from the top or the side.

GENERAL REFERENCES

Garvin, G. S., "Compounding and Processing Rubbers and Resins," Chap. XVI in C. E. Schildknecht, ed., *Polymer Processes*, Interscience Publishers, New York, 1956.

McKelvey, J. M., *Polymer Processing*, John Wiley & Sons, New York, 1962.

Meeks, P. J., "Injection Molding," Chap. 21 in A. Renfrew and P. Morgan, eds., *Polythene*, 2nd ed., Interscience Publishers, New York, 1960.

Modern Plastics Magazine, Encyclopedia Issue, Breskin Publications, New York, annually.

Thayer, G. B., J. W. Mighton, R. B. Dahl, and C. E. Beyer, "Injection Molding," Chap. 5 in E. C. Bernhardt, ed., *Processing of Thermoplastic Materials*, Reinhold Publishing Corp., New York, 1959.

BIBLIOGRAPHY

1. Gilmore, G. D., and R. S. Spencer, *Modern Plastics* **27,** #8, 143 (April, 1950); **28,** #8, 117 (April, 1951).
2. Keiser, N., L. H. Cirker, and P. D. Kohl, *SPE J.* **17,** 340 (1961).
3. Maier, R., and A. Thierstein, *SPE J.* **17,** 331 (1961).
4. Morse, A. R., *SPE J.* **17,** 539 (1961).

B. Other Processing Methods

Blowing

Although the blowing of plastics has evolved from standard glass-blowing techniques, procedures have been markedly modified to take advantage of the unique properties of thermoplastic polymers.

Blow molding In this operation, a section of tubing (*parison*) is extruded into an open mold (Fig. 20-4). By means of compressed air or steam the plastic is then blown into the configuration of the mold. This technique is widely used for the manufacture of bottles and similar articles.

Blowing film and tubing Continuous tubing can be made by extruding the plastic through an annular die, and maintaining gas pressure within the tube. In this fashion, rather thin-walled, uniform-diameter tubing can be manufactured at a rapid rate. By properly controlling the air rate, it is possible to increase the tube diameter to a point where the tube is essentially

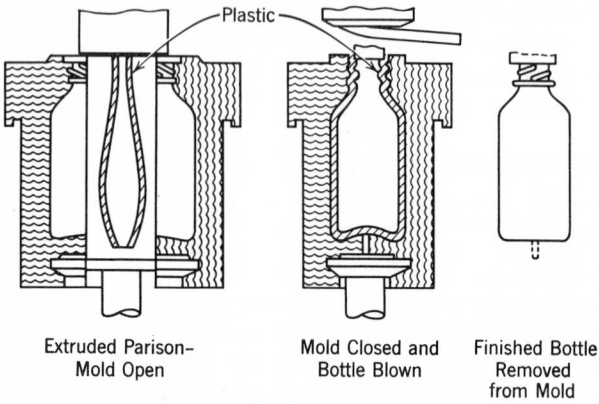

Fig. 20-4. Sketch of blow-molding process. Courtesy of the Society of the Plastics Industry, Inc.

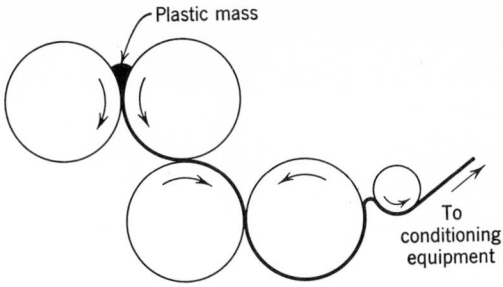

Fig. 20-5. Diagram of a four-roll calender (Winding and Hiatt, gen. ref.).

a cylindrical film. This cylinder can be slit and laid flat, yielding a continuous film of excellent uniformity. In some film-blowing installations, the tube, after blowing and cooling, is nipped between a pair of rolls before being slit; this operation confines the air as a bubble between the extruder and the rolls, making continuous addition of air unnecessary.

Calendering

Calendering is a process used for the continuous manufacture of sheet or film. Granular resin, or thick sheet, is passed between pairs of highly polished heated rolls under high pressure (Fig. 20-5). For the production of thin film, a series of pairs is used, with a gradual reduction in roll separation as the stock progresses through the unit. Proper calendering requires precise control of roll temperature, pressure, and speed of rotation. By maintaining a slight speed differential between a roll pair, it is often possible to impart an exceedingly high gloss to the film or sheet surface. An embossed design can be produced on the surface by means of a calender roll, appropriately engraved. By calendering a mixture of granular resin chips of varying color, it is possible to produce unusual decorative effects (e.g., marblization) in the product; this technique is widely employed in the manufacture of flooring compositions.

Casting

In casting processes, a liquid material is poured into a mold and solidified by physical (e.g., cooling) or chemical (e.g., polymerization) means, and the solid object is removed from the mold. Casting utilizes low-cost equipment (molds can be made out of soft, inexpensive materials such as rubber and plaster) but is a relatively slow process.

Casting of thermosetting resins Thermosetting resins are cast by stopping the polymerization at the A-stage, where the resin is still fusible and fluid. After the mold is filled, the resin is cured in an oven. More details are given in Chapter 19*A*.

Casting of vinyl polymers Vinyl polymers, primarily the acrylic resins, are cast by preparing a syrup of monomer and polymer, and polymerizing the monomer in the mold. Sheets, rods, and tubes are prepared as described in Chapter 15*B*.

Casting of cellulosics Cellulose acetate-butyrate and mixtures of ethyl cellulose and wax are cast by plasticizing the resins to the point where they flow freely when hot, but solidify on cooling. The process is analogous to the casting of metals.

Centrifugal casting Cylindrical articles such as pipe and drums can be made by rotating a cylindrical metal mold charged with granular polymer. The granules cover the walls of the mold and are fused in place by the application of heat from the outside. The process is similar to thermofusion, described on page 497.

Film casting Films, including photographic film and cellophane, are made by flowing a solution of the polymer onto an extremely smooth surface in the form of a large polished wheel or, occasionally, a metal belt or band. After the solvent has evaporated (or, in the case of cellophane, the polymer has coagulated) the film is stripped from the casting surface.

Coating

The technology of coating fabrics and paper, a major outlet for some plastics, is a subject whose magnitude and complexity place it outside the scope of this book. The plastic may be utilized as a melt, solution, latex, paste, or enamel or lacquer. It may be applied to the substrate by spreading with a knife, brushing, using a roller, calendering, casting, or extrusion. Coating processes include *dipping*, in which a form (such as for a rubber glove) is dipped into a suspension or latex of polymer and then into a bath of coagulating agent. After several dips to build up the desired thickness, the film is stripped from the form and subjected to heat treatments to cure or crosslink the resin. In "*slush*" *molding*, a hollow mold is filled with a more viscous latex or "slush" of partly plasticized material such as a vinyl plastisol (Chapter 15*D*). The excess is poured out, leaving a film which is heat treated and removed. The processes are very flexible and involve inexpensive equipment.

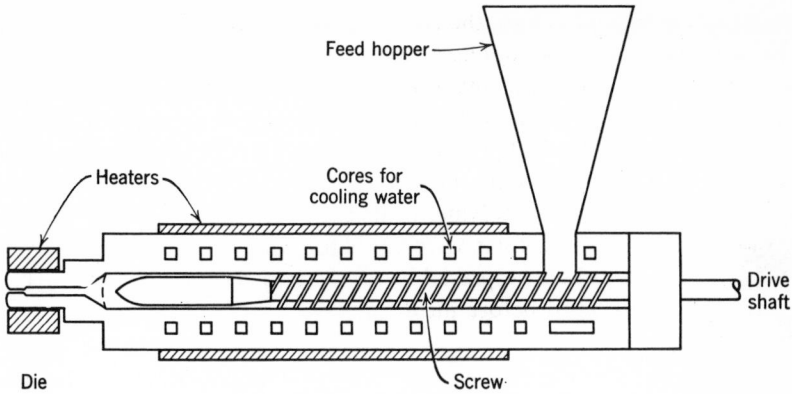

Fig. 20-6. Diagram of a plastics extruder.

Extrusion

In the extrusion process, polymer is propelled continuously along a screw through regions of high temperature and pressure where it is melted and compacted, and finally forced through a die shaped to give the final object (Fig. 20-6). A wide variety of shapes can be made by extrusion, including rods, channels and other structural shapes, tubing and hose, sheeting up to several feet wide and $\frac{1}{4}$ in. or more thick, and film of similar width down to a few thousandths of an inch in thickness.

The screw of an extruder is divided into several sections, each with a specific purpose (Fig. 20-7). The feed section picks up the finely divided polymer from a hopper and propels it into the main part of the extruder.

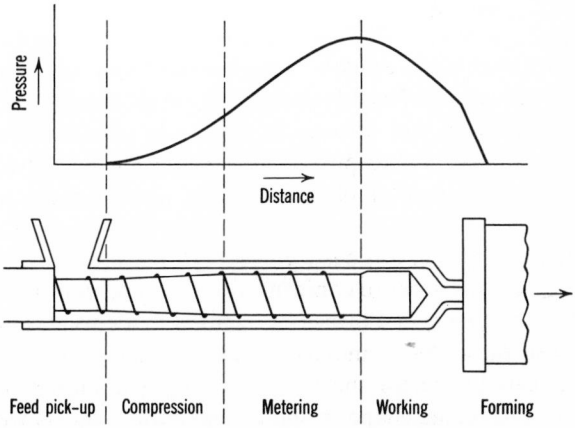

Fig. 20-7. Details of an extruder screw (1).

In the compression section, the loosely packed feed is compacted, melted, and formed into a continuous stream of molten plastic. Some external heat must be applied, but much is generated by friction. The metering section contributes to uniform flow rate, required to produce uniform dimensions in the finished product, and builds up sufficient pressure in the polymer melt to force the plastic through the rest of the extruder and out of the die. Since viscous polymer melts can be mixed only by the application of shearing forces (their viscosity is too high to allow turbulence or diffusion to contribute appreciably to mixing), an additional working section may be needed before the die.

Foaming

The production of plastic foams is accomplished by generating a gas in a fluid polymer, usually at an elevated temperature. The chemical production of a gas during polymerization to form urethan foams is described in Chapter 19*E*. Thermoplastics are foamed by incorporating either a blowing agent, which decomposes to a gas at elevated temperature, or an inert gas. Rubber latices are foamed by whipping mechanically to a froth.

Forming

Postforming thermosetting resins Laminated sheets of thermosetting resins are formed into various shapes by a process resembling that used for sheet metal. The sheet is heated before the final thermosetting reaction, shaped quickly in a mold or around a form, and held in place with light pressure until it sets up.

Forming thermoplastic sheets Vacuum forming is used widely for the manipulation of cellulose acetate and acrylic resin sheeting. A sheet of the plastic is warmed and laid across a hollow mold cavity, and a vacuum is drawn on the cavity. Atmospheric pressure forces the sheet to conform with the mold; after cooling of the sheet, the vacuum is released, and the formed object removed.

Many modifications of this basic process are in use. In one, an inflatable rubber bag forces the sheet to conform with the mold contour. In another variation the sheeting is first dished by vacuum to a curvature greater than ultimately desired, a *convex* mold is placed into the mold cavity, and the vacuum is released while the sheet is still warm. The sheet recovers sufficiently to conform to the shape of the convex mold, and is frozen in that shape by cooling.

Laminating and low-pressure molding

Both lamination, a high-pressure process, and low-pressure molding involve (*a*) the impregnation of sheets (wood, paper, fabric) with a liquid or dissolved thermosetting resin which acts as an adhesive; (*b*) assembly of the individual sheets; and (*c*) compression and curing.

In low-pressure molding or laminating, the sheets of impregnated material are laid over a mold and held in place by a rubber mattress or bag which is inflated with steam to provide heat and pressure to hold the laminate in place and effect the cure. Many variations of the process are practiced.

Lamination differs from low-pressure molding in that standard shapes are produced in sufficient quantity to allow economical application of hot-press methods. Resins requiring higher temperatures for curing can also be used.

Thermofusion

This technique consists of fusing finely divided thermoplastic resin, having about the particle size of table salt, in inexpensive sheet-metal molds. Fusion is accomplished by placing the filled mold in an oven. Polyethylene is the resin most frequently used. Most articles powder molded in this way are quite large, a weight of 100 lb being not unusual. They include barrels, hoppers, small boats, and similar products.

GENERAL REFERENCES

Bernhardt, E. C., ed., *Processing of Thermoplastic Materials*, Reinhold Publishing Corp., New York, 1959, Sec. II.

Garvin, G. S., *loc. cit.*

Griff, A. L., *Plastics Extrusion Technology*, Reinhold Publishing Corp., New York, 1962.

McKelvey, J. M., *loc. cit.*

Modern Plastics Magazine, *loc. cit.*

Renfrew, A., and P. Morgan, eds., *Polythene*, 2nd ed., Interscience Publishers, New York, 1960, Part II.

Winding, C. C., and G. D. Hiatt, *Polymeric Materials*, McGraw-Hill Book Co., New York, 1961, Chaps. 4, 5.

BIBLIOGRAPHY

1. McKelvey, J. M., *SPE J.* **9**, #3, 12 (1953).

C. Fillers

Many plastics are virtually useless alone but are converted into highly serviceable products by combining them with particulate or fibrous solids. Phenolic and amino resins are almost always compounded, or filled, with substances like wood flour, pure short-fiber cellulose, powdered mica, and asbestos. These materials greatly enhance dimensional stability, impact resistance, tensile and compressive strength, abrasion resistance, and thermal stability. The use of glass fiber as a reinforcing filler for polyester resins is another important illustration. Soft thermoplastics like coumarone-indene and hydrocarbon resins are usually blended with very large amounts (over 80 per cent by weight) of mineral solids such as crushed quartz, limestone, or clay. In compounds of this type, the resin functions as an interparticle adhesive; these products often have poor tensile properties but excellent compressive strength, abrasion resistance, and dimensional stability. A final and extremely important example of the beneficial effect of fillers is the reinforcement of rubber, discussed in Chapter 22*B*.

The principal fillers used for thermosetting molding compounds are the following (1, 2):

a. Wood flour, made by grinding soft wood to about 100 mesh. This is the least expensive and most widely used filler for phenolic resins.

b. Cotton flock, obtained from cotton linters or clippings. This filler improves impact resistance somewhat.

c. α-Cellulose, obtained from wood pulp or high-quality paper. This filler has the advantage of being white, but effects little improvement in properties. It is widely used with amino resins to produce light colored objects.

d. Macerated fabric and cord. These fillers lead to compounds with ten or more times the impact strength of corresponding wood flour filled compositions.

e. Synthetic fibers. These materials, in the form of flock, chopped fiber, or dipped fabric, are used with newer products such as allylic resins, rather than the low-cost phenolics.

f. Rubber. Nitrile rubber, in combination with more conventional fillers, improves flexibility as well as impact strength in phenolic resins.

g. Carbon. Although beneficial in rubbers (Chapter 22*C*), carbon black does not improve the properties of thermosetting resins significantly. It is used as a pigment and for obtaining electrical conductivity.

h. Asbestos. Asbestos improves the heat, chemical, and moisture resistance of thermosetting resins.

i. Quartz. Quartz-filled phenolics are used in missile nose cones because of their good resistance to extreme temperatures.

j. Mica. This filler leads to excellent electrical properties and low moisture absorption.

k. Glass. In the form of short ($\frac{1}{4}$–$\frac{1}{2}$ in.) fibers, glass leads to good impact strength, electrical properties, and temperature resistance in many thermosetting compositions. This use is not to be confused with its function as long fibers or woven fabric to reinforce "polyester" resins, as described in Chapter 19*C*.

GENERAL REFERENCE

Modern Plastics Magazine, *loc. cit.*

BIBLIOGRAPHY

1. Seymour, R. B., *Modern Plastics* **38**, #1A (Encyclopedia Issue), 522 (1960).
2. Winding, C. C., and G. D. Hiatt, *op. cit.*, pp. 226–227.

D. Plasticizers and Other Additives

Plasticizers

Plasticizers (1, 3) are added to plastics to improve flow and, therefore, processability, and to reduce the brittleness of the product. This is achieved by lowering the glass transition temperature below room temperature, thus achieving a change in properties from those of a hard, brittle, glasslike solid to those of a soft, flexible, tough material (Chapter 7*D*). An example is the plasticization of poly(vinyl chloride) and vinyl chloride-acetate copolymers (Chapter 15*D*). Similar changes in properties can, of course, be brought about by altering the molecular structure of the polymer (e.g., by copolymerization, sometimes called *internal plasticization*).

The basic requirements which must be met by a plasticizer are compatibility and permanence. The plasticizer must be miscible with the polymer. This implies a similarity in the intermolecular forces active in the two components, and explains why compatibility is difficult to achieve with a nonpolar polymer such as polyethylene. Permanence requirements demand low vapor pressure and a low diffusion rate of the plasticizer within the polymer, both of which are obtained by the use of high-molecular-weight plasticizers.

The efficiency of the plasticizer in bringing about the desired changes in properties is important in determining the proportion in which the plasticizer must be added to the resin. Plasticizer efficiency may be evaluated by a number of different semiempirical tests. Some of these measure the amount of nonsolvent needed to cause phase separation when added to polymer-plasticizer solutions (dilution ratio), the viscosity of dilute solutions of the polymer in the plasticizer, polymer-solvent interaction constants measured on these solutions, the depression of the glass transition temperature, the melt viscosity of the plasticized polymer, the electrical or mechanical properties of the plasticized polymer, or the molecular size and shape or viscosity of the plasticizer itself. Not all these tests, needless to say, rate plasticizer candidates in the same order. The selection of a particular plasticizer still depends to a large extent upon empirical results rather than theoretical predictions.

The following types of plasticizers are in common use (2, *Modern Plastics*, *loc. cit.*):

a. Phthalate esters, accounting for over half of the total volume of plasticizers used.

b. Phosphate esters, chiefly tricresyl phosphate, valued primarily for their flameproofing characteristics.

c. Adipates, azelates, oleates, and sebacates, used chiefly in vinyl resins for improving low-temperature flexibility.

d. Epoxy plasticizers, produced by reacting hydrogen peroxide with unsaturated vegetable oils and fatty acids.

e. Fatty acid esters from natural sources, used primarily as extenders to reduce cost (*secondary plasticizers*).

f. Glycol derivatives, employed mainly as lubricants and mold-release agents, and as plasticizers for poly(vinyl alcohol).

g. Sulfonamides, used to plasticize cellulose esters, phenolic and amino resins, and amide and protein plastics.

h. Hydrocarbons and hydrocarbon derivatives, serving as secondary plasticizers.

Other additives

Antioxidants The role of antioxidant in preventing or inhibiting the oxidation of polymers is usually filled by a substance which itself is readily oxidized, although in some cases the antioxidant may act by combining with the oxidizing polymer to form a stable product. Common antioxidants fall in the classes of phenols, aromatic amines, and salts and condensation products of amines and aminophenols with aldehydes, ketones, and thio compounds.

Colorants Depending on their solubility in the polymer, colorants may be classified as dyes (soluble) and pigments (insoluble). Dyes yield transparent colored materials and hence are used primarily with the transparent resins, such as polystyrene and poly(methyl methacrylate). Since dyes are organic substances and are, as a class, not as heat stable as pigments, dyes are less frequently used in high-melting crystalline polymers. Dyes which are primarily produced for use with textiles in processes requiring water solubility (Chapter 21*C*) are not compatible with plastics; the selection of oil-soluble dyes is much smaller by comparison.

Pigments may be either organic or inorganic. With a few important exceptions (such as the phthalocyanines), organic pigments are the less heat stable and are restricted in use to plastics processable at relatively low temperatures. Because of their low refractive index, organic pigments sometimes, when dispersed finely enough, appear transparent in plastics. This is not true of the inorganic pigments, however, and their use is restricted to translucent or opaque compositions.

In contrast to the technique of dyeing (Chapter 21*C*), the coloring of plastics is normally carried out by adding the colorants to the powdered plastic, tumbling, and compounding on hot rolls or in an extruder (see page 495). Other techniques are occasionally used: e.g., colored castings of poly(methyl methacrylate) are produced by dissolving or dispersing the colorants in the syrup before polymerizing.

Flame retardants The most useful material imparting flame retardance to plastics is antimony trioxide. It must be used with a source of available chlorine to be effective; it is presumed that antimony oxychloride is the active flame-retarding agent. Aside from this compound and other antimony derivatives, the phosphate ester plasticizers are widely used for reducing flammability, especially in vinyl resins.

The major factors in reducing the flammability of materials appear to be (*a*) elimination of volatile fuel, as by cooling; (*b*) production of a thermal barrier, as by charring, thus eliminating fuel by reducing heat transfer; and (*c*) quenching the chain reactions in the flame, as by adding suitable radical scavengers.

Stabilizers In addition to antioxidants and flame retardants, certain polymers require stabilizers to achieve and maintain utility. An important example is the vinyl resins, whose stabilization is discussed in Chapter 15*D*. Other applications of stabilizers include the use of carbon black to prevent photochemical degradation (by excluding light), as in polyethylene, and of ultraviolet light absorbers, such as hydroxybenzophenones, to improve the light stability of both plastics and the dyes which serve as their colorants.

*TABLE 20-1. Typical properties of commercial thermoplastic polymers**

Property \ Polymer	Acetal Resin	Poly(methyl Methacrylate)	Cellulose Acetate	Chlorinated Polyether	66 Nylon	Linear Polyethylene	Branched Polyethylene	Polypropylene
Specific gravity, g/cm³	1.43	1.17–1.20	1.23–1.34	1.4	1.09–1.14	0.941–0.965	0.910–0.925	0.90–0.91
Refractive index, n_D^{25}	1.48	1.49	1.46–1.50	...	1.53	1.54	1.51	1.49
Tensile strength, psi	10,000	7000–11,000	1900–8500	6000	7000–11,000	3100–5500	1000–2300	4300–5700
Elongation, %	15–75	3–10	6–70	60–160	90	15–100	90–650	250–700
Tensile modulus, 10^5 psi	4.1	4.5	0.6–4.0	1.6	2.6–4.0	0.8–1.5	0.17–0.35	1.3–2.0
Impact strength, ft-lb/in. of notch	0.4–2.3	0.3–0.5	0.4–5.2	0.4	1.0	1.5–12	>16	0.6–6.0
Heat-distortion temp., °F, 66 psi	338	160–195	110–209	300	300–360	140–180	105–121	210–230
Dielectric constant, 1000 cycles	3.7	3.0–3.5	3.5–7.0	3.0	4.0–4.5	2.30–2.35	2.25–2.35	2.0–2.1
Dielectric loss, 1000 cycles	0.004	0.03–0.05	0.01–0.06	0.01	0.02–0.04	<0.0002	<0.0005	<0.0003
Water absorption, $\frac{1}{8}$ in. bar, 24 hr, %	0.12	0.3–0.4	1.9–6.5	0.01	0.4–1.5	<0.01	<0.015	<0.01
Burning rate	Slow	Slow	Slow to self-exting.	Self-exting.	Self-exting.	Very slow	Very slow	Slow
Effect of sunlight	Chalks	None	Slight	Slight	Discolors, embrittles	Requires carbon black	Surface crazing	Requires carbon black
Effect of strong acids or bases	Attacked	Attacked	Decomposes	Attacked, acids	Attacked, acids	Resistant	Resistant	Resistant
Effect of organic solvents	Resistant	Soluble	Soluble	Resistant	Resistant	Resistant below 80°C	Soluble above 80°C	Resistant below 80°C
Clarity	Opaque	Transparent	Transparent	Opaque	Opaque	Opaque	Opaque	Opaque

Property \ Polymer	Polycarbonate	Polytetra-fluoroethylene	Polystyrene	Impact Polystyrene	Styrene-Acrylonitrile Copolymer	ABS and Polyblend	Rigid Vinyl	Plasticized Vinyl
Specific gravity, g/cm³	1.2	2.1–2.2	1.04–1.06	0.98–1.10	1.08–1.10	0.99–1.10	1.35–1.45	1.16–1.35
Refractive index, n_D^{25}	1.58	1.35	1.59–1.60	...	1.57	...	1.52–1.55	...
Tensile strength, psi	8500–9500	2000–4500	5000–9000	3500–6500	9500–12,000	2500–9000	5000–9000	1500–3500
Elongation, %	60–100	200–400	1.0–2.5	5–80	1.5–3.5	10–140	2.0–4.0	200–450
Tensile modulus, 10^5 psi	3.2	0.6	4–5	3–4.5	4–5.6	1.0–4.1	3.5–6	...
Impact strength, ft-lb/in. of notch	12–16	3.0	0.25–0.4	0.5–11	0.35–0.5	0.7–12	0.4–20	Varies
Heat-distortion temp., °F, 66 psi	280–290	250	150–195	148–200	195–200	165–225	130–165	...
Dielectric constant, 1000 cycles	3.0	2.0	2.4–2.65	2.4–4.5	2.5	2.7–4.8	3.0–3.3	4.0–8.0
Dielectric loss, 1000 cycles	0.0011	<0.0002	0.0001–0.0003	0.0004–0.002	0.007–0.010	0.002–0.012	0.009–0.017	0.07–0.16
Water absorption, ⅛ in. bar, 24 hr, %	0.3	0.00	0.03–0.05	0.1–0.3	0.2–0.3	0.1–0.3	0.07–0.4	0.15–0.75
Burning rate	Self-exting.	None	Slow	Slow	Slow	Slow	Self-exting.	Slow to self-exting.
Effect of sunlight	Yellows	None	Yellows	Loses strength	Yellows	Yellows	Darkens	Darkens
Effect of strong acids or bases	Attacked slowly	Very resistant	Attacked, acids	Attacked, acids	Attacked, acids	Attacked, acids	Resistant	Resistant
Effect of organic solvents	Soluble	Very resistant	Soluble	Soluble	Soluble	Soluble	Soluble	Soluble
Clarity	Transparent	Opaque	Transparent	Opaque	Transparent	Opaque	Transparent	Transparent

* Source: *Modern Plastics* Magazine, Encyclopedia Issue, Breskin Publications, New York, 1961.

*TABLE 20-2. Typical properties of commercial thermosetting resins**

Property \ Resin and Filler	Diallyl Phthalate	Melamine-Formaldehyde			Phenol-Formaldehyde			Polyester
	Glass	α-Cellulose	Glass Fiber	Wood Flour	Glass Fiber	Fabric or Cord	Rubber-Wood Flour	Glass
Specific gravity, g/cm³	1.55–1.68	1.47–1.52	1.9–2.0	1.32–1.45	1.75–1.95	1.36–1.43	1.24–1.35	1.35–2.0
Refractive index, n_D^{25}	...	...	...	...	...	...	...	...
Tensile strength, psi	6500–7000	7000–13,000	6000–10,000	6500–9000	5000–10,000	3300–9000	4500–9000	10,000–25,000
Elongation, %	...	0.6–0.9	...	0.4–0.8	0.2	0.37–0.57	0.75–2.25	0.5–5.0
Tensile modulus, 10^5 psi	15–20	12–14	...	8–12	33	9–13	4.0–6.0	8–20
Impact strength, ft-lb/in. of notch	0.5–15	0.24–0.35	8.0–16.0	0.24–0.60	10–50	0.75–8.0	0.35–1.0	5–30
Heat distortion temp., °F, 66 psi	325–500	400	400	260–340	>600	250–300	230–260	...
Dielectric constant, 1000 cycles	4.1–4.5	7.8–9.2	...	4.4–9.3	6.9	5.0–11	...	4.0–6.0
Dielectric loss, 1000 cycles	0.004–0.008	0.015–0.036	...	0.04–0.20	0.02	0.04–0.20	...	0.01–0.05
Water absorption, ⅛ in. bar, 24 hr, %	0.12–0.35	0.1–0.6	0.3–0.6	0.3–1.0	0.1–1.2	0.4–1.75	1–2	0.01–1.0
Burning rate	Self-exting.	None	None	Very slow	None	Almost none	Slow	Slow to self-exting.
Effect of sunlight	None	Color change	Slight	Darkens	Darkens	Darkens	Darkens	Slight
Effect of strong acids or bases	Slight	Attacked	Attacked, acids	Attacked	Attacked	Attacked	Attacked	Attacked
Effect of organic solvents	None	None	None	None	None	None	None	Slight
Clarity	Opaque	Translucent	Opaque	Opaque	Opaque	Opaque	Opaque	Opaque

Resin and Filler / Property	Silicone	Urea-Formaldehyde	Epoxy		Allyl Carbonate	Phenolic	Polyester, Cast	
	Glass	α-Cellulose	Mineral	None (cast)	None (cast)	None (cast)	None (liquid)	None (flexible)
Specific gravity, g/cm³	1.68–2.0	1.47–1.52	1.6–1.9	1.11–1.4	1.30–1.40	1.30–1.32	1.10–1.46	1.01–1.20
Refractive index, n_D^{25}	...	1.54–1.56	...	1.61	1.50–1.575	1.58–1.66	1.52–1.57	1.537–1.55
Tensile strength, psi	4000–5000	6000–13,000	5000–7000	4000–13,000	5000–6000	6000–9000	6000–10,000	800–1800
Elongation, %	...	0.5–1.0	...	3–6	...	1.5–2.0	<5	40–310
Tensile modulus, 10^5 psi	...	10–15	...	3.5	3.0	4–5	3.0–6.4	...
Impact strength, ft-lb/in. of notch	3–15	0.25–0.40	0.25–0.35	0.2–1.0	0.2–0.4	0.25–0.40	0.2–0.4	>7.0
Heat distortion temp., °F, 66 psi	>900	260–290	250–350	115–550	140–190	165–175	140–400	...
Dielectric constant, 1000 cycles	3.2–5.0	7.0–7.5	4.2	3.5–4.5	3.35–5.0	5.5–6.0	2.8–5.2	4.5–7.1
Dielectric loss, 1000 cycles	0.0035–0.02	0.025–0.035	0.019	0.002–0.02	0.01	0.01–0.05	0.005–0.025	0.016–0.05
Water absorption, ⅛ in. bar, 24 hr, %	0.1–0.2	0.4–0.8	0.1	0.08–0.15	0.03–0.44	0.3–0.4	0.15–0.60	0.50–2.5
Burning rate	None to slow	Very slow	Self-exting.	Slow	Slow to self-exting.	Very slow	Slow to self-exting.	...
Effect of sunlight	Slight	None	Slight	None	Yellows	Colors fade	Yellows	...
Effect of strong acids or bases	Slight attack	Attacked	Slight attack	Attacked	Attacked, acids	Attacked	Attacked	Attacked
Effect of organic solvents	Attacked	None to slight	None	Resistant	Resistant	Attacked	Attacked	Attacked
Clarity	Opaque	Transparent	Opaque	Transparent	Transparent	Transparent	Transparent	Transparent

*Source; *Modern Plastics* Magazine, Encyclopedia Issue, Breskin Publications, New York, 1961.

Compounding

The term *compounding* is applied both to the selection of additives to modify the properties of a polymer, and to their incorporation with the polymer to give a homogeneous mixture, in a form suitable for efficient use in the subsequent processing or fabrication step. It is in the latter sense that compounding is here considered.

The traditional compounding device in the plastics industry is the two-roll mill, which looks and operates much like the top half of a calender (Fig. 20-5). By proper selection of temperatures and speeds of rotation, the plastic is made to adhere to the front roll, except as it is cut off by the operator. The compounding ingredients are added to the plastic mass as it passes between the rolls.

The roll mill has been supplanted in many operations by the compound- or mixer-extruder, an extruder in which the function of the mixing section of the screw is emphasized. The use of the extruder for compounding has many advantages: contamination is reduced, inert atmospheres or vacuum may be utilized, continuous processes are more readily achieved, etc.

Other compounding devices in common use are internal mixers such as kneaders, masticators, and paddle blenders; tumblers; and blenders.

GENERAL REFERENCES

Buttery, D. N., *Plasticizers*, Franklin Publishing Co., Palisades, N.J., 2nd ed., 1960.
Garvin, G. S., *loc. cit.*
Modern Plastics Magazine, *loc. cit.*
Schildknecht, C. E., "Stabilization of Polymers," Chap. XIII in C. E. Schildknecht, ed., *op. cit.*

BIBLIOGRAPHY

1. Boyer, R. F., *J. Appl. Phys.* **20,** 540 (1949); *Tappi* **34,** 357 (1951).
2. Reid, W. J., *Modern Plastics* **39,** #1, 165 (September, 1961).
3. Stickney, P. B., and L. E. Cheyney, *J. Polymer Sci.* **3,** 231 (1948).

E. Tables of Plastics Properties

Tables 20-1 and 20-2 list comparative properties for typical examples of the major commercial thermoplastic and thermosetting resins, respectively. It should be noted that the properties of a given plastic can vary widely, depending on compounding, fabrication, thermal history, and many other variables. The numbers in the tables may represent only a fraction of the total range of values attainable and should be used only for comparative purposes.

21

Fiber Technology

The wide range of materials classified as fibers includes natural and synthetic, organic and inorganic products. Some polymers used as fibers, such as nylon and cellulose acetate, serve equally well as plastics. In the final analysis, the classification of a substance as a fiber depends more on its shape than on any other property. One common definition of a fiber requires that its length be at least 100 times its diameter. Artificial fibers can usually be made in any desired ratio of length to diameter. Among the natural fibers, cotton, wool, and flax are often found with lengths 1000–3000 times their diameter: coarser fibers such as jute, ramie, and hemp have lengths 100–1000 times their diameter.

To be useful as a textile material, a synthetic polymer must have suitable characteristics with respect to several physical properties. These include a high softening point to allow ironing, adequate tensile strength over a fairly wide temperature range, solubility or meltability for spinning, a high modulus or stiffness, and good textile qualities such as those defined on page 509. In addition to these primary requirements, many other properties of the material are important if it is to be suitable for textile applications. Some of these are listed in Table 21-1. The desirable values of these properties and the molecular structure needed to obtain them are discussed in the following sections of this chapter.

The textile industry is quite complex, and many of the technologies which comprise it are intricate and obscure. A new fiber must conform to a set of desirable properties which have been largely decided upon in advance. Manipulation of molecular structure and synthesis variables does not allow freedom to achieve all the desired properties, so compromise is inevitable. Even a successful compromise in physical properties does not ensure success for a new fiber. Other factors which must be favorable include all the processes leading to some knitted or woven cloth ready for wearing trials, a sound economic situation and competitive price structure,

TABLE 21-1. Fiber properties important in textile uses

Chemical	Physical	Biological	Fabric Properties
Stability toward	Mechanical	Toxicological	Appearance
Acids	Tenacity	Dermatological	Drape
Bases	Elongation	Resistance to	Hand
Bleaches	Stiffness	Bacteria	Luster
Solvents	Flex life	Molds	Comfort
Heat	Abrasion resistance	Insects	Warmth
Sunlight	Work recovery		Water sorption
Aging	Tensile recovery		Moisture retention
Flammability	Thermal		Wicking
Dyeability	Melting point		Stability
	Softening point		Shape
	Glass transition temperature		Shrinkage
	Decomposition temperature		Felting
	Electrical		Pilling
	Surface resistivity		Crease resistance
			Crease retention

a good supply of raw materials, and, eventually, a good record of customer acceptance.

Information on price and production volume of the major fibers is found in Chapter 18 and in the sections in Part IV describing the fiber-forming polymers.

A. Textile and Fabric Properties

Definitions of textile terms

Types of fibers Most artificial fibers may be obtained either as a very long *continuous filament*, or as *staple*, made by cutting continuous filament into relatively short lengths. Natural fibers, with the exception of silk, are obtained only in the form of staple.

Denier The *denier* of a fiber, a measure of its size, is defined as the weight in grams of 9000 meters of the fiber. It is thus proportional to the density of the fiber and to its cross-sectional area. At least ten other measures of the size of fibers are in use, but the denier is the most widely accepted.

Tenacity The tensile strength of a fiber is usually expressed in terms of *tenacity*. Tenacity is defined as the strength per unit size number, such as the denier, where the size number is expressed as a weight per unit length.

Tenacity is thus a function of the density of the fiber as well as its tensile strength. Two fibers having the same denier and the same breaking strength have the same tenacity, usually expressed in grams per denier; if they have the same density, they have the same tensile strength as well. Tensile strength (pounds per square inch) = tenacity (grams per denier) × density × 12,791.

Creep *Creep* in a fiber is defined as in a plastic material, i.e., viscoelastic deformation under load.

Moisture content and moisture regain Most textile fibers absorb some moisture from their surroundings. If the amount of moisture present at equilibrium under standard conditions (65% relative humidity and 70°F) is expressed as a percentage of the total weight of the (moist) fiber, it is known as the *moisture content*; if it is expressed as a percentage of the oven-dry weight (110°C) of the fiber, it is known as the *moisture regain.*

Crimp *Crimp* is the waviness of a fiber, a measure of the difference between the length of the unstraightened and that of the straightened fiber. Some naturally occurring fibers, notably certain wools, have a natural crimp. Crimp can be artificially produced in fibers by suitable heat treatment or by rolling them between heated, fluted rolls.

Fabric property terms The esthetic qualities of fabrics are defined in terms such as *appearance*, *hand* (or *handle*), and *drape*. Although the qualitative meanings of these terms are obvious, their quantitative definitions and interpretations in terms of fiber properties are quite difficult. Hand and drape are largely determined by the tensile and elastic behavior of the fiber.

Properties of textile fibers

Electrical properties Since fibers are not normally used in electrical applications, their only electrical property of great interest is their resistivity. Too high resistivity leads to the development of static electrical charges, which cause the fabric to cling unpleasantly and to be difficult to clean. Some of the synthetic fibers, including the nylons, polyesters, and acrylics, are poor in this respect.

Mechanical properties The mechanical properties of fibers are quite complex and have been the subject of much experimental work. A stressed textile fiber is a complicated viscoelastic system, in which a number of irreversible processes can take place. Some typical fiber stress-strain curves

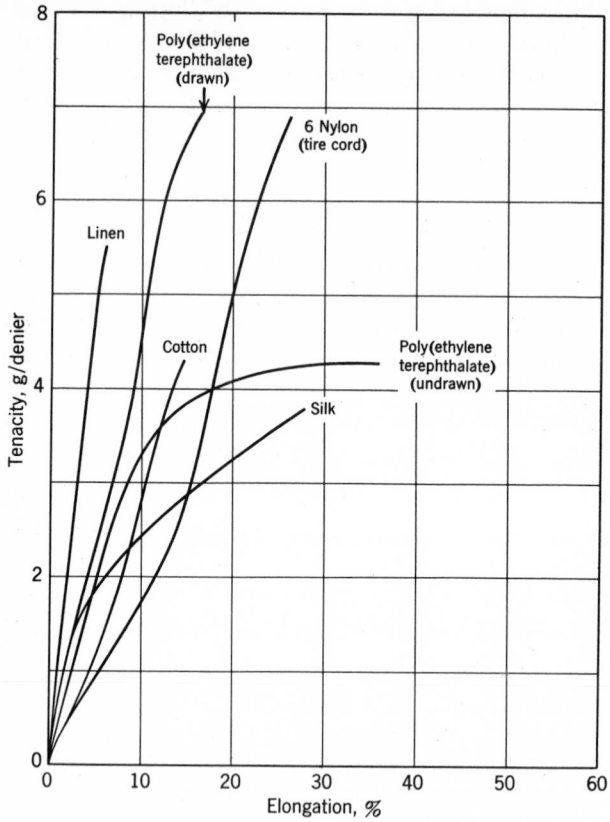

Fig. 21-1. Stress-strain curves of silklike fibers (1).

are shown in Figs. 21-1 and 21-2. They may be divided roughly into two groups: the silklike curves, featuring almost constant high modulus to the break point, and the woollike curves, featuring a sharp drop in modulus at low stress followed by a long period of elongation at almost constant stress. Intermediate characteristics are shown by some of the synthetics such as nylon: the first region of reversible elasticity is followed by an irreversible region in which drawing takes place. When orientation is complete, another reversible elastic region is found. This may end at the point of failure, or plastic flow may take place.

The tenacity at break of typical fibers ranges from about 1 g/denier for acetate rayon up to 5 or more g/denier for nylon and some of the natural fibers. A fiber weaker than acetate is not a practical textile, and one stronger than nylon would find use for its great strength only in exceptional cases.

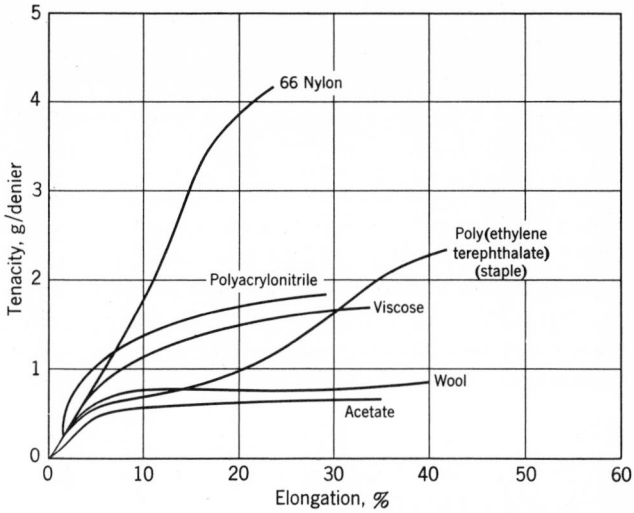

Fig. 21-2. Stress-strain curves of woollike fibers (1).

Moisture regain The moisture regain of the synthetic polyesters, nylons, and acrylics is so much lower than that of any of the natural fibers that the synthetics may be classed as hydrophobic materials. This is an advantage for rapid drying.

Dyeability Hydrophobic fibers are in general difficult to dye. The acrylics are particularly bad in this respect, and modification of the polymer structure by copolymerization is resorted to in order to obtain materials which can be dyed. Nylon is intermediate in dyeability, while cellulose and the cellulosic fibers are eminently dyeable.

Chemical stability All textile fibers must be stable to water, dry-cleaning solvents, and dilute acids, alkalis, and bleaches, and all are reasonably adequate in this respect. Silk and nylon are the least satisfactory fibers from the standpoint of weatherability, whereas the acrylics are outstanding in this respect.

Fabric properties

Esthetic factors (See preceding definitions of textile terms.) Silk is the outstanding example of a fiber with good esthetic properties. Filament acetate, which possibly ranks next, is replacing filament viscose rayon in many applications for which luster, hand, and drape are more important than good mechanical properties. Among the staple fibers, the polyesters appear to be most nearly woollike and most pleasing.

Comfort The comfort of a fabric is a highly important but little understood property. It is related to the structure of the fabric in determining ventilation and heat-insulating characteristics and wicking ability. Comfort is not, however, dependent upon high moisture absorption in the fiber.

Crease resistance and crease retention These two properties are related in that a fabric which can be creased only with difficulty tends to retain its crease. The thermoplastic fibers such as nylon, polyesters, and acrylics are very good in both respects. Creasing in wear becomes more important at high moisture contents. Cellulosics are naturally poor in crease resistance, although they may be modified by resin finishes to give good crease resistance.

Fabric stability The stability of shape and dimensions of the synthetic fibers is outstandingly good. Wool retains its shape well in garments, but felts and shrinks under wet treatment.

The phenomenon of "pilling" has become important in some of the synthetics. A small nodule or pill can be formed on some fabrics by gentle rubbing, as the surface fibers are raised and tangled. In wool these pills are not important because they wear off rapidly, but in nylon and the polyesters the pills do not break away and may become quite unsightly.

Wear resistance There is no laboratory test for abrasion in a fabric which correlates with actual use. There is evidence that two or more independent parameters are required to define wear in a fabric. In general, acetate rayon is most easily abraded, with viscose, cotton, and wool following as a group (except that viscose when wet is not much better than acetate), while the polyesters, acrylics, and nylon are progressively better.

GENERAL REFERENCES

Mauersberger, H. R., ed., *Matthew's Textile Fibers*, 6th ed., John Wiley & Sons, New York, 1954, Chaps. 1–3.

Moncrieff, R. W., *Man-Made Fibres*, 3rd ed., John Wiley & Sons, New York, 1957, Part I.

Press, J. J., ed., "Textile Engineering Principles," Chapter XIII in J. J. Press, ed., *Man-Made Textile Encyclopedia*, Interscience Publishers, New York, 1959.

BIBLIOGRAPHY

1. Heckert, W. W., in S. B. McFarlane, ed., *Technology of Synthetic Fibers*, Fairchild Publications, New York, 1953, Chap. 5.

B. Spinning

The conversion of bulk polymer to fiber form is accomplished by *spinning*. In most cases, spinning processes require solution or melting of the polymer (an exception is the spinning of fibers from an aqueous dispersion, as in the case of polytetrafluoroethylene, followed by removal of liquid and sintering).

If a polymer can be melted under reasonable conditions, the production of a fiber by *melt spinning* is preferred over solution processes. When melt spinning cannot be carried out, a distinction as to type of process is made depending upon whether the solvent is removed by evaporation (*dry spinning*) or by leaching out into another liquid which is miscible with the spinning solvent but is not itself a solvent for the polymer (*wet spinning*). Dry spinning has some advantages over wet spinning. The three processes have many features in common.

To obtain solubility in a polymer which cannot be melted, recourse may be had to modification of the polymer to allow solubility in common solvents, or to a search for unusual solvents. Thus the first commercial artificial fiber, Chardonnet silk, was made by spinning cellulose nitrate containing 10.5–12.2% nitrogen from a mixture of ether and alcohol. Viscose and cellulose acetate (Chapter 18) are examples of the modification of cellulose to obtain solubility in common solvents, but the cuprammonium process is based on the discovery of a solvent for cellulose itself. Vinyl chloride may be copolymerized with vinyl acetate or chlorinated to obtain solubility: vinylidene chloride has been copolymerized with ethyl acrylate for the same reason. On the other hand, only an intensive search for unusual solvents has allowed the commercial exploitation of polyacrylonitrile fiber.

The conversion of the spun polymer melt or solution to a solid fiber involves cooling, solvent evaporation, or coagulation, depending on the type of spinning used. The rates of these processes decrease in the order listed. Cooling of a fine filament is normally very rapid and can be controlled within relatively narrow limits. Solvent evaporation involves simultaneous outward mass transfer and inward heat transfer, the rate-controlling step invariably being outward diffusion of solvent. Coagulation involves two-way mass transfer, the coagulating agent (e.g., acid) diffusing inward, and the products of coagulation (e.g., salts, H_2S) diffusing out. As a consequence of these facts, it is very easy to obtain a melt-spun fiber which possesses uniform properties throughout its cross section, but almost impossible to do so with solvent-spun or coagulated fibers. In addition, the rapid cooling of melt-spun fibers tends to produce

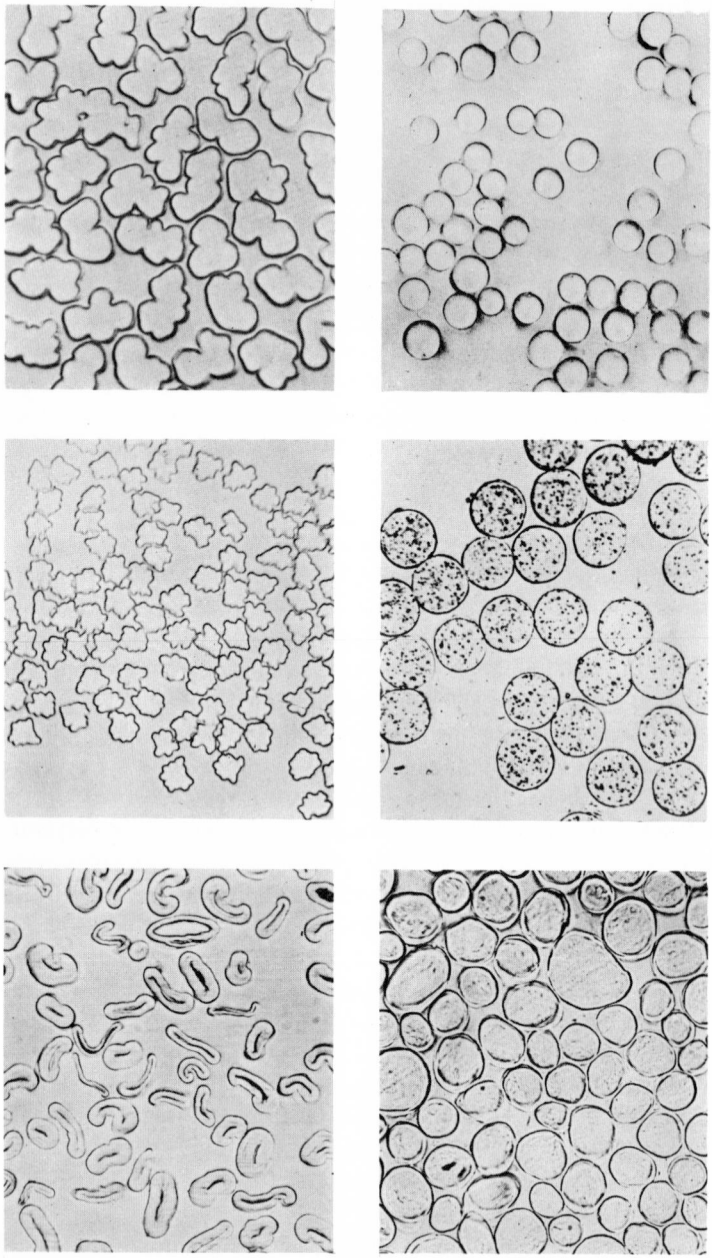

Fig. 21-3. Photomicrographs, X250, showing cross-sectional shapes of fibers (Riley, gen. ref.). Top, left to right: cotton, viscose, acetate. Bottom, left to right: wool, 66 nylon (pigmented), poly(ethylene terephthalate). Courtesy of J. L. Riley.

an almost circular cross section, whereas with solvent-spun fibers the cross sections are usually elliptical, and with coagulated fibers (because of the absence of strong surface-tension effects) the cross sections are ordinarily highly convoluted, as shown in Fig. 21-3.

Melt spinning

The process of melt spinning is inherently simple. Molten polymer is pumped at a constant rate under high pressure through a plate called a *spinneret* containing a large number of small holes. The liquid-polymer streams emerge downward from the face of the spinneret, usually into air. They solidify and are brought together to form a thread and wound up on bobbins. A subsequent drawing step is necessary to orient the fibers.

The polymer is melted by contacting a hot grid in the form of steel tubing, which is heated by electric current or by hot vapors (Fig. 21-4*a*). It is usually necessary to protect the polymer melt from oxygen by blanketing it with steam or an inert gas such as carbon dioxide or nitrogen. If the viscosity of the molten polymer is low, it may pass directly to the metering (constant-rate) pump. For melts of higher viscosity a booster pump may be used. Other methods of melting have been proposed, including the use of an extrusion-type screw, one section of which can serve as its own metering pump. Methods in which the metering pump is replaced, e.g., by a source of gas pressure or other device, do not appear to offer sufficiently precise control to hold the denier of fine yarns constant.

While the polymer is molten, certain physical and chemical changes can take place. Molecular weight may change in condensation polymers, since moisture may be removed by the nitrogen bleed needed to prevent oxidation, or may be introduced if steam is used for the same purpose. Interchange reactions can take place, changing weight-average molecular weight and melt viscosity. Thermal decomposition may lead to the formation of gas bubbles, which interfere badly with spinning, or to branching and crosslinking reactions. In the case of 66 nylon this decomposition leads to the formation of gel particles which must be removed from the polymer melt by a filter pack consisting of layers of graded sand between wire gauze placed just before the spinneret.

The spinneret itself may consist of a 2- to 3-in. diameter steel disk about $\frac{1}{4}$ in. thick with 50–60 countersunk holes 0.010 in. or less in diameter. The denier of the filament is determined not by the diameter of the holes but by the rate at which polymer is pumped through the spinneret and the rate at which the filaments are wound up.

The filaments emerge from the spinneret face into air and begin to cool. An air blast may be used to speed up the cooling process. After the

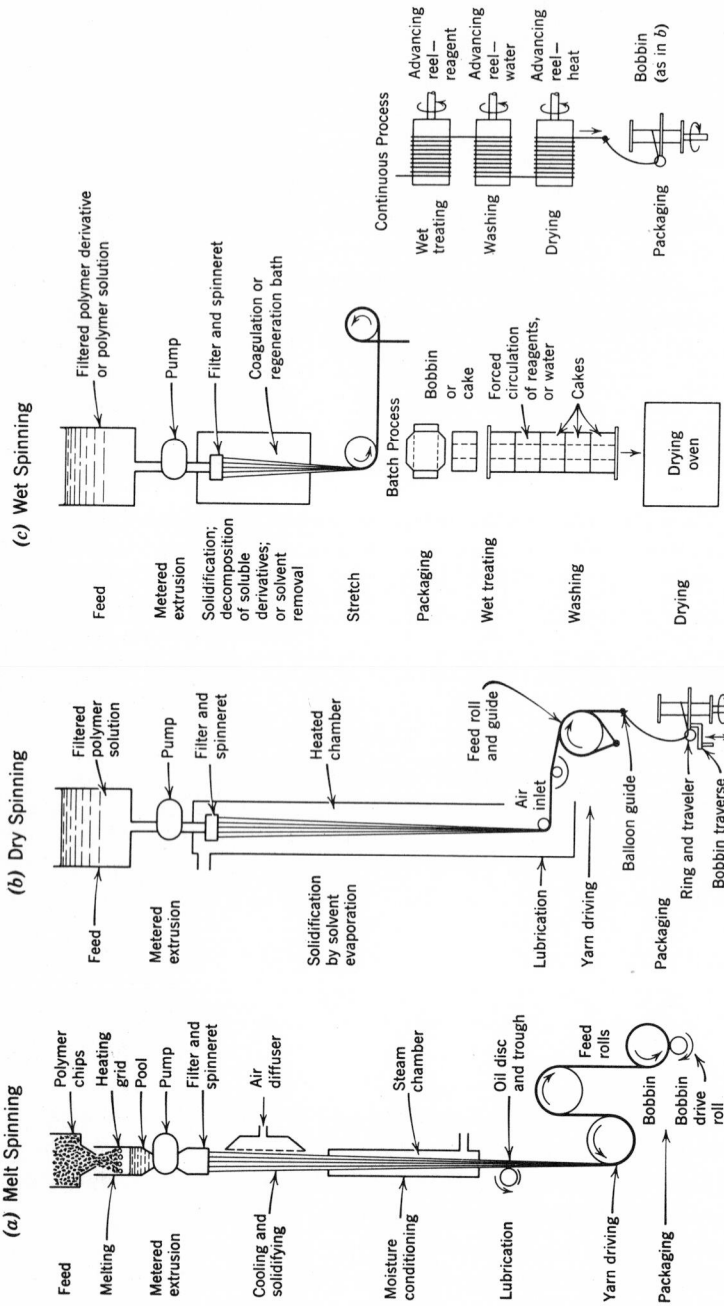

Fig. 21-4. Schematic diagrams of the three principal methods of spinning fibers (Riley, gen. ref.).

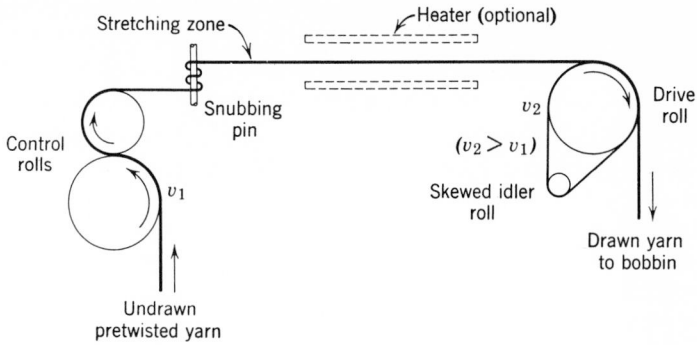

Fig. 21-5. Schematic diagram of a typical drawing process (Riley, gen. ref.).

filaments have traveled far enough to become solid (about 2 ft) they are brought together and wound up. Speeds of about 2500 ft/min are usually employed.

The filaments as spun are almost completely unoriented. Most of the stretching that occurs between the spinneret and windup does so while the filament is still molten, and there is sufficient time for molecular orientation to relax before the fiber cools and crystallizes. Consequently a separate drawing step is necessary to produce the orientation of the crystallites necessary for optimum physical properties. In the drawing step somewhat lower speeds are required than can be achieved in spinning, so the two steps are usually done separately. The drawing machine consists of two sets of rolls (Fig. 21-5), one to feed the undrawn yarn from a supply package at velocity v_1 and the other, moving about four times as fast, to collect the drawn yarn at velocity v_2. The filaments may pass over a metal pin between the two sets of rolls; drawing is localized in the neighborhood of the pin. The yarn is then collected on a strong metal bobbin. Freshly drawn yarn has a tendency to contract somewhat in length. Heat is evolved during the drawing process, since work is done on the polymer.

Dry spinning

In the dry-spinning process, the filament is formed by the evaporation of solvent from the polymer solution into air or an inert gas atmosphere. The solvent must, of course, be volatile. Dry spinning has been used for many years for spinning cellulose acetate from acetone solution, and is now employed for several vinyl fibers including poly(vinyl chloride) and its copolymers, and polyacrylonitrile spun from solution in organic solvents.

So far no process is used in which the polymer is made in the same solvent from which it is dry spun. Therefore, the first step in dry spinning is always the preparation of the polymer solution. The choice of solvent for spinning is based on considerations of solvent power, boiling point, heat of evaporation, stability, toxicity, ease of recovery, etc. Nonpolar solvents are preferred because of convenient boiling points and nonhygroscopic nature, but may cause hazardous buildup of static charge. Low-boiling solvents with high heats of evaporation may cause polymer precipitation or coagulation on the surface of the fiber, in part due to condensation of moisture as the fiber cools when solvent evaporates. This leads to loss of luster and strength. Acetone is preferred for many dry-spinning operations. High solution concentrations are desired, and elevated temperatures are often used to keep viscosities from becoming too high. Cellulose acetate is spun from 20–45% solutions with 400–1000 poise viscosity at 40°C. As in other types of spinning, polymer solutions are usually filtered just before dry spinning.

Spinning is done in a vertical tubular cell (Fig. 21-4*b*), jacketed for temperature control, in which air, steam, or inert gas may be passed through either concurrently or countercurrently as required. Downward spinning is preferred for small-denier fibers and upward spinning for high deniers, to control draw better by eliminating the influence of gravity. Dry spinning is carried out at rates as high as 2500–3000 ft/min, about the same as melt spinning.

Wet spinning

In the wet-spinning process, a solution of a polymer or polymer derivative is spun through a spinneret into a liquid which can coagulate the polymer or derivative (Fig. 21-4*c*). The chemical reaction to convert a derivative to the final polymer can take place simultaneously or later. Wet spinning is commonly used for viscose, cellulose, and some synthetic fibers such as polyacrylonitrile spun from salt solutions.

The essential feature in wet spinning is the transfer of the mass of the solvent from the polymer to the coagulating bath. This transfer is not accompanied by heat of solidification of the polymer, as in melt spinning, or heat of evaporation of the solvent, as in dry spinning. A heat of chemical reaction may be present but is not an essential part of the process.

As in other types of spinning, the surface or interfacial tension forces around the filament are quite strong so that it is difficult to produce at the spinneret a filament with a cross section other than circular. These forces

also tend to make the filament break up into drops. This tendency is opposed by the viscosity of the polymer; hence more viscous solutions are easier to spin.

It is also advantageous to work with as high concentration of polymer as possible, for obvious reasons. This again leads to high viscosity, and the practical limit may be reached above which the polymer cannot be filtered, pumped, or extruded. For this reason some polymers are wet spun at elevated temperatures where viscosity is lower. Higher temperatures usually promote crystallization and lead to denser, stronger fibers. Cellulose fibers may be spun at 50°C, acrylonitrile fibers as high as 160–185°C.

During coagulation several processes take place simultaneously, including diffusion, osmosis, and salting out. Because of the interplay of these processes, coagulation occurs in rather different ways for different fiber-solvent systems. Usually, coagulation is rapid and the fibers cannot be stretched greatly during this step. The skin of the fiber sets to a gel, and later the volume of material in the core is reduced as the rest of the solvent is removed. The skin then has to fold to accommodate the reduced volume. This gives a characteristic wrinkled cross section to some fibers, such as viscose rayon (Fig. 21-3). It is difficult to produce by wet spinning fibers whose final cross section is circular, despite their initial production at the spinneret in this shape. As a result of the shrinkage, the skin of the fibers is oriented much more than the core.

In other cases, coagulation is much slower and the filaments may be stretched to as much as 30 times their original length before coagulation (as in the spinning of cellulose from cuprammonium solutions into water).

The need for sufficient time for coagulation and other treatments depending on diffusion, and the considerable viscous drag of the coagulating baths, limit the rates of wet spinning to 150–300 ft/min in the usual cases. These speeds are low enough that drawing can be done immediately after spinning in a continuous operation.

Wet-spun yarn may be collected on a bobbin or as a loose cake, which is subjected to chemical treatments, washing, and drying. Alternately, these steps may be carried out continuously as indicated in Fig. 21-4*c*.

GENERAL REFERENCES

Frey, W., and A. Sippel, "Fibres by Dry Spinning Processes," Chap. 16 in R. Hill, ed., *Fibres from Synthetic Polymers*, Elsevier Publishing Co., New York, 1953.

Lodge, R. M., "Fibres by Melt Extrusion," Chap. 14 in R. Hill, ed., *op. cit.*

Preston, J. M., "Fibres by Wet Spinning Processes," Chap. 15 in R. Hill, ed., *op. cit.*

Riley, J. L., "Spinning and Drawing Fibers," Chap. XVIII in C. E. Schildknecht, ed., *Polymer Processes*, Interscience Publishers, New York, 1956.

C. Fiber After-treatments

Although it is beyond the scope of this book to treat in detail the production of fabrics from fibers, there are certain intermediate steps between spinning and weaving which are ultimately dependent on the physical, chemical, and molecular structure of the fiber.

All natural and some synthetic fibers must be washed or scoured to free them of natural oils, dirt, chemicals, and other foreign impurities. They must be lubricated or sized or both for proper processing into cloth. They will very likely be dyed for pleasing appearance, and they may have any of a variety of treatments applied to impart or control crease resistance, softening, water repellency, slipping, dimensional stability, shrinkage, or many other properties.

After some or all of these treatments, the fiber is ready to be transformed into a fabric. The most important method of doing this is weaving, in which a set of yarns running lengthwise (warp) is interlaced with a second set at right angles (filling). Other methods of producing fabrics include knitting, in which a series of yarns is looped together, and the manufacture of nonwoven fabrics, papers, and felts, whereby the fibers are bonded together into flat sheets by heat, pressure, and, possibly, bonding agents. Variations of the weaving process lead to pile fabrics, laces, braids, etc. The properties of the fabric depend on the fiber properties, the yarn construction (lengths and diameters of the fibers, size of yarns, amount of twist), the fabric construction (number of yarns per unit area and patterns in which they are combined), and the finishes applied to the fibers and the fabric.

Scouring

The removal of impurities from textile materials, called *scouring*, is carried out by the use of surface-active agents such as soaps and synthetic detergents. These materials have the property of reducing the surface tension of water, and their main function in scouring is to reduce the interfacial tension of their solution toward fats and oils. Most textile fibers contain such fats, oils, or waxes, either naturally or inadvertently or purposely added during various operations. The principal task of scouring is to remove these substances, since most of the other ingredients of the soil are embedded in the fatty material. On repeated laundering, however, some fine soil may be redeposited on the fat-free surface of the fiber, find its way into cavities on the surface, and become extremely resistant to removal.

In contrast to initial scouring, one important function of detergents in household and laundry use is to prevent this redeposition of soil.

Lubrication

Lubrication of fibers is necessary to reduce their friction against themselves and against elements of the processing machinery. In the lubrication of yarns, it is necessary to preserve a high static coefficient of friction to keep the yarn in place on its spool or cone, while reducing the dynamic coefficient so as to obtain high speed of movement of the yarn without the generation of heat.

Lubricants may be vegetable or mineral oils or suitably refined petroleum products. Vegetable oils have been preferred because of easy and complete removal by saponification. The best lubricants are water soluble, such as the poly(alkylene glycols).

Another function of the lubricant, which is especially important in the case of the moisture-resistant synthetic fibers, is to reduce the static electric charge on the fibers by lowering their surface resistivity.

Sizing

A *size* is a surface coating used to protect the yarn during weaving. It makes the yarn smooth by binding protruding fibers onto the core of the yarn. Since the fabric is usually dyed after weaving, the size must be easily removed.

Starch is used almost exclusively to size cotton and, to some extent, rayon and wool. It is unsuitable for continuous-filament yarns of rayon, acetate, or the synthetics because of poor adhesion. Gelatin, poly(vinyl alcohol), or other polymeric materials are used to size these yarns.

Dyeing

Dyeing consists in placing a fiber in an (aqueous) solution of a dye and leaving it there until an equilibrium is established in which most of the dye is adsorbed on the fiber and only a small part remains in the dye bath. When equilibrium is reached, the dye concentration is uniform throughout the cross section of the fiber. To attain these conditions, the dye must be preferentially adsorbed in the fiber by means of some type of intermolecular bond such as a van der Waals or hydrogen bond. It is clear also that because of the size of the dye molecules they can penetrate only the amorphous regions of the polymer. Two requirements for successful dyeing are,

then, sites to which dye molecules can bond and amorphous regions in the fiber.

Natural cellulosic and protein fibers were known for centuries to dye well with acid dyes which bonded to the hydroxyl or amino groups in the fibers. When cellulose acetate was introduced, a new problem arose which was solved by the use of so-called dispersed dyes containing terminal amine or hydroxyl groups. They are water insoluble and are used as dispersions, hence the name. It seems probable that these dyes undergo hydrogen bonding to the carbonyl oxygen of the acetate groups.

Synthetic fibers have often been very difficult to dye. Polyamide, polyester, and acrylic fibers contain sites for bonding dye molecules, but have such compact structures that dye absorption is extremely slow. Rate of dyeing has been increased by going to higher temperatures, or by utilizing swelling agents for the polymers. The dye sites can be modified, as, e.g., in the treatment of acrylic fibers with copper salts so as to form sites having affinity for anionic dyes. Such fibers as poly(vinyl chloride), poly(vinylidene chloride), and polyethylene contain no sites for dye absorption. It is necessary to copolymerize these monomers with small amounts of materials which do provide dyeing sites in order to obtain satisfactory coloration.

Finishing

Finishes which affect hand Under the heading of finishes which affect the feel or hand of fabrics come those which increase or decrease the natural friction between the fibers. Thus lubricants make the fabrics feel softer or have increased pliability. On the other hand, antislip finishes such as rosin, carboxymethyl cellulose, and hydroxyethyl cellulose impart a harsh feel to the fabric.

Conditioners Ethylene and propylene glycols and their low polymers, with the ability to absorb and retain moisture, act as softening, plasticizing, and antistatic agents for the hydrophilic textile fibers, and can counteract to some extent the harshness or stiffness which accompanies the use of antislip agents.

Water-repellent finishes Treatment with waxes such as paraffin emulsions imparts water repellency to cellulosic fabrics, but the effect is not permanent. Permanence can be enhanced by subsequent treatment of the fabric with aluminum or zirconium salts in solution, but the best water repellents at present are those which use stearoylamide pyridinium chloride. This molecule contains the stearoyl radical to give the desired water repellency. the amidomethyl radical which can form a semiacetal

with a hydroxyl group of cellulose, and the pyridinium group to solubilize the molecule in water so as to bring it into intimate contact with the fiber. Baking the fiber causes the molecule to split off pyridine hydrochloride, leaving the stearoyl amidomethyl group attached to the cellulose.

Silicones also impart water repellency to fabrics.

Wash-wear finishes One of the major advances in textile technology in recent years has been the introduction of resin finishes which impart the wrinkle- and crease-resistant properties associated with "wash and wear" fabrics. Ordinarily used with cotton and viscose rayon, resin finishes are currently applied to the majority of shirts, blouses, dresses, trousers, and retail piece goods produced.

The resin finishes consist of aqueous solutions of urea-formaldehyde or melamine-formaldehyde precondensates, or of cyclic ureas such as dimethylol ethylene urea. It is thought that wrinkle resistance is imparted through the crosslinking of adjacent cellulose chains in the fiber, rather than from resin formation in or between the fibers. The deposition of the crosslinked polymer between fibers is specifically avoided, since it leads to stiffness and increased body of the fabric.

The fabric is treated with an aqueous solution of the monomers containing an acid or acid-generating catalyst. Excess liquid is removed, and the fabric is dried and cured by heat treatment. Crosslinking is demonstrated by the reduced swelling of the fibers. The treatments function by reducing the irreversible elongation of the fiber: a greater force is needed to stretch the fiber, but a larger percentage of the elongation is recovered.

Textured yarns

The ability to stabilize the physical form of the thermoplastic fibers by heat setting has been utilized in the production of a number of "textured" yarns which have unusually high elasticity. These yarns are made by imparting a tight twist to continuous-filament yarn (usually nylon), heat treating to "set" the twist, and then untwisting the yarn. The resulting coils and twists in the individual fibers can be reversibly pulled out to 200–400% elongation. Variations of the process exist, and other textured yarns are made by many different processes.

Nonwoven fabrics

A nonwoven fabric is a web or continuous sheet of staple-length fibers, aldidown mechanically. The fibers may be deposited in a random manner

*TABLE 21-2. Properties of fibers**

Property \ Fiber	Acetate	Triacetate	Viscose: Regular	Viscose: High Tenacity	Nylon 66: Continuous Filament	Nylon 66: Staple	Vinyl Chloride-Acetate Staple	Polyester: Continuous Filament	Polyester: Staple	Vinylidene Chloride Continuous Filament	Acrylic Staple	Cotton	Silk	Wool
Specific gravity g/cm^3	1.30	1.30	1.50	1.50	1.14	1.14	1.33	1.38	1.38	1.70	1.14	1.50	1.25	1.30
Tensile strength, 10^3 lb/in.^{2}	18–23	20–23	29–47	58–97	66–88	58–105	10–12	77–88	56–78	25–60	33–38	42–125	45–83	17–28
g/denier	1.1–1.4	1.2–1.4	1.5–2.4	3.0–5.0	4.5–6.0	4.0–7.2	0.6–0.7	4.4–5.0	3.2–4.4	1.1–2.9	2.3–2.6	2.1–6.3	2.8–5.2	1.0–1.7
Wet strength, % dry	60–70	67–70	44–54	55–75	85–90	85–90	100	100	100	100	80	110–130	75–95	76–97
Ultimate elongation, %	25–45	25–40	15–30	9–20	26	16–45	10–13	19–23	30–36	20–35	20–28	3–10	13–31	20–50
Recovery, % from	23	43	30	37	100	...	...	80	...	95	...	45	33	63
elongation, %	20	10	20	15	8	...	...	8	...	10	...	5	20	20
Stiffness, 10^{10} dynes/cm^2	3.0–4.8	3.5–5.2	6.4–9.1	9.9–23.6	3.0	1.5	3.3–4.5	12.2	6.1	0.3–1.7	5.0	5.7–11.2	8.4–12.9	2.7–3.9
g/denier	26–41	35–45	48–68	74–176	30	1.5	28–38	100	50	2–11	50	42–82	76–117	24–34
Moisture regain, % at 21°C, 65% R.H.	6.3–6.5	3.2	11.5–16.6	11.5–16.6	4.0–4.5	4.0–4.5	0	0.4	0.4	0	1.5	8.5	11.0	17.0
Melting or softening temperature, °C	260M	300M	...	...	265M	265M	127M	265M	265M	125S	245S	...	...	...
Loss in strength in sun, prolonged exposure	Yes	Slight	Yes	Yes	Yes	Yes	No	Yes	Yes	No	No	Yes	Yes	Yes

* *Source:* *Modern Plastics* Magazine, Encyclopedia Issue, Breskin Publications, New York, 1961.

or preferentially oriented in one direction. The sheet is then bonded together with an adhesive.

Almost all fibers can be used in the production of nonwoven fabrics, those most widely employed being cotton, rayon, and nylon. Binders include poly(vinyl acetate), rubber lattices, dispersions of other vinyl polymers, amino resins, and water-soluble cellulose derivatives.

The major use for nonwoven fabrics was once in disposable products, such as tea bags, diapers, and sanitary goods, and industrial wiping cloths. However, their application in household items such as draperies and in interfacings for garments, backings for coated fabrics, etc., has increased rapidly.

GENERAL REFERENCES

Coke, C. E., "Non-Woven Fabrics," pp. 1–38 in J. E. Lynn and J. J. Press, eds. *Advances in Textile Processing*, Vol. 1, Interscience Publishers, New York, 1961.
Moncrieff, R. W., *op. cit.*, Part V.
Press, J. J., ed., *op. cit.*, especially Chaps. V, VIII–XVI.
Vickerstaff, T., *The Physical Chemistry of Dyeing*, Oliver and Boyd, London, 1950.

D. Table of Fiber Properties

Table 21-2 lists comparative properties for typical examples of the major textile fibers. It should be noted that the properties of fibers of the same chemical type can vary widely depending on precise composition, heat treatment, yarn structure, and many other variables. The values in the table should be used only for comparative purposes.

22

Elastomer Technology

The final class of high polymers to be considered in Part V is elastomers. Like fibers, elastomers are considered apart from other polymeric materials because of their special properties. Unlike fibers, elastomers do not in general lend themselves to plastics uses; elastomers must be amorphous when unstretched and must be above their glass transition temperature to be elastic, whereas plastics must be crystalline or must be used below this temperature to preserve dimensional stability.

In review of earlier chapters, it may be recalled that the unique properties of elastomers include their ability to stretch and retract rapidly, exhibit high strength and modulus while stretched, and recover fully on release of the stress. To obtain these properties, certain requirements are placed upon the molecular structure of the compounds: they must be high polymers, be above their glass transition temperatures, be amorphous in the unstretched state (but preferably develop crystallinity on stretching), and contain a network of crosslinks to restrain gross mobility of the chains.

A. Compounding and Elastomer Properties

The process by which a network of crosslinks is introduced into an elastomer is called *vulcanization*. The chemistry of vulcanization (Section *B*) is complex and has not been well understood throughout the century of practice of the process since its discovery by Goodyear in 1839. The profound effects of vulcanization, however, are clear: it transforms an elastomer from a weak thermoplastic mass without useful mechanical properties into a strong, elastic, tough rubber (Table 22-1). The tensile strength, stiffness, and hysteresis (representing loss of energy as heat) of natural rubber before and after vulcanization are shown in Fig. 22-1.

TABLE 22-1. Properties typical of raw and vulcanized natural rubber

Property	Raw Rubber	Vulcanized Rubber*
Tensile strength, psi	300	3000
Elongation at break, %	1200	800
Permanent set	Large	Small
Rapidity of retraction (snap)	Good	Very good
Water absorption	Large	Small
Solvent resistance (hydrocarbons)	Soluble	Swells only

* Not reinforced.

For many uses, however, even vulcanized rubbers do not exhibit satisfactory tensile strength, stiffness, abrasion resistance, and tear resistance. Fortunately, these properties can be enhanced by the addition of certain fillers to the rubber before vulcanization. Fillers for rubber can be divided into two classes: *inert fillers*, such as clay, whiting, and barytes, which may make the rubber mixture easier to handle before vulcanization but have little effect on its physical properties; and *reinforcing fillers*

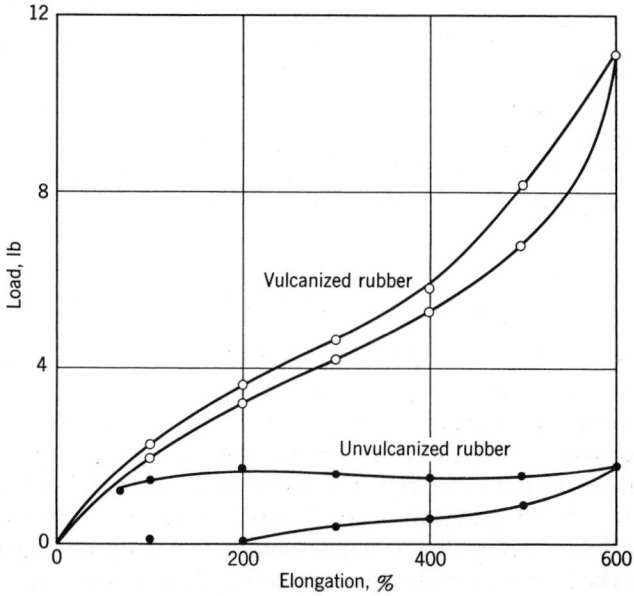

Fig. 22-1. Stress-strain curves to 600% elongation and back, typical of unvulcanized and vulcanized natural rubber (1).

TABLE 22-2. Effect of vulcanization and reinforcement on properties of typical elastomers

Property	Elastomer		
	Natural Rubber	SBR	Butyl
Tensile strength, psi			
Vulcanized	3000	400	3000
Reinforced	4500	3000	3000
Elongation at break, %			
Vulcanized	700	800	1000
Reinforced	600	500	400
400% modulus,* psi			
Vulcanized	400	200	300
Reinforced	2500	2000	1000

* Tensile stress at 400% elongation; a measure of stiffness.

(Section *D*), which do improve the above-named unsatisfactory properties of the vulcanized rubber. Carbon black is the outstanding reinforcing filler for both natural and synthetic rubbers. The added effects of reinforcement with carbon black on the properties of rubbers over the effects of vulcanization alone are illustrated in Table 22-2 and Fig. 22-2. Although the nature of reinforcement is not completely understood, it appears to add a network of many relatively weak "fix points" to the more diffuse network of strong primary bond crosslinks introduced by vulcanization. Vulcanization restrains the long-range movements of the polymer molecules but leaves their local segmental mobility high; reinforcement stiffens the mass and improves its toughness by restricting this local freedom of movement.

Mechanical properties of elastomers

The stress-strain behavior of pure gum (i.e., nonreinforced) vulcanizates of the various elastomers depends markedly on the molecular structure, polarity, and crystallizability of the polymers. Nitrile rubber and SBR, being copolymers of complex and irregular structure, do not crystallize at all, but natural rubber and butyl rubber crystallize on stretching at room temperature. In these elastomers the crystallites have a stiffening effect like that of a reinforcing filler with the result that their stress-strain curves turn up markedly at higher elongations (see Fig. 22-2, lower curves). The similarity of these stress-strain curves to those predicted from the kinetic theory of rubber elasticity (Chapter 6*B*, Fig. 6-5) should be noted.

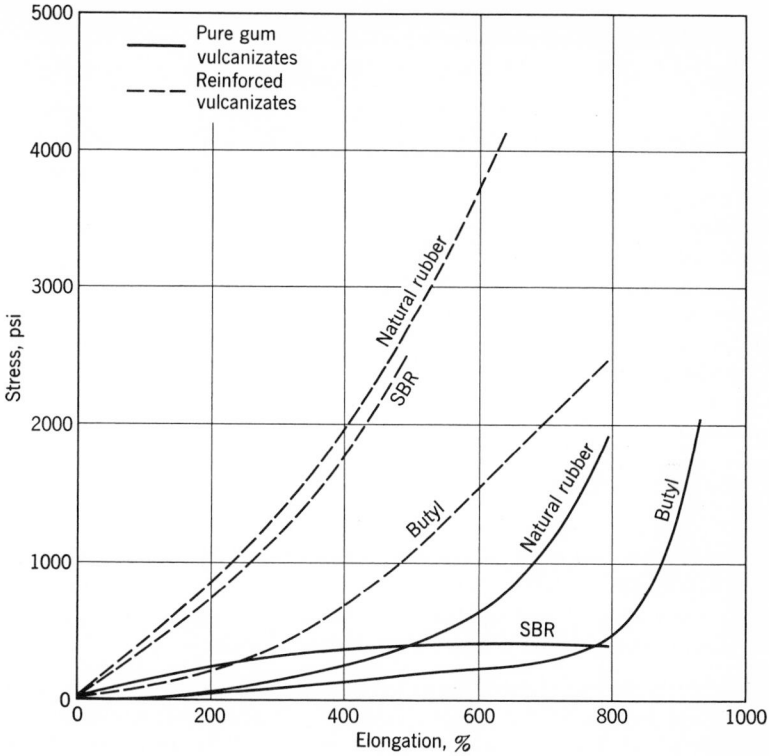

Fig. 22-2. Stress-strain curves typical of several vulcanized and reinforced elastomers (Schmidt and Marlies, gen. ref.).

Unvulcanized rubber undergoes a very large amount of mechanical conditioning when first exposed to tensile stress. Figure 22-3 shows the first two cycles of loading and unloading of a tensile specimen of unvulcanized natural rubber. As successive cycles take place, the changes in resistance to stretching, tensile strength, energy absorption, and permanent set become smaller. This mechanical conditioning is not accompanied by molecular weight degradation. Such degradation does occur, however, in the process of milling raw rubber to reduce its melt viscosity to the desired range before the incorporation of fillers, etc. Here the process is complicated by the presence of oxygen. Vulcanized rubber undergoes a similar but much less extensive change in stress-strain properties on successive cycles of test.

The mechanical properties of elastomers depend greatly upon the rate of testing. Since service conditions for automobile tires (by far the most important end use of elastomers) involve rapid cyclic stresses, dynamic test

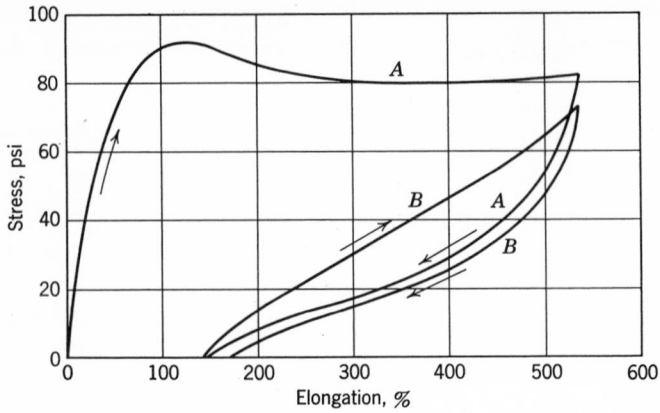

Fig. 22-3. Mechanical conditioning during the first (*AA*) and second (*BB*) cycles of stress-strain, typical of unvulcanized natural rubber (1).

methods are essential if the results are to correlate with the performance of the finished article.

Oxidative aging of elastomers

Rubbers which retain double bonds in their vulcanized structure, such as natural rubber, SBR, and nitrile rubber, are sensitive to heat, light, and particularly oxygen (4). Unless protected with antioxidants (of which the phenyl α- and β-naphthylamines are the most common) the rubbers age by an autocatalytic process accompanied by an increase in oxygen content. The results of the aging may be either softening or embrittlement, suggesting competing processes of degradation and crosslinking of the polymer chains. The first step in the process is presumed to be the same as that in the oxidation of any olefin, i.e., the formation by free radical attack of a peroxide at a carbon atom next to a double bond:

$$-CH_2CH{=}CH- \longrightarrow -\underset{\cdot}{C}H-CH{=}CH- + H\cdot$$

$$H\cdot + -\underset{\cdot}{C}H-CH{=}CH- + O_2 \longrightarrow -\underset{\underset{\underset{OH}{|}}{O}}{\underset{|}{C}H}-CH{=}CH-$$

Subsequent steps may include crosslinking or chain cleavage or involvement of the double bond, probably initially through an epoxide group. Many of the steps in the reaction resemble closely those of certain types of vulcanization.

The relative susceptibility of elastomers to oxidative aging is measured by the rate at which they absorb oxygen at elevated temperatures. Some typical results (3) are shown in Fig. 22-4.

Compounding

As described in Section 20*D*, the term *compounding* describes the selection of additives and their incorporation into a polymer so as to give a homogeneous mixture ready for subsequent processing steps. Except for differences in the nature of the compounding ingredients, especially as required for vulcanization and reinforcement and described under those headings, the compounding of rubbers is similar to that of most plastics. Typical equipment described in Chapter 20*D* is used for rubber as well as plastics compounding.

In the case of natural rubber, SBR, and the other synthetic rubbers produced by emulsion polymerization, the polymer is first available in the form of a latex. As described in Chapter 14*A* for natural rubber, the normal procedure in compounding is to coagulate and dry the latex, and then to masticate and compound the rubber on mills or in other equipment. With SBR, however, two important compounding steps, *oil extending* and *masterbatching*, are commonly carried out before coagulation of the latex.

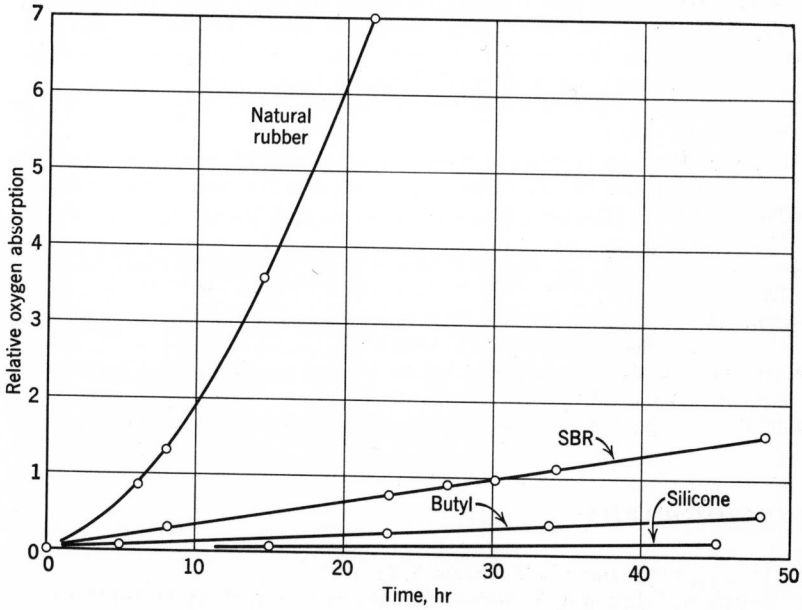

Fig. 22-4. Rate of absorption of oxygen at 130°C by various elastomers (3).

Oil extending (5) The use of hydrocarbon oils to dilute, or extend, rubbers has been known for many years. Since 1950, this practice has been applied to SBR polymerized to high molecular weight. The oil serves as a plasticizer and softener, reducing the melt viscosity of the rubber to that normally required in compounding. The process, which leads to a less expensive final product, is widely used: about 90% of the tire tread stock produced in this country is oil extended to some extent. The oil is added as an emulsion to the latex before coagulation.

Masterbatching (2) A convenient method of mixing rubber and carbon black is the coprecipitation of a mixture of the rubber latex and an aqueous slurry of the black. This method gives adequate mixing when the particle size of the black is similar to that of the latex particle and the number of each type of particle per cubic centimeter before the coagulation can be made nearly the same. This situation holds for SBR and the most useful blacks. Latex masterbatching has now become an established technique in the rubber industry.

GENERAL REFERENCES

Barnhart, R. R., and T. H Newby, "Antioxidants and Antiozonants," Chap. 6 in M. Morton, ed., *Introduction to Rubber Technology*, Reinhold Publishing Corp., New York, 1959.

Bueche, F., *Physical Properties of Polymers*, Interscience Publishers, a Division of John Wiley & Sons, New York, 1962, Chaps. 10, 14.

Garvey, B. S., Jr., "History and Summary of Rubber Technology," Chap. 1 in M. Morton, ed., *op. cit.*

Garvin, G. S., "Compounding and Processing Rubbers and Resins," Chap. XVI in C. E. Schildknecht, ed., *Polymer Processes*, Interscience Publishers, New York, 1956.

McMillan, F. M., "Rubber Plasticizers, Softeners, and Extenders," Chap. 7 in M. Morton, ed., *op. cit.*

Naunton, W. J. S., ed., *The Applied Science of Rubber*, St. Martin's Press, New York, 1961.

Schmidt, A. X., and C. A. Marlies, "Principles of High-Polymer Theory and Practice," McGraw-Hill Book Co., New York, 1948, Chaps. 12, 13.

Whitby, G. S., C. C. Davis, and R. F. Dunbrook, eds., *Synthetic Rubber*, John Wiley & Sons, New York, 1954.

Wolf, R. W., "Nonblack Compounding Ingredients," Chap. 9 in M. Morton, ed., *op. cit.*

BIBLIOGRAPHY

1. Hock, L., and S. Bostrom, *Kautschuk* **2,** 130 (1926).
2. Janssen, H. J. J., and K. V. Weinstock, *Rubber Chem. Tech.* **34,** 1485 (1961).
3. Mesrobian, R. B., and A. V. Tobolsky, *J. Polymer Sci.* **2,** 463 (1947).

4. Shelton, J. R., *Rubber Chem. Tech.* **30,** 1251 (1957).
5. Storey, E. B., *Rubber Chem. Tech.* **34,** 1402 (1961).

B. Vulcanization

Ever since Goodyear's first experiments in heating rubber with small amounts of sulfur, this process has been the best and most practical method for bringing about the drastic property changes described by the term *vulcanization*, not only in natural rubber but also in the diene synthetic elastomers such as SBR, butyl, and nitrile rubbers. It has been found since, however, that neither heat nor sulfur is essential to the vulcanization process. Rubber can be vulcanized or *cured* without heat by the action of sulfur chloride, for example. A large number of compounds which do not contain sulfur can vulcanize rubber; these fall generally into two groups, oxidizing agents (selenium, tellurium, organic peroxides, nitro compounds) and generators of free radicals (organic peroxides, azo compounds, many accelerators, etc.). Thus it is clear that there is no single method or chemical reaction of vulcanization.

Since the chemical reactions associated with vulcanization are varied and involve only a few atoms in each polymer molecule, a definition of vulcanization in terms of the physical properties of the rubber is necessary. In this sense, vulcanization may be defined as any treatment that decreases the flow of an elastomer, increases its tensile strength and modulus, but preserves its extensibility. There is little doubt that these changes are due primarily to chemical crosslinking reactions between polymer molecules. As might be expected, such properties as tensile strength are relatively insensitive to the onset of these crosslinking reactions; tensile strength does indeed change ten-fold during curing, but this is evidence of the profound alteration of polymer properties by the process. Tests based on melt flow are more sensitive to initial crosslinking reactions and are widely used in the rubber industry.

Although vulcanization takes place by heat in the presence of sulfur alone, the process is relatively slow. It can be speeded many-fold by the addition of small amounts of organic or inorganic compounds known as *accelerators*. Many accelerators require the presence of still other chemicals known as *activators* or *promoters* before their full effects are realized. These activators are usually metallic oxides, such as zinc oxide. They function best in the presence of a rubber-soluble metallic soap, which may be formed during the curing reaction from the activator and a fatty acid. The most efficient combination of chemicals for sulfur vulcanization includes sulfur, an organic accelerator, a metallic oxide, and a soap.

Chemistry of vulcanization

It is convenient to group vulcanization reactions into two categories: nonsulfur vulcanizations, in which peroxides, nitro compounds, quinones, or azo compounds are the curing agents; and vulcanizations brought about by sulfur, selenium, or tellurium.

The understanding of vulcanization has been considerably advanced by the recognition of the following two concepts which arise from the analogy of vulcanization with the crosslinking and (oxidative) degradation of simple olefins:

a. The point of initial attack in most vulcanization reactions is a hydrogen atom on a methylene group alpha to the double bond itself.

b. Free radical mechanisms are involved in almost all cases.

Nonsulfur vulcanization From the two concepts just given, the following scheme has been developed for nonsulfur vulcanizations:

a. A free radical R· is formed by the decomposition or oxidation of the curing agent, or as a step in the oxidative degradation of the rubber.

b. This free radical initiates vulcanization by abstracting a hydrogen atom from one of the α-methylene groups (in natural rubber the methyl side group directs the attack to the methylene group nearest it):

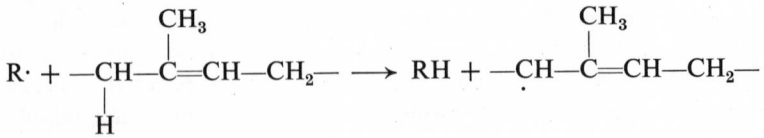

c. The rubber free radical then attacks a double bond in an adjacent polymer chain. This results in the formation of a crosslink and the regeneration of a free radical in a reaction analogous to propagation in an addition polymerization (Chapter 9):

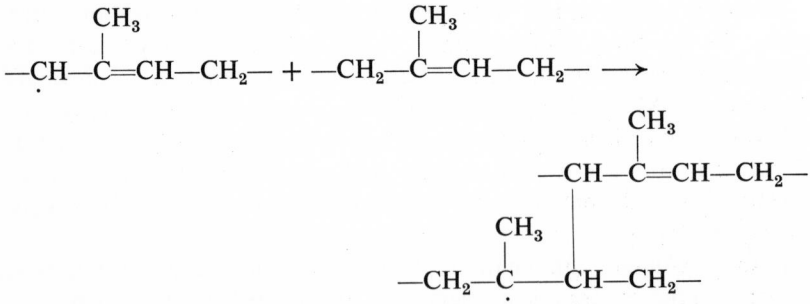

Vulcanization may continue by several such propagation steps.

d. Chain transfer may occur:

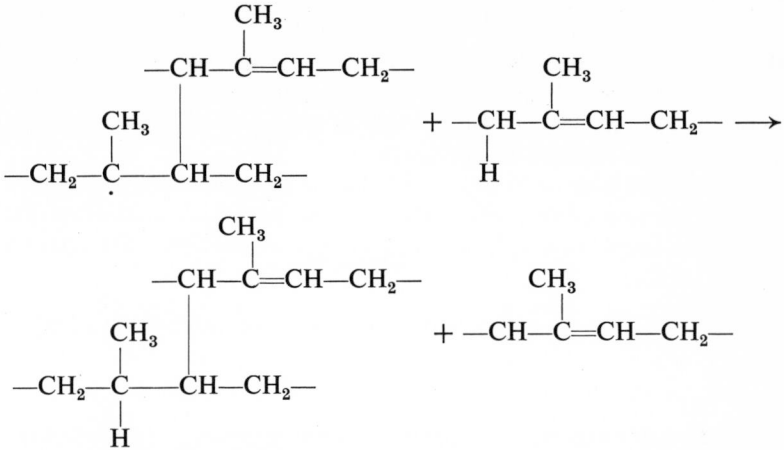

e. Termination probably occurs by reaction of the rubber free radical with a free radical fragment of the curing agent:

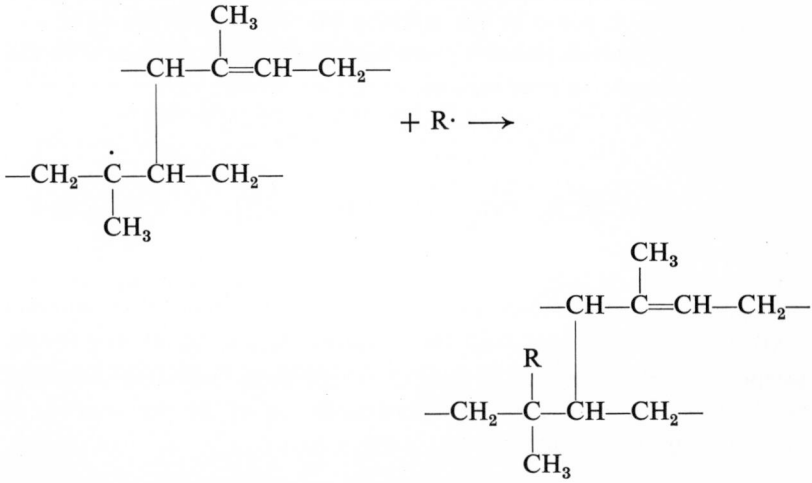

In contrast to addition polymerization in a fluid system, the termination reaction between two rubber free radicals:

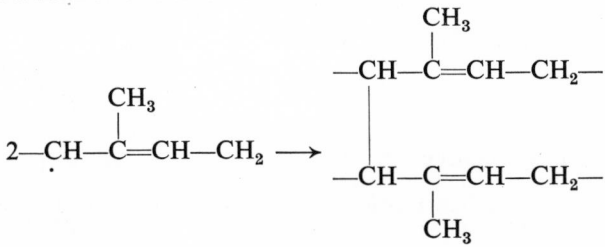

is considered unlikely because of the low probability of two such radicals coming into position to react, owing to the high viscosity of the medium.

Sulfur vulcanization Not all the reactions which take place in sulfur curing are well understood. It is clear that the sulfur can undergo a variety of reactions. Some, but not all, of it goes to form sulfide or disulfide crosslinks between chains, e.g.,

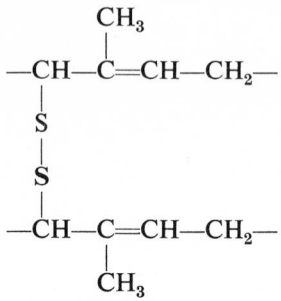

Some double bonds are saturated by reactions analogous to step *c* of the mechanism above, some by the addition of hydrogen sulfide (a known product of sulfur vulcanization—see next paragraph), and some by the formation of cyclic sulfides such as

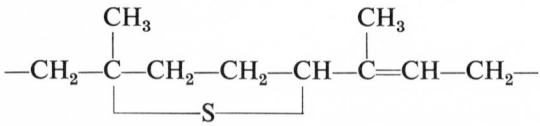

Some sulfur reacts with an activator such as zinc oxide to form the corresponding sulfide.

Most accelerators are either free radical formers or strong proton acceptors (as is sulfur itself). Thus C—C crosslinks may also be formed by the free radical mechanism described above, or by the removal of hydrogens by the proton acceptors, e.g.,

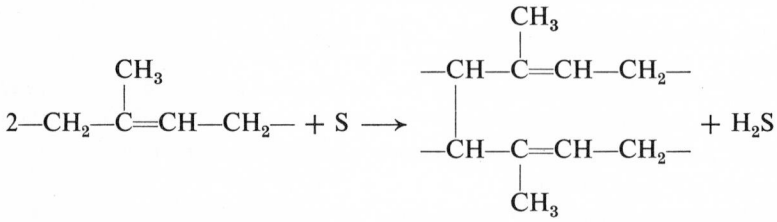

Many other possibilities exist.

Other types of vulcanization Some reagents, such as mercaptans,

thiol acids, and mercapto acids, can add directly to diene double bonds. Bisthiol acids can produce direct interchain crosslinks by this mechanism:

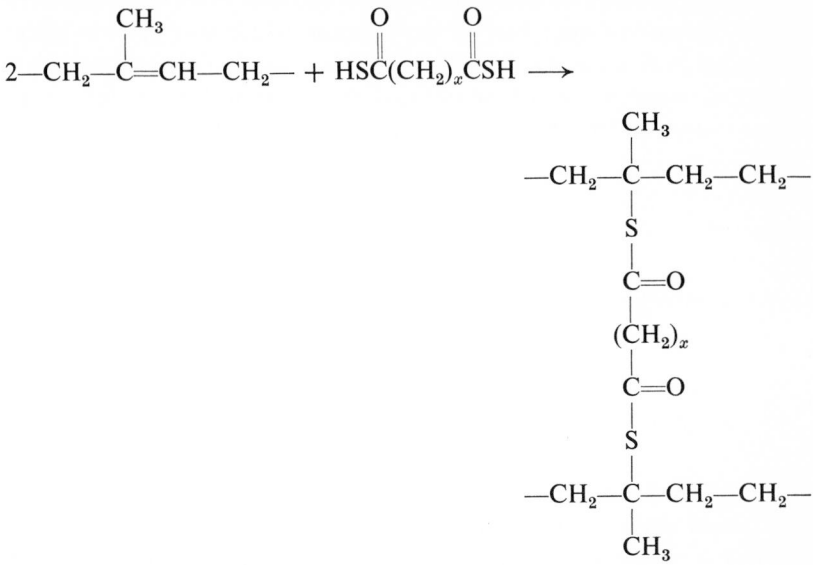

Physical aspects of vulcanization

Vulcanization of the type just described takes place when 0.5–5 parts (by weight) of sulfur is combined with 100 parts of rubber. If the reactions are allowed to continue until considerably more sulfur has combined (say 30–50 parts per 100 parts rubber), a rigid, nonelastomeric plastic known as *hard rubber* or *ebonite* is formed. The changes in physical properties accompanying increasing amounts of combined sulfur are shown in Fig. 22-5. The region of maximum tensile properties in the neighborhood of 5–10 parts combined sulfur is known as the region of *optimum cure* for this *mix* (100 parts rubber and 50 parts sulfur, no accelerator; not a typical commercial mix). Beyond this point the *stock* (i.e., mixture) is said to be *overcured* and is likely to be leathery, i.e., stiffer and harder than at the optimum cure but weaker and less extensible. At high proportions of combined sulfur the strength of the material increases again, and the elongation becomes very low as the hard rubber region is reached.

The properties of soft vulcanizates vary with cure time in the general way depicted in Fig. 22-6. The effect of increasing temperature is, as expected, to move the curves to shorter times. Experiments in which the

composition of the mix is varied show that the initial rise in tensile strength is accompanied by the rapid incorporation of sulfur in the rubber, and that, as the rate of sulfur addition decreases, the rate of increase of tensile strength begins to decline. It is also known that the total amount of combined sulfur does not measure the extent of change in physical properties during cure, and that the ratio of sulfur combined to double bonds used up varies widely from one mix to another.

These facts suggest that processes of crosslinking and degradation occur simultaneously during vulcanization. Additional evidence for this view arises from the fact that badly overcured stocks may undergo *reversion*, i.e.,

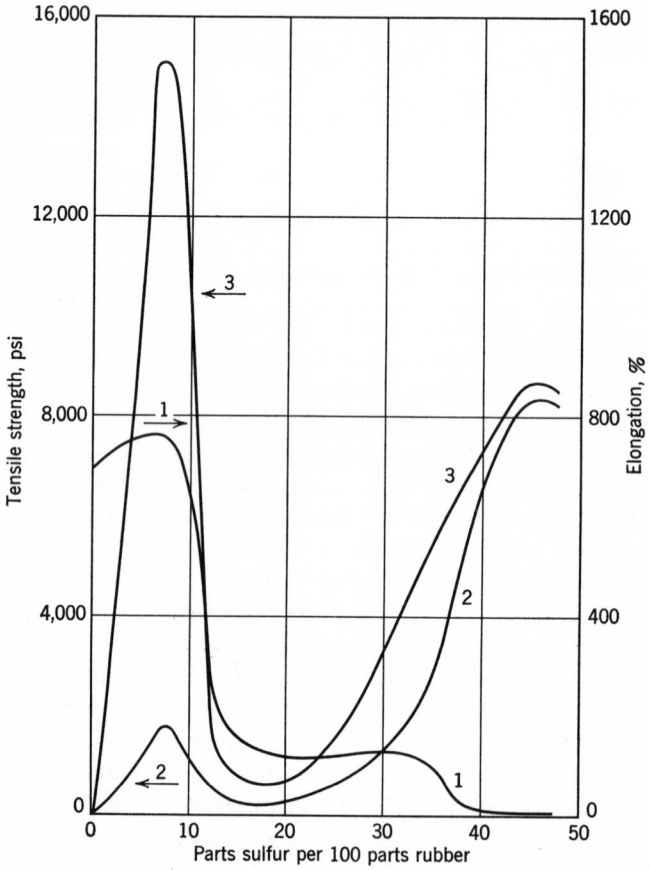

Fig. 22-5. Physical properties as a function of extent of vulcanization, measured by amount of combined sulfur (Schmidt and Marlies, gen. ref.). Natural rubber vulcanized with sulfur only. Curve 1, elongation; curve 2, tensile strength based on initial cross section; curve 3, tensile strength based on cross section at break.

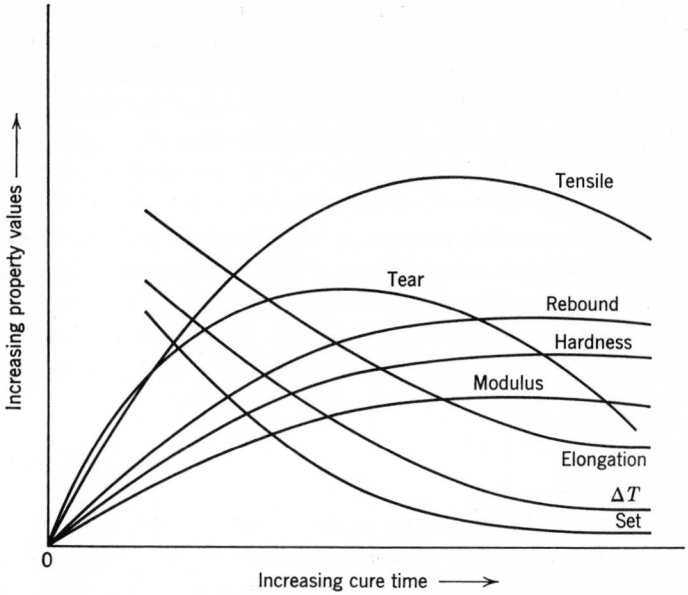

Fig. 22-6. Effect of cure time on the physical properties of elastomers (Garvey, gen. ref.). ΔT is heat buildup in a compression-flex test.

become soft and tacky with notable loss in strength. Such stocks resemble oxidatively degraded rubber. Reversion differs from overcuring in that it usually takes place on prolonged heating when not enough sulfur is present to get into the region of leathery cures.

Accelerators

Until about 1900, rubber vulcanization was a slow process. Large excesses of sulfur over the amount required for optimum cure were used. The remaining free sulfur diffused to the surface of the stock and caused *bloom*. The curing time was several hours, despite the fact that lead, calcium, magnesium, or zinc oxides (inorganic accelerators) speeded the reaction slightly.

The rubber industry was revolutionized by the introduction of organic accelerators for vulcanization. These include sulfur-containing compounds such as thioureas, thiophenols, mercaptans, dithiocarbamates, xanthates, trithiocarbamates, dithio acids, mercaptothiazoles and mercaptobenzothiazoles, and thiuram sulfides, plus a few nonsulfur types such as ureas, guanidines, and aldehydeamines.

Except for the generalization that accelerators are likely to contribute

to vulcanization by initiating free radical chains or by abstracting protons, little is known about their specific action in speeding up vulcanization. Some accelerators, such as thiuram disulfides, are known to dissociate thermally at vulcanization temperatures:

$$R_2N{-}\overset{\overset{\displaystyle S}{\|}}{C}{-}S{-}S{-}\overset{\overset{\displaystyle S}{\|}}{C}{-}NR_2 \longrightarrow CS_2 + S + R_2N\overset{\overset{\displaystyle S}{\|}}{C}NR_2$$

The sulfur is consumed in promoting vulcanization. Since the mechanism may involve free radical steps, further action is possible. Zinc salts of mercaptobenzothiazole and the dithiocarbamic acids may act similarly. On the other hand, an accelerator such as mercaptobenzothiazole itself, which is known not to decompose or to react with olefins at curing temperatures, must always be used with an activator.

Activators

The presence of basic metallic oxides or salts of lead, calcium, zinc, or magnesium appears to be necessary to obtain the full effect of almost all organic accelerators. Solubility of the compound is important; for this reason the oxides are used with organic acids (stearic acid, rosin), or the soaps (stearates, laurates) of the metals are utilized instead. Aside from the specific need for an activator with an acidic accelerator such as mercaptobenzothiazole, as just noted, the presence of the oxide appears to be important in determining the type of crosslinking reactions which take place. These appear to be strong ionic bridges of the type

$$\text{rubber—(accelerator} + \text{S)}^{-}\ Zn^{+2}\ {}^{-}\text{(accelerator} + \text{S)—rubber}$$

formed during vulcanization. In cases in which the activator is important, these bridges may contribute substantially to the crosslinking in the vulcanizate.

TABLE 22-3. Typical composition of pure-gum vulcanizate of natural rubber

Material	Parts by Weight
Rubber	100
Sulfur	0.25–1.5
Accelerator	0.25–1.5
Zinc oxide (activator)	1–10
Stearic acid or zinc laurate	1–5
Antioxidant	0–1.5

Typical vulcanization mix

The majority of sulfur-vulcanized products of natural rubber are based on a mix with the composition range shown in Table 22-3. If no other ingredients are added, the product is known as a *pure-gum vulcanizate. Loaded stocks*, made by adding carbon black or other fillers to this recipe, are discussed in Section *C*.

GENERAL REFERENCES

Craig, D., *Rubber Chem. Tech.* **30,** 1291 (1957).
Garvey, B. S., Jr., "History and Summary of Rubber Technology," Chap. 1, and "Accelerators of Vulcanization," Chap. 5, in M. Morton, ed., *op. cit.*
Oneacre, L. E., "Vulcanization," Chap. 4 in M. Morton, ed., *op. cit.*
Scheele, W., *Rubber Chem. Tech.* **34,** 1306 (1961).
Schmidt, A. X., and C. A. Marlies, *loc. cit.*

C. Reinforcement

Pure-gum vulcanizates compounded according to Table 22-3 are relatively soft, pliable, and extensible, and are most useful for such items as rubber bands, tubing, and gloves. Rubber would be of little value in modern industry were it not for the fact that certain fillers enhance such mechanical properties as tensile strength, stiffness, tear resistance, and abrasion resistance of the stocks into which they are incorporated. Such loading agents are called *reinforcing fillers*, and the improvement in service life of a heavy-duty rubber item, such as a belt or tire tread, resulting from the enhancement of these properties, is called *reinforcement*.

Types of fillers

Only a few fillers are reinforcing for even one or two of the important mechanical properties of natural rubber, while some weaken the vulcanizates in one or more important respects. The latter are known as *inert fillers*. In general it is found that the reinforcing action of a filler depends upon its nature, the type of elastomer with which it is used, and the amount of filler present.

Carbon black is the only important reinforcing filler for most elastomers, although silica and silicate fillers (1) can be used in some cases, as with

the silicone elastomers (Chapter 19*F*). Important nonreinforcing fillers include talc, zinc oxide, and magnesium carbonate.

Nature of the filler The chemical composition of the filler is a primary determinant of its reinforcing action. Carbon blacks are always more efficient reinforcing agents than, e.g., talc, even though particle size, surface condition, and other properties can be varied widely for both. Within a single chemical class, particle size and surface condition are important variables. Their effects on reinforcement by carbon black are discussed on page 544.

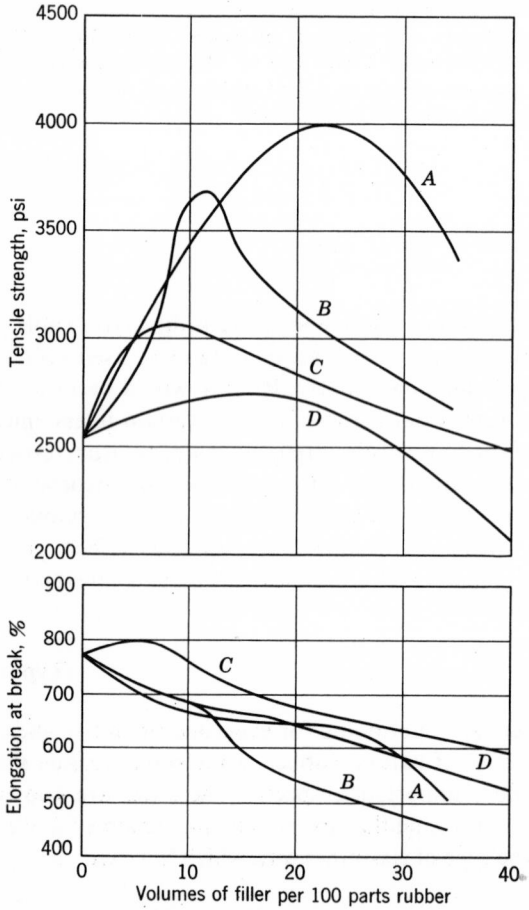

Fig. 22-7. Effect of filler loading on tensile properties typical of natural rubber (2). Curve *A*, carbon black; curve *B*, china clay; curve *C*, zinc oxide; curve *D*, whiting.

Nature of the elastomer Although a good reinforcing filler such as carbon black increases the tear and abrasion resistance of all the current major elastomers, it has no effect in enhancing the tensile strength of neoprene or butyl rubber, in contrast to the marked tensile reinforcement it causes in natural rubber, SBR, nitrile rubber, and even the polysulfide elastomers. These differences are ascribed to a conflict between two effects of the filler. (*a*) A reinforcing filler stiffens and strengthens the structure by introducing a network of many relatively weak fix points. (*b*) Simultaneously, it may interfere with the ability of the polymer to crystallize at high elongations, simply by its bulkiness in the system. Since stiffness and tensile strength are greatly enhanced by crystallization, the specific effect of the filler depends upon which of these actions predominates.

Amount of filler Increasing amounts of reinforcing filler cause continuing improvement in properties until a maximum is reached, representing the optimum loading for this composition. Beyond this point, additional filler merely acts as diluent and the properties of the vulcanizate again deteriorate. This is illustrated in Fig. 22-7.

Carbon black

Various colloidal forms of carbon constitute the most important reinforcing filler for rubber. Carbon blacks are the only inexpensive materials that reinforce all three of the important properties—tensile strength, tear resistance, and abrasion resistance. Many different carbon blacks are used, differing mainly in particle size, surface condition, and degree of agglomeration. Roughly, the degree of reinforcement increases with decreasing particle size of the black down to the practical lower limit at which blacks can be made, about 100 A diameter.

Types of carbon black *Channel blacks* are made by burning natural gas in a limited supply of air and allowing the carbon to deposit on cold iron channels. They are sometimes designated by processability, as easy-processing channel, etc. Particle diameters range from 200 to 300 A. *Furnace blacks* are replacing channel blacks in many uses. They are made by burning gas or oil in a furnace in limited air and removing the carbon from the off-gases by a centrifugal and electrostatic precipitation. They range from 300 to 800 A in particle diameter, and are often designated by their effects on the vulcanizates, as high-abrasion, high-modulus, semi-reinforcing, or conductive furnace blacks. Coarser blacks, such as *thermal blacks* (from hydrocarbons decomposed in the absence of air), *acetylene blacks*, and *lamp blacks*, are less commonly used as reinforcing fillers.

Structure of carbon blacks The electron microscope shows that the carbon blacks are essentially spherical, with a tendency in some cases to align in continuous chainlike structures. These chains form electrically conducting paths through the rubber, hence the names conductive channel and conductive furnace blacks. Surface areas calculated from electron microscope measurements of particle diameter and by nitrogen absorption agree well, indicating that the particles are relatively nonporous. They have a crystal structure similar to that of graphite except that the sheets of hexagonally close-packed carbon atoms are arranged randomly above one another. This suggests that unsatisfied bonds are present which may be important in reinforcement. The surfaces of the particles contain much adsorbed material, including hydrogenated and oxygenated structures. As a result the *p*H of a water slurry of the black may range from 2.5 to 11.

Effects of carbon black structure on reinforcement The three parameters, particle size, *p*H, and *structure index*, i.e., tendency to agglomerate or form chains, appear to be predominant in determining the reinforcing behavior of blacks. Tensile strength, abrasion and tear resistance, hardness, and toughness increase with decreasing particle size, whereas rebound and ease of processing become poorer. Blacks with a high structure index are difficult to disperse and give high stiffness and hardness but low tensile strength, toughness, and electrical resistivity. The *p*H influences the rate of vulcanization, depending upon the acidic or basic nature of the accelerator.

The stiffness of a reinforced vulcanizate rises steeply with the loading of carbon black and is nearly independent of the particle size of the black. These two features are unique for this property. Other properties, such as tensile strength, go through a maximum as loading is increased (Fig. 22-7). They also depend upon particle size, as mentioned above.

The tensile strength of natural rubber can be increased about 40% by reinforcement. In SBR, however, the properties of the pure-gum vulcanizates are very poor, since the elastomer does not crystallize. The tensile strength of SBR can be raised about ten-fold by reinforcement, making it equivalent to natural rubber for fully reinforced stocks. The tear resistance of SBR is also poor for unloaded vulcanizates, but is equivalent to that of natural rubber after reinforcement (depending somewhat on the SBR; see Chapter 14*B*).

Abrasion resistance is highly important for heavy-duty rubber such as tire tread stock. The abrasion resistance of both natural rubber and SBR can be improved at least five-fold by proper reinforcement. Unfortunately, the resilience of the rubber decreases with increasing loading of filler. As a consequence hysteresis loss and heat buildup increase. Thus the use of a

reinforcing filler represents a compromise between adequate abrasion and tear resistance and abnormal heat buildup.

Theories of reinforcement Theoretical approaches to reinforcement can be divided into efforts to ascertain what types of bonds exist between the carbon black particles and the elastomer chains, and attempts to explain the dependence of the various physical properties of vulcanizates on loading and on particle size of the blacks.

Although the exact nature of the rubber-carbon black bonds has not been ascertained, their presence has been demonstrated in several ways. The viscosity of suspensions of carbon black in solutions of SBR in hydrocarbon solvents is considerably greater than expected, and corresponds to the adsorption of a layer of rubber hydrocarbon 100–200 A thick around each carbon particle. It is also known that, when a mixture (unvulcanized) of carbon black and rubber is dissolved in benzene, not all the rubber will enter into solution. The amount which remains adsorbed on the carbon black is proportional to the surface area of the particles. There was also a linear relation between the amount of adsorbed rubber and the abrasion resistance of the tread stocks made from the mixes used in these experiments.

Since this adsorption appeared to involve bond forces stronger than the typical secondary bond polymer-solvent interactions, the heats of adsorption of a number of compounds on carbon black were measured. Simple gases and low-molecular-weight hydrocarbons, both saturated and unsaturated, gave heats of adsorption corresponding to typical secondary bonds, but dihydromyrcene, a low polymer of isoprene, gave values suggesting a few points of primary bond crosslinking. It is now recognized (3, 4) that these primary bonds are responsible for the phenomenon of reinforcement. They result from the reaction of rubber free radicals, formed during milling and vulcanization, with active sites on the surface of the carbon black particles.

The stiffening effect of both reinforcing and inert fillers has been moderately well accounted for by combining the equations of the kinetic theory of rubber elasticity (Chapter 6*B*) with those for the viscosity of solid spheres or rodlike particles in solution.

GENERAL REFERENCES

Schmidt, A. X., and C. A. Marlies, *loc. cit.*
Snyder, J. W., and M. H. Leonard, "Carbon Black," Chap. 8 in M. Morton, ed., *op. cit.*
Studebaker, M. L., *Rubber Chem. Tech.* **30,** 1400 (1957).

*TABLE 22-4. Typical properties of commercial elastomers**

Property \ Elastomer	Natural Rubber	SBR Normal	SBR Cold	Neoprene	Butyl	Urethan	Nitrile	Polysulfide	Acrylate	Silicone	Fluorocarbon	*cis*-1,4 Isoprene
Tensile strength, psi	4500	3000	3800	4000	3000	2000	2500	1200	2500	1500	1000	4500
Elongation, %	600	500	550	800	400	300	550	400	400	600	600	600
Modulus, psi, 300–400% elongation	2500	2000	2500	1000	1000	1200	1500	1400	...	...	250	2500
Dynamic properties	Excel.	Fair	Good	Fair	Poor	Good	Poor	Poor	Good	Poor	Poor	Excel.
Permanent set	Low	Low	Low	Moder.	Moder.	Moder.	Moder.	High	Low	High	High	Low
Tear resistance	Good	Poor	Fair	Good	Excel.	Good	...	...	Good	Poor	...	Good
Abrasion resistance	Fair	Good	Good	Good	Good	Excel.	...	Poor	Good	Poor	...	Fair
Gas permeability	High	High	High	Moder.	Low	...	Moder.	Low	Low	...	...	High
Upper use temp., °C	100	100	100	80–100	120	100	140	80–150	200	250	250	100
Lower use temp., °C	−60	−55	−55	−45	−50	−45	−15 to −55	−40	−25	−90	−15 to −40	−60
Oxidative and ozone resistance	Poor	Fair	Fair	Good	Good	Good	Good	Good	Excel.	Excel.	Excel.	Fair
Solvent resistance	Poor	Poor	Poor	Good	Poor	Good	Good	Good	Good	Excel.	Excel.	Poor

* Sources: Morton, gen. ref., Section *A*; Whitby *et al.*, gen. ref., Section *A*; and references listed in other chapters for the corresponding elastomers.

BIBLIOGRAPHY

1. Bachmann, J. H., J. W. Sellers, M. P. Wagner, and R. F. Wolf, *Rubber Chem. Tech.* **32,** 1286 (1959).
2. Barron, H., and F. H. Cotton, *Trans. Inst. Rubber Ind.* **7,** 209 (1931–32).
3. Garten, V. A., and G. K. Sutherland, *Proceedings of the 3rd Rubber Technology Conference*, W. Heffer and Sons, Cambridge, England, 1954, pp. 536–552.
4. Watson, W. F., *Proceedings of the 3rd Rubber Technology Conference*, W. Heffer and Sons, Cambridge, England, 1954, pp. 553–564.

D. Table of Elastomer Properties

Table 22-4 lists values of some physical and chemical properties of major commercial elastomers. In comparison with the corresponding tables in Chapters 20 and 21, it should be noted that the properties of elastomers depend in very large extent on the details of vulcanization, reinforcement, and compounding. It is, consequently, almost futile to assign "typical" values of properties according to polymer type. Therefore, most of the data in Table 22-4 are of no more than qualitative significance, and comparisons should be made with extreme caution.

Appendix I

List of Symbols

Symbols in dimensional formulas have the following significance:

M = mass T = time Q = electric charge
L = length θ = temperature

Symbol	Dimension	Common Unit	Definition	Chapter Where First Defined
A	L	...	Angstrom unit	
A	ML^2T^{-2}	kcal	Helmholz free energy, work content	1*C*
A_1	...	mole/g	First virial coefficient	3*B*
A_2	$M^{-1}L^3$	ml mole/g^2	Second virial coefficient	3*B*
A_3	$M^{-2}L^6$	ml^2 mole/g^3	Third virial coefficient	3*B*
A	T^{-1}	sec^{-1}	Collision frequency factor (first-order reaction)	9*F*
A	$M^{-1}L^3T^{-1}$	liter/mole sec	Collision frequency factor (second-order reaction)	9*F*
a	...	...	Arbitrary constant	
a	...	...	Exponent in modified Staudinger equation	3*E*
a	L	A	Length of crystal unit cell axis	5*B*
a_T	...	...	"Shift factor" for time-temperature superposition	6*C*
b	...	...	Arbitrary constant	
b	L	A	Length of crystal unit cell axis	5*B*
b	L	cm or A	Size parameter of Gaussian distribution	6*B*

Symbol	Dimension	Common Unit	Definition	Chapter Where First Defined
C	...	...	Arbitrary constant	
C	...	...	Transfer constant in chain polymerization	9*C*
C_m	...	$mole^{1/2}/g^{1/2}$	Thermodynamic constant of the Flory-Huggins dilute solution theory	2*C*
C_p	$L^2T^{-2}\theta^{-1}$	cal/g °C	Specific heat	4*E*
c	ML^{-3}	g/cm^3	Concentration	
c	ML^{-3}	g/deciliter	Concentration	3*E*
c	L	A	Length of crystal unit cell axis	5*B*
D	L^2T^{-1}	cm^2/sec	Diffusion constant	3*D*
d	...	...	Signifies the total derivative	
d	L	cm	Diameter of spherical particle	3*C*
E	ML^2T^{-2}	kcal	Energy content	1*C*
E	ML^2T^{-2}	kcal	Activation energy	6*A*
e	...	...	Base of natural logarithms	
e	...	...	Polarity factor in the Alfrey-Price equation	11*C*
F	ML^2T^{-2}	kcal	Gibbs free energy	1*C*
F_1,F_2	...	...	Mole fractions in copolymer	11*A*
f	...	...	Fraction of polymer in given phase	2*D*
f	MT^{-1}	dyne sec/cm	Frictional coefficient	3*D*
f	MLT^{-2}	dyne	Force	6*B*
f	...	...	Volume fraction	6*D*
f	...	...	Functionality	8*C*
f	...	...	Initiator efficiency	9*C*
f_1,f_2	...	...	Mole fractions in monomer feed	11*A*
G	$ML^{-1}T^{-2}$	psi or $dyne/cm^2$	Modulus of elasticity	6*B*
g	...	...	Ratio of sizes of branched and unbranched molecules	2*B*
g	...	...	Coefficient of Γ^2c^2 in expansion of osmotic pressure	3*E*

Symbol	Dimension	Common Unit	Definition	Chapter Where First Defined
g	...	...	Symmetry tensor in EPR analysis	4*D*
g	...	...	Abbreviation for *gauche* conformation	5*B*
H	ML^2T^{-2}	kcal	Heat content or enthalpy	1*C*
H	$M^{-1}L^2$	mole cm^2/g^2	Light scattering calibration constant	3*C*
H	$L^{-1}T^{-1}Q^{-1}$	gauss	Magnetic field strength	4*D*
h	ML^2T^{-1}	erg sec	Planck's constant	
I	ML^2T^{-2}	lumen	Transmitted light intensity	3*C*
I_0	ML^2T^{-2}	lumen	Incident light intensity	3*C*
i	...	...	Summation index	
i	$ML^{-1}T^{-2}$	lumen/cm^3	Scattered light intensity per unit volume of scatterer	3*C*
J	$M^{-1}LT^2$	cm^2/dyne	Elastic compliance	6*C*
j	...	...	Summation index	
K	$M^{-1}L^2$	mole cm^2/g^2	Light scattering calibration constant	3*C*
K	$M^{-1}L^3$	deciliter $mole^{1/2}/g^{3/2}$	Constant in Flory viscosity equation	3*E*
K'	...	...	Empirical constant in modified Staudinger viscosity equation	3*E*
k, k'	...	...	Arbitrary constants	
k	$ML^2T^{-2}\theta^{-1}$	erg/deg molecule	Boltzmann's constant	
k	$M^{-1}L^3T^{-1}$	liter/mole sec	Rate constant (second order)	8*C*, 9*C*
k_d	T^{-1}	sec^{-1}	Rate constant for decomposition of initiator (first order)	9*C*
k', k''	...	...	Coefficients in Huggins viscosity equation	3*E*
l	L	cm or A	Length	
ln	...	...	Abbreviation for natural logarithm	
log	...	...	Abbreviation for logarithm to the base 10	

Symbol	Dimension	Common Unit	Definition	Chapter Where First Defined
M	...	g/mole	Molecular weight	
$\bar{M}_n$	...	g/mole	Number-average molecular weight	3*B*
$\bar{M}_{os}$	...	g/mole	Osmotic molecular weight	3*B*
$\bar{M}_v$	...	g/mole	Viscosity-average molecular weight	3*E*
$\bar{M}_w$	...	g/mole	Weight-average molecular weight	3*C*
$\bar{M}_z$	...	g/mole	*z*-Average molecular weight	3*D*
m	L^{-2}	cm^{-2}	Number of elastic chains per unit area	6*B*
N	...	...	Number of items	
N_o	M^{-1}	$mole^{-1}$	Avogadro's number	
n	...	...	Mole fraction	2*C*
n	...	...	Refractive index	3*C*
n	...	...	Exponent in Avrami equation	5*E*
$\tilde{n}$	...	...	Refractive increment	3*D*
P	...	...	Reactivity of radical in Alfrey-Price equation	11*C*
$P(\theta)$	...	...	Angular light-scattering function	3*C*
p	$ML^{-1}T^{-2}$	psi or dyne/cm^2	Pressure	
p	...	...	Extent of reaction	8*C*
Q	...	...	General reactivity factor in Alfrey-Price equation	11*C*
q	...	...	Fraction of cross-linkable monomers on a chain	12*C*
R	$ML^2T^{-2}\theta^{-1}$	erg/mole deg or cal/mole deg	Gas constant	
R	...	...	Ratio of volumes of dilute and precipitated phases	2*D*
R_θ	L^{-1}	cm^{-1}	Rayleigh ratio	3*C*
r	L	cm or A	Distance from origin	
r	...	...	Ratio of number of *A* and *B* groups present in stepwise polymerization	8*C*

Symbol	Dimension	Common Unit	Definition	Chapter Where First Defined
r, $(\overline{r^2})^{1/2}$	L	cm or A	Root-mean-square end-to-end distance	2*B*
r_1, r_2	...	...	Monomer reactivity ratios	11*A*
S	$ML^2T^{-2}\theta^{-1}$	kcal/deg	Entropy	1*C*
s, $(\overline{s^2})^{1/2}$	L	cm or A	Radius of gyration or root-mean-square distance of chain segments from center of gravity	2*B*
s	T	sec	Sedimentation constant	3*D*
s	$ML^{-1}T^{-2}$	psi or dyne/cm^2	Stress or applied force per unit area	6*A*
T	θ	°K or °C	Temperature	
T_g	θ	°K or °C	Glass transition temperature	6*D*
T_m	θ	°K or °C	Crystalline melting point	5*F*
t	T	sec	Time, efflux time	3*E*
t	...	...	Abbreviation for *trans* conformation	5*B*
u	L^3	cm^3	Excluded volume	2*C*
u	...	...	Size parameter for light scattering of spheres	3*C*
V	L^3	cm^3	Volume, specific volume, molar volume	
v	...	...	Volume fraction	2*A*
v	...	...	Size parameter for light scattering of random coils	3*C*
v	LT^{-1}	cm/sec	Velocity	6*A*
v	$ML^{-3}T^{-1}$	mole/liter sec	Rate of reaction	9*C*
v_f	L^3	cm^3	Free volume	6*D*
$\bar{v}$	$M^{-1}L^3$	cm^3/mole	Partial molar or specific volume	2*C*
W	...	...	Probability function	2*C*
w	...	...	Weight, weight fraction	
x	L	cm or A	Distance, space coordinate	

Symbol	Dimension	Common Unit	Definition	Chapter Where First Defined
x	...	...	Chain length, number of segments in chain, degree of polymerization	2*B*
x	...	...	Size parameter for light scattering of rods (Fig. 3-4 only)	3*C*
$\bar{x}_n$	...	...	Number-average degree of polymerization	8*C*
$\bar{x}_w$	...	...	Weight-average degree of polymerization	8*C*
y	...	...	Space coordinate	
y	...	...	Size parameter for light scattering of discs (Fig. 3-4 only)	3*C*
y	...	...	Parameter of distribution functions	9*E*
Z	...	...	Number of atoms in polymer chain	6*A*
z	...	...	Space coordinate	
z	...	...	z-Average (see $\overline{M}_z$)	
z	...	...	Parameter of Zimm treatment of excluded volume	2*C*
z	...	...	Lattice parameter of Flory-Huggins theory (Fig. 2-7 only)	2*C*
z	...	...	Dissymmetry of scattered light	3*C*
z	...	...	Parameter of distribution functions	9*E*
α	...	...	Denotes first in a series	
α	...	...	Expansion factor of dissolved polymer chain	2*B*, 2*C*
α	L	cm or A	Dimension	6*B*
α	...	...	Branching coefficient	8*D*
α_f	$L^3\theta^{-1}$	cm^3/deg	Volume expansion coefficient	6*D*
β	L	cm or A	Dimension	6*B*

Symbol	Dimension	Common Unit	Definition	Chapter Where First Defined
β	...	...	Denotes second in a series	
β	$ML^3T^{-1}Q^{-1}$	Bohr magneton	Magnetic moment of electron spin	4*D*
β	(angle)	degree	Angle of crystal unit cell	5*B*
Γ	$M^{-1}L^3$	cm^3/g	Coefficient of c in expansion of osmotic pressure	3*B*
Γ	...	...	Gamma function	9*E*
γ	L^{-3}	cm^{-3}	Number of particles per cm^3	3*C*
γ	...	...	Activity coefficient	3*D*
γ	L	cm or A	Dimension	6*B*
γ	...	...	Strain or deformation in response to applied stress	6*C*
$\dot{\gamma}$	T^{-1}	sec^{-1}	Shear rate	6*A*
Δ	...	...	Signifies a change in or difference between the values of a thermodynamic function for two states	
δ	(angle)	degree	Phase difference due to energy absorption	4*G*
δ	...	...	Ratio of termination to propagation rate constants in copolymerization	11*A*
δ^2	$ML^{-1}T^{-2}$	$calorie/cm^3$	Cohesive energy density	2*A*
ϵ	...	...	$[(1/\epsilon) - 1]$ is kinetic chain length in polymer degradation	12*E*
ζ	...	...	Size parameter of crystallization theory	5*F*
η	$ML^{-1}T^{-1}$	dyne sec/cm^2 poise	Viscosity	
η_{inh}	$M^{-1}L^3$	deciliter/g	Inherent viscosity	3*E*
η_r	...	...	Relative viscosity	3*E*
η_{red}	$M^{-1}L^3$	deciliter/g	Reduced viscosity	3*E*
η_{sp}	...	...	Specific viscosity	3*E*
$[\eta]$	$M^{-1}L^3$	deciliter/g	Intrinsic viscosity	3*E*

Symbol	Dimension	Common Unit	Definition	Chapter Where First Defined
Θ	θ	°K	Flory temperature, where polymer-solvent interactions are zero	2*C*
θ	(angle)	degree	Angle of observation	3*C*
κ	...	...	Thermodynamic constant of Flory-Huggins dilute solution theory	2*C*
κ	$M^{-1}LT^2$	cm^2/dyne	Compressibility of liquid	3*C*
λ	L	cm or A	Wavelength of light in air	3*C*
λ_s	L	cm or A	Wavelength of light in solution $= \lambda/n$	3*C*
μ	L	micron	Abbreviation for micron	
μ	$ML^3T^{-1}Q^{-1}$	Bohr magneton	Magnetic moment of nucleus	4*D*
μ_0	MLQ^{-2}	henry/meter	Magnetic permeability of vacuum	4*D*
ν	...	...	Number of items	
ν	T^{-1}	sec^{-1}	Frequency	4*A*
ν	...	...	Kinetic chain length	4*C*
π	$ML^{-1}T^{-2}$	atmosphere	Osmotic pressure	2*C*
ρ	ML^{-3}	g/cm^3	Density	3*B*
ρ	...	...	Ratio of number of functional groups on branch units to total number of such groups	8*D*
ρ	...	...	Parameter of distribution functions	9*E*
Σ	...	...	Summation sign	
σ	...	...	Parameter of fractionation theory	2*D*
σ	...	...	Parameter of distribution functions	9*E*
σ	...	...	Chain transfer constant in polymer degradation	12*E*
τ	L^{-1}	cm^{-1}	Turbidity	3*C*

Symbol	Dimension	Common Units	Definition	Chapter Where First Defined
τ	T	sec	Relaxation time, retardation time	6*C*
τ_s	T	sec	Mean lifetime of a free radical	9*E*
Φ	M^{-1}	deciliter/mole cm^3	Universal constant of Flory viscosity theory	3*E*
ϕ	...	...	Cross termination probability constant in copolymerization	11*A*
χ	...	...	Polymer-solvent interaction constant of Flory-Huggins theory	2*C*
ψ	...	...	Thermodynamic constant of Flory-Huggins dilute solution theory	2*C*
ω	...	...	Denotes last in a series	
ω	T^{-1}	sec^{-1}	Angular velocity	3*D*
∂	...	...	Signifies partial derivative	
[ ]	...	...	Denote concentration of substance within brackets (except in symbol $[\eta]$)	
°	...	...	Degree sign	
×	...	...	Multiplication sign	

Appendix II

Table of Physical Constants

Symbol	Name	Value
A	Angstrom unit	$1\ \text{A} = 10^{-8}$ cm
e	Base of natural logarithms	2.718
h	Planck's constant	6.625×10^{-27} erg sec
k	Boltzmann's constant	1.3804×10^{-16} erg/deg
N_0	Avogadro's number	6.025×10^{23} mole^{-1}
R	Gas constant	8.317×10^{7} erg/mole °K 1.9864 cal/mole °K 82.06 cm^3 atm/mole °K
°K	Absolute temperature	0°C = 273.16°K
μ	Micron	$1\mu = 10^{-3}$ mm $= 10^{-4}$ cm $= 10^{4}$ A

Appendix III

Trademarks, Generic Names, and Manufacturers

This list is intended not to be exhaustive but to cover the more common trademarks of polymeric materials. Products are arranged according to the chapter in which they are discussed. More complete information on trademarks and manufacturers may be found in the following:

Guidebook to Man-Made Fibers, United Piece Dye Works, New York, 1960.

Modern Plastics Magazine, Encyclopedia Issue, Breskin Publications, New York, annually.

Morton, M., ed., *Introduction to Rubber Technology*, Reinhold Publishing Corp., New York, 1959.

Press, J. J., ed., *Man-Made Textile Encyclopedia*, Interscience Publishers, New York, 1959.

Simonds, H. R., ed., *Source Book of the New Plastics*, Reinhold Publishing Corp., New York, Vol. 1, 1959; Vol. 2, 1961; other volumes to follow.

TABLE 1. Trademarks and probable compositions

Trademark	Probable Composition	Manufacturer*
	Chapter 13A, B. Polyethylene	
A-C	Polyethylene and polyethylene wax	Allied
Alathon	Polyethylene	Du Pont
Bakelite	Polyethylene	U. Carbide
———	Polyethylene	Catalin
Dylan, Super Dylan	Polyethylene	Koppers
El Rex	Polyethylene	Rexall
Epolene	Polyethylene	Eastman
Fortiflex	Polyethylene	Celanese
Hi-fax	Polyethylene	Hercules
Grex	Polyethylene	Grace
Marlex	Polyethylene	Phillips
———	Polyethylene	Olin
Petrothene	Polyethylene	U.S.I.

Trademark	Probable Composition	Manufacturer
Poly-Em, Poly-Eth	Polyethylene	Spencer
Tenite	Polyethylene	Eastman
	Chapter 13C. Polypropylene	
Alathon	Polypropylene	Du Pont
Escon	Polypropylene	Enjay
Herculon	Polypropylene (fiber)	Hercules
Moplen	Polypropylene	Chemore
Olane, Olefane	Polypropylene	Avisun
Poly-Pro	Polypropylene	Spencer
Pro-fax	Polypropylene	Hercules
Tenite	Polypropylene	Eastman
	Chapter 13D. Other Olefin Polymers	
Alathon	Ethylene copolymers	Du Pont
Cryovac L	Irradiated polyethylene	Grace
Elvax	Poly(ethylene *co* vinyl acetate)	Du Pont
Fortiflex	Poly(ethylene *co* butene-1)	Celanese
Grex	Ethylene copolymers	Grace
Hypalon	Chlorosulfonated polyethylene	Du Pont
Irrathene	Irradiated polyethylene	G.E.
Marlex	Ethylene copolymers	Phillips
Nordel	Poly(ethylene *co* propylene)	Du Pont
Parlon P	Chlorinated polypropylene	Hercules
Poly-Eze	Ethylene copolymers	Spencer
Zetafin	Poly(ethylene *co* ethyl acrylate)	Dow
	Chapter 14. Diene Polymers	
Ameripol	Polyisoprene (*cis*-1,4)	Goodrich
Budene, Tufsyn	Polybutadiene (*cis*-1,4)	Goodyear
Bucron	Poly(butadiene *co* isoprene)	Enjay
Butaprene	Butadiene copolymers	Firestone
Chemigum	Poly(butadiene *co* acrylonitrile)	Goodyear
Cis-4	Polybutadiene (*cis*-1,4)	Phillips
Coral	Polyisoprene (*cis*-1,4)	Firestone
Diene	Polybutadiene (*cis*-1,4)	Firestone
Hycar	Butadiene copolymers	Goodrich
Natsyn	Polyisoprene (*cis*-1,4)	Goodyear
Paracryl	Poly(butadiene *co* acrylonitrile)	U.S. Rubber
Parlon	Chlorinated rubber	Hercules
Pliofilm	Rubber hydrochloride	Goodyear
Plioflex	Poly(butadiene *co* styrene)	Goodyear
Trans-4	Polybutadiene (*trans*-1,4)	Phillips
Vistanex	Polyisobutylene	Enjay

Trademark	Probable Composition	Manufacturer
	Chapter 15A. Polystyrene and Related Polymers	
Abson	ABS resin	Goodrich
Bakelite	Polystyrene	U. Carbide
Bakelite C-11	Poly(styrene *co* acrylonitrile)	U. Carbide
———	Polystyrene	Catalin
Cycolac	ABS resin	Borg-Warner
Dylene	Polystyrene	Koppers
Fosta Tuf-Flex	Styrene grafted on synthetic rubber	Foster-Grant
Fostacryl	Poly(styrene *co* acrylonitrile)	Foster-Grant
Fostarene	Polystyrene	Foster-Grant
Kralastic	ABS resin	U.S. Rubber
Lustran	ABS resin	Monsanto
Lustrex	Polystyrene	Monsanto
Royalite	ABS resin	U.S. Rubber
Styron	Polystyrene	Dow
Tyril	Poly(styrene *co* acrylonitrile)	Dow
	Chapter 15B. Acrylic Polymers	
Acrilan	Poly(acrylonitrile *co* vinyl acetate)	Chemstrand
Acrylite	Poly(methyl methacrylate)	Allied
Bavick-11	Poly(methyl methacrylate *co* α-methylstyrene)	Baker
Creslan	Poly(acrylonitrile *co* acrylic esters, and others)	Amer. Cyanamid
Darvan	Poly(vinylidene cyanide *co* vinyl acetate, 1:1)	Celanese
Evr-kleer	Poly(methyl methacrylate)	Cast Optics
Hycar	Acrylate elastomer	Goodrich
Implex	Poly(methyl methacrylate) blended with rubber, olefin, or ABS polymer	Rohm and Haas
Lucite	Poly(methyl methacrylate)	Du Pont
Orlon	Polyacrylonitrile	Du Pont
Plexiglas	Poly(methyl methacrylate)	Rohm and Haas
Verel	Poly(acrylonitrile *co* vinyl acetate, vinyl chloride, and/or vinylidene chloride)	Eastman
Zefran	Poly(acrylonitrile *co* vinyl acetate) with poly(vinyl pyrrolidone) grafted on	Dow
Zerlon	Poly(methyl methacrylate *co* styrene)	Dow
	Chapter 15C. Poly(vinyl Esters) and Derived Polymers	
Butacite	Poly(vinyl butyral)	Du Pont
Butvar	Poly(vinyl butyral)	Shawinigan

Trademark	Probable Composition	Manufacturer
Elvacet	Poly(vinyl acetate)	Du Pont
Elvanol	Poly(vinyl alcohol)	Du Pont
Formvar	Poly(vinyl formal)	Shawinigan
Gelva	Poly(vinyl acetate)	Shawinigan
Gelvatol	Poly(vinyl alcohol)	Shawinigan
Lemac	Poly(vinyl acetate)	Borden
Lemol	Poly(vinyl alcohol)	Borden
Saflex	Poly(vinyl butyral)	Monsanto
Vinac	Poly(vinyl acetate)	Air Reduction
———	Poly(vinyl alcohol)	Air Reduction
Vinylite	Poly(vinyl acetate), poly(vinyl alcohol)	U. Carbide
	Chapter 15D. Chlorine-Containing Polymers	
Bakelite	Poly(vinyl chloride) and copolymers	U. Carbide
———	Poly(vinylidene chloride) and copolymers (saran)	Dow
Dynel	Poly(vinyl chloride *co* acrylonitrile, 60:40)	U. Carbide
———	Poly(vinyl chloride) and copolymers	Escambia
Exon	Poly(vinyl chloride) and copolymers	Firestone
Geon, Koroseal	Poly(vinyl chloride) and copolymers	Goodrich
Marvinol	Poly(vinyl chloride) and copolymers	U.S. Rubber
Opalon, Ultron	Poly(vinyl chloride) and copolymers	Monsanto
Pliovic	Poly(vinyl chloride) and copolymers	Goodyear
Rovana	Poly(vinylidene chloride) and copolymers	Dow
Tygon	Poly(vinyl chloride) and copolymers	U.S. Stoneware
———	Poly(vinyl chloride *co* vinyl acetate, 88:12) (vinyon)	U. Carbide
Velon	Poly(vinyl chloride *co* vinylidene chloride)	Firestone
Vinylite	Poly(vinyl chloride) and copolymers	U. Carbide
Vygen	Poly(vinyl chloride) and copolymers	Gen. Tire
	Chapter 16. Fluorocarbon Polymers	
Aclar, Halon	Polychlorotrifluoroethylene and copolymers	Allied
Fluorel	Poly(vinylidene fluoride *co* hexafluoropropylene) (elastomer)	3M
Genetron	Polychlorotrifluoroethylene, modified	Allied
Kel-F	Polychlorotrifluoroethylene, poly(chlorotrifluoroethylene *co* vinylidene fluoride) and others	3M

Trademark	Probable Composition	Manufacturer
Kynar	Poly(vinylidene fluoride)	Pennsalt
Tedlar	Poly(vinyl fluoride)	Du Pont
Teflon FEP	Poly(tetrafluoroethylene *co* hexafluoropropylene)	Du Pont
Teflon TFE	Polytetrafluoroethylene	Du Pont
Viton	Poly(vinylidene fluoride *co* hexafluoropropylene) (elastomer)	Du Pont
Vydax	Polytetrafluoroethylene (low molecular weight)	Du Pont
	Chapter 17A. Polyamides	
———	Poly(hexamethylene adipamide), 66 nylon	Chemstrand
———	Poly(hexamethylene adipamide), 66 nylon	Du Pont
Zytel 42, 101, 103, 105	Poly(hexamethylene adipamide), 66 nylon (plastics)	Du Pont
Zytel 31, 33, 37X, 38	Poly(hexamethylene sebacamide), 610 nylon (plastics)	Du Pont
Caprolan	Polycaprolactam, 6 nylon	Allied
———	Polycaprolactam, 6 nylon	Catalin
———	Polycaprolactam, 6 nylon	Chemstrand
Nylenka	Polycaprolactam, 6 nylon	Amer. Enka
———	Polycaprolactam, 6 nylon	Firestone
Fosta-Nylon	Polycaprolactam, 6 nylon	Foster-Grant
Plaskon	Polycaprolactam, 6 nylon	Allied
———	Polycaprolactam, 6 nylon	Spencer
Zytel 211	Polycaprolactam, 6 nylon (plastic)	Du Pont
BCI 800	Alkoxy substituted 66 nylon	B.C.I.
BCI 1100	Poly(ω-amino undecanoic acid), 11 nylon	B.C.I.
Versamid	Vegetable oil polyamides	Gen. Mills
Herox	(Undisclosed, monofilament)	Du Pont
Tynex	(Undisclosed, monofilament)	Du Pont
Zytel 61, 63, 69, 109	(Copolymers, undisclosed, plastics)	Du Pont

Trademark	Probable Composition	Manufacturer
	Chapter 17B. Polyesters	
Cronar, Dacron, Mylar	Poly(ethylene terephthalate)	Du Pont
Fortrel	(Undisclosed)	Celanese
Kodel, Tenite	Poly(1,4-cyclohexylene dimethylene terephthalate)	Eastman
Videne, Vitel, Vycron	Polyester of terephthalic and isophthalic acids (70:30) plus ethylene glycol	Goodyear
	Chapter 17C. Polyurethans	
Lycra	(Undisclosed, fiber)	Du Pont
Vyrene	(Undisclosed, fiber)	U.S. Rubber
Adiprene	Polyether-based (elastomer)	Du Pont
Chemigum	Polyester-based (elastomer)	Goodyear
Estane	Polyester-based (elastomer)	Goodrich
Uscothane	Polyether- or polyester-based (elastomer)	U.S. Rubber
Vulkollan	Polyester-based (elastomer)	Mobay
	Chapter 17D-F. Other Heterochain Thermoplastics	
Celcon	Acetal copolymer	Celanese
Delrin	Acetal polymer (polyoxymethylene)	Du Pont
Lexan	Polycarbonate from bisphenol A	G.E.
Merlon	Polycarbonate	Mobay
Penton	Poly(2,2-di(chloromethyl)propylene oxide)	Hercules
Polyox	Poly(ethylene oxide)	Carbide
Thiokol	Polysulfides	Thiokol
	Chapter 18B. Regenerated Cellulose	
———	Cuprammonium rayon	Amer. Bemberg
———	Viscose rayon	Amer. Enka
Avisco	Viscose rayon	Amer. Viscose
———	Viscose rayon	Celanese
Cordura	Viscose, high-tenacity tire cord	Du Pont
———	Viscose rayon	Du Pont
Fortisan	Saponified cellulose acetate	Celanese
I.R.C.	Viscose rayon	Ind. Rayon
Narco	Viscose rayon	N.A. Rayon
	Chapter 18C. Cellulose Esters	
Acele	Cellulose (di)acetate	Du Pont

Trademark	Probable Composition	Manufacturer
Arnel	Cellulose triacetate	Celanese
Avisco	Cellulose (di)acetate	Amer. Viscose
———	Cellulose (di)acetate	Celanese
———	Cellulose (di)acetate	Du Pont
Forticel	Cellulose propionate	Celanese
Hercocel	Cellulose (di)acetate	Hercules
Kodapak I, Tenite I	Cellulose (di)acetate	Eastman
Kodapak II, Tenite II, Uvex	Cellulose acetate-butyrate	Eastman
Kodapak IV	Cellulose triacetate	Eastman
Lumarith	Cellulose (di)acetate	Celanese
Tenite	Cellulose acetate-propionate	Eastman
	Chapter 18D. Other Cellulose Polymers	
Ethocel	Ethyl cellulose	Dow
Hercocel	Ethyl cellulose	Hercules
Methocel	Methyl cellulose	Dow
	Chapter 19A. Phenolic Resins	
Bakelite	Phenol-formaldehyde	U. Carbide
———	Phenol-formaldehyde	Catalin
Durez	Phenol-formaldehyde	Hooker
Plaskon	Phenol-formaldehyde	Allied
Resinox	Phenol-formaldehyde	Monsanto
	Chapter 19B. Amino Resins	
Avisco	Urea-formaldehyde	Amer. Viscose
Beetle	Urea-formaldehyde	Amer. Cyanamid
———	Urea- and melamine-formaldehyde	Catalin
Cymel	Melamine-formaldehyde	Amer. Cyanamid
Plaskon	Urea- and melamine-formaldehyde	Allied
Resimene	Urea- and melamine-formaldehyde	Monsanto
	Chapter 19C. Unsaturated Polyester Resins	
Atlac	Polyester-styrene	Atlas
Durez	Polyester-styrene	Hooker
Hetron	Polyester-styrene	Hooker
Laminac	Polyester-styrene	Amer. Cyanamid
Marco	Polyester-styrene	Celanese
———	Polyester-styrene	Mobay
Paraplex	Polyester-styrene	Rohm and Haas

Trademark	Probable Composition	Manufacturer
Plaskon	Polyester-styrene	Allied
Polylite	Polyester-styrene	Reichold
Selectron	Polyester-styrene	P.P.G.
Vibrin	Polyester-styrene	U.S. Rubber
	Chapter 19D. Epoxy Resins	
Araldite	Epoxy	Ciba
Epon	Epoxy	Shell
	Chapter 19F. Silicone Polymers	
Silastic	Silicone	Dow
	Chapter 19G. Miscellaneous Thermosetting Resins	
CR-39	Allyl carbonate	Cast Optics
Dapon	Diallyl phthalate	F.M.C.
Duralon	Furane	U.S. Stoneware
Durez	Diallyl phthalate	Hooker
Plaskon	Alkyd, diallyl phthalate	Allied

* See Table 3 for complete names of manufacturers.

TABLE 2. Generic names and sources

Generic Name	Composition	Source
acetate	Cellulose acetate (fiber)	FTC*
acrylic	At least 85% by wt. polyacrylonitrile (fiber)	FTC
BR	Butadiene-based rubbers	ASTM†
butyl	Poly(isobutylene *co* isoprene)	Enjay
cellophane	Regenerated cellulose film	Industrywide
celluloid	Cellulose (di)nitrate	Industrywide
CR	Chloroprene-based rubbers	ASTM
glyptal	Alkyd resin	Industrywide
IIR	Isobutylene-isoprene-based rubbers	ASTM
IR	Isoprene-based rubbers, synthetic	ASTM
modacrylic	At least 35% but not more than 85% by wt. polyacrylonitrile (fiber)	FTC
NBR	Nitrile-butadiene-based rubbers	ASTM
NCR	Nitrile-chloroprene-based rubbers	ASTM
neoprene	Polychloroprene	Du Pont
NR	Natural rubber	ASTM
nylon	Polyamide	Industrywide
nytril	At least 85% by wt. poly(vinylidene dinitrile) (fiber)	FTC
olefin	At least 85% by wt. polypropylene or other polyolefin (fiber)	FTC
polyester	At least 85% by wt. polyester of a glycol and terephthalic acid (fiber)	FTC
rayon	Regenerated cellulose, substituents replacing no more than 15% of the hydroxyl groups (fiber)	FTC
saran	At least 80% by wt. poly(vinylidene chloride) (fiber)	FTC
SBR	Styrene-butadiene-based rubbers	ASTM
spandex	At least 85% polyurethan (fiber)	FTC
triacetate	Cellulose acetate, when not less than 92% of the hydroxyls are acetylated (fiber, optional to acetate)	FTC
tyrex	Regenerated cellulose, high-tenacity tire cord	Industrywide
vinal (formerly vinylon)	At least 50% by wt. poly(vinyl alcohol), with poly(vinyl alcohol) and poly(vinyl acetal) totaling over 85% by wt. (fiber)	FTC
vinyon	At least 85% by wt. poly(vinyl chloride)	FTC
viscose	Regenerated cellulose (fiber)	Industrywide

* Federal Trade Commission.

† American Society for Testing Materials, *Tentative Recommended Practice for Nomenclature for Synthetic Elastomers*, ASTM Designation: D1418-56T.

TABLE 3. Manufacturers cited in list of trade names

Air Reduction Chemical Co.	New York, N.Y.
Allied Chemical Corp.	New York, N.Y.
American Bemberg	New York, N.Y.
American Cyanamid Co.	New York, N.Y.
American Enka Corp.	New York, N.Y.
American Viscose Corp.	Marcus Hook, Pa.
Atlas Chemical Industries, Inc.	Wilmington, Del.
AviSun Corp.	Marcus Hook, Pa.
The J.T. Baker Chemical Co.	Phillipsburg, N.J.
Belding Corticelli Industries	New York, N.Y.
The Borden Chemical Co.	New York, N.Y.
Borg-Warner Corp., Marbon Chemical Division	Washington, W. Va.
Cast Optics Corp.	Hackensack, N.J.
Catalin Corp. of America	New York, N.Y.
Celanese Corp. of America	New York, N.Y.
Chemore Corp.	New York, N.Y.
Chemstrand Corp.	New York, N.Y.
Ciba Products Corp.	Fairlawn, N.J.
The Dow Chemical Co.	Midland, Mich.
E. I. du Pont de Nemours and Co., Inc.	Wilmington, Del.
Eastman Kodak Co., Eastman Chemical Products, Inc.	Kingsport, Tenn.
Enjay Chemical Co., Division of Humble Oil and Refining Co.	New York, N.Y.
Escambia Chemical Corp.	Wilton, Conn.
Firestone Tire and Rubber Co., Firestone Plastics Division	Pottstown, Pa.
Firestone Synthetic Rubber and Latex Co.	Akron, Ohio
FMC Corp, Chemicals and Plastics Division	New York, N.Y.
Foster Grant & Co., Inc.	Leominster, Mass.
General Electric Co., Chemical Materials Dept.	Pittsfield, Mass.
General Mills, Inc.	Minneapolis, Minn.
General Tire and Rubber Co.	Akron, Ohio
The B. F. Goodrich Co., B. F. Goodrich Chemical Co. Division	Cleveland, Ohio
Goodyear Tire and Rubber Co., Goodyear Chemical Division	Akron, Ohio
W. R. Grace and Co., Polymer Chemical Division	Clifton, N.J.
Hercules Powder Co.	Wilmington, Del.

Hooker Chemical Corp., Durez Plastics Division	North Tonawanda, N.Y.
Industrial Rayon Corp.	New York, N.Y.
Koppers Co., Inc., Plastics Division	Pittsburgh, Pa.
Minnesota Mining and Manufacturing Co., Chemical Division	St. Paul, Minn.
Mobay Chemical Co.	Pittsburgh, Pa.
Monsanto Chemical Co., Plastics Division	Springfield, Mass.
North American Rayon Corp.	New York, N.Y.
Olin Mathieson Chemical Corp., Chemicals Division	Baltimore, Md.
Pennsalt Chemicals Corp.	Philadelphia, Pa.
Phillips Chemical Co.	Bartlesville, Okla.
Pittsburgh Plate Glass Co.	Pittsburgh, Pa.
Reichold Chemicals, Inc., Alsinyte Division	San Diego, Calif.
Rexall Chemical Co.	Paramus, N.J.
Rohm and Haas Co.	Philadelphia, Pa.
Shawinigan Resins Corp.	Springfield, Mass.
Shell Oil Co., Shell Chemical Co. Division	New York, N.Y.
Spencer Chemical Co.	Kansas City, Mo.
Thiokol Chemical Corp.	Trenton, N.J.
Union Carbide Corp.	New York, N.Y.
U.S. Industrial Chemicals Co., Division of Nat'l Distillers and Chemical Corp.	New York, N.Y.
U.S. Rubber Co., Naugatuck Chemical Division	Naugatuck, Conn.
The U.S. Stoneware Co.	Tallmadge, Ohio

Author Index*

Abitz, W., 146, **154**
Aelion, R., 129, **140**
Agar, A. W., 144, **145**
Akin, R. B., **444**
Albrecht, A. C., 39, **40**
Alcock, T. C., 148, **154**
Alexander, L. E., **99**
Alexander, P., 358, **359**
Alfrey, T., 47, **51**, 192, 194, **198**, 311, 312, 318, **318**, 319, **319**, **325**, 329, **333**, **353**
Allen, P. W., **26**, **51**, **54**, **63**, **72**, **85**, **336**
Ambrose, E. J., 170, **171**
Amos, J. L., 395, **400**
Anagnostopoulos, C. E., 414, **416**
Anderson, A. W., 302, 305, **306**
Anderson, D. H., **96**
Angelo, R. J., 353, **353**
Angier, D. J., 335, 336, **336**
Anson, M. L., 72, **73**
Anthony, R. L., 188, **198**
Archibald, W. J., 77, **79**
Arcus, C. L., 123, **123**
Arnett, L. M., 274, **280**
Arnold, R. G., 437, **438**, **479**
Ashby, C. E., 58, 61, **63**, 302, 305, **306**
Asmussen, F., 26, **27**
Atkins, J. T., 70, **73**, 210, **211**, 295, **297**
Avrami, M., 156, **157**

Bachmann, J. H., 541, **547**
Badgley, W. J., 48, **51**, 62, **63**
Bagchi, S. N., 99, **100**, 147, **154**
Bailey, E. D., **79**
Bailey, J., 171, **171**
Bailey, K., 72, **73**
Baker, C. A., 49, **51**
Baker, W. O., 129, **140**, **207**
Baldwin, R. L., 77, **79**
Balwit, J. S., 358, **359**, 366, **369**
Bamford, C. H., 42, **47**, **96**, **99**, **280**, 282, **283**, **286**, **318**, 334, **336**, **347**
Banks, R. L., 307, **309**
Bannerman, D. G., 217, **220**, **434**, **436**
Bannister, D. W., 49, **51**
Barb, W. G., **280**, 318, **318**, **347**
Barnhart, R. R., **532**
Barnhart, W. S., 425, **426**
Barron, H., 542, **547**
Bartlett, P. D., 54, **54**
Bartovics, A., 47, **51**
Baskett, A. C., 365, 366, **369**
Bassett, D. C., **108**, 144, **145**
Bassi, I. W., 119, **124**, 302, **307**
Batchinski, A. J., 178, **179**
Baum, F. J., 70, **73**
Bawn, C. E. H., 36, 37, **40**, **209**, 219, **220**, 303, **306**
Baxendale, J. H., 54, **54**, **269**, 347, **348**
Bayer, O., 437, **438**
Beaman, R. G., 212, **220**, 297, **301**
Beasley, J. K., 285, **286**
Beati, E., 336, **337**
Beckmann, C. O., 49, **51**
Beevers, R. B., 213, 215, **220**
Bekkedahl, N., 154, **157**, 199, **204**

* Page numbers in **boldface** refer to references at end of sections.

Bengough, W. I., 341, **348**
Bernhardt, E. C., **491**, **497**
Berry, G. C., 209, **212**
Berry, K. L., 347, **348**, 419, **424**
Bevington, J. C., 13, **15**, 54, **54**, 265, **269**, **270**, 408, **411**
Beyer, C. E., **491**
Billmeyer, F. W., Jr., 50, **51**, 59, 61, **63**, 69, 70, 72, **72**, **73**, 148, 150, **154**, 161, **164**, 210, **211**, 276, **280**, 284, **286**, 295, **297**, 366, **369**
Birshtein, T. M., 126, 136, **140**
Bjorksten, J., **436**, **475**
Blais, J. F., **472**
Blaschke, F., 274, **280**
Bockhoff, F. J., 375, **376**
Bohn, C. R., 125, **140**
Bohrer, J. J., 311, 312, **318**, 319, **325**, **333**, **353**
Bonart, R., 99, **100**, 147, **154**
Bonnar, R. U., **63**
Bostrom, S., 527, 530, **532**
Boundy, R. H., **400**
Bovey, F. A., 119, **123**, **347**, **359**
Boyd, D. R. J., 365, **369**
Boyd, R. H., 161, **164**, 178, **179**, 217, **220**
Boyer, R. F., 50, **51**, 52, **52**, 199, **204**, 212, 213, **220**, **400**, 499, **506**
Boyer-Kawenoki, F., 43, **47**
Bragaw, C. G., 111, **116**
Bragg, J. K., 139, **140**
Branson, H. R., 136, 138, **140**, 311, **319**
Breil, H., 302, **307**
Breslow, D. S., 303, 304, **306**
Bressan, G., 120, 121, 122, 123, **124**
Brewer, L., 30
Brice, B. A., 66, 70, 71, 72, **73**
Briggs, E. R., 318, **319**
Bristow, G. M., 26, **27**
Broda, A., 50, **51**
Brodnyan, J. G., 82, **85**
Brookman, E. F., 275, **280**
Brooks, R. E., 166, **171**, 221, 222, 224, **226**
Brower, F. M., 210, **212**, 222, **226**
Brown, D. W., 421, **424**
Brown, J. F., Jr., 122, 123, **123**
Brown, W. B., 299, **301**
Bruce, J. M., Jr., 302, **306**
Bryant, W. M. D., 16, **21**, 70, **73**, 138, **140**, 146, **154**, 163, **164**, 210, **211**, 289, **290**, 365, 369, **369**, **371**
Buchdahl, R., 146, 150, **154**
Buckley, D. J., **392**
Bueche, A. M., 83, **85**, 112, **116**
Bueche, F., **30**, 83, **85**, **104**, 178, **179**, **187**, 197, **198**, 201, **204**, 210, **211**, **220**, **532**
Bungenberg de Jong, H. G., 41, **47**
Bunn, C. W., **99**, 125, 126, 127, 129, 130, 131, 132, 133, 135, 137, 138, **139**, **140**, 146, 148, 151, **154**, 165, 169, **171**, 202, **204**, 215, **220**
Burbank, R. D., 144, **145**
Burk, R. E., 62, **63**, 247, 248, 249, 250, **252**, 256, 257, 258, **259**
Burlant, W. J., **229**, **336**
Burnett, B. B., 149, **154**
Burnett, G. M., **252**, **259**, **269**, **272**, **280**, 282, **283**, **286**, **290**, **293**, **297**, **301**, 311, 312, 316, **318**, 322, 325, **325**, 329, **333**, **336**, **340**, 344, **347**, **348**, **357**
Burton, R. L., 168, **171**
Busse, W. F., 175, **179**, 180, **187**
Butler, G. B., 353, **353**
Buttery, D. N., **506**

Cabannes, J., 66, **73**
Campbell, T. W., 219, **220**, **244**, **261**, **272**, **293**
Cantow, H.-J., 78, **79**
Cantow, M., 299, **301**
Capron, N., 425, **426**
Carey, R. H., 115, **116**
Carlin, E. R., 336, **337**
Carlson, B. C., 375, **376**
Carothers, W. H., 5, **11**, 13, **15**, 237, **239**, 430, 434
Carpenter, D. K., 72, **73**
Carr, C. I., Jr., 72, **73**
Carswell, T. S., 110, **116**, 466, **469**
Casassa, E. F., **30**, 39, **40**, 57, **64**
Cassel, J. M., **90**
Catsiff, E., 188, 189, **198**
Ceresa, R. J., 335, **336**
Chandrasekhar, S., 27, **30**
Chapiro, A., 336, **336**, **359**

Charlesby, A., 209, **212**, 358, 359, **359**, 366, **369**
Cheyney, L. E., 499, **506**
Chiang, Y., **108**
Chien, J. C. W., 305, **306**
Christopher, W. F., 446, **447**
Ciferri, A., 182, **187**
Cirker, L. H., 491, **492**
Claesen, M., 336, **337**
Claesson, S., 82, **85**
Clark, A., 307, **309**
Clark, E. S., 98, 99, **100**, 126, 138, **140**, 169
Clarke, J. T., 408, **411**
Claver, G. C., 146, 150, **154**
Cleland, R. L., 300, **301**
Cobbold, A. J., 151, **154**
Cochran, W., 98, **100**
Coffman, D. D., 425, **426**
Cogdell, J. F., 442, **444**
Cohen, S. G., 54, **54**, 336, **337**
Cohn, E. S., 300, **301**
Coke, C. E., **525**
Colborne, R. S., **229**
Colclough, R. O., 295, **297**
Cole, O. D., **383**
Coleman, L. E., **428**
Coles, H. W., 170, **171**, **187**, **198**, **204**, **207**
Collette, J. W., 374, **376**
Collins, E., 335, **337**
Combs, R. L., 89, **90**
Conix, A., 54, **54**
Conway, B. E., **87**
Cooke, H. G., Jr., 276, **280**
Cooke, R. C., 48, **51**
Cooper, W., **123**, **306**, **309**
Coover, H. W., Jr., 372, **373**, 404, **406**
Coran, A. Y., 414, **416**
Corey, R. B., 134, 136, 138, **140**
Corradini, P., 118, 119, 121, **124**, 128, 135, **140**, 302, **307**
Cotton, F. H., 542, **547**
Cottrell, T. L., **21**
Cragg, L. H., **51**, 80, **85**
Craig, D., **541**
Cram, D. J., 304, 305, **306**
Crawford, G. H., 428, **428**
Crawshaw, A., 353, **353**
Crick, F. H. C., 98, **100**, 123, **123**
Critchfield, F. E., 88, **90**
Cronin, E. W., 444, **444**
Cross, L. H., 94, 95, **96**
Crozier, R. N., 408, **411**
Cruickshank, J., 60, **64**
Curry, B. A., 425, **426**

Dahl, R. B., **491**
Dainton, F. S., 288, **290**, 294, **297**, 356, **358**
D'Alelio, G. F., **209**
Dal Nogare, S., **90**
Daniels, W. W., 54, **54**
Dauben, W. G., 120, 121, **124**
Daum, G. A., **388**, 403, **406**
Davidson, H. R., 119, **124**
Davies, B. L., **383**
Davies, D. R., 98, **100**
Davis, C. C., **348**, **388**, **532**, 546
Debye, P., 17, **21**, 65, 66, **73**, 83, **85**, 178, **179**
Debye, P. P., 66, **73**
Dec, J., 276, **280**
DeCoste, J. B., 115, **116**
De Dáni, A., **475**
De Hoff, G. R., 424, **424**
de Jong, J. I., 471, **472**
de Jonge, J., 471, **472**
Dennis, L. M., 112, **116**
Denoon, C. E., Jr., 269, **270**
Desreux, V., 48, **51**, 117, **124**
de Than, C. B., 69, **73**
Dewar, M. J. S., 282, **283**
D'Ianni, J. D., 374, **376**, **383**, **388**
Dickhauser, E., 310, **319**
Di Marzio, E. A., 202, **204**
Dimbat, M., **63**
Dixon, S., 426, **426**
Doak, K. W., **369**, **371**, **373**
Doban, R. C., 420, 421, **424**
Dobry, A., 43, **47**, 62, **63**
Dobry-Duclaux, A., **87**
Dole, M., 105, 106, **107**, **108**, 159, 160, 161, 162, **164**, 215, 217, **220**, 358, 359, **359**
Dombrow, B. A., **438**, **479**
Donati, M., 121, 122, **124**
Doolittle, A. K., 201, **204**
Doolittle, D. B., 201, **204**

Doremus, R. H., 141, **145**, 148, **153**, **154**, **157**
Dostal, H., 311, **318**
Doty, P., 29, **30**, 67, 72, **73**
Dukert, A. A., 425, **426**
Dunbrook, R. F., **348**, **388**, **532**, 546
Dunkel, W. L., **392**
Dunkelberger, D. L., 300, **301**
Dunn, A. S., 334, 335, **337**

Eaborn, C., **482**
Eastham, A. M., **261**, **297**
Edgar, O. B., 214, **220**
Edsall, J. T., 72, **73**
Einstein, A., 65, **73**
Eirich, F. R., 83, **86**, **87**, 174, 175, 176, 178, **179**, **198**, **207**
Eisenschitz, R., 18, **21**
Eldridge, J. E., 70, **73**, 210, **211**
Eley, D. D., 296, **297**
Elias, H.-G., 60, **63**
Eliassaf, J., 41, **47**
Elliott, A., **96**, **99**, 170, **171**
Elliott, J. H., 48, **51**
Elving, P. J., **72**
Endres, G. F., 295, **297**
Eppe, R., 141, **145**, 146, 150, 151, **154**
Epprecht, A. G., 88, **89**
Ernst, S. L., 408, **411**
Evans, A. G., 294, 295, **297**
Evans, M. G., 54, **54**, **290**, 327, 328, **333**, 347, **348**
Evans, R. D., 158, **164**
Evering, B. L., 307, **309**
Ewart, R. H., 81, **85**, 282, **283**, 345, **348**
Eyring, H., 174, 178, **179**, 201, **204**

Farina, M., 120, 121, 122, 123, **124**
Faucette, W. A., 219, **220**
Faucher, J. A., 206, 207, **207**
Fawcett, E. W., 363, **369**
Fawcett, M. S., 374, **376**
Fedor, W. S., 377
Fekete, F., 482, **482**
Feller, M., 307, **309**
Fellman, R. P., 300, **301**
Ferren, R. A., 425, **426**
Ferry, J. D., 192, 196, 197, **198**, 201, **204**, **205**, **207**
Fettes, E. M., **446**
Field, E., 307, **309**
Fikentscher, H., 310, **319**
Fineman, M., 316, **319**
Fischer, E. W., 141, **145**, 146, 150, 151, **154**
Fischer, M., 302, **306**
Fisher, B. S., 374, **376**
Fisher, C. H., 219, **220**
Fisher, H. L., **361**
Fishman, M. M., **72**
Fixman, M., 39, **40**, 85, **86**
Flom, D. G., 112, **116**
Florin, R. E., **104**, 421, **424**
Flory, P. J., 5, **10**, 13, **14**, **15**, **30**, 34, 35, **40**, 41, 42, 43, 44, 45, 46, **46**, 47, **47**, **51**, 57, 59, **63**, 78, **79**, 82, 83, **85**, **86**, **87**, 126, 136, **140**, **152**, 158, 159, **164**, 175, **179**, 181, 182, 186, 187, **187**, 199, 202, **204**, 210, **212**, 213, 214, **220**, 237, **239**, 247, 248, 249, 250, **252**, 255, 256, 257, 258, **259**, 261, **261**, 262, 266, **269**, **270**, **280**, **283**, **286**, **290**, **293**, **297**, **301**, **318**, 333, **348**, 349, 351, **353**, 410, **411**
Floyd, D. E., **434**
Ford, T. A., 425, **426**
Fortner, C. P., 171, **171**
Fox, D. W., **447**
Fox, J. C., 83, **85**
Fox, J. J., 94, **96**, 364, **369**
Fox, T. G., Jr., 82, 83, **85**, 174, 175, 176, **179**, 199, **204**, 213, **220**, 300, **301**, 305, **306**
Fraenkel, G. K., **105**
Francis, P. S., 48, **51**
Frank, F. C., 141, 144, **145**, 151, **154**
Franklin, R. E., 98, **100**
Franklin, R. L., 423, **425**
Franta, W. A., 158, **164**, 210, 211, **212**, 221, 222, 223, **226**, 227, **230**, 366, **369**
Fraser, R. D. B., 95, **96**
Freeman, R. F. J., 36, 37, **40**
Frey, W., **519**
Friedlander, H. N., 308, **309**
Frilette, V. J., 48, **51**
Fryling, C. F., **348**
Fuchs, O., 48, **51**
Fujita, H., 76, **79**
Fuller, C. S., 129, **140**

Funt, B. L., 335, **337**
Fuoss, R. M., 59, **63**, **87**
Furukawa, J., **261**, **444**

Gailey, J. A., 168, **171**
Gamrath, H. R., 414, **416**
Gans, R., 67, **73**
Gardon, J. L., 61, **63**
Garner, E. V., 129, **140**
Garrett, A. B., 419, **425**
Garton, V. A., 545, **547**
Garvey, B. S., Jr., **532**, 539, **541**
Garvin, G. S., **491**, **497**, **506**, **532**
Gaskins, F. H., 82, **85**
Gaylord, N. G., 135, **139**, **301**, 303, **306**, **309**, **444**, **446**, **477**
Gee, G., 33, 35, 36, 37, **40**, 62, **63**, 182, 183, **187**
Geil, P. H., 141, 142, 143, 144, **145**, 148, 149, 150, 152, 153, **154**, 168
George, D., 115, **116**
Gerngross, O., 146, **154**
Geymer, D. O., 36, **40**
Giachetti, E., 302, 303, **307**
Giannini, U., 302, **307**
Gibbs, J. H., 202, **204**
Gibson, M. E., Jr., 88, **90**
Gibson, R. O., 363, **369**
Gilman, L., 110, **116**
Gilmore, G. D., 490, **492**
Gladding, E. K., 374, **376**
Glasstone, S., 30, 174, **179**
Glazer, J., **383**
Gloor, W. E., **371**
Glover, C. A., 58, **63**
Glusker, D. L., 300, 301, **301**
Goldberg, A., 48, **51**
Goldberg, E. J., 374, **376**
Goldblum, K. B., 446, **447**
Goldfinger, G., 311, 318, **318**, **319**
Golding, B., **361**
Goode, W. E., 300, **301**
Goodwin, P. A., 375, **376**
Gordon, M., 285, **286**, 306, **306**
Gordy, W., 104, **105**
Gould, D. F., **469**
Grafflin, M. W., **15**, **452**, **455**, 457, **460**, **462**
Graham, R. K., 300, **301**, 336, **337**
Grassie, N., **357**
Grassie, V. R., **373**
Gratch, S., 174, 175, 176, **179**
Green, J. S., 48, 49, **51**
Greensmith, H. W., 112, **116**
Gregg, R. A., 274, 278, **280**
Grievson, B. M., 214, **220**
Griff, A. L., **497**
Griffis, C. B., 428, **428**
Grimley, T. B., **40**
Gross, S. T., 119, **124**
Grosser, F., 119, **124**
Gruben, A., 88, **90**
Grummitt, O., 62, **63**, 247, 248, 249, 250, **252**, 256, 257, 258, **259**
Grün, F., 185, **188**
Guillet, J. E., 89, **90**, 336, **337**, 372, **373**
Gurnee, E. F., **198**
Guth, E., 183, 184, 185, **187**, 188, **188**, **198**
Gutowsky, H. S., **104**
Guzmán, G. M., **51**, 408, **411**

Haas, H. C., 336, **337**
Haber, F., 346, **348**
Hall, C. E., **108**
Hall, R. W., 47, **51**
Halwer, M., 66, 70, 71, 72, **73**
Hamm, F. A., **108**
Hammer, C. F., 58, 61, **63**, 222, 224, **226**
Hammerschlag, H., **51**
Hammett, L. P., 331, **333**
Hampton, R. R., 385, **388**
Hanby, W. E., **96**, **99**
Hanford, W. E., 13, **15**, 54, **55**, 433, **434**
Hanna, R. D., **373**
Hansen, J. E., 166, **171**, 221, 222, **226**
Happoldt, W. B., Jr., 420, 423, **424**
Harborth, G., 276, **280**
Hardesty, R. H., 442, **444**
Harding, S. W., 178, **179**
Harker, B., **436**, **475**
Harkins, W. D., 345, **348**
Harlen, F., 365, **369**
Harper, D. A., 437, **438**
Harmon, J., 13, **15**, 54, **55**
Harrington, J. P., 110, **116**
Harris, M., 210, **212**
Hart, R., **336**
Hart, V. E., 421, **424**

Hartley, F. D., 149, **154**, 155, **157**
Hartshorne, N. H., **108**
Hathaway, C. T., 171, **171**
Hausdorff, H., 92, 93, **96**
Hauter, R. L., 115, **116**
Haven, A. C., Jr., 219, **220**
Heckert, W. W., 510, 511, **512**
Heffelfinger, C. J., 168, **171**
Heidner, R. H., 88, **90**
Henglein, A., 335, **337**
Hengstenberg, J., 310, **319**
Henning, J., **436**, **475**
Herman, D. F., 302, **306**
Hermans, P. H., 433, **434**
Herrmann, K., 146, **154**
Hettinger, W. P., Jr., 161, **164**
Heuer, W., 13, **15**, 79, 82, **86**
Hiatt, G. D., 109, 110, **116**, **362**, 396, **400**, 467, **469**, 493, **497**, 498, **499**
Hicks, J. A., 335, **337**
Higginson, W. C. E., 299, **301**
Hildebrand, J. H., 26, **27**
Hill, F. B., 437, **438**
Hill, J. W., 430
Hill, R., 129, 131, **139**, 165, **171**, 214, 215, 216, **220**, **229**, **290**, **434**, **436**, **519**
Hines, R. A., 369, **371**
Hirshfelder, J. O., 201, **204**
Hittmair, P., 115, **116**
Hobbs, L. M., 209, **212**
Hock, L., 527, 530, **532**
Hoeve, C. A. J., 182, **187**
Hoff, E. A. W., 204, **205**
Hoffman, A. S., **229**, **336**
Hoffman, J. D., 151, **154**
Hogan, J. P., 307, **309**
Hohenstein, W. P., 48, **51**
Holmes, D. R., 125, 129, 135, 137, **140**, 202, **204**
Holmes, J., 163, **164**
Holmes, L. G., 48, **51**
Holtzer, A., 29, **30**
Holzkamp, E., 302, **307**
Honeyman, J., **451**, **455**, **460**, **462**
Hookway, H. T., 49, **51**, **63**
Hopff, H., 310, **319**
Hopkins, I. L., **207**
Horn, M. B., **406**
Horne, S. E., Jr., 121, **124**, 302, **306**, 382, **383**
Horowitz, E., **96**
Hosemann, R., 99, **100**, 147, **154**
Houtz, R. C., 405, **406**
Houwink, R., **89**, **115**, **229**
Howard, G. J., **252**, **259**
Howard, J. E., 115, **116**
Howard, R. O., 408, **411**
Howells, E. R., 138, **140**
Howlett, F., 48, **51**
Hsaio, C. C., 110, **116**
Huggins, M. L., 34, 36, **40**, 81, **85**, 117, **124**
Hughes, R. E., 126, **140**
Husemann, E., 275, **280**
Hymo, L. A., 421, **424**

Immergut, E. H., v, 59, 60, **63**, **64**, **336**
Inderfurth, K. H., **434**
Ingram, D. J. E., **105**
Inskeep, G. C., 454, **455**
Iserson, H., 425, **426**
Ivin, K. J., 288, **290**, 356, **358**
Iwayanagi, S., 204, **205**

Jaffe, J. A., 423, **424**
James, H. M., 183, 184, 185, 187, **188**
Janssen, H. J. J., 532, **532**
Jelley, E. E., **108**
Jellinek, H. H. G., **357**
Jenckel, E., 214, **220**
Jenkins, A. D., **280**, **318**, **347**
Jilk, L. T., 60, **64**
Johnson, D. H., 277, **280**
Johnson, D. P., 88, **90**
Johnston, F. L., 425, **426**
Jolly, R. E., 175, **179**
Jorczak, J. S., **446**
Joyce, R. M., 13, **15**, 54, **55**, 433, **434**
Joyner, F. B., 404, **406**

Kalb, G. H., 425, **426**
Kamaliddin, A. R., 36, 37, **40**
Kamath, P. M., 336, **337**
Kämpf, G., 150, **154**
Kane, R. P., 426, **426**
Kargin, V. A., 129, **140**, 166, **171**
Kast, W., 162, **164**
Katchalsky, A., 41, **47**, 58, **64**

Katz, M., 262, **270**
Kauzmann, W., 198, **205**
Kaye, W., 166, **171**
Ke, B., **108**
Kedzie, R. W., 148, **154**
Keeley, F. W., 374, **376**
Keeling, C. D., 359, **359**
Keesom, W. H., 17, **21**
Kegeles, G., 75, 77, **79**
Keiser, N., 491, **492**
Kell, R. W., 13, **15**
Keller, A., **108**, 141, 142, 144, **145**, 149, 150, 151, **153**, **154**, 163, **164**, 166, 167, **171**
Keller, R. A., **90**
Kenyon, A. S., 48, **51**
Keskkula, H., 112, **116**
Ketelaar, J. A. A., **21**
Khoury, F., 144, **145**
Kienle, R. H., 255, **259**
Kiessling, J. C., **400**
Kilb, R. W., 85, **86**
Kin, L., 210, **212**, 222, **226**
Kirkwood, J. G., 83, **85**
Kissinger, H. E., **108**
Kitchen, W., 451, **452**, 454, **455**
Kitson, R. E., 54, **54**, 58, **63**
Klainer, S. M., 77, **79**
Kline, G. M., **89**, **90**, **96**, **100**, **104**, **108**
Klug, A., 98, **100**
Klug, H. P., **99**
Knight, A. C., 420, **424**
Koch, T. A., 222, 224, **226**
Kockelbergh, G., 336, **337**
Kohl, P. D., 491, **492**
Kohle, H., 275, **280**
Kokle, V., 59, 61, **63**, 366, **369**
Kolthoff, I. M., **72**, **90**, **108**, **347**
Kopecky, K. R., 304, 305, **306**
Kovacs, A. J., 157, **157**, 201, 202, **205**
Kraemer, E. O., 13, **15**, 81, **85**, 285, **286**
Králiček, J., 297, **301**
Krassig, H., 451, **452**, 454, **455**
Kratky, O., 29, **30**, 162, **164**
Krebs, E., 13, **15**
Kresser, T. O. J., **369**, **373**
Krigbaum, W. R., 36, 37, **40**, 41, **47**, 48, **51**, 57, **63**, 72, **73**, 83, **85**
Krimm, S., 95, 96, **96**, 163, **164**
Kruissink, Ch. A., 61, **64**
Kuhn, W., 28, **30**, 185, **188**
Kurata, M., 39, **40**, 84, **85**
Kurz, J. E., 48, **51**
Kwolek, S. L., 243, **244**

Lagally, P., 275, **280**
Laidler, K. J., 174, **179**
Lambert, J. M., 119, **124**
Landel, R. F., 196, 197, **198**, 201, **205**
Landler, L. L., 115, **116**
Landler, Y., 332, **333**
Lanning, W. C., 307, **309**
Lansing, W. D., 13, **15**, 285, **286**
Larchar, A. W., 369, 370, **371**
Larsen, H. A., 424, **424**
Larson, N. R., 161, **164**
Lauritzen, J. I., Jr., 151, **154**
Lavin, E., 408, **411**
Lawrence, J. R., **475**
Lawton, E. J., 358, **359**, 366, **369**
Lazan, B. J., 111, **116**
Leaderman, H., **198**
Lebedev, Y. S., 104, **105**
Lee, H., **477**
Leeper, H. M., 140, **145**
Lehrle, R. S., **63**
Leiner, H. H., 88, **90**
Leonard, M. H., **545**
Lester, G. R., 149, **154**
Letsinger, R. L., 300, **301**
Levesque, C. L., 269, **270**
Levy, M., 299, **301**
Lewis, E. E., 421, **425**
Lewis, F. M., 312, 316, **319**, 320, **325**, 329, **333**, **482**
Lewis, G. N., 30
Liang, C. Y., 95, 96, **96**
Lipscomb, W. N., **99**
Liquori, A. M., 136, 137, **140**
Loconti, J. D., 48, **51**, 170, **171**, **187**, **198**, **204**, **207**
Lodge, R. M., **519**
Lohmander, U., 82, **85**
London, F., 18, **21**
Long, V. C., 209, **212**
Longworth, R., 175, **179**
Lontz, J. F., 210, **212**, 221, 222, 224, **226**, 420, 423, **424**
Lord, F. W., 149, **154**, 155, **157**
Lord, S. S., Jr., **89**

Loshaek, S., 174, 175, 176, **179**, 213, **220**
Lou, R. L. H., 72, **73**
Lovelace, A. M., **428**
Ludlum, D. B., 302, 305, **306**
Luongo, J. P., 365, **369**, 370, **371**
Lynn, J. E., **525**
Lysloff, I., 300, 301, **301**

MacDonald, R. N., 442, **444**
MacDougall, F. H., 30
Mackie, J. S., 108, **109**
Madorsky, S. L., 89, **90**, 357, **358**, 421, **424**
Magat, E. E., 217, **220**, 300, **301**, **434**, **436**
Maier, R., 491, **492**
Malcolm, G. N., 41, **47**
Mallouk, R. S., 424, **425**
Malm, F. S., 115, **116**
Maloney, D. E., 115, **116**
Mandelkern, L., 76, **79**, **157**, 159, 161, **164**, 219, **220**
Mandell, E. R., 101, 102, **105**
Manley, R. St. J., 141, **145**
Mark, H. F., 5, **11**, **14**, 47, 48, **51**, 60, 62, **63**, 81, **85**, 115, **116**, 117, **124**, 129, 132, 133, 135, **139**, **140**, 154, 155, 156, **157**, 183, 184, 185, **187**, 199, **204**, 219, **220**, 229, **229**, 237, 239, **239**, **244**, 256, **259**, **301**, 303, **306**, **309**, 311, 312, **318**, 319, **325**, 333, **336**, **353**, 430, **434**
Marlies, C. A., 529, **532**, 538, **541**, **545**
Maron, S. H., 72, **73**
Marshall, I., 166, **171**, **436**
Martens, C. R., **484**
Martin, A. E., 94, **96**, 364, **369**
Martin, H., 302, **307**
Martin, R. W., **469**
Martin, T. J., 115, **116**
Marvel, C. S., **244**, 269, **269**, **270**, 276, **280**, **293**, **333**, 353, **353**, 354, **354**
Mason, S. G., 61, **63**
Masson, C. R., 60, **64**
Matheson, M. S., 274, **280**
Matsuo, H., 359, **359**
Matthews, J. L., 163, **164**
Mauersberger, H. R., **442**, **512**
Maxwell, B., 110, **116**
Mayo, F. R., 54, **55**, 275, 278, **280**, 312, 316, 318, **318**, **319**, 320, 321, 322, 323, 324, 325, **325**, 329, **333**
Mayo, L. R., **393**
Mazzanti, G., 302, **307**
McClellan, A. L., **21**
McCormick, H. W., 78, **79**, 210, **212**, 222, **226**, 299, **301**
McCrum, N. G., 190, 191, **198**, 203, 204, **205**, 372, **373**
McDevit, W. F., 149, **154**
McFarlane, S. B., 510, 511, **512**
McGregor, R. R., **482**
McGrew, F. C., 7, **11**, **15**, 204, **205**, 366, 367, **369**, 372, **373**
McKelvey, J. M., **491**, 495, **497**
McMahon, E. J., 115, **116**
McMillan, F. M., **532**
McPherson, J. L., 210, **212**, 221, 222, 224, **226**, 420, 421, **425**
Mead, D. J., 59, **63**
Meadows, G. W., 295, **297**
Meals, R. N., **482**
Medalia, A. I., **347**
Meehan, E. J., **347**
Meeks, P. J., **491**
Megson, N. J. L., **469**
Melville, H. W., **11**, 13, **15**, 54, **54**, 60, **64**, 265, **270**, 282, **283**, 334, 335, 336, **337**, 408, **411**
Menzies, R. F., 60, **64**
Merckling, N. G., 302, **306**
Merrett, F. M., 225, **226**
Merrill, E. W., 82, **85**, 177, **179**
Merz, E. H., 47, **51**, 318, **319**
Meselson, M., 77, **79**
Mesrobian, R. B., 336, **337**, 391, **392**, 531, **532**
Metz, D. J., 336, **337**
Meyer, K. H., 129, **140**
Meyerhoff, G., **85**, 299, **301**
Michaels, A. S., 341, **348**
Mickley, H. S., 82, **85**, 341, **348**
Mie, G., 67, **73**
Mighton, H. R., 158, **164**
Mighton, J. W., **491**
Miles, F. D., **460**
Milhaelsen, J. D., 421, **424**
Milkovich, R., 299, **301**
Miller, A. A., 358, **359**

Miller, R. G. J., 163, **164**
Miller, R. L., 97, **100**, 119, **123**, **139**, 146, 150, **154**
Miller, W. L., 353, **353**
Milner, D. C., 359, **359**
Milville, M. E., 425, **426**
Mitchell, J., Jr., 58, **63**, 88, **89**, **90**, **108**
Mitsuhashi, S., 144, **145**
Miyazawa, T., 96, **96**
Moncrieff, R. W., **362**, **406**, **411**, **416**, **434**, **436**, **438**, **442**, **451**, **455**, **460**, **462**, **512**, **525**
Montermoso, J. C., **426**, **427**, 428, **428**, **482**
Montonna, R. E., 60, **64**
Mooney, M., 175, **179**
Mooney, R. C. L., 127, **140**
Moore, A. L., 341, **348**
Moore, G. E., 166, **171**, 221, 222, **226**
Moore, J. E., 300, **301**
Moore, L. D., Jr., **63**, 209, **212**
Moran, A. L., 426, **426**
Morehead, F. F., 141, **145**
Morey, D. R., 50, **51**
Morgan, L. B., **145**, **153**, 155, **157**, **171**
Morgan, P., 169, **171**, **369**, **371**, **491**, **497**
Morgan, P. W., 243, **244**
Morse, A. R., 491, **492**
Morton, A. A., 300, **301**
Morton, M., **362**, 374, **376**, **383**, **388**, **392**, **393**, 403, **406**, **446**, **482**, **532**, 539, **541**, **545**, 546, **559**
Mosely, J. R., 161, **164**
Müller, A., 126, **140**, 161, **164**
Müller, E., 437, **438**, **439**
Munden, A. R., **434**
Munn, J. P., **393**
Muus, L. T., 59, 61, **63**, 70, **73**, 98, 99, **100**, 138, **140**, 204, **205**, 210, **211**, 285, **286**, 366, **369**, 372, **373**
Mulvaney, J. E., **301**
Myers, W. L., 300, **301**
Myerson, I., 59, 60, **64**

Nagasawa, M., **87**
Nason, H. K., 110, **116**
Natta, G., 13, **15**, 54, **55**, 117, 118, 119, 120, 121, 122, 123, **123**, **124**, 128, 135, **140**, 269, **270**, 302, 303. **307**, 308, 309, **309**, 336, **337**, 372, **373**, 374, **376**
Naunton, W. J. S., **532**
Naylor, M. A., 276, **280**
Nazarian, G. M., 77, **79**
Neal, A. M., **393**
Nelson, J. A., 437, **438**, **479**
Nelson, W. K., 302, **306**
Neu, R. F., **392**
Neumann, J. A., 375, **376**
Neville, K., **477**
Newbury, N. R., 303, 304, **305**
Newby, T. H., **532**
Newing, M. J., 36, 37, **40**
Newitt, E. J., 59, 61, **63**, 366, **369**
Newman, M. S., 120, 121, **124**
Newman, S. B., **108**
Nichols, J. B., **79**, 163, **164**
Niegish, W. D., 144, **145**
Nielsen, L. E., 97, **100**, 119, **123**, **139**
Niwinska, T., 50, **51**
Noe, R. W., 141, **145**
Nordt, E., 49, **51**
Norrish, R. G. W., 275, **280**, 294, **297**, 336, **337**, 341, **348**
North, A. M., 275, **280**
Norton, J. R., Jr., 112, **116**

Oakes, W. G., 204, **205**
O'Conner, A., 141, **145**, 151, **154**
O'Driscoll, K. F., 298, **301**, 332, **333**
O'Dwyer, J. J., 115, **116**
Oemler, A. N., 58, **63**
Ogg, C. L., 54, **55**
Oglesby, A. C., 336, **337**
Ohlberg, S. M., 222, 224, **226**
Oita, K., 308, **309**
Oka, S., 174, **179**
Oncley, J. L., **87**
Oneacre, L. E., **541**
Onyon, P. F., **85**, **280**, 282, **283**, **318**, **347**
Orr, W. J. C., 35, 36, 37, **40**
Osgan, M., 123, **124**
Oster, G., **72**
Ostwald, W., 3, **11**
Oth, A., 50, **51**
Ott, E., **15**, **452**, **455**, 457, **460**, **462**
Overberger, C. G., 295, **297**, **301**
Owens, F. H., 300, **301**

Padden, F. J., Jr., 144, **145**
Paist, W. D., **460**
Pake, G. E., **105**
Palmer, R. P., 151, **154**
Pals, D. T. F., 61, **64**
Pao, Y.-H., 82, **85**, 178, **179**
Park, G. S., 54, **54**, 347, **348**
Parks, G. S., 161, **164**
Parrette, R. L., 58, **64**
Pasquon, I., 302, 303, **307**, 372, **373**
Pasternak, R. A., 77, **79**
Patat, F., 60, **63**
Paton, J. G., 363, **369**
Patton, T. C., **484**
Pauling, L., **21**, 134, 136, 138, **140**
Peaker, F. W., **40**, 49, **51**, **63**, **72**, **123**, **145**, **153**, **157**, **171**, **252**, **259**, **306**, **309**, 336, **337**
Pease, D. C., 302, **306**, 369, 370, **371**
Peck, V., 166, **171**
Pedersen, K. O., **79**
Peiser, H. S., 125, 127, **140**, 163, **164**
Pepper, D. C., 296, **297**, 332, **333**
Peraldo, M., 120, 121, 123, **124**
Perkins, J. R., 115, **116**
Perkinson, G., 82, **85**
Perrin, M. W., 363, **369**
Peterlin, A., 151, **154**
Peters, E. F., 307, **309**
Peterson, J. H., 347, **348**, 419, 420, **424**
Peterson, S., 437, **438**
Petke, F. E., 255, **259**
Philippoff, W., 82, **85**
Phillips, C. S. G., 49, **51**
Piepenbrink, H.-F., 437, **438**
Pierce, R. H. H., Jr., 138, **140**, 163, **164**
Pike, R. A., 482, **482**
Pimentel, G. C., **21**
Pinner, S. H., 366, **369**
Pino, P., 302, **307**
Pitzer, K. S., 30, 120, 121, **124**
Plesch, P. H., **297**, 336, **337**
Plunkett, R. J., 419, **425**
Pohl, H. A., 54, **55**
Polanyi, M., 294, **297**
Polowinski, S., 50, **51**
Porod, G., 29, **30**, 97, **100**, 162, **164**
Porter, W. L., 54, **55**
Posner, A. S., **100**
Postelnek, W., **428**
Press, J. J., **362**, **512**, **525**, **559**
Preston, J. M., **519**
Price, C. C., 13, **15**, 54, **55**, 123, **124**, **270**, 329, 331, **333**
Price, F. P., 148, 151, **154**, 155, **157**
Price, G. F., **54**
Proskauer, E. S., **90**, **108**
Ptitsyn, O. B., 126, 136, **140**
Purves, C. B., **15**

Quinn, F. A., Jr., 161, **164**

Raetz, R. W., 47, **51**
Raff, R. A. V., 222, 224, **226**, **369**, **371**, **373**
Ram, A., 82, **85**
Rånby, B. G., 141, **145**, 228, **229**, **452**, **455**, **460**, **462**
Randall, M., 30
Ray, N. H., 58, **64**
Rayleigh, Lord, 27, **30**, 64, 67, **73**
Reding, F. P., 108, **109**
Ree, T., 178, **179**
Reed, G. A., 275, **280**
Rehberg, C. E., 219, **220**
Rehner, J., Jr., 63, **64**
Reid, W. J., 500, **506**
Reitenour, J. S., 58, 61, **63**
Reneker, D. H., 142, 143, 144, **145**, 148, **154**
Renfrew, A., 169, **171**, **369**, **371**, **491**, **497**
Renfrew, M. M., 421, **425**
Rexford, D. R., 426, **426**
Rexroad, H. N., 104, **105**
Rice, S. A., **87**
Rich, A., 98, **100**
Richards, A. W., 296, **297**
Richards, R. B., 94, 95, **96**, 163, **164**
Riley, J. L., 514, 516, 517, **519**
Riseman, J., 83, **85**
Ritscher, T. A., 60, **63**
Rivers, J. T., 423, **425**
Robb, J. C., **40**, 49, **51**, **63**, **123**, **145**, **153**, **157**, **171**, **252**, **259**, **306**, **309**
Robb, L. E., 423, **424**, 425, **426**
Roberts, B. W., 141, **145**, 148, **153**, **154**, **157**
Roberts, D. E., 161, **164**
Roberts, J. D., **104**

Robinson, D. W., 204, **205**
Rocard, Y., 66, **73**
Rochow, E. G., **482**
Roder, T. M., 166, **171**, 221, 222, **226**
Roe, R.-J., 285, **286**, 306, **306**
Roebuck, A. K., 307, **309**
Roedel, M. J., 302, **306**, 365, **369**
Rogers, S. S., 219, **220**
Roha, M., **306**
Roig, A., 39, **40**, 84, **85**
Rollin, S., 60, **63**
Rose, D. G., 359, **359**
Rosenberg, J. L., 49, **51**
Ross, M., 358, **359**
Ross, S. D., 316, **319**
Roth, F. L., 183, **188**
Roth, J., 222, 224, **226**
Rouse, P. E., 178, **179**, 197, **198**
Rowlinson, J. S., 41, **47**
Roy, M. F., 269, **270**, 354, **354**
Rudd, J. F., 177, **179**, 211, **212**
Rudin, A., 108, **109**
Rudner, M. A., **424**, **427**
Rugg, J. S., 426, **426**
Rushman, D. F., **63**
Russell, E. W., 115, **116**
Russell, K. E., 294, **297**
Rydholm, S. A., 228, **229**, **452**, **455**, **460**, **462**

Saegusa, T., **261**, **444**
Safranski, L. W., **90**
Salem, W. J., 77, **79**
Salkind, A., 60, **63**
Salyer, I. O., 48, **51**
Sample, J. H., 269, **270**, 354, **354**
Sandeman, I., 163, **164**
Sandt, B. W., 421, **424**
Sauer, J. A., 110, **116**
Saunders, J. H., 438, **439**, **479**
Sauter, E., 140, **145**
Schachman, H. K., **79**
Schaefgen, J. R., 82, **86**, 125, **140**
Scheele, W., **541**
Schellman, J. H., 139, **140**
Schildknecht, C. E., 119, **124**, 217, **220**, 228, **229**, **269**, 291, 292, **293**, **297**, **301**, 307, **309**, **340**, 342, **347**, **348**, **411**, **434**, **446**, **452**, **455**, **460**, **469**, **472**, **477**, **491**, **497**, **506**, 514, 516, 517, **519**, **532**
Schiller, A. M., **301**
Schlesinger, W., 140, **145**
Schmidt, A. X., 529, **532**, 538, **541**, **545**
Schmieder, K., 190, **198**
Schneider, N. S., 29, **30**, 48, **51**, 170, **171**, **187**, **198**, **204**, **207**
Schneiders, J., 310, **319**
Schoknecht, G., 99, **100**, 147, **154**
Schreyer, R. C., 302, **306**
Schuler, N. W., 336, **337**
Schultz, A. R., 42, 43, **47**
Schulz, G. V., 46, **47**, 49, **51**, 252, **253**, 269, **270**, 274, 275, 276, **280**
Schweitzer, C. E., 442, **445**
Scott, R. G., **108**, 141, **145**
Scott, R. L., 26, **27**
Sebban-Dannon, J., 336, **337**
Šebedna, J., 297, **301**
Sedlak, V. A., 421, **424**
Seeger, N. U., 437, **439**
Sella, C., 152, **154**, 166, **171**
Sellers, J. W., 541, **547**
Semon, W. L., **388**
Severini, F., 336, **337**
Sewell, P. R., 408, **411**
Seymour, R. B., 498, **499**
Shearer, N. H., Jr., 372, **373**, 404, **406**
Shelton, J. R., 530, **533**
Shelton, R. S., 112, **116**
Signer, R., 140, **145**
Silberberg, A., 41, **47**
Simha, R., 311, **319**, 356, **358**
Simonds, H. R., **362**, **373**, **559**
Simpson, W., 365, **369**
Simril, V. L., 425, **426**
Sinnott, K. M., 204, **205**
Sippel, A., **519**
Skeist, I., 320, **325**, **477**
Slichter, W. P., 101, 102, **104**, **105**
Small, P. A., 21, **21**, 26, **27**
Smets, G., 335, 336, **336**, **337**
Smith, C. W., 105, 106, **108**
Smith, D. J., 129, 130, **140**
Smith, H., 58, **64**
Smith, J. F., 426, **426**
Smith, W. F., 345, **348**, 437, **438**
Smith, W. M., **411**, **416**
Smith, W. V., 282, **283**

Smoluchowski, M. S., 65, **73**
Smothers, W. J., **108**
Snyder, J. W., **545**
Snyder, W. H., 300, **301**
Sokolava, E. A., 126, 136, **140**
Sood, S. P., 295, **297**
Sookne, A. M., 210, **212**
Sorenson, W. R., **244**, **261**, **272**, **293**
Sparks, W. J., **392**
Speiser, R., 70, 71, 72, **73**
Spencer, R. S., 52, **52**, 199, **204**, 490, **492**
Sperati, C. A., 158, **164**, 210, 211, **212**, 221, 222, 223, 224, **226**, 227, 228, **230**, 366, **369**, 420, 421, **424**, **425**
Spiegels, M. C., 48, **51**
Spurlin, H. M., **15**, **452**, **455**, 457, **460**, **462**
Stabin, J. V., 59, **64**
Stacey, K. A., **73**
Stahl, F. W., 77, **79**
Stanley, R. R., 58, **63**
Stannett, V., 295, **297**
Starkweather, H. W., Jr., 158, 161, **164**, 166, **171**, 210, 211, **212**, 217, **220**, 221, 222, 223, 224, **226**, 227, **230**, 366, **369**, 420, 421, **424**, **425**
Statton, W. O., 97, **100**, 125, **140**, 141, **145**, 152, 153, **154**, 163, **164**, 166, 167, **171**
Staudinger, H., 13, **15**, 79, 82, **86**, 117, **124**, 140, **145**, 262, 269, **270**, 310, **319**, 363, **369**, 442, **445**
Staudinger, M., 140, **145**
Stavely, F. W., 121, **124**, 382, **383**
Staverman, A., 61, **64**
Stead, B. D., 334, **337**
Stein, R. S., 67, **73**, 95, **96**, 170, **171**, **187**, **198**, **204**, **207**
Steinbrunn, G., 310, **319**
Steinhofer, A., 13, **15**, 269, **270**
Stevenson, A. C., 426, **426**
Stewart, D. M., **89**
Stickney, P. B., 499, **506**
Stiles, E., 300, 301, **301**
Stille, J. K., **11**, **306**, **309**
Stivala, S. S., **477**
Stober, K. E., 395, **400**
Stockmayer, W. H., 29, **30**, 39, **40**, 45, **47**, 50, **51**, 57, **64**, 84, **85**, **86**, 284, 285, **286**, 300, **301**, 319, **325**, 408, **411**
Storey, E. B., 532, **533**
Straus, S., 89, **90**, 421, **424**
Strella, S., 110, **116**
Stross, F. H., **63**
Stroupe, J. D., 136, **140**
Strutt, J. W., *see* Rayleigh, Lord
Stuart, A., **108**
Stuart, H. A., 141, **145**, 146, 150, 152, **154**, 157, **157**, 160, 161, **164**, 168, **171**
Studebaker, M. L., **545**
Sturtevant, J. M., **107**
Suen, T. J., **469**, **472**
Supnik, R. H., 110, **116**
Sutherland, G. B. B. M., 95, 96, **96**, 294, **297**
Sutherland, G. K., 545, **547**
Svedberg, T., **79**
Swan, P. R., 144, **145**
Symons, N. K. J., 141, **145**
Szwarc, M., 295, **297**, 299, **301**, 335, **337**

Takemura, T., 82, **86**
Tamblyn, J. W., 50, **51**
Tanford, C., **30**, **73**, **79**, **85**, **87**, **100**, **179**
Tate, B. E., 54, **55**
Taylor, G. B., 54, **55**, 250, **253**
Taylor, R. P., 13, **15**, 54, **54**, 265, **270**
Teach, W. C., **400**
Teeple, J. H., **115**
Temple, R. B., 170, **171**
Thayer, G. B., **491**
Thierstein, A., 491, **492**
Thomas, P. E., 210, **212**, 221, 222, 224, **226**, 420, **425**
Thomas, R. M., **392**
Thomas, W. M., 341, **348**
Thompson, A. B., 166, **171**
Thompson, H. W., **96**
Thompson, M. S., **400**
Thompson, R. J., 446, **447**
Thompson, W. B., Jr., 424, **425**
Throssell, J. J., 295, **297**
Tiers, G. V. D., 119, **123**
Till, P. H., Jr., 141, **145**
Tobin, M. C., 95, **96**
Tobolsky, A. V., 163, **164**, 188, 189,

198, 207, **207**, 239, **239**, 256, **259**, 277, **280**, 298, **301**, 391, **392**, 531, **532**
Todd, N. W., 424, **424**
Tompa, H., 30, **30**, **40**, 42, 44, **46**, **47**, **63**, **85**
Tordella, J. P., 163, **164**, 175, **179**
Tosi, M., 151, **154**
Tovey, H., **436**, **475**
Treiber, E., 453, **455**
Treloar, L. R. G., 33, **40**, 182, 184, **187**
Trillat, J.-J., 152, **154**, 166, **171**
Trommsdorff, E., 275, **280**, 342, **347**
Truett, W. L., 302, **306**
Tryon, M., **96**
Tsvetkov, I. D., 104, **105**
Tuckett, R. F., 204, **205**
Tung, L. H., 285, **286**
Turnbull, D., 141, **145**, 148, 153, **154**, **157**
Turner, D. T., 336, **336**

Ubbelohde, A. R., 161, **164**
Ueberreiter, K., 26, **27**, 157, **157**, 214, **220**
Ullman, R., 115, **116**
Ullrich, R. D., **373**
Urquhart, A. R., 48, **51**

Vale, R. L., 336, **337**
Van Amerongen, G. J., **229**
Vand, V., 98, **100**
Van de Hulst, H. C., 72, **73**
Van der Meulen, P. A., 255, **259**
Van der Weg, K. R., 148, 150, **154**
Van Holde, K. E., 77, **79**
Van Horn, P., 454, **455**
Vassallo, D. A., 88, **90**
Vaughan, M. F., 48, 49, **51**
Verbanc, J. J., 374, **376**, **438**, **479**
Vickerstaff, T., **525**
Vincent, P. I., 166, **171**, 221, **226**
Vinograd, J. R., 77, **79**
Voevodskii, V. V., 104, **105**
Vofsi, D., 58, **64**
Vogl, O., 443, **445**
Voorn, M. J., **46**
Voss, A., 310, **319**
Voter, R. C., 365, **369**
Wack, P. E., 188, **198**
Waddington, F. B., 365, **369**
Wagner, M. P., 541, **547**
Wagner, R. H., **63**
Wagner-Jauregg, T., 310, **319**
Waldron, J. D., 365, **369**
Wales, M., 77, **79**
Walker, E. E., 21, **21**, 26, **27**, 216, **220**, **229**
Wall, F. T., 30, 311, **319**
Wall, L. A., **89**, **104**, 356, **357**, 358, **358**, **359**, 421, **424**
Wallder, V. T., 115, **116**
Walling, C., **270**, **280**, **283**, **286**, **290**, 312, 317, 318, **318**, **319**, 320, 321, 322, 323, 324, 325, **325**, 328, 329, **333**, 351, **353**
Walter, E. R., 108, **109**
Walter, N. M., 141, **145**
Waltz, J. E., 54, **55**
Ward, I. M., 54, **55**
Watson, M. T., 436, **436**
Watson, W. F., 26, **27**, 335, **336**, 545, **547**
Webber, A. C., 114, 116
Weinstock, K. V., 532, **532**
Weiss, P., 346, **348**
Weissberg, S. G., 76, **79**
Weissberger, A., **63**, **72**, **79**, **89**, **90**, **96**, **99**, **104**, **107**, **108**
Werber, F. X., **309**
Wesslau, H., 48, **51**, 285, **286**
West, W., **96**
Wethington, J. A., Jr., 161, **164**
Wetzel, J. W., 375, **376**
Wheeler, O. L., 408, **411**
Whinfield, J. R., **436**
Whitby, G. S., 5, **11**, **14**, 81, **85**, 129, 132, 133, **139**, 154, 155, 156, **157**, 183, 184, 185, **187**, 199, **204**, 237, **239**, **244**, 262, **270**, **348**, **388**, **392**, **393**, 430, **434**, **532**, 546
White, E. F. T., 213, 215, **220**, 334, **336**
White, H. G., 437, **438**
Whitmore, F. C., 294, **297**
Whitney, J. F., 138, **140**, 222, 224, **226**
Wichterle, O., 297, **301**
Wicker, T. H., Jr., 404, **406**

Wijga, P. W. O., 211, **212**, 221, 222, 224, **226**
Wilfong, R. E., **220**, **436**
Willbourn, A. H., 204, **205**, 365, **369**
Williams, E. G., 363, **369**
Williams, H. L., **348**
Williams, J. W., **79**
Williams, L. C., 76, **79**
Williams, M. L., 196, 197, **198**, 201, **205**
Williams, R. J. P., 49, **51**
Williams, T. C., 482, **482**
Williams, T. F., 359, **359**
Williamson, A. T., 58, **64**
Willis, H. A., 94, 95, **96**, 163, **164**
Willits, C. O., 54, **55**
Wilson, A., 428, **428**
Wilson, F. C., 163, **164**
Winchester, C. T., 382, **383**
Windemuth, E., 437, **438**
Winding, C. C., 109, 110, **116**, **362**, 396, **400**, 467, **469**, 493, **497**, 498, **499**
Winkler, D. E., 413, **416**
Wittbecker, E. L., 243, **244**
Wolf, K., 190, **198**
Wolf, R. F., 541, **547**
Wolf, R. W., **532**
Wolock, I., 115, **116**
Wood, D. L., 365, **369**, 370, **371**
Wood, L. A., 154, 155, 156, **157**, 183, **188**, 215, **220**
Woodall, N. B., **96**
Wooding, N. S., 299, **301**
Woodward, A. E., 335, **337**
Wooten, W. C., 89, **90**
Wren, W. G., **383**
Wunderlich, B., 161, **164**, 215, 217, **220**

Yamakawa, H., 39, **40**
Yoncoskie, B., 300, **301**
Young, C. A., 437, **438**
Young, L. J., **318**
Yorgiadis, A., 111, **116**

Zambelli, A., 372, **373**
Zapp, R. L., **392**
Zaukelies, D. A., 167, **171**
Zemany, P. D., 366, **369**
Ziegler, E. E., 115, **116**
Ziegler, K., 297, **301**, 302, **307**
Zimm, B. H., 29, **30**, 39, **40**, 59, 60, **64**, 67, 69, 72, **73**, 83, 84, 85, **86**, 139, **140**, 197, **198**, 284, **286**
Zlets, A., 307, **309**
Zoss, A. O., 119, **124**

Subject Index

Ablation, 468
Abrasion resistance, elastomers, 375, 445, 527, 541, 543, 544
  fibers, 405, 431, 437, 512
  measurement, 112
  plastics, 402, 431
ABS resins, 399, 503
Accelerators, 539; *see also* Vulcanization
Acetal resins, *see* Polyoxymethylene
Acetate rayon, 456–8
Acetylation, cellulose, 456
  cotton, 451
  poly(vinyl alcohol), 409
Acrylate elastomers, 387, 403
Acrylic esters, polymerization mechanism, 292
Acrylic fibers, 405
Acrylic plastics, 400–5
Acrylonitrile, fiber, 405
  polymerization mechanism, 292, 341
Activation energy, degradation, 287
  polymerization rate, 286
  viscous flow, 173, 200
Activators, 540; *see also* Vulcanization
Addition polymerization, 262–300; *see also* Chain polymerization
Addition polymers by step-reaction mechanisms, 238
Adhesives, acrylate, 403
  amino resin, 472
  cyanoacrylate, 404
  epoxy, 477
  phenolic, 469

Adhesives, polyisobutylene, 390
  poly(vinyl acetate), 407
Aldehydes, polymers of, 443
"Alfin" polymerization, 300
Alfrey-Price equation, 329
Alginate fiber, 461
Alkali cellulose, 452
Alkyd resins, 242, 483
Allyl resins, allyl diglycol carbonate, 484, 505
  diallyl phthalate, 484
Alternation, 313; *see also* Copolymerization
Amide interchange, 242
Amino resins, 470–2, 505
Analysis, chemical, of polymers, 88
Angular dissymmetry, 67–9
Anionic polymerization, 292, 297–301
Antioxidants, 500
Arc resistance, 114
Arrhenius equation, 173, 286
Asbestos, 498
Association, 12
Atactic configuration, 119
Athermal solutions, 32
Autoacceleration, 275
Avrami equation, 156
Āzeotropic copolymerization, 315

Bag molding, 475
Balata, 382
Batch polymerization, 339

Bead polymerization, *see* Suspension polymerization
Binoidal curve, 42
Birefringence, 158, 170
Block copolymers, 225, 229, 334
Block methods, 460
Block osmometer, 59
Bloom, 539
Blow molding, 368, 371, 492
Blowing, 492
Boiling point elevation, *see* Ebulliometry
Bonds, chemical, hydrogen, 18
  primary, 15
  secondary, 17
Branching, 7, 29
  in chain polymerization, 285
  detection, by infrared spectroscopy, 90, 364
    by intrinsic viscosity and molecular weight, 84
  in diene and divinyl polymers, 348
  and melt viscosity, 176, 209
  molecular size and, 29
  and molecular weight distribution, 255, 285
  multichain polymers, 252
  in polyethylene, 364
  and polymer properties, 209, 218
  in poly(vinyl acetate) and poly(vinyl alcohol), 408
  in SBR, 385
  in stepwise polymerization, 253–9
  in styrene-divinyl benzene copolymers, 398
  transfer to polymer, 278, 285
Brittle point, temperature, 111
Brittle strength, 221; *see also* Impact resistance, strength
Bulk polymerization, 337–9
Buna rubber, 384
Butadiene, 383
  polymerization mechanism, 292
Butyl rubber, 388–92

Calender, calendering, 493
Calorimeter, calorimetry, 105
Carbanion polymerization, 292, 297–301
Carbon black, 498, 543; *see also* Reinforcement
Carbonium ion polymerization, 292–6
Casein, 441
Casting, 493–4
  acrylic resins, 401
  cellulose plastics, 494
  centrifugal, 494
  film, 494
  phenolic resins, 468
  thermosetting resins, 494
Catalyst, 242, 246–7, 291–309, 370, 372, 382, 389, 395, 406, 412, 456, 464–8
Cationic polymerization, 292–6
Ceiling temperature, 288
Cellophane, 60, 454
Celluloid, 14, 459
Cellulose, 241, 449–51, 498
  acetate, fiber, 14, 456–8, 517, 524
    plastic, 14, 456–8, 502
  acetate-butyrate, 458
  acetate-propionate, 459
  alkali, 452
  bacterial, 60
  crystal structure, 129
  esters, 456–60
  ethers, 461
  history, 14
  nitrate, 14, 60, 459–60
  propionate, 459
  regenerated, 60, 452–5; *see also* Viscose
  solubility of derivatives, 228
  triacetate, 456–8, 524
  xanthate, 452
Centrifugation, *see* Ultracentrifugation
Chain branching, *see* Branching
Chain expansion factor, 29, 38
Chain folding, 142, 151
Chain flexibility, 28
  and polymer properties, 217
Chain polymerization
  activation energy, 286–7
  autoacceleration, 275
  configuration, 268
  degradation, 354–7
  diene, divinyl compounds, 348–53
  end group analysis, 54
  free energy, 288
  free radicals, 264; *see also* Radicals

Chain polymerization, heat, 288
induction period, 268
inhibitors, inhibition, 267
initiators, initiation, 264, 271, 273, 294, 298
initiator efficiency, 264
kinetic chain length, 277
kinetics, 272–80, 295–6
laboratory methods, 270
molecular weight, 279, 283–5
molecular weight distribution, 283–5
photoinitiation, 271
propagation, 265, 273, 294, 298
radical lifetime, 281
rate constants, 274, 281, 283
retarders, retardation, 267
rotating sector, 281
steady state, 273, 295, 300, 305
and step polymerization, 238
termination, 265, 273, 294, 299
thermal, 277
transfer, 260, 277, 295, 299
See also Anionic, Batch, Bulk, Cationic, Continuous, Emulsion, Solution, Suspension polymerization; Copolymerization
Chain reaction polymers, configuration, 268
end group analysis, 54
Chain transfer, *see* Transfer, Chain polymerization
Chemical properties, measurement, 115
Chemical reactions of polymers, 353
elastomers, 380–2, 533–7
fibers, 452–4, 456–7
plastics, 374–5, 398, 407–10, 413, 456–7, 459, 461
Chemical resistance of polymers, 367, 372, 397, 402, 413, 421, 425–8, 432, 435, 443–4
Chlorinated polyether, 444, 502
Chlorinated polyethylene, 374
Chlorinated polypropylene, 374
Chlorinated poly(vinyl chloride), 415
Chlorinated rubber, 380
Chlorine-containing polymers, 374, 380, 412–6, 444, 502, 503, 524
Chlorosulfonated polyethylene, 374
Chlorotrifluoroethylene, 426
Chromatographic absorption, 49
Clarification of polymer solutions, 80
Coacervation, 41, 48
Coatings, 368, 438, 494
Cocatalyst, 291, 294–6
Cohesive energy (density), 19, 26
Coining, 422
Cold drawing, 165
Cold molding, 484, 488
Cold rubber, 385
Colligative properties, 31, 55–63
Collision factor, frequency, 245, 287
Colloid, colloidal state, 11
Color, measurement, 113
Colorants, 501; *see also* Dyeing
Combination, 265; *see also* Chain polymerization
Comfort of fabrics, 512
Compliance, elastic, 195
Compounding, elastomers, 386–90, 531
plastics, 506
Compression molding, 487
Condensation polymerization, 237–61; *see also* Step-reaction polymerization
Condensation polymers by chain mechanism, 238
Configurations, 9, 13, 117–123
in chain polymerization, 268, 294
cis vs. *trans*, 120, 352
definition, 9
in diene polymers, 352
Conformations, 9, 27, 117–123
definition, 9
in dissolved polymers, 27
trans vs. *gauche*, 125, 135
Consolute temperature, 41
Contact resins, *see* Polyester resins
Continuous filament, 508
Continuous polymerization, 339
Contour length, 27
Coordinate bond, 16
Coordination catalysts, 302–6, 370, 372, 382
Coordination complexes, 12, 302–6
Coordination polymerization, 292, 302–6, 370, 372, 382
Copolymer equation, 311, 320, 350–2
Copolymerization, 310–336
Alfrey-Price equation, 329
alternating, 313, 329

Copolymerization, azeotropic, 315
chemistry, 326–33
coordination, of dienes, 353
cross termination, 316
of dienes, divinyl compounds, 350–2
history, 310
ideal, 312
initiation, 316
ionic, 331
kinetics, 310–18
monomer reactivity, 326
penultimate unit, 318
polarity, 329
and polymer properties, 224
radical reactivity, 328
rate, 316
reactivity ratios, 312
resonance, 327
step-reaction polymers, 332
steric effects, 329
termination, 316
ternary systems, 318
Copolymers, block, 334
composition, 313, 319–25
and properties, 224
diene and divinyl, 350–2
graft, 335
homogeneity, attainment, 325
solubility, 229
See also names of monomers
Corn protein fiber, 442
Cotton, 451–2, 524
Cotton flock, 498
Coumarone resins, 418
Covalent bond, 16
Crazing, 115, 397, 402
Crease resistance, retention, 512, 523
Creep, 190, 509
Crepe rubber, 379
Crimp, 509
Critical miscibility temperature, 41
Crosslinked polymers, 8, 225, 348–52, 358, 375, 398, 463–84
Crosslinking, chemical, polyethylene, 375
diene and divinyl polymers, 348–52
kinetics, 253–5
and molecular weight and distribution, 255–9
and polymer properties, 225
Crosslinking, by radiation, 358, 375
Cross termination, 316
Cryoscopy, 55–6, 58–9, 61
Crystal defects, 147–8
Crystal structures, 98–9, 124–40, 146–7
Crystalline melting, 157–8
Crystalline melting point, determination, 157–9, 215
effect of structure on, 159–61, 212–4
and glass transition, 212
melting range, 158
property changes at, 157–8
solubility and, 25
Crystallinity, 9, 157–64
degree, 162–3
lamellae, 141
paracrystals, 147
and polymer properties, 221, 227
single crystals, 140–4
spherulites, 148–51
structural requirements for, 124
Crystallization, 9, 151–62
Crystals, single, 140–4
Cure, curing, 537; *see also* Vulcanization
Cyanoethylation of cotton, 451
Cyclic compounds, polymerization, 260
Cyclized rubber, 381

Dashpot, 192–5
Debye equation, 66
Degradation, 354–7
Degree of crystallinity, 162–3; *see also* Crystallinity
Degree of polymerization, 4, 277, 287; *see also* Molecular weight
Denier, 508
Density, and degree of crystallinity, 162
of polymers, 363–4, 370–2, 420, 432, 502–5, 524
See also Specific volume
Density gradient, sedimentation in, 77
Depolymerization, 354–7
Depropagation, 354–7
Deswelling, 52
Diallyl phthalate resins, 484, 504
Dialysis, 49
Dielectric properties, measurement, 114
of polymers, 226–9, 232, 367, 397, 421, 502–5
Diene polymers, 292, 348–52, 377–93

Differential thermal analysis, 106–7, 159
Diffusion, autoacceleration, 275
  fractionation by, 49
  measurement, 115
  osmotic membranes, 60
  sedimentation, 78
Diisotactic, 120
Dilatometer, 271
Dilution, free energy, 31, 34
Dip coating, dipping, 494
Dipole, dipole force, dipole moment, 17
Diradicals, 277
Directional order, 125
Dispersion force, 18
Dispersion polymerization, 342
Disproportionation, 265; *see also* Chain polymerization
Dissymmetry, 68
Disyndiotactic, 120
Ditactic, 120
Divinyl polymers, 348–52
Double bond, 263
Drape, 509
Drawing, 165
Dry spinning, 517
Drying resins, 483
DTA, 106–7, 159
Dyeing, 511, 521
Dynamic mechanical properties, of elastomers, 386–93, 546
  measurement, 111, 190, 529

Ebonite, 380
Ebulliometry, 55–8
Economics of polymer uses, 232
Efficiency, initiator, 264
  plasticizer, 500
Efflux time, 79
Elasticity, kinetic theory of rubberlike, 180–7
Elastomers, 526–47; *see also* Rubbers
  chain structure, 185
  defining properties, 180
  ideal, 180
  mechanical properties, 528
  molecular requirements, 180, 230
  oxidative aging, 530
  price and production volume, 377
  stress-strain behavior, 183–5
  table of properties, 546
Electrical properties, measurement, 114–5
  of polymers, 226–9, 232, 367, 397, 421, 502–5
  and structure, 227–8
Electron microscopy, 108
Electron paramagnetic resonance spectroscopy, 103
Electronic structure of compounds, 263
Electrostatic forces, 263
Elongation, 109
Elution, 48–9
Emulsion polymerization, 343–7, 384
End group analysis, 13, 53
End-to-end distance, 27, 38
Energy, *see* Free energy
Entropy, definition, 20
  elasticity, 181
  of fusion, 160
  of mixing, 31–2
  of polymerization, 289
Epoxy resins, 475–7, 505
EPR, 103–4
Equilibrium, esterification, 456–60
  polymerization-depolymerization, 268
  step-reaction polymerization, 246, 435
Ester interchange, 242, 435
Ethyl acrylate rubbers, 403
Ethyl cellulose, 461
Ethylene, 292, 363
  polymers, 363–71
Ethylene oxide, 260, 443
Excluded volume, 29, 37–9
Extent of reaction, 246; *see also* Step-reaction polymerization
Extruders, extrusion, 495

Fabrication, elastomers, 375, 379, 386, 390, 526–45
  fibers, 513–25
  plastics, 367, 370, 401, 411, 414, 422, 460, 474, 478, 487–506
Fabrics, nonwoven, 523
  properties, 509, 511
  textured, 523
Fatigue tests, 111
Felting, felts, 436, 439
Fiber, 507–25
  after treatment, 520
  definition, 507

Fiber, finishing, 522
molecular requirements, 231
photomicrographs, 514
price and production volume, 429, 448
properties, 509–11, 524
spinning, 513
textile terms, 508
Fibers, acetate, 457
acrylic, 405
alginate, 461
casein, 441
cellulose, 448–62
cellulose acetate, 458
chlorine-containing, 415
cotton, 450
fluorocarbon, 423
modacrylic, 405, 415
nylon, 432
nytril, 405
poly(ethylene terephthalate), 435
polyurethan, 437
poly(vinyl alcohol), 409
protein, 441
rayon, 454, 457
regenerated cellulose, 454
saran, 415
silk, 440
triacetate, 457
vinal, 409
vinyon, 415
viscose, 454
vulcanized, 455
wool, 439
zein, 442
Filament, nylon, 433
polyethylene, 371
polypropylene, 373
Fillers, elastomers, 527, 541; *see also* Reinforcement
plastics, 498
Film, blown, 492
cast, 494
cellophane, 454
polyethylene, 367
poly(ethylene terephthalate), 436
polypropylene, 373
polystyrene, 399
poly(vinyl fluoride), 425
Finishes, 521–3
First-order transition, definition, 10
Flame retardants, 501
Flammability, measurement, 113
Flash, 488
Flex life, measurement, 111
molecular structure and, 221
See also Dynamic mechanical properties
Flory constant Φ, 84
Flory-Huggins theory, 34
Flory-Krigbaum theory, 37
Flory temperature Θ, 38, 63, 76, 78, 84
Flow, melt, viscous, 173–9
Fluorine-containing polymers, 229, 419–28, 503
Foams, 477, 481, 496
Folded chain structures, 142
Forming, 496
Fractionation, 47–51
Free energy, of mixing, 31, 34
of polymerization, 289
of solution, 25
of stretching elastomers, 181
Free radicals, *see* Radicals
Free volume, 201
Freezing point depression, 55–6, 58–9, 61
Frequency factor, 245, 287
Frictional coefficient, 78
Friedel-Crafts catalysts, polymerization, 293
Fringed micelle, 145
Functional group determination, 53, 92–4
Functionality, 239
Furan resins, 484
Furfural, 469
Fusion, heat and entropy, 160–2, 215
See also Crystalline Melting point

Gas chromatography, 89
Gas-phase polymerization, 340
Gases, light scattering in, 64
Gate, 490
Gaussian distribution, 28, 181
Gel, gel point, gelation, 253–5; *see also* Crosslinking, Step-reaction polymerization
Gel cellophane, 60
Gel effect, 275

Gel volume, 50
Gels, swelling of, 52
Glass, 499
  reinforced plastics, 472–5
Glass transition, glass transition temperature, glassy state, 10, 198–204
  measurement, 199
  molecular interpretation, 201
  and polymer properties, 212–14
  time effects in, 212–14
Gloss, 113
Gradient elution, 48
Graft copolymers, 225, 229, 335–6
Granular polymerization, *see* Suspension polymerization
GR-S, 384–5; *see also* SBR
Gum rubber, 540–1
Gutta percha, 129, 161, 382

Hammett equation, 331
Hand, handle, 509
Hard resins, 483
Hard rubber, 380
Hardness, 113, 221, 402
Haze, 113
Head-to-head, tail-to-tail configuration, 269
Head-to-tail configuration, 13, 269
Heat, buildup, 388, 391, 393, 526, 529, 544–6
  capacity, 105
  content, 105
  distortion temperature, test, 112
  and elongation of elastomers, 182
  of fusion, 160
  of mixing, 31, 34
  of polymerization, 288
  resistance, stability, elastomers, 375, 380, 387, 391, 393
    fibers, 432, 435
    plastics, 367, 370–1, 373, 397, 402, 408, 413, 421, 425, 427, 432, 435, 443–4, 468, 471, 481–2
  shrinking, 371, 376, 415
Helix-coil transition, 139
Heterochain polymers, 6, 429–84
Hevea brasiliensis, Hevea rubber, *see* Rubbers, natural
Hobbing, 422
Homochain polymers, 6, 363–428
Hooke's law, Hookean elasticity, 172, 192
Huggins equation, 81
Hydrogen bond, 18–19

Ideal copolymerization, 312
Ideal elastomer, 180
Ideal solutions, 31
Impact resistance, strength, 111, 367, 372, 397, 399, 402, 421, 431, 436, 443–4, 446, 458, 460, 471, 498–9, 502–5
Indene resins, 418
Induction force, 17
Induction period, 268
Infrared spectroscopy, 54, 90–6, 170, 366
Inherent viscosity, 80
Inhibition, inhibitors, 267
Initiation, initiator, copolymerization, 316
  degradation, 355
  ionic chain polymerization, 294–9
  radical chain polymerization, 264–72
  vulcanization, 534
Initiator efficiency, 264
Injection molding, 368, 370, 373, 397, 402, 424, 432, 444, 446, 458, 489
Insulation, *see* Wire and cable insulation
Interchange, 242, 251, 435
Interfacial polymerization, 243
Intermolecular forces, 17–21
Internal friction, 190, 204
Internal plasticization, 499
Intrinsic viscosity, 79–85
Ion-exchange resins, 398, 469
Ionic bond, 15
Ionic polymerization, 293–301, 389
Ionization, 86
Irradiation, 358, 375
Irregular solutions, 32
Isobutylene, 292, 388
Isopiestic molecular weight determination, 58
Isoprene, 12, 388
Isotactic configuration, 119

Isotactic polymers, 119, 135–6, 372, 396, 402
Isothermal distillation, 58

Jump frequency, 174

Kelvin (Voigt) element, 193
Keratin, 133, 136, 439–41
Kinetic chain length, 277
Kinetic theory of elasticity, 180–7
Kinetics, of coordination polymerization, 305
  of copolymerization, 310–8
  of crystallization, 154–7
  of degradation, 355–6
  of ionic chain polymerization, 295–6, 299
  of radical chain polymerization, 272–80
  of step-reaction polymerization, 245–7, 253–5
Knitting, 520

Lactams, lactones, 260
Lamellae, lamellar crystals, 141–4, 148–53, 166
Laminating resins, lamination, 411, 468, 472, 474, 476, 481, 497
Latex, 379, 384, 392, 531–2
Lattice defects, 147–8
Lattice model, 32
Lifetime, radical, 281
Light microscopy, 108
Light scattering, 64–72
Lignin, 462
Limiting viscosity number, 80
Linear polyethylene, 302, 307, 369–71
Linear polymer, 3
Liquids, light scattering in, 64
"Living" polymers, 299
Loading, loaded stock, 541; *see also* Reinforcement
Logarithmic normal distribution, 285
Logarithmic viscosity number, 80
Lubrication, 521

Macromolecular concept, 13
Mass spectrometry, 89
Masterbatching, 532
Mastication, 379, 387
Maxwell element, 192
Mechanical properties, of amorphous polymers, 172–204
  of crystalline polymers, 205–7
  measurement, 109–12
  and structure, 221–6
  See also Dynamic mechanical properties, Tensile properties, *and under the various polymers by name*
Melamine-formaldehyde polymers, 470–2, 505
Melt flow, 172–9, 209, 421
Melt fracture, 177
Melt index, 175
Melt spinning, 515
Melt viscosity, 172–9, 209, 421
Melting point, *see* Crystalline melting point
Membranes, osmotic, 59–61
Mercerization, 451
Mesomerism, *see* Resonance
Metal oxides, 307–9
Metallic bond, 17
Methacrylic esters, 292, 403
Methyl acrylate, 403
Methyl cellulose, 461
Methyl methacrylate, 401
Mica, 499
Micelle, 343
Microscopy, 108
Mixing, liquids, 25–6
  polymer melts, 491, 496
  polymer solutions, 31–4
Modulus of elasticity, *see* Stiffness
Moisture content, regain, 509
Moisture sensitivity, elastomers, 446
  fibers, 409, 415, 432, 435, 439, 441–2, 450–1, 454, 458, 524
  plastics, 372, 407–9, 418, 421, 432, 442–3, 454, 458, 461, 471, 478, 502–5
Mold, molding, 368, 370, 373, 397, 402, 414, 422–6, 432, 458, 468, 471, 474, 481, 487–91, 494, 497
Molecular motion, 152, 201–3
Molecular weight, in chain polymerization, 277, 294
  control, 256, 279
  correlation, with intrinsic viscosity, 82–3
    with melt viscosity, 175

Molecular weight, definition, 4, 6
in degradation, 355
determination, cryoscopy, 58
deswelling of gels, 52
ebulliometry, 58
end group analysis, 53
light scattering, 69–72
osmometry, 59–63
ultracentrifugation, 76–8
determined by random processes, 6
diene and divinyl polymers, 349–51
early estimates, 11
and polymer properties, 221, 226–7
and solubility, 226
in step-reaction polymerization, 247–59
temperature dependence, chain polymerization, 287
See also Molecular weight averages, Molecular weights of polymers
Molecular weight averages, number-, 7, 56
osmotic-, 61
significance of weight- and number-, 66
viscosity-, 82
weight-, 7, 66
z-, 76
Molecular weight distribution, 6, 7, 13
branched polymers, 257, 285, 349–51
in chain polymerization, 283–5
cumulative, 50
determination, distribution of sedimentation constants, 78
fractionation, 49–50
sedimentation equilibrium, 77
two molecular weight averages, 51
differential, 50
early recognition, 6, 13
graphical representation, 6, 50, 249–57
integral, 50
multichain polymer, 251
polyethylene, 366
in ring-scission polymers, 261
in step-reaction polymers, 249–57
Molecular weights of polymers, cellulose, 450
natural rubber, 378
polyethylene, 366
polyisobutylene, 389
Molecular weight of polymers, polytetrafluoroethylene, 420
poly(vinyl acetate) and derived polymers, 409
Monomer reactivity ratios, 326
Monomers, 3, 263, 292
Mooney viscosimeter, 175
Multichain polymer, 251
"Mutual" recipe, 343, 385

Natural fibers, *see* Cotton, Protein fibers, Silk, Wool
Natural rubber, *see* Rubbers, natural
N-carboxyanhydrides, 260
Neoprene (polychloroprene), 392, 546
Network polymers, *see* Crosslinked polymers
Network structure of elastomers, 184–7
Newtonian flow, liquid, 173
Nitrile rubber, 386–7, 498, 546
Nitrocellulose, 14, 60, 459–60
Nitroethylenes, 292
NMR, 100–3, 119, 148
Nonsulfur vulcanization, 534
Nonwoven fabrics, 523
Novolac, 464
Nuclear magnetic resonance spectroscopy, 100–3, 119, 148
Nucleation, 151
Number-average molecular weight, 7, 56
Nylon, alkoxy substituted, 433
history, 430
infrared spectrum, 93
soluble, 433
See also Polyamides
Nylon 6, *see* Polycaprolactam
Nylon 11, *see* Poly(ω-aminoundecanoic acid)
Nylon 66, *see* Poly(hexamethylene adipamide)
Nylon 610, *see* Poly(hexamethylene sebacamide)

Octet, 15
Oil extension, 532
Olefinic groups, infrared spectra, 94
α-Olefins, polymers, 292, 374
Optical activity, 123
Optical properties, 113, 227, 232, 367, 397, 402

Order-disorder transitions, 137–9
Organisols, 414
Orientation, 164–71, 224, 231–2
Osmometer, 59–60
Osmometry, 55–63
Osmotic membranes, 60
Osmotic molecular weight, 61
Osmotic pressure, 35, 55–63
Ostwald-Fenske viscometer, 80
Overcure, 537
Oxidative aging, elastomers, 375, 379, 387, 389, 391, 393, 530–1
  plastics, 367, 373, 413, 432
Oxidized rubber, 381
Ozone resistance, 375, 391, 393

Paracrystallinity, 147
Partial molar quantities, 35, 36, 38
Partial specific volume, 39, 76
Particle light scattering functions, 67
Pearl polymerization, *see* Suspension polymerization
Penultimate unit, 318
Peptide linkage, 134, 439
Permanent set, 180
Permeability, 60–1, 115
Peroxides, 264
Phase separation, 41–6
Phenol-aldehyde polymers, 241, 463–9
Phenol-formaldehyde resins, 14, 241, 463–9, 504–5
Phenolic resins, with formaldehyde, 14, 241, 463–9, 504–5
  with furfural, 469
  with resorcinol, 469
p-Phenylene group, 218
Photoinitiation, photopolymerization, 271
Photometer, 69–71
Pilling, 512
Pipe, 368, 371, 414, 443
Plastication, 491
Plasticization, plasticizers, 499–500
  cellulose acetate, 458
  cellulose nitrate, 14, 460
  and polymer properties, 225
  poly(methyl methacrylate) sheet, 401
  poly(vinyl butyral), 411
  vinyl resins, 414
Plasticizer efficiency, 500
Plasticizing resins, 483
Plastics, prices and production volumes, 363, 369, 372, 394, 401, 407, 412–3, 418, 419, 429, 432, 435, 443, 449, 458, 460, 463, 470, 473, 477, 483
  properties, requirements, 232
    tables, 502–5
Plastisol, 414
Plunger, 490
Polarity, 17, 20, 227–8, 291
Polarizability, 18
Polarized light, 66, 95, 108
Polyacetals, 240
Polyacrylamide, 404
Polyacrylates, 403
Poly(acrylic acid), 86, 404
Polyacrylonitrile, 405
  cohesive energy density, 21
  crystal structure, 125
  polymerization mechanism, 341
  properties, table, 524
  solubility parameter, 26
  spinning, 517
Polyamides, 429–34; *see also* Nylon
  copolymers, 214, 433
  crystal structure, 128
  end group analysis, 54
  heat and entropy of fusion, 215
  interfacial polymerization, 243
  melt viscosity, 176
  melting points, 214–7
  numbering system, 429
  vegetable oil, 234
Poly(ω-aminoundecanoic acid), 433
Polyanhydrides, 240
Polyblends, 399, 503
Polybutadiene, 119, 357, 384, 387–8
Poly(1-butene), 136–7
Polycaprolactam, 4, 128, 141, 144, 433
Polycarbonates, 446, 503
Polychloroprene, 129, 392, 546
Polychlorostyrenes, 229, 399
Polychlorotrifluoroethylene, 60, 426
Poly(cyclohexyl methacrylate), 403
Poly(decamethylene adipate), 176
Poly(decamethylene sebacate), 158
Poly(dimethyl butadiene), 384
Poly(dimethyl siloxane), 176

Polyelectrolytes, 86
Polyester resins, 472–5, 504–5
Polyesters, 434–6
  crystal structures, 129
  end group analysis, 54
  heat and entropy of fusion, 215
  polymerizable, styrenated, *see* Polyester resins
  polymerization, 237, 240, 242, 435
Polyethers, 442–4, 502
Polyethylene, 4, 363–71, 502
  branched, 363–8, 502
  chain folding, 142
  chlorinated, 374
  chlorosulfonated, 374
  cohesive energy density, 21
  crosslinked, 375
  crystal structure, 126
  crystallinity, 162–3
  degradation products, 357
  heat, and entropy of fusion, 161
    and free energy of polymerization, 289
  infrared spectrum, 93–4
  irradiated, 375
  linear, 369–71, 502
  mechanical properties, 206
  melt viscosity, 175
  molecular motion, 204
  single crystals, 142
  softening temperature, 221
  solubility parameter, 26
  solution viscosity, 84
  stiffness, 221
  stress-strain curve, 109
Poly(ethylene adipate), 5
Poly(ethylene glycol), 12
Poly(ethylene terephthalate), 434–6, 524
  cohesive energy density, 21
  crystal structure, 129
  DTA curve, 107
  polymerization, 242
  solubility parameter, 26
  specific heat, 105
Polyglycols, 12, 443
Poly(hexamethylene adipamide), 431–2, 502, 524
  cohesive energy density, 21
Poly(hexamethylene adipamide), crystal structure, 128
  infrared spectrum, 93
  molecular weight stabilization, 431
  polymerization, 242, 431
  solubility parameter, 26
  spherulites, 224
  spinning, 515
Poly(hexamethylene sebacamide), 128, 433
Polyisobutylene, 4, 388–90
  cohesive energy density, 21
  crystal structure, 136
  degradation products, 357
  melt viscosity, 176
  polymerization, 294, 389
  solution viscosity, 83
  stress relaxation, 189
Polyisoprene, 12, 378–83, 546; *see also* Rubbers, natural
  cohesive energy density, 21
  crystal structure, 129
  degradation products, 357
  stereoregular, 382
Poly(lauryl methacrylate), 403
Polymer, definition, 3
Polymer melts, 173–9, 209–10
Polymer properties, determinants, 208
  and structure, 208–29
  tables, 502–5, 524, 546
  and utilization, 230–3
  See also names of polymers
Polymerization, alternating intra-intermolecular, 353
  classification of methods, 5–6, 237–9, 388–9
  diene and divinyl, 348–52
  with metal oxides, 307–9
  monomers and mechanisms, 291–3
  rates, 246, 270, 286–8
  See also Anionic, Batch, Bulk, Cationic, Chain, Continuous, Coordination, Emulsion, Interfacial, Solution, Step-reaction, Suspension polymerization; Copolymerization
Polymerization-depolymerization equilibrium, 288
Polymer-solvent interaction constant, 34–44, 57, 160

Polymethacrylates, 204, 219, 400–3
Poly(methacrylic acid), 12, 404
Poly(methyl cyanoacrylate), 404
Poly(5-methyl-hexene-1), 119
Poly(methyl methacrylate), 400–3, 502
  anionic polymerization, 300
  autoacceleration of polymerization, 275
  cohesive energy density, 21
  copolymers, 225, 404
  degradation products, 357
  glass transition, 213
  graft copolymers, 225
  infrared spectrum, 93–4
  melt viscosity, 176
  solubility parameter, 26
  suspension polymerization, 342
  tacticity, 119, 300
  termination of polymerization, 265
Poly(α-methyl styrene), 357
Poly(methyl styrenes), 398
Poly(α-olefins), 119, 219, 374
Polyoxymethylene, 13, 442–3, 502
  heat and entropy of fusion, 161
  lattice defects, 148
  spherulites, 149, 224
  x-ray diffraction pattern, 169
Polypeptides, 133, 136, 139, 439–42
Polyperfluoroacrylates, 427
Polypropylene, 371–4, 502
  chain stiffness, 217
  crystal structure, 136
  glass transition, 204
  heat and entropy of fusion, 217
  molecular motion, 204
  NMR spectrum, 102
  spherulites, 224
Poly(propylene oxide), 123
Polysiloxanes, 241, 479–81
Polystyrene, 4, 12–14, 394–9, 503
  anionic polymerization, 299
  cohesive energy density, 21
  crystal structure, 136
  degradation products, 357
  glass transition, 213
  infrared spectrum, 93–4
  "living" polymer, 299
  melt viscosity, 176
  solubility parameter, 26
  tacticity, 119
Polystyrene, termination of polymerization, 265
Polysulfides, 241, 445
Polysulfonamides, 243
Polytetrafluoroethylene, 419–24, 503
  chain stiffness, 217
  crystal phase changes, 138
  crystal structure, 138
  EPR spectrum, 104
  flex life, 222
  glass transition, 204
  heat and entropy of fusion, 161, 217
  infrared spectrum, 93–4
Poly(trimethylene glutarate), 150
Polyureas, 240, 243
Polyurethans, 60, 240, 243, 437–8, 477
Poly(vinyl acetals), 410
Poly(vinyl acetate), 406–7
  amorphous nature, 125
  chain branching, 409
  cohesive energy density, 21
  dispersion polymerization, 342
  glass transition, 202
Poly(vinyl alcohol), 60, 407–9
  chain branching, 409
  crystal structure, 127
  crystallizability, 125
Poly(vinyl amines), 416
Poly(4-vinyl N-butyl pyridinium bromide), 86
Poly(vinyl butyral), 60, 410–1
Poly(vinyl carbazole), 416
Poly(vinyl chloride), 4, 412–4
  cohesive energy density, 21
  crystal structure, 128
  infrared spectrum, 93–4
  mechanical properties, 206
  polymerization mechanism, 341
  solubility parameter, 26
Poly(vinyl esters), 406
Poly(vinyl ethers), 417
Poly(vinyl fluoride), 425
Poly(vinyl formal), 410
Poly(vinyl isobutyl ether), 119
Poly(vinyl ketones), 417
Poly(vinyl oxazolidinone), 417
Poly(vinyl pyrrolidone), 417
Poly(vinylidene fluoride), 425
Popcorn polymerization, 341

Postforming, 496
Precipitation, 47–8, 340–1
Preform, preforming, 475, 488
Premix, 475
Prepreg, 475
Prices, *see* Prices and production volumes *under* Elastomers, Fiber, and Plastics
Primary bonds, 15–7
Principal sections, segments, 186
Processing, *see* Fabrication
Production volumes, *see* Prices and production volumes *under* Elastomers, Fiber, and Plastics
Promoters, 533, 540
Propagation, 265; *see also* Chain Polymerization
Protein fibers, plastics, 441
Pyramidal single crystals, 142

Q-e scheme, 329, 331
Quartz, 499

Radiation, high energy, crosslinking of polyethylene, 358, 375
  effect on polymers, 358–9
Radical polymerization, *see* Chain polymerization
Radicals (free), 6, 264; *see also* Chain polymerization
  addition to double bonds, 265
  concentration in polymerization, 265, 281, 345–6
  detection by EPR, 103–4
  generation, 264
  lifetime, 281
  monomers polymerized by, 292
  reactivity in copolymerization, 328
  stability, 264, 328
Radioactive initiators, 265, 420
Radius of gyration, 27, 72, 84
Ram, 490
Raman spectra, 95
Random chain, coil, flight, walk, 27
Random degradation, 354–5
Raoult's law, 31–2
Rate constants, polymerization, *see* Chain, Step-reaction polymerization
Rates of initiation, polymerization, propagation, termination, *see* Chain, Step-reaction polymerization
Rayleigh ratio, scattering, 63
Rayon, acetate, 456–8
  cuprammonium, 455
  triacetate, 458
  viscose, 454
Reactivity, and molecular size, 245
  monomer, 326
  radical, 328
  ratios, 312–15
  See also Copolymerization
Rebound, 180
Redox initiation, initiator, 347
Reduced viscosity, 80
Reflectance, 113
Refractionation, 48
Refractive increment, 76
Refractive index, 64–6, 75–6, 397, 403, 421
Refractive index gradient, *see* Specific refractive increment
Regenerated cellulose, 452–5
Regular solutions, 32
Regulator, 279
Reinforced plastics, 468, 472–5, 476, 483, 484
Reinforcing agents, fillers, 527, 541
Reinforcement, 541–5
  butyl rubber, 390
  by carbon black, 543–4
  neoprene, 393
  nitrile rubber, 387
  and polymer properties, 225
  polysulfide rubber, 445
  SBR, 386
  silicone rubber, 482
  theories, 545
Relative viscosity, 80
Relaxation time, 195–6
Release resins, 481
Repeat unit, 4, 98
Resilience, 180
Resin finishing, 436, 451, 523
Resistivity, 114
Resole, 465
Resonance, 216, 263, 327
Resorcinol-formaldehyde resins, 469

Retardation, retarders, 267
Retardation time, 193
Retarded elasticity, 193
Reversion, 538
Rheology, 172–207
Rheometer, 175
Ring formation, 244
Ring-scission polymerization, 259
Root-mean-square end-to-end distance, 27, 38
Rotating sector, 281
Rotating spreader, 491
Rubber, crystal structure, 129
  crystallization and melting, 155–6
  energy, entropy, and heat of solution, 35–7
  fillers, 541
  heat and entropy of fusion, 161
  infrared spectrum, 93–4
  See also Elasticity, Elastomers
Rubbers, acrylate, 403–4, 546
  butadiene-acrylic acid, 387
  butyl, 388–91, 546
  chlorosulfonated polyethylene, 374
  ethylene-propylene, 374
  fluorocarbon, 426, 546
  fluorosiloxane, 428
  natural, 12–14, 60, 378–9, 526–46
    chlorinated, 380
    cyclized, 380
    graft copolymer with methyl methacrylate, 225
    hard, 14, 381
    hydrochloride, 381
    oxidized, 381
  neoprene, 392–3, 546
  nitrile, 386, 546
  nitrosofluoro, 428
  polybutadiene, stereoregular, 387
  polyisoprene, stereoregular, 382
  polysulfide, 445, 546
  SBR, 383–6, 546
  silicone, 482, 546
  urethan, 437, 546

Safety glass, 411
Salt, polymerization of nylon, 431
SBR, 383–6, 546
  degradation products, 357
  emulsion polymerization, 343
Schlieren optics, 75
Scouring, 520
Seaweed fiber, 461
Second-order transition, 10
Second virial coefficient, 39, 57, 61–3, 66, 76
Secondary bond, 17–21
Sector method, 281
Sedimentation, approach to equilibrium, 77
  constant, 78
  equilibrium, 76–7
  transport, 77–8
  velocity, 77–8
Semipermeable membrane, 59–61
Semipolar bond, 16
Shared-electron bond, 16
Shear rate, and melt viscosity, 175–7
  and solution viscosity, 82
Shear thickening, thinning, 173
Sheeting, 367, 401, 411, 414, 427, 458, 460
Shift factor, 196
Side chains and polymer properties, 219; *see also* Branching
Silicones, 479–82, 505
Silk, 240, 440, 524
  crystal structure, 133
Single crystals, 140–4, 166
Single phase concept, 147
Size, sizing, 521
"Slush" molding, 494
Smoked sheet rubber, 379
Snap, 180
Soap, 343–6
Soda cellulose, 452
Sodium polyacrylate, 86
Softening-point, temperature, effect on properties, 221
  measurement, 112
  See also Crystalline melting point
Sol, 253
Solubility, 25–7
  parameter, 26
  polymer properties and, 226–9
  rate, 26
Solution polymerization, 340
Solution viscosity, 79–85
Solvent resistance, elastomers, 387, 391, 393, 438, 445, 482
  fibers, 432, 435, 440, 450, 457–8, 462

Solvent resistance, plastics, 367, 374, 397–8, 404, 408–9, 412, 420–1, 427, 432, 443–4, 461, 471
and structure, 20, 226–9, 231
Specific cohesive energy, 19, 26
Specific gravity, *see* Density
Specific heat, 105, 182
Specific refractive increment, 66, 75
Specific viscosity, 80
Specific volume, crystallization and melting, 155–8
degree of crystallization, 162
glass transition, 199, 202
See also Density
Spectrometer, infrared, 91–2
Spectrophotometer, 113
Spherulites, 148–51
and properties, 222
ringed, 150
Spinneret, 515
Spinning, 513–9
Spring, 192
Stabilizers, 501
Stabin osmometer, 59
Staple, 508
Statistics of polymerization, *see* Step-reaction polymerization
Staudinger equation, 82–3
Steady state, 273, 295, 300, 305; *see also* Chain polymerization
Step-reaction polymerization, 5, 237–61
branching, branching coefficient, branch point, unit, 253–4
and chain polymerization, 238
control of molecular weight, 250
end group analysis, 54
extent of reaction, 246
functionality, 239
gel, gelation, gel point, 253–4
interchange, 242, 251
interfacial, 243
kinetics, 246–7
linear, 245–52
molecular weight and distribution, linear, 249
multichain, 252
polyfunctional, 255
ring scission, 261
multichain, 251–2
nylon, 431
Step-reaction polymerization, poly-(ethylene terephthalate), 435
polyfunctional, 253–9
polymer types in, 239–41
rate constants, 246–7
reactivity and molecular size, 245
ring formation, 244
ring scission, 259–61
statistics, 248–50
Stereospecific polymerization, 10; *see also* Coordination polymerization
Stiffness, elastomers, 528, 541, 546
fibers, 450, 507, 523–4
and glass transition, 198, 206
measurement, 109
plastics, 320, 502–5
rubber elasticity, 184–7
and structure, 221
Stoichiometric balance, 250, 431
Stokes' law, 78
Stress cracking, 115, 366–7, 373, 374
Stress relaxation, 188
Stress-strain curves, properties, elastomers, 183–6
fibers, 509–11
measurement, 109
Structure index, 544
Styrene, 395
Styrene-butadiene rubber, *see* SBR
Sulfur vulcanization, 536
Summative fractionation, 50
Superpolymers, 430
Suspension polymerization, 341, 395, 401, 403, 412
Swelling, 25, 52, 387
Syndiotactic, 119
Synthetic boundary cell, 75
Synthetic elastomers, rubbers, *see* Rubbers
Synthetic fibers, *see* Fibers

Tacticity, 119
Tear strength, 366–7, 381, 436, 455, 541
Temperature gradient elution, 49
Tenacity, 455, 458, 508
Tensile properties, strength, elastomers, 386–7, 391, 445, 541, 546
fibers, 405, 415, 431, 441, 454, 458, 524
measurement, 109

Tensile properties, strength, plastics, 366, 370, 372, 376, 397, 399, 402, 431, 443, 455, 471, 502–5
and structure, 210
Terminal sections, segments, 186
Termination, 265; *see also* Chain polymerization
Tetrafluoroethylene, 419
Textiles, 508
Thermal initiation, polymerization, 277
Thermal properties, tests for, 112–13; *see also* Heat resistance, stability
Thermodynamics, of polymer solutions, 32–46
of polymerization, 286–90
of rubber elasticity, 180–3
of simple liquids, 31–2
Thermofusion, 497
Thermoplastics, definition, 8
Thermosetting resins, definition, 8
Time-temperature superposition, 188, 196
Tire cord, 432–3, 454
Tires and tubes, 386, 392, 393, 438, 445, 529
Torsion pendulum, 190
Toughness, 367, 372, 375, 397, 399–400, 402, 404–5, 407, 421, 425, 431, 443–4, 446, 455, 458, 460–1
Transfer, transfer agent, constant, 266, 278–9; *see also* Chain polymerization
Transfer molding, 489
Transmittance, 113
Transparency, 114
Tritactic, 121
Trommsdorff effect, 275
Turbidimetric titration, 50
Turbidity, 63–4
Two-phase concept, 146

Ubbelohde viscometer, 80
Ultracentrifugation, 50, 73–8; *see also* Sedimentation
Ultracentrifuge, 74–5
Unit cell, 98
Unzippering, 355
Upper Newtonian region, viscosity, 82, 177
Urea-aldehyde polymers, 241, 470–2
Urea-formaldehyde resins, 470–2, 505
Uses of polymers, 230–3
elastomers, 374–5, 377–93, 426–8, 437–8, 445–6, 482
fibers, 405, 409–10, 415–6, 432–3, 435, 437, 439–42, 448–62
plastics, 367–76, 394–418, 419–27, 429–47, 458–61, 463–84

van der Waals bond, force, 17
Vapor pressure lowering, 55–6, 58
Vicat test, 112, 221
Vinyl acetate, 406
Vinyl chloride, 412
Vinyl monomers, 263
polymerization mechanism, 292
Vinyl polymerization, *see* Chain polymerization
Vinyl resins, 14, 413, 503
Vinylidene monomers, polymerization mechanism, 292
Virial coefficient (second), 39, 57, 61–3, 66, 76
Viscoelasticity, 10, 188–98, 200
Viscometers, 80–1, 174
Viscose, 454, 518, 524
Viscosity, melt, 173–9
and structure, 209–10
solution, 79–85
Viscosity-average molecular weight, 82
Viscosity number, 80
Viscosity ratio, 80
Voigt element, 193
Volatility, 19
Volumes, production, *see* Prices and production volumes *under* Elastomers, Fiber, and Plastics
Vulcanization, 14, 533–41
of elastomers, 374–5, 386–7, 390, 393, 426, 445, 482
kinetics, 348–50
Vulcanized fiber, 455

Wash and wear, 436, 523
Water repellents, 522
Water sensitivity, *see* Moisture sensitivity
Water-soluble polymers, 86, 404, 408, 417, 443, 461–2
Wear resistance, 512; *see also* Abrasion resistance

Weathering, elastomers, 375, 386–7, 391, 393, 445
fibers, 405, 413, 432, 440, 524
measurement, 115
plastics, 367, 373, 397, 400, 402, 413, 432, 438, 460, 502–5
Weaving, 520
Weight-average molecular weight, 7, 66–7
Wet spinning, 518
Wire and cable insulation, 232, 368, 392–3, 414, 423–5, 427, 432
WLF equation, 178, 196, 200
Wood flour, 498
Wool, 439, 524
Work content, 20

Xanthates, xanthation, 453
X-ray diffraction, 97–9, 124, 152, 169

Yarns, textured, 523
Yield stress, 109, 173, 221

z-Average molecular weight,76
Zein, 442
Zero strength temperature, 112
Zimm-Myerson osmometer, 59
Zimm plot, 69, 72
Zone melting, 49

PENSACOLA JUNIOR COLLEGE LIBRARY
QD281.P6 B43 1962
Billmeyer, Fred W. 000
Textbook of polymer scien 210101
3 5101 00041400 8

DATE DUE

DEC 2 4 1997			

Pensacola Junior College
Library

QD281.P6B43 29,581
1962

Billmeyer, Fred W.
Textbook of polymer science

PENSACOLA JUNIOR COLLEGE LIBRARY